The COVID-19 Pandemic

Joseph M. Rosen • Rita R. Colwell
Editors

The COVID-19 Pandemic

Science, Technology, and the Future of Healthcare Delivery

Springer

Editors
Joseph M. Rosen
Geisel School of Medicine
Dartmouth-Hitchcock Medical Center
Hanover, NH, USA

Rita R. Colwell
Institute for Advanced Computer Studies
University of Maryland
College Park, MD, USA

ISBN 978-3-031-62771-2 ISBN 978-3-031-62772-9 (eBook)
https://doi.org/10.1007/978-3-031-62772-9

This Springer imprint is published by the registered company Springer Nature Switzerland AG
The registered company address is: Gewerbestrasse 11, 6330 Cham, Switzerland

Preface

The COVID-19 pandemic caused millions of deaths and upended the global economy. It is critical that we learn from this tragic period of time and provide a view of the landscape through which the pandemic drove, an overview of the science and technology that helps us understand the trajectory of the pandemic. Genomics, epidemiology, and health care throughout the COVID-19 pandemic, its virus, and pathogenicity are included. How the virus impacted and continues to affect our healthcare system and economy is discussed. Simulation modeling used to predict spread of the virus and risk of infection in America and globally, and the critical role of communication in pandemic response, provides an understanding of the landscape of the pandemic.

Synthetic biology was employed to develop and produce effective diagnostics that were rapidly deployed, with vaccine development in unlimited quantity a colossal achievement. US and global production of vaccine was unparalleled in human history. Most importantly, treatment of patients infected with the virus using telemedicine and virtual reality for remote intensive care settings introduced a whole new layer of medical care. Biotechnology played a major role in the COVID-19 pandemic response, with many advances in engineering sciences introduced.

The virus itself was analyzed and described in exacting detail. A broad overview of biodefense, closely coupled to this and earlier pandemics in human history, lays bare the cascading of potential subsequent pandemics threatening future human populations and provides a foundation for understanding the relentlessness of pandemics. Internationally, individual countries responded to COVID-19 uniquely, providing more information about pandemic response, including Vietnam, Afghanistan, and China, all of which present unique roles that they played. As would be expected, variants of the virus created the need for variable responses, hence chronology of the pandemic is detailed, highlighting the many factors at play.

Robotics and artificial intelligence, new technologies still developing, were introduced during COVID-19's relentless global spread and now are being incorporated into contemporary practice of medicine and applications for improving public health. The development of mRNA vaccines and their role in synthetic biology have become contemporary technologies. All of these novel applications are key for future healthcare systems, based essentially on the defense highway system, the Koop healthcare system, named for the former surgeon general who played a significant role in the conception of this dynamic system that is adaptable to future pandemics. It integrates health care across all 50 states, territories, and tribal areas, helping us prepare a more effective response to the next, inevitable pandemic. The economics of the pandemic, from the perspective of the world bank, and the resilience and its role in preparing communities to respond to the pandemic are included.

Finally, disinformation having played a crucial role in the COVID-19 pandemic is discussed in this overview of COVID-19. In summary, the chapters in this book, as intended, provide a comprehensive and multidimensional view of the COVID-19 pandemic. We hope it will prove informative for our readers.

Hanover, NH, USA Joseph M. Rosen
College Park, MD, USA Rita R. Colwell

The COVID-19 Timeline

Dec. 31, 2019:
The World Health Organization (WHO) Country Office in China translated a statement released by the Wuhan Municipal Health Commission regarding cases of "viral pneumonia." The office contacted the International Health Regulations (IHR) center at the WHO Western Pacific Regional Office.

Around the same time, evidence of cases of "pneumonia of unknown cause" in Wuhan was discovered through a media report on ProMED, a program by the International Society for Infectious Diseases, by WHO's Epidemic Intelligence from Open Sources platform ("Pneumonia of Unknown Cause—China" 2020).

Dec. 31, 2019:
WHO China Country Office was directly informed of cases of pneumonia of unknown cause in Wuhan, and all the cases were connected to the Huanan Seafood Wholesale Market in Wuhan ("CDC Museum COVID-19 Timeline" 2022).

Jan. 11, 2020:
The first death from "viral pneumonia" was a 61-year-old man in China. The Chinese media's preliminary lab tests led to the discovery of a new type of coronavirus (Allam 2020).

Jan. 13, 2020:
The first case of the novel coronavirus outside of China was reported in Thailand by the Ministry of Public Health ("WHO Statement on Novel Coronavirus in Thailand" 2020) ("CDC Museum COVID-19 Timeline" 2022).

Jan. 20, 2020:
The first US laboratory-confirmed case of COVID-19 was officially confirmed in the state of Washington through samples taken on January 18 by the Centers for Disease Control and Prevention (CDC). To respond to this outbreak, the CDC activated the Emergency Operations Center ("CDC Museum COVID-19 Timeline" 2022).

Jan. 22, 2020:
During a WHO emergency meeting in Geneva, Director-General Dr. Tedros Adhanom Ghebreyesus stated, "Make no mistake, though, this is an emergency in China. But it has not yet become a global health emergency. It may yet become one." Human-to-human spread of the novel coronavirus was confirmed by the WHO ("Statement on the Meeting of the International Health Regulations…" 2020).

Jan. 23, 2020:
China announced a lockdown in Wuhan (Feng 2021).

Jan. 23, 2020:
WHO held an Emergency Committee meeting under the International Health Regulations where representatives from South Korea, Thailand, China, and Japan discussed the coronavirus cases in their countries. China reported 557 cases total on this date ("Statement on the Meeting of the International Health Regulations…" 2020).

Jan. 27, 2020:
A Level 3 travel advisory was issued by the U.S. embassy in China to reconsider travel due to the coronavirus. Health officials in China had reported thousands of cases throughout the country ("China Travel Advisory: Level 3 – Reconsider Travel (28 Jan 2020)" 2020).

Jan. 27, 2020:
A press release was issued by the WHO regional director for Southeast Asia, urging countries in the region to prepare for increased detection of cases to prevent further spread ("WHO Calls for Urgent, Aggressive Actions…" 2020).

Jan. 27, 2020:
The CDC advised all travelers to cease nonessential travel to China ("CDC Advises Travelers to Avoid All Nonessential Travel to China" 2020).

January 27, 2020—12,545 cases globally
January 28, 2020—100 deaths globally
("WHO Coronavirus (COVID-19) Dashboard" 2022)

Jan. 30, 2020:
WHO Director-General Ghebreyesus officially declared a public health emergency of international concern from the coronavirus outbreak ("Statement on the Second Meeting of the International Health Regulations…" 2020).

Feb. 1, 2020:
Four countries—the United States, Singapore, Russia, and Australia—all banned foreign travelers that recently traveled in China ("Timeline-How the Global Coronavirus Pandemic Unfolded" 2020).

Feb. 2, 2020:
The first coronavirus death outside of China was reported in the Philippines, and China's total cases rise to over 14,000 (Schnirring 2020).

Feb. 2, 2020:
The first Polymerase Chain Reaction (PCR) lab kits were shipped to the WHO Regional Offices ("Timeline: WHO's COVID-19 Response" 2022) ("Coronavirus disease (COVID-19) Situation Report 1 Philippines" 2020, p. 3).

Feb. 5, 2020:
The Diamond Princess cruise ship held 3700 passengers that had to quarantine for nearly a month after over 700 passengers tested positive and 14 died ("Timeline-How the Global Coronavirus Pandemic Unfolded" 2020).

Feb. 5, 2020:
WHO held the first daily media briefings by Director-General Dr. Tedros Adhanom Ghebreyesus ("Timeline: WHO's COVID-19 Response" 2022).

Feb. 7, 2020:
A Chinese whistleblower, Dr. Li Wenliang, died from the coronavirus (Lee 2020).

February 10, 2020—1,000 deaths globally
("WHO Coronavirus (COVID-19) Dashboard" 2022)

Feb. 11, 2020:
The coronavirus was officially named COVID-19 by the WHO. Director-General Tedros Adhanom Ghebreyesus explained that the "CO" stands for "corona," the "VI" stands for "virus," the "D" for "disease," and the "19" is for the identified year of 2019 ("WHO Director-General's Remarks…" 2020) ("Naming the Coronavirus Disease…" 2019).

Feb. 25, 2020:
An update on COVID-19 as well as the CDC's COVID-19 response to America was presented by the CDC ("Transcript for the CDC Telebriefing Update on COVID-19" 2020).

Feb. 25, 2020:
A COVID-19 case was reported in Egypt making it the first in Africa. Following this, the second case in WHO's African Region was confirmed in Algeria. The regional director for Africa urged all countries in the continent to step up their readiness for the virus ("A Second COVID-19 Case Is Confirmed in Africa" 2020).

February 29, 2020—1st death in the United States
("WHO Coronavirus (COVID-19) Dashboard" 2022)

March 1, 2020:
Both King County and the United States marked their second COVID-19 death (Hudetz 2020) ("Public Health Confirms Two More COVID-19 Cases…" 2020).

March 7, 2020:
WHO released a statement of action to delay, contain, control, reduce, and stop the impact of COVID-19 at every opportunity as the number of global cases surpasses 100,000 ("WHO Statement on Cases of COVID-19 Surpassing 100 000" 2020).

March 19, 2020:
California governor Gavin Newsom issued a stay-at-home order ("Governor Gavin Newsom Issues Stay at Home Order" 2020).

March 20, 2020:
Governor Andrew Cuomo of New York issued guidance on essential services under his "New York State on PAUSE" executive order ("Governor Cuomo…" 2020).

March 20, 2020—10,000 deaths globally
("WHO Coronavirus (COVID-19) Dashboard" 2022)

March 24, 2020:
The 2020 Olympics were delayed in a joint statement from the International Olympic Committee and the Tokyo 2020 Organizing Committee ("Joint Statement…" 2022).

March 26, 2020:
The CARES Act, which stands for Coronavirus Aid, Relief, and Economic Security, was passed by the U.S. Senate, providing $2 trillion in aid to small businesses, hospitals, and both local and state governments while eliminating the Medicare sequester from May 2020 to December 31, 2020 ("CDC Museum COVID-19 Timeline" 2022) ("H.R. 748…" 2021).

April 2, 2020: Worldwide cases reach 1 million
("WHO Coronavirus (COVID-19) Dashboard" 2022)

April 7, 2020:
A report from the Chicago Tribune noted that 68% of the COVID-19-related deaths in Chicago occurred among the city's African American community, illuminating racial disparities of the pandemic in the United States ("CDC Museum COVID-19 Timeline" 2022) (Kim and Bostwick 2020) (Unruh et al. 2022).

April 8, 2020:
The 11 million citizens of Wuhan were able to leave their homes after months as the city reopens ("Timeline-How the Global Coronavirus Pandemic Unfolded" 2020) ("Wuhan's Lockdown Lifted" 2020).

April 8, 2020:
The United Nations created a COVID-19 Supply Chain Task Force to scale up the production of personal protective equipment, oxygen, and lab tests for countries in need ("WHO Director-General's Opening Remarks…" 2022).

April 8, 2020:
The U.S. Health and Human Services announced the first contract for ventilator production goes to General Motors under the Defense Production Act ("CDC Museum COVID-19 Timeline" 2022).

April 9, 2020—100,000 deaths globally
("WHO Coronavirus (COVID-19) Dashboard" 2022)

April 13, 2020:
Restrictions began to ease in multiple European countries. Manufacturing and construction resumed in Spain, stores began to reopen in Italy and Austria, and schools returned for Danish children (Brandon 2020).

April 13, 2020:
One hundred thirty scientists, manufacturers, funders, and other stakeholders published a statement committing to working together with the WHO to develop a vaccine quickly against COVID-19 ("Public Statement for Collaboration on COVID-19 Vaccine Development" 2020).

April 13, 2020:
President Trump announced that the United States will stop funding to the World Health Organization, alarming the public health community ("CDC Museum COVID-19 Timeline" 2022).

April 24, 2020:
President Trump said that scientists should explore whether injecting disinfectant might help COVID-19 patients. Physicians worry people will poison themselves with bleach ("President Trump on 'Injecting' Disinfectants" 2020).

April 24, 2020:
Despite health experts warning against reopening states too early, the states of Georgia, Alaska, and Oklahoma began to reopen ("CDC Museum COVID-19 Timeline" 2022).

May 11, 2020:
A team of patients published the results of the first survey of "long COVID" symptoms. They reported more than 50 symptoms, including cognitive impairment and fatigue (McCorkell et al. 2020).

May 13, 2020:
The WHO warned that COVID-19 could become endemic like HIV and never go away. WHO emergencies expert Mike Ryan told an online briefing: "It is important to put this on the table: this virus may become just another endemic virus in our communities" (COVID-19 Virtual Press conference" 2020, p. 21).

May 14, 2020:
The United Nations warned of a looming mental illness crisis as millions of people were surrounded by death and disease and forced into isolation, poverty, and anxiety ("COVID-19 and the Need for Action on Mental Health" 2020).

May 15, 2020:
The White House announced Operation Warp Speed as a national effort to accelerate the development, manufacturing, and distribution of COVID-19 medical countermeasures (MCMs) ("Remarks by President Trump on Vaccine Development" 2020).

May 21, 2020: Worldwide cases reach 5 million
("WHO Coronavirus (COVID-19) Dashboard" 2022)

May 22, 2020:
Brazil overtook Russia to become the No. 2 in global cases while the United States stayed at the No. 1 spot ("Brazil…" 2022) ("India…" 2022) ("United States of America…" 2022).

May 28, 2020—100,000 deaths in the United States
("WHO Coronavirus (COVID-19) Dashboard" 2022)
("CDC Museum COVID-19 Timeline" 2022)

May 29, 2020:
President Donald Trump claimed he was ending the United States' relationship with WHO because of the organization's poor handling of the COVID-19 outbreak. He also accused the UN agency of becoming a puppet organization of China ("Timeline-How the Global Coronavirus Pandemic Unfolded" 2020) ("Remarks by President Trump on Actions Against China" 2020).

May 29, 2020:
The COVID-19 Technology Access Pool (C-TAP) was launched by 30 countries and international partners, to increase access to vaccines, tests, and treatments for all people ("WHO and MPP Welcome NIH's Offer…" 2022).

June 8, 2020:
New Zealand lifted all social and economic restrictions except border controls, one of the first countries to return almost to pre-pandemic normality (Menon 2020).

June 8, 2020:
The World Bank predicted that COVID-19 will cause the worst global economic recession since World War II ("COVID-19 to Plunge Global Economy…" 2020).

June 10, 2020:
The number of confirmed cases of COVID-19 hit 2 million in the United States as new infections continue to rise in 20 states (Chappell 2020).

June 15, 2020:
After 83 days of lockdown, England allowed retail stores to reopen (Aspinall 2022).

June 19, 2020:
Brazil's cases topped 1 million, and deaths approached 50,000, marking its status as the world's second worst-hit country ("Brazil…" 2022).

June 28, 2020: Worldwide cases reach 10 million
("WHO Coronavirus (COVID-19) Dashboard" 2022)
June 29, 2020—500,000 deaths globally
("WHO Coronavirus (COVID-19) Dashboard" 2022)

July 6, 2020:
India reported 700,000 infections, becoming the country with the third highest number of cases, behind the United States and Brazil. ("India…" 2022).

July 6, 2020:
A WHO survey found that 73 countries were at risk of running out of antiretroviral medications as a result of COVID-19. Twenty-four countries had critically low stock ("WHO: Access to HIV Medicines Severely Impacted…" 2020).

July 8, 2020:
Australia's second-largest city, Melbourne, went back into lockdown (Bungard 2020).

Aug. 5, 2020:
As the global death toll topped 700,000, the WHO said young people must limit their partying instincts to help prevent new outbreaks during part of a live chat with Dr. Michael Ryan ("Live Q&A on #COVID19 with @DrMikeRyan and Dr @ mvankerkhove. #AskWHO", 12:50).

Aug. 6, 2020:
Africa's cases surpassed 1 million, with South Africa accounting for more than half of the caseload ("WHO Coronavirus (COVID-19) Dashboard" 2022) ("South Africa…" 2022).

Aug. 10, 2020:
Brazil's death toll reached 100,000, continuing to climb as cities reopen shops and dining ("Brazil…" 2022).

Aug. 11, 2020:
After less than 2 months of human testing, Russia became the first country to approve a COVID-19 vaccine, called Sputnik V (Burki 2020).

Sept. 7, 2020: WHO
India passed Brazil as the second-worst hit country, with more than 4.2 million cases, and the United States remained the worst affected country ("Brazil…" 2022) ("India…" 2022) ("United States of America…" 2022).

September 22, 2020—200,000 deaths in the United States
("WHO Coronavirus (COVID-19) Dashboard" 2022)
("CDC Museum COVID-19 Timeline" 2022)

Sept. 23, 2020:
Johnson & Johnson (J&J) began trials of a single-shot vaccine that could simplify distribution. Rivals from Moderna Inc, Pfizer Inc, and AstraZeneca all require two shots separated by several weeks ("Johnson & Johnson" 2020) (Routh 2020).

September 28, 2020—One million deaths globally
("WHO Coronavirus (COVID-19) Dashboard" 2022)

Sept. 29, 2020:
COVID-19-related deaths doubled from only half a million in the previous 3 months. The United States, Brazil, and India led in the burden of fatalities ("WHO Coronavirus (COVID-19) Dashboard" 2022).

Oct. 2, 2020:
President Trump tested positive for COVID-19 ("CDC Museum COVID-19 Timeline" 2022).

October 19, 2020—Global cases top 40 million
("WHO Coronavirus (COVID-19) Dashboard" 2022)

Oct. 22, 2020:
The U.S. Food and Drug Administration (FDA) approved Remdesivir, an antiviral, as the first COVID-19 drug ("FDA Approves First Treatment for COVID-19" 2020).

The U.S. reported an unprecedented 100,000 cases in one single day ("United States—COVID-19 Overview—Johns Hopkins" 2022).

Nov. 9, 2020:
The FDA issued Emergency Use Authorization (EUA) for Eli Lilly's antibody treatment Bamlanivimab. It is authorized for treatment of recently diagnosed, mild to moderate COVID-19 in high-risk patients ("Coronavirus (COVID-19) Update: FDA Revokes…" 2021, para. 2).

November 9, 2020: Worldwide cases reach 50 million
("WHO Coronavirus (COVID-19) Dashboard" 2022)

Nov. 16, 2020:
Moderna announced their preliminary phase three trial data which showed that their coronavirus vaccine has 94% efficacy in preventing COVID-19 ("Moderna's COVID-19 Vaccine Candidate…" 2020).

Nov. 18, 2020:
The Pfizer-BioNTech vaccine was announced to be 95% effective ("Pfizer and BioNTech Conclude Phase 3 Study…" 2020).

Nov. 21, 2020:
The FDA granted EUA for monoclonal antibody treatment, the second treatment to be approved ("Coronavirus (COVID-19) Update: FDA Authorizes Monoclonal…" 2020).

Dec. 11, 2020:
FDA granted EUA for a COVID-19 vaccine made by Pfizer-BioNTech ("Pfizer and BioNTech Announce New Agreement…" 2022, para. 7).

Dec. 14, 2020:
Nurse Sandra Lindsay became the first person in the United States to receive a COVID-19 vaccine ("Senior UN Official…" 2022).

December 14, 2020—300,000 deaths in the United States
("WHO Coronavirus (COVID-19) Dashboard" 2022)

January 27, 2021: Worldwide cases reach 100 million
("WHO Coronavirus (COVID-19) Dashboard" 2022)

Dec. 18, 2020:
The FDA issued an Emergency Use Authorization for the second COVID-19 vaccine, the Moderna vaccine ("FDA Takes Additional Action…" 2020).

January 15, 2021—Two million deaths globally
January 19, 2021—400,000 deaths in the United States
("WHO Coronavirus (COVID-19) Dashboard" 2022)

Feb. 15, 2021:
The WHO granted emergency use for the AstraZeneca/Oxford COVID-19 vaccine ("WHO Lists Two Additional COVID-19 Vaccines…" 2021).

February 21, 2021—500,000 deaths in the United States
("WHO Coronavirus (COVID-19) Dashboard" 2022)

Feb. 27, 2021:
FDA approved EUA for Johnson and Johnson's one-shot COVID-19 vaccine ("Emergency Use Authorization" 2022).

April 13, 2021:
FDA and CDC announced a recommended pause in the administration of the Johnson and Johnson vaccine to investigate six reported cases of blood clots ("Coronavirus (COVID-19) Update: FDA Limits…" 2022, para. 6) ("CDC Museum COVID-19 Timeline" 2022).

April 17, 2021—Three million deaths globally
("WHO Coronavirus (COVID-19) Dashboard" 2022)

May 10, 2021:
WHO labeled the Delta variant, originally discovered in India in late 2020, as a variant of concern ("COVID-19 Virtual Press Conference Transcript—10 May 2021," 2021).

June 14, 2021—600,000 deaths in the United States
July 3, 2021—Four million deaths globally
("WHO Coronavirus (COVID-19) Dashboard" 2022)

Aug 4, 2021: Worldwide cases reach 200 million
("WHO Coronavirus (COVID-19) Dashboard" 2022)

Aug. 23, 2021:
The FDA officially approved the Pfizer-BioNTech vaccine for those 16 and older ("FDA Approves First COVID-19 Vaccine" 2021).

Sept. 22, 2021:
FDA approved a booster dose of Pfizer-BioNTech following the two-shot sequence ("FDA Authorizes Booster…" 2021, para. 1).

October 1, 2021—700,000 deaths in the United States
October 5, 2021: Worldwide cases reach 235 million
("WHO Coronavirus (COVID-19) Dashboard" 2022)

Oct. 8, 2021:
Brazil reported more than 600,000 deaths from COVID-19, the second country to do so after the United States ("Brazil…" 2022) ("United States of America…" 2022).

Oct. 11, 2021:
The World Health Organization recommended an additional dose of the vaccine for severely or moderately immunocompromised people ("WHO Advisory Group…" 2021).

Oct. 26, 2021:
FDA voted 17-0 recommending Pfizer-BioNTech's COVID-19 vaccine in children ages 5 to 11 ("Vaccines and Related Biological Products Advisory Committee Meeting" 2021).

November 1, 2021—Five million deaths globally
("WHO Coronavirus (COVID-19) Dashboard" 2022)

November 9, 2021: Worldwide cases reach 250 million
("WHO Coronavirus (COVID-19) Dashboard" 2022)

Nov. 16, 2021:
The WHO announced that almost all global COVID-19 cases (99%) originated from the delta strain (Towey 2021).

Nov. 26, 2021:
WHO announced that the Omicron variant is a variant of concern. It was first sequenced in South Africa ("Tracking Sars-COV-2 Variants" 2021).

December 15, 2021—800,000 deaths in the United States
("WHO Coronavirus (COVID-19) Dashboard" 2022)

Dec. 15, 2021:
The WHO UNITY Studies Collaborator Group published a preprint of a systematic review of COVID-19 seroprevalence studies. They concluded that true infections were underestimated, with a quarter of the global population having COVID-19 antibodies in April 2021 (Bergeri 2021).

Dec. 20, 2021:
CDC estimated that the Omicron variant comprised 73% of U.S. COVID-19 cases ("CDC COVID Data Tracker" 2022).

Dec. 27, 2021:
The CDC shortened the recommended isolation period for COVID-positive Americans from 10 days to 5 days if symptoms were improving. People who were fully vaccinated and boosted did not need to quarantine following exposure but should wear masks ("CDC Updates and Shortens Recommended Isolation…" 2021).

Dec. 22, 2021:

The FDA granted EAU for Pfizer's Paxlovid, the first oral antiviral COVID-19 drug ("Coronavirus (COVID-19) Update: FDA Authorizes First Oral Antiviral for Treatment of COVID-19" 2021).**Jan. 3, 2022:**
The FDA amended the EUA of the Pfizer-BioNTech vaccine to expand the single booster use for children 12–15. They also allowed for a third dose for immunocompromised children 5–11 ("Coronavirus (COVID-19) Update: FDA Takes Multiple Actions…" 2022).

Jan. 19, 2022
The Biden Administration launched a program to mail COVID-19 test kits directly to American households ("Fact Sheet: The Biden Administration to Begin…" 2022).

Jan. 21, 2022:
The FDA officially approved the Moderna vaccine for individuals 18 and older ("Coronavirus (COVID-19) Update: FDA Takes Key Action…" 2022).

February 4, 2022—900,000 deaths in the United States
("WHO Coronavirus (COVID-19) Dashboard" 2022)

Feb. 24, 2022:
A report by the Lancet estimated that 6.7 million children had suffered from COVID-19-associated orphanhood and caregiver death in the past 2 years ("COVID-19-Associated Orphanhood" 2022).

Feb. 25, 2022:
The CDC recommended wearing a mask based on individual comfort preferences and regional COVID-19 levels ("Use and Care of Masks" 2022).

March 2, 2022:
The U.S. federal government launched the "Test to Treat" response wherein people that test positive for COVID-19 at a pharmacy can receive antiviral pills immediately for free ("Remarks by President Biden in State of the Union Address" 2022).

March 7, 2022—Global COVID-19 deaths exceed 6 million
("WHO Coronavirus (COVID-19) Dashboard" 2022)

March 30, 2022:
FDA authorized a second booster dose for Americans ages 50 and older, along with some immunocompromised individuals, at least 4 months after their initial booster ("Coronavirus (COVID-19) Update: FDA Authorizes Second…" 2022, para. 2).

April 8, 2022:
A Commonwealth Fund study estimated that the U.S. vaccination campaign prevented 2.2 million deaths, 17 million hospitalizations, and 66 million infections (Schneider et al. 2022).

April 12, 2022: Worldwide cases reach 500 million
May 12, 2022—1 million deaths in the United States
("WHO Coronavirus (COVID-19) Dashboard" 2022)

May 17, 2022:
The FDA expanded eligibility for the Pfizer-BioNTech vaccine booster dose for children 5–11 ("Coronavirus (COVID-19) Update: FDA Expands Eligibility…" 2022).

June 17, 2022:
Moderna and Pfizer-BioNTech vaccines for children as young as 6 months were authorized by the FDA ("Coronavirus (COVID-19) Update: FDA Authorizes Moderna…" 2022).

June 18, 2022:
The CDC recommended COVID-19 vaccines for children 6 months to 5 years old ("CDC Recommends COVID-19 Vaccines for Young Children" 2022).

June 26, 2022:
The CDC announced that data from the Household Pulse Survey found that 19% of American adults that had COVID-19 still had symptoms (long COVID) ("Nearly One in Five American Adults…" 2022).

Aug. 17, 2022:
CDC director Dr. Rachel Walensky sent messages within the CDC admitting missteps during the pandemic. She noted that the agency must develop better

procedures to respond more efficiently to the next public health emergency (Edwards 2022).

Aug. 31, 2022
The FDA authorized bivalent COVID-19 vaccines from Moderna and Pfizer-BioNTech for use as a booster. These vaccines protect against both the original strain and the Omicron variant ("Coronavirus (COVID-19) Update: FDA Authorizes Moderna, Pfizer-BioNTech Bivalent..." 2022).

Sept. 1, 2022:
The *New England Journal of Medicine* published a study on Pfizer's Paxlovid use for severe COVID-19 outcomes, announcing that mortality for those above 65 years old was reduced by 81%. There was no evidence of benefits for younger adults (Arbel et al. 2022).

Oct. 19, 2022:
The CDC authorized Novavax Monovalent boosters for adults 18 and older ("CDC Allows Novavax Monovalent..." 2022).

Statistics

Global Status as of April 2022
Total cases: **500 million**
Total deaths: **6.21 million**
("WHO Coronavirus (COVID-19) Dashboard" 2022)

US Status as of June 2022
Total cases: **85.7 million**
Total deaths: **1.01 million**
("WHO Coronavirus (COVID-19) Dashboard" 2022)

US Status as of November 2022
Total cases: **96 million**
Total deaths: **1.06 million**
("United States of America..." 2022)

Global Status as of December 2022
Total cases: **641.4 million**
Total deaths: **6.62 million**
("WHO Coronavirus (COVID-19) Dashboard" 2022)

Bibliography

Allam, Zaheer. "The First 50 Days of COVID-19: A Detailed Chronological Timeline and Extensive Review of Literature Documenting the Pandemic." *Surveying the COVID-19 Pandemic and Its Implications*, U.S. National Library of Medicine, 2020, https://www.ncbi.nlm.nih.gov/pmc/articles/PMC7378494/

Arbel, Ronen, et al. "Nirmatrelvir Use and Severe COVID-19 Outcomes during the Omicron Surge: Nejm." New England Journal of Medicine, 1 Sept. 2022, https://www.nejm.org/doi/full/10.1056/NEJMoa2204919

"A Second COVID-19 Case Is Confirmed in Africa." *World Health Organization*, World Health Organization, 2020, https://www.afro.WHO.int/news/second-COVID-19-case-confirmed-africa

Aspinall, Evie. "COVID-19 Timeline." British Foreign Policy Group, 28 Mar. 2022, https://bfpg.co.uk/2020/04/COVID-19-timeline/

Bergeri, Isabel, et al. "Global epidemiology of SARS-CoV-2 infection: a systematic review and meta-analysis of standardized population-based seroprevalence studies, Jan 2020-Oct 2021." medRxiv (2021)

Brandon, Simon. "Spain, Italy and France Are Starting to Lift Their Lockdowns." World Economic Forum, 2020, https://www.weforum.org/agenda/2020/04/three-of-the-worlds-hardest-hit-nations-are-preparing-to-end-their-lockdowns/

"Brazil: WHO Coronavirus Disease (COVID-19) Dashboard with Vaccination Data." World Health Organization, World Health Organization, 2022, https://COVID19.WHO.int/region/amro/country/br

Bungard, Matt, and Mary Ward. "As the Day Unfolded: Act Records New COVID-19 Cases All Linked to Victorian Outbreak; Melbourne Lockdown Preparation Begins as NSW Border Closed." The Sydney Morning Herald, The Sydney Morning Herald, 8 July 2020, https://www.smh.com.au/national/coronavirus-updates-live-melbourne-victorian-regional-areas-prepare-for-COVID-19-lockdown-nsw-border-closed-indefinitely-20200707-p559w3.html

Burki, Talha Khan. "The Russian Vaccine for COVID-19." The Lancet. Respiratory Medicine, U.S. National Library of Medicine, Nov. 2020, https://www.ncbi.nlm.nih.gov/pmc/articles/PMC7837053/

Callard, Felicity, and Perego, Elisa. "How and why patients made Long COVID." Social Science & Medicine, vol. 268, 2021, p. 113426, https://doi.org/10.1016/j.socscimed.2020.113426

"CDC Advises Travelers to Avoid All Nonessential Travel to China." Centers for Disease Control and Prevention, Centers for Disease Control and Prevention, 28 Jan. 2020, https://www.cdc.gov/media/releases/2020/s0128-travelers-avoid-china.html

"CDC Allows Novavax Monovalent COVID-19 Boosters for Adults Ages 18 and Older." Centers for Disease Control and Prevention, Centers for Disease Control and Prevention, 19 Oct. 2022, https://www.cdc.gov/media/releases/2022/s1019-novavax.html

"CDC COVID Data Tracker." Centers for Disease Control and Prevention, Centers for Disease Control and Prevention, 2022, https://COVID.cdc.gov/COVID-data-tracker/#variant-proportions

"CDC Museum COVID-19 Timeline." Centers for Disease Control and Prevention, Centers for Disease Control and Prevention, 16 Aug. 2022, https://www.cdc.gov/museum/timeline/COVID19.html

"CDC Recommends COVID-19 Vaccines for Young Children." Centers for Disease Control and Prevention, Centers for Disease Control and Prevention, 18 June 2022, https://www.cdc.gov/media/releases/2022/s0618-children-vaccine.html

"CDC Updates and Shortens Recommended Isolation and Quarantine Period for General Population." Centers for Disease Control and Prevention, Centers for Disease Control and Prevention, 29 Dec. 2021, https://www.cdc.gov/media/releases/2021/s1227-isolation-quarantine-guidance.html

Chappell, Bill, and Rob Stein. "U.S. Hits 2 Million Coronavirus Cases as Many States See a Surge of Patients." NPR, NPR, 11 June 2020, https://www.npr.org/sections/coronavirus-live-updates/2020/06/10/873473805/u-s-hits-2-million-coronavirus-cases-as-many-states-see-a-surge-of-patients

"China Travel Advisory: Level 3 – Reconsider Travel (28 Jan 2020)." U.S. Embassy & Consulates in China, 28 Jan. 2020, https://china.usembassy-china.org.cn/china-travel-advisory-level-3-reconsider-travel-28-jan-2020/

"Coronavirus (COVID-19) Update: FDA Authorizes First Oral Antiviral for Treatment of COVID-19." U.S. Food and Drug Administration, FDA, 2021, https://www.fda.gov/news-events/press-announcements/coronavirus-COVID-19-update-fda-authorizes-first-oral-antiviral-treatment-COVID-19

"Coronavirus (COVID-19) Update: FDA Authorizes Moderna and Pfizer-BioNTech COVID-19 Vaccines for Children down to 6 Months of Age." U.S. Food and Drug Administration, FDA, 2022, https://www.fda.gov/news-events/press-announcements/coronavirus-COVID-19-update-fda-authorizes-moderna-and-pfizer-biontech-COVID-19-vaccines-children

"Coronavirus (COVID-19) Update: FDA Authorizes Moderna, Pfizer-Biontech Bivalent COVID-19 Vaccines for Use as a Booster Dose." U.S. Food and Drug Administration, FDA, 2022, https://www.fda.gov/news-events/press-announcements/coronavirus-COVID-19-update-fda-authorizes-moderna-pfizer-biontech-bivalent-COVID-19-vaccines-use

"Coronavirus (COVID-19) Update: FDA Authorizes Monoclonal Antibodies for Treatment of COVID-19." U.S. Food and Drug Administration, FDA, 2020, https://www.fda.gov/news-events/press-announcements/coronavirus-COVID-19-update-fda-authorizes-monoclonal-antibodies-treatment-COVID-19

"Coronavirus (COVID-19) Update: FDA Authorizes Second Booster Dose of Two COVID-19 Vaccines for Older and Immunocompromised Individuals." U.S. Food and Drug Administration, FDA, 2022, https://www.fda.gov/news-events/press-announcements/coronavirus-COVID-19-update-fda-authorizes-second-booster-dose-two-COVID-19-vaccines-older-and

"Coronavirus (COVID-19) Update: FDA Expands Eligibility for Pfizer-Biontech COVID-19 Vaccine Booster Dose to Children 5 through 11 Years." U.S. Food and Drug Administration, FDA, 2022, https://www.fda.gov/news-events/press-announcements/coronavirus-COVID-19-update-fda-expands-eligibility-pfizer-biontech-COVID-19-vaccine-booster-dose

"Coronavirus (COVID-19) Update: FDA Limits Use of Janssen COVID-19 Vaccine to Certain Individuals." U.S. Food and Drug Administration, FDA, 2022, https://www.fda.gov/news-events/press-announcements/coronavirus-COVID-19-update-fda-limits-use-janssen-COVID-19-vaccine-certain-individuals#:~:text=On%20April%2013%2C%202021%2C%20the,plan%20for%20proper%20recognition%20and

"Coronavirus (COVID-19) Update: FDA Revokes Emergency Use Authorization for Monoclonal Antibody Bamlanivimab." U.S. Food and Drug Administration, FDA, 2021, https://www.fda.gov/news-events/press-announcements/coronavirus-COVID-19-update-fda-revokes-emergency-use-authorization-monoclonal-antibody-bamlanivimab

"Coronavirus (COVID-19) Update: FDA Takes Key Action by Approving Second COVID-19 Vaccine." U.S. Food and Drug Administration, FDA, 2022, https://www.fda.gov/news-events/press-announcements/coronavirus-COVID-19-update-fda-takes-key-action-approving-second-COVID-19-vaccine

"Coronavirus (COVID-19) Update: FDA Takes Multiple Actions to Expand Use of Pfizer-Biontech COVID-19 Vaccine." U.S. Food and Drug Administration, FDA, 2022, https://www.fda.gov/news-events/press-announcements/coronavirus-COVID-19-update-fda-takes-multiple-actions-expand-use-pfizer-biontech-COVID-19-vaccine

"Coronavirus disease (COVID-19) Situation Report 1 Philippines." World Health Organization, 9 March 2020, https://www.WHO.int/docs/default-source/wpro%2D%2D-documents/countries/philippines/emergencies/COVID-19/WHO-phl-sitrep-1-COVID-19-9mar2020.pdf?sfvrsn=2553985a_2

"COVID-19 and the Need for Action on Mental Health." UN Executive Office of the Secretary-General (EOSG) Policy Briefs and Papers, 2020, https://doi.org/10.18356/13fff923-en

"COVID-19 to Plunge Global Economy into Worst Recession since World War II." World Bank, World Bank Group, 8 June 2020, https://www.worldbank.org/en/news/press-release/2020/06/08/COVID-19-to-plunge-global-economy-into-worst-recession-since-world-war-ii

"COVID-19 Virtual Press Conference Transcript – 10 May 2021." World Health Organization, World Health Organization, 2021, https://www.WHO.int/publications/m/item/COVID-19-virtual-press-conference-transcript%2D%2D-10-may-2021

"COVID-19 Virtual Press conference." World Health Organization, 13 May 2020, https://www.WHO.int/docs/default-source/coronaviruse/transcripts/WHO-pressconference-13may2020.pdf?sfvrsn=ee0d2cde_2

"COVID-19-Associated Orphanhood." The Lancet Child & Adolescent Health, 2022, https://www.thelancet.com/infographics-do/COVID-19-associated-caregiver-deaths

Edwards, Erika. "CDC Director Calls for Drastic Changes to the Agency Following Pandemic Missteps." NBCNews.com, NBCUniversal News Group, 17 Aug. 2022, https://www.nbcnews.com/health/health-news/cdc-director-calls-drastic-changes-agency-rcna43536

"Emergency Use Authorization." U.S. Food and Drug Administration, FDA, 2022, https://www.fda.gov/emergency-preparedness-and-response/mcm-legal-regulatory-and-policy-framework/emergency-use-authorization#vaccines

"Fact Sheet: The Biden Administration to Begin Distributing at-Home, Rapid COVID-19 Tests to Americans for Free." The White House, The United States Government, 14 Jan. 2022, https://www.whitehouse.gov/briefing-room/statements-releases/2022/01/14/fact-sheet-the-biden-administration-to-begin-distributing-at-home-rapid-COVID-19-tests-to-americans-for-free/

"FDA Approves First COVID-19 Vaccine." U.S. Food and Drug Administration, FDA, 2021, https://www.fda.gov/news-events/press-announcements/fda-approves-first-COVID-19-vaccine

"FDA Approves First Treatment for COVID-19." U.S. Food and Drug Administration, FDA, 2020, https://www.fda.gov/news-events/press-announcements/fda-approves-first-treatment-COVID-19

"FDA Authorizes Booster Dose of Pfizer-Biontech COVID-19 Vaccine for Certain Populations." U.S. Food and Drug Administration, FDA, 2021, https://www.fda.gov/news-events/press-announcements/fda-authorizes-booster-dose-pfizer-biontech-COVID-19-vaccine-certain-populations

"FDA Takes Additional Action in Fight against COVID-19 by Issuing Emergency Use Authorization for Second COVID-19 Vaccine." U.S. Food and Drug Administration, FDA, 2020, https://www.fda.gov/news-events/press-announcements/fda-takes-additional-action-fight-against-COVID-19-issuing-emergency-use-authorization-second-COVID

Feng, Emily. "Wuhan's Lockdown Memories 1 Year Later: Pride, Anger, Deep Pain." NPR, NPR, 23 Jan. 2021, https://www.npr.org/sections/goatsandsoda/2021/01/23/959618838/wuhans-lockdown-memories-one-year-later-pride-anger-deep-pain

"Governor Cuomo Issues Guidance on Essential Services under the 'New York State on Pause' Executive Order." Governor Kathy Hochul, 2020, https://www.governor.ny.gov/news/governor-cuomo-issues-guidance-essential-services-under-new-york-state-pause-executive-order

"Governor Gavin Newsom Issues Stay at Home Order." California Governor, 19 Mar. 2020, https://www.gov.ca.gov/2020/03/19/governor-gavin-newsom-issues-stay-at-home-order/

"H.R. 748 — 116th Congress: H.R. 748: Coronavirus Aid, Relief, and Economic Security Act." www.GovTrack.us. 2019. November 10, 2022 https://www.govtrack.us/congress/bills/116/hr748

Hudetz, Mary, et al. "A Second Person in U.S. – and King County – Dies of Novel Coronavirus; More Cases Likely." The Seattle Times, The Seattle Times Company, 1 Mar. 2020, https://www.seattle-times.com/seattle-news/health/officials-2-men-in-king-county-being-treated-for-COVID-19/

"India: WHO Coronavirus Disease (COVID-19) Dashboard with Vaccination Data." World Health Organization, World Health Organization, 2022, https://COVID19.WHO.int/region/searo/country/in

"Johnson & Johnson." Content Lab U.S., 2020, https://www.jnj.com/johnson-johnson-initiates-pivotal-global-phase-3-clinical-trial-of-janssens-COVID-19-vaccine-candidate

"Joint Statement from the International Olympic Committee and the Tokyo 2020 Organising Committee – Olympic News." International Olympic Committee, IOC, 21 May 2022, https://olympics.com/ioc/news/joint-statement-from-the-international-olympic-committee-and-the-tokyo-2020-organising-committee

Kim, Sage J., and Wendy Bostwick. "<? COVID19?> Social vulnerability and racial inequality in COVID-19 deaths in Chicago." Health Education & Behavior 47.4 (2020): 509–513

Lee, Keng. "Coronavirus Kills Chinese Whistleblower Ophthalmologist." American Academy of Ophthalmology, 11 Feb. 2020, https://www.aao.org/headline/coronavirus-kills-chinese-whistleblower-ophthalmol

"Live Q&A on #COVID19 with @DrMikeRyan and Dr @mvankerkhove. #AskWHO" Periscope, uploaded by WHO, 2020, https://www.pscp.tv/w/1mnxelYQVMoJX

McCorkell, L., Assaf, G., Davis, H., Wei, H., Akrami, A., 2020. Patient-Led Research for COVID-19: Embedding Patients in the Long COVID Narrative [WWW Document]. https://doi.org/10.31219/osf.io/n9e75

Menon, Praveen. "New Zealand Says Coronavirus 'Eliminated' and Life Can Resume without Restrictions." World Economic Forum, 2020, https://www.weforum.org/agenda/2020/06/new-zealand-coronavirus-eliminated-restrictions-COVID19-jacinda-arden/

"Moderna's COVID-19 Vaccine Candidate Meets Its Primary Efficacy Endpoint in the First Interim Analysis of the Phase 3 Cove Study." Moderna's COVID-19 Vaccine Candidate Meets Its Primary Efficacy Endpoint in the First Interim Analysis of the Phase 3 COVE Study, 2020, https://investors.modernatx.com/news/news-details/2020/Modernas-COVID-19-Vaccine-Candidate-Meets-its-Primary-Efficacy-Endpoint-in-the-First-Interim-Analysis-of-the-Phase-3-COVE-Study/default.aspx

"Nearly One in Five American Adults WHO Have Had COVID-19 Still Have 'Long COVID.'" Centers for Disease Control and Prevention, Centers for Disease Control and Prevention, 22 June 2022, https://www.cdc.gov/nchs/pressroom/nchs_press_releases/2022/20220622.htm

"Pfizer and BioNTech Announce New Agreement with U.S. Government to Provide Additional Doses of COVID-19 Vaccine." Pfizer, 2022, https://www.pfizer.com/news/press-release/press-release-detail/pfizer-and-biontech-announce-new-agreement-us-government#:~:text=Pfizer%2DBioNTech%20COVID%2D19%20Vaccine%20is%20FDA%20authorized%20under%20Emergency,months%20of%20age%20and%20older

"Pfizer and BioNTech Conclude Phase 3 Study of COVID-19 Vaccine Candidate, Meeting All Primary Efficacy Endpoints." Pfizer, 2020, https://www.pfizer.com/news/press-release/press-release-detail/pfizer-and-biontech-conclude-phase-3-study-COVID-19-vaccine

"Pneumonia of Unknown Cause – China." *World Health Organization*, World Health Organization, 2020, https://www.WHO.int/emergencies/disease-outbreak-news/item/2020-DON229#:~:text=On%2031%20December%202019%2C%20the,the%20national%20authorities%20in%20China

"President Trump on 'Injecting' Disinfectants." C-SPAN, 2020, https://www.c-span.org/video/?c4871173%2Fpresident-trump-injecting-disinfectants

"Public Health Confirms Two More COVID-19 Cases, Bringing the Total in King County to Six." Public Health News and Blog, 2020, https://kingcounty.gov/depts/health/news/2020/March/1-COVID19-cases.aspx

"Public Statement for Collaboration on COVID-19 Vaccine Development." *World Health Organization*, World Health Organization, 2020, https://www.who.int/news/item/13-04-2020-public-statement-for-collaboration-on-covid-19-vaccine-development

"Remarks by President Biden in State of the Union Address." The White House, The United States Government, 2 Mar. 2022, https://www.whitehouse.gov/briefing-room/speeches-remarks/2022/03/02/remarks-by-president-biden-in-state-of-the-union-address/

"Remarks by President Trump on Actions Against China." National Archives and Records Administration, National Archives and Records Administration, 2020, https://trumpwhitehouse.archives.gov/briefings-statements/remarks-president-trump-actions-china/

Routh, Jennifer. "Fourth Large-Scale COVID-19 Vaccine Trial Begins in the United States." National Institutes of Health, U.S. Department of Health and Human Services, 23 Sept. 2020, https://www.nih.gov/news-events/news-releases/fourth-large-scale-COVID-19-vaccine-trial-begins-united-states

Schneider, Eric C, et al. "Impact of U.S. COVID-19 Vaccination Efforts: An Update on Averted Deaths, Hospitalizations, and Health Care Costs through March 2022." Impact COVID Vaccination Efforts: Update Through March 2022 | Commonwealth Fund, 8 Apr. 2022, https://www.commonwealthfund.org/blog/2022/impact-us-COVID-19-vaccination-efforts-march-update

Schnirring, Lisa. "Philippines Has First NCoV Death Outside of China as Cases Top 14,000." CIDRAP, 2 Feb. 2020, https://www.cidrap.umn.edu/news-perspective/2020/02/philippines-has-first-ncov-death-outside-china-cases-top-14000

"Senior UN Official Meets First Person in U.S. Vaccinated against COVID | | 1UN News." United Nations, United Nations, 2022, https://news.un.org/en/story/2022/02/1112642

"South Africa: WHO Coronavirus Disease (COVID-19) Dashboard with Vaccination Data." World Health Organization, World Health Organization, 2022, https://COVID19.WHO.int/region/afro/country/za

"Statement on the Meeting of the International Health Regulations (2005) Emergency Committee Regarding the Outbreak of Novel Coronavirus 2019 (N-COV) on 23 January 2020." World Health Organization, World Health Organization, 2020, https://www.WHO.int/news/item/23-01-2020-statement-on-the-meeting-of-the-international-health-regulations-(2005)-emergency-committee-regarding-the-outbreak-of-novel-coronavirus-(2019-ncov)

"Statement on the Second Meeting of the International Health Regulations (2005) Emergency Committee Regarding the Outbreak of Novel Coronavirus (2019-Ncov)." *World Health Organization*, World Health Organization, 30 Jan. 2020, https://www.WHO.int/news/item/30-01-2020-statement-on-the-second-meeting-of-the-international-health-regulations-(2005)-emergency-committee-regarding-the-outbreak-of-novel-coronavirus-(2019-ncov)

"Timeline-How the Global Coronavirus Pandemic Unfolded." Reuters, Thomson Reuters, 29 Sept. 2020, https://www.reuters.com/article/health-coronavirus-timeline/timeline-how-the-global-coronavirus-pandemic-unfolded-idUSL1N2GN04J

"Timeline: WHO's COVID-19 Response." World Health Organization, World Health Organization, 2022, https://www.WHO.int/emergencies/diseases/novel-coronavirus-2019/interactive-timeline#!

Towey, Robert. "WHO Says Delta Variant Accounts for 99% of COVID Cases around the World." CNBC, CNBC, 16 Nov. 2021, https://www.cnbc.com/2021/11/16/WHO-says-delta-variant-accounts-for-99percent-of-COVID-cases-around-the-world.html

"Tracking Sars-COV-2 Variants." World Health Organization, World Health Organization, 2021, https://www.WHO.int/activities/tracking-SARS-CoV-2-variants

"Transcript for the CDC Telebriefing Update on COVID-19." Centers for Disease Control and Prevention, Centers for Disease Control and Prevention, 26 Feb. 2020, https://www.cdc.gov/media/releases/2020/t0225-cdc-telebriefing-COVID-19.html

"United States – COVID-19 Overview – Johns Hopkins." Johns Hopkins Coronavirus Resource Center, 2022, https://coronavirus.jhu.edu/region/united-states

"United States of America: WHO Coronavirus Disease (COVID-19) Dashboard with Vaccination Data." World Health Organization, World Health Organization, 2022, https://COVID19.WHO.int/region/amro/country/us

Unruh, Larissa H et al. "Health disparities and COVID-19: A retrospective study examining individual and community factors causing disproportionate COVID-19 outcomes in Cook County, Illinois." PloS one vol. 17,5 e0268317. 16 May. 2022, https://doi.org/10.1371/journal.pone.0268317

"Use and Care of Masks." Centers for Disease Control and Prevention, Centers for Disease Control and Prevention, 2022, https://www.cdc.gov/coronavirus/2019-ncov/prevent-getting-sick/about-face-coverings.html

"Vaccines and Related Biological Products Advisory Committee Meeting." Food and Drug Administration, 26 October 2021, https://www.fda.gov/media/153447/download

"WHO Advisory Group Recommends Extra COVID-19 Vaccine Dose for Immuno compromised | | UN News." United Nations, United Nations, 2021, https://news.un.org/en/story/2021/10/1102732

"WHO and MPP Welcome NIH's Offer of COVID-19 Health Technologies to C-TAP." *World Health Organization*, World Health Organization, 2022, https://www.WHO.int/news/item/03-03-2022-WHO-and-mpp-welcome-nih-s-offer-of-COVID-19-health-technologies-to-c-tap

"WHO Calls for Urgent, Aggressive Actions to Combat COVID-19, as Cases Soar in South-East Asia Region." *World Health Organization*, World Health Organization, 2020, https://www.WHO.int/southeastasia/news/detail/17-03-2020-WHO-calls-for-urgent-aggressive-actions-to-combat-COVID-19-as-cases-soar-in-south-east-asia-region

"WHO Coronavirus (COVID-19) Dashboard." World Health Organization, World Health Organization, https://COVID19.WHO.int/

"WHO Director-General's Opening Remarks at the Media Briefing on COVID-19 – 8 April 2020." *World Health Organization*, World Health Organization, 2020, https://www.WHO.int/director-general/speeches/detail/WHO-director-general-s-opening-remarks-at-the-media-briefing-on-COVID-19%2D%2D8-april-2020

"WHO Director-General's Remarks at the Media Briefing on 2019-NCoV on 11 February 2020." *World Health Organization*, World Health Organization, 2020, https://www.WHO.int/director-general/speeches/detail/WHO-director-general-s-remarks-at-the-media-briefing-on-2019-ncov-on-11-february-2020

"WHO Lists Two Additional COVID-19 Vaccines for Emergency Use and COVAX Roll-Out." World Health Organization, World Health Organization, 2021, https://www.WHO.int/news/item/15-02-2021-WHO-lists-two-additional-COVID-19-vaccines-for-emergency-use-and-covax-roll-out

"WHO Statement on Cases of COVID-19 Surpassing 100 000." *World Health Organization*, World Health Organization, 2020, https://www.WHO.int/news/item/07-03-2020-WHO-statement-on-cases-of-COVID-19-surpassing-100-000

"WHO Statement on Novel Coronavirus in Thailand." World Health Organization, World Health Organization, 2020, https://www.WHO.int/news-room/detail/13-01-2020-WHO-statement-on-novel-coronavirus-in-thailand

"WHO: Access to HIV Medicines Severely Impacted by COVID-19 as AIDS Response Stalls." *World Health Organization*, World Health Organization, 2020, https://www.WHO.int/news/item/06-07-2020-WHO-access-to-hiv-medicines-severely-impacted-by-COVID-19-as-aids-response-stalls

"Wuhan's Lockdown Lifted." South China Morning Post, 2020, https://www.scmp.com/video/scmp-originals/3079078/lockdown-wuhan-officially-lifted-life-still-far-normal

Contents

Introduction to Pandemics and Biodefense

David R. Franz

1 Introduction

History can be known to all of us; some of it we have lived and some we have learned from others or read about. While we can't change history, we can learn from history. Our understanding of the present depends to a great extent on where we have been, what we have done and observed, and our resulting frame of reference; unfortunately, today it also depends on who we choose to believe [1]. The present is changing constantly, and the future is unknown. For the most part, we are "driving in the dark" regarding the future [2]. It's likely that the more we know about the past, the better we will be at imagining the future. Finally, we all will have a different perspective regarding what the future might hold, depending on which part of "the historical infectious disease, human behavior, and technology elephant" we understand the best. The authors who have contributed to this volume, as a group, have seen and understand a significant part of our complex "pandemic and biodefense" history. Some have spent careers focused on reducing the negative impact on society of pandemics, epidemics, bioterrorism, biological warfare, and even laboratory accidents or technical surprises. This introductory chapter contains a brief overview of my perspective and understanding of our history regarding *pandemics and biodefense*.

2 The Big Ones We Remember

I grew up in Kansas, roughly midway, as the crow flies, between the small town of Satanta, in southwest Kansas, and Fort Riley in northeast Kansas, both locations often mentioned in the story of the "great influenza" of the early twentieth century

D. R. Franz (✉)
SBDGlobal, Gettysburg, PA, USA

J. M. Rosen, R. R. Colwell (eds.), *The COVID-19 Pandemic*,
https://doi.org/10.1007/978-3-031-62772-9_1

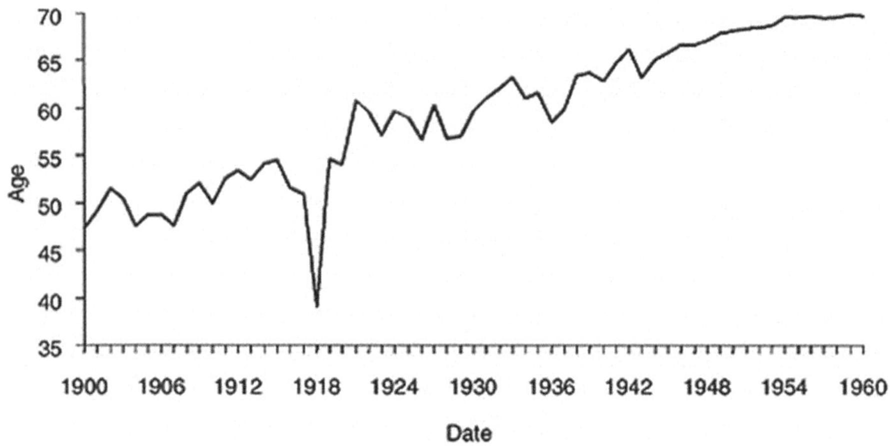

Fig. 1 Prof. Joshua Lederberg carried a viewgraph with this *x–y* plot in his briefcase for many of his informal "lecture/discussions" before groups of scientists and citizens. It depicts the average age at death in the United States and clearly demonstrates the impact of the 1918 pandemic flu virus

[3]. I recall my mother introducing me to her elementary school classmate who survived the flu. The rest I've had to get from books, like most of my peers. Medicine, technologies, transportation, communication, and knowledge of microbes and their workings were much different at that time and the global impact of WWI probably enhanced spread and worked counter to response and control. It is thought that 50 million humans died globally and more than one-half million in the USA. That pandemic drastically changed the life expectancy of Americans, as depicted in a viewgraph chart (Fig. 1), which the late Nobel Laureate Joshua Lederberg used in many of his presentations. The high mortality in the healthy 20- to 40-year-old age group was a surprise. Our predecessors fought that pandemic, not with mRNA vaccines and antiviral drugs, but to a great extent with nonpharmaceutical interventions (NPI). It's interesting that, like our ancestors a century before our time, we started our current COVID-19 pandemic with little more than NPI in our armamentarium as well.

In the background, while the great influenza burned through the younger population, we had come to live with smallpox. This disease of approximately 30% mortality was endemic and pandemic at various times in history. It may have killed 300 million people globally in the twentieth century [4]. Its DNA has been sequenced from mummies that are thousands of years old. Records of inoculations for smallpox in China date back to the tenth century. In a huge accomplishment, the disease was eradicated in 1977—declared officially so in 1979—with a massive vaccination program led by the World Health Organization and in cooperation with many nations. In the end, a vaccine, essentially a "weakened" orthopoxvirus called vaccina, and global cooperation were what made eradication possible. In the final stages, a process called "ring vaccination" was used to build walls of immunity around the few positive cases remaining in the world. While there are likely still some of the extremely stable smallpox "scabs" in existence [5], with most humans vaccinated, the virus ran out of places to hide. Fortunately, smallpox was a "human-only disease," so when *enough* humans were protected, it had no place to go. Eradication may have suggested to the public health community that…"we can do this!" *Now let's just take these endemic viral diseases one at a time and eradicate them.* We would globally eradicate only one more, Rinderpest, a disease of cattle and related wild species, in 2011 [6]. It's just those two that we've put behind us. Along with smallpox, measles devastated native American populations five centuries ago. We've greatly reduced measles incidence at times over the last century, but it's still here and has been in the news during the current pandemic in the context of increased vaccine hesitancy—regarding all vaccines—at home and even unrest and war, in other parts of the world; both have reduced vaccine coverage [7].

And the great influenza wasn't the last "infectious earthquake"; we've had outbreaks and epidemic "aftershocks" ever since: natural, accidental, and even intentional. Some of us remember the polio outbreak of 1952 impacting tens of thousands, mostly children. We also remember the effective vaccine, quickly developed and distributed; for many of us, it amounted to just a trip to our grammar school lunchroom and some red liquid on a sugar cube. I knew one young man in my small hometown who contracted polio; he was confined to wearing aluminum leg braces for the rest of his life. My sister was a young Registered Nurse at the time, caring for paralyzed patients in iron lungs, big stainless-steel tubes into which the patient was placed so the machine could pull a negative pressure around the chest to inflate the lungs, doing the work of the patients' own paralyzed "breathing muscles." We didn't change our patterns of life a lot because of polio in those, my elementary school years, as I recall; at least not for long. I do remember that swimming pools were off-limits in my town, considered likely sources of spread. I also had one childhood friend whose parents negotiated with him and his siblings, in the end agreeing to buy the family their first television set, if the kids would agree to NOT attend the Kansas State Fair that year. There may have always been misinformation in pandemics and epidemics, with so much being new and unknown every time, but in those days there was probably less disinformation than we see today. A friend told me that watching TV while lying on one's stomach with chin on one's hands—thereby overextending the neck and back—was one way to "catch" the paralytic disease. Most of us just swallowed that sugar cube and moved on…but those were simpler times.

We've tried to keep up or even get ahead of infectious diseases since. We have learned a lot from every intervention in our communities by an aggressive virus or bacterium, but not enough to eliminate pandemics, epidemics, or even outbreaks for that matter. The biotech revolution was still 20 or 30 years away, but some of the fundamental research that would support it began to show itself.

3 Early Technical Breakthroughs

In the early 1950s, Rosalind Franklin, using X-ray crystallography, discovered the helical shape of DNA, then Watson and Crick described the double-helix—like a twisted ladder—made up of nucleotide pairs, and demonstrated that it contained genetic information coding "life." This breakthrough doesn't sound like a solution to the problem of infectious disease, but these discoveries were critical in laying the foundation for what would become molecular biology and the powerful tools of a distant biotechnology revolution that would follow.

In the mid-1950s, shortly after the characterization of DNA, oligonucleotide synthesis was introduced as a new and powerful tool. This was the ability to artificially synthetize short lengths of DNA and RNA which eventually supported powerful laboratory capabilities and laid the early groundwork for DNA sequencing developed by Frederick Sanger in 1977. Sanger sequencing would become the most widely used sequencing technique through the last quarter of the twentieth century. Then, next-generation sequencing (NGS), such as is done today globally on Illumina machines (Illumina, San Diego, CA) would be discovered in the mid- to late 1990s and become widely used worldwide, ever since [8]. The powerful polymerase chain reaction (PCR), developed in 1983 [9], gave us the ability to identify specific segments of DNA or RNA. Along with synthesis and sequencing, PCR has had an enormous positive impact on the science necessary to deal with both pandemics and biological security. These tools have allowed us to identify the genome of an agent very specifically, and with the addition of powerful informatics, to detect the agent in a haystack of human, animal, or contaminant nucleic acids.

4 Transition, New Challenges, and Still Learning from History

Most of us remember the HIV pandemic that the world became aware of in the early 1980s. It is still ongoing globally, now often called an epidemic. This was an example of a viral illness for which antiviral development has been underway for many years. Nonpharmaceutical interventions were all we had for nearly 10 years after the discovery of the disease. HIV also carried a stigma, impacting predominantly the illicit sex and drug-user community. We remain, today, heavily dependent on chronic drug treatments to control the secondary disease, AIDS, and keep patients relatively free from clinical signs. Approximately 4000 people, globally, become infected daily [10].

So, now we've begun talking about natural disease and the development of tools to diagnose and, hopefully, control some of the diseases. Like almost everything that is natural or man-made, bacteria and viruses have been used to do intentional harm.

Bacteria and viruses were discovered a long time ago. Antony von Leeuwenhoek first saw and described bacteria in 1676, but it wasn't until the mid-1800s that Semmelweis demonstrated disinfection could cure disease. Then, four decades later in 1892, Dmitri Ivanovsky discovered the tobacco mosaic virus by demonstrating that there was something smaller than a bacterium, an infectious particle which was not filtered out of a solution when all bacteria were.

It typically doesn't take long, when a new discovery or technology comes along, for someone with malintent to use it for harm; that has been true for a long time. Improvised biological weapons are said to have been used in centuries past, the most common early mention is of the Mongols catapulting plague (*Yersinia pestis*) infected human bodies over the walls of Caffa, in what is now Ukraine, in the 1300s. We, of course, don't remember any of that. Biologicals were not used offensively in WWI; only chemicals. Perhaps circulating H1N1 was a deterrent reminder that bacteria, or even more so a virus, used as a weapon might be complicated and outcomes difficult to control.

5 Misuse in the Modern Era

The USA developed biological weapons of war from roughly 1943 to 1969. It is my understanding, based on reviewing old literature and numerous personal discussions with Mr. William (Bill) C. Patrick, III and Mr. Tommy Dashiell, both key scientists in what we call the "old program" at Ft Detrick, that the USA had developed potentially highly effective biological weapons by the time the program ended in 1969: possibly of nuclear equivalence.[1] The several field tests conducted, namely, simulant trials in the Pacific and San Francisco Bay area and in the New York City subway system, all in the 1960s, were convincing. In 1969, President Nixon visited Ft Detrick, MD, and announced that the US offensive program had ended [11]. The USA then took the lead in supporting, signing, and ratifying the Biological Weapons Convention (BWC) which went into force in 1975. The BWC has been ratified by 184 states as of 2022 [12].

The Soviet Union, which had begun an offensive program in the 1920s, signed the BWC in 1972 and ratified it in 1975. However, the Soviets continued their offensive efforts on a massive scale, increasing the scope of their program eventually to 30–40 thousand scientists and engineers between 1972 and 1992. Their offensive work continued, even after a 1979 industrial accident at an anthrax plant in Sverdlovsk [13] killed more than 60 people downwind of the plant. The world's discovery of the program led to President Yeltsin's admission of the 1979 accident in 1992, which was a turning point in Western awareness of the enormous, highly secret modern-era Soviet offensive program.

In 1989, Vladimir Pasechnik, a senior Soviet biological weapons scientist defected to the UK, bringing with him knowledge of the huge science and industrial base of approximately 20 institutes called Biopreparat, a network of nominally civilian laboratories and production facilities, established in 1974 to support the Soviet Ministry of Defense (MoD) biological warfare program. That core offensive

[1] Franz, DR. (1995–2009) Personal communication with Mr. William C Patrick III. ....AND... Franz's Statement to House Committee on Foreign Affairs, May 1, 2014.

MoD program was centered in facilities in Sergiyev Posad, Sverdlovsk (now Ekataterinburg), and Kirov region, possibly along with other sites. There was also a large network of Anti-Plague Institutes, apparently a "legitimate" public health network, but a ready source of interesting pathogens passed to the MoD for potential weaponization. The Soviets had denied this massive program, even after the Sverdlovsk anthrax accident exposed part of it. Details provided by Pasechnik [14] and, a few years later by a second defector from Biopreparat named Alibekov, would aid the West significantly in understanding what had gone on in these highly secret facilities, some in so-called "closed cities," which were typically in remote locations and not shown on maps of the time.

Also in 1989, during what was believed to be a lethal outbreak of Simian Hemorrhagic Fever in a research monkey colony in Reston, VA, an apparent filovirus was discovered through electron microscopic examination of tissues from dead monkeys. Further examination of tissue samples in a BSL-4 laboratory at Ft. Detrick, MD, concluded that it was, in fact, an Ebola virus. Investigating scientists had to assume what became known as Ebola Reston was as lethal as African strains; therefore, the animal colony was depopulated and the entire facility was eventually raised. In the end, several animal handlers seroconverted, demonstrating that they had been exposed to the virus, but none became ill. The monkeys had been imported from the Philippines and represented the first introduction of an Ebola virus to north America. A national best-selling book, *The Hot Zone*, told the story and found great interest, particularly among young science students and their teachers [15].

About this time, we also learned of an Iraqi BW program. While Saddam Hussein's scientists and engineers began working on biological weapons in the early 1980s, it was not until the first half of the decade of the 1990s that we understood the scale and details of the program.

During the last few years of the Cold War, now with the knowledge of both the enormous Soviet offensive program and the much smaller Iraqi one, the focus of the US research and development of medical and nonmedical defenses was located inside the Department of Defense (DoD). Field force leadership, typically with a much better understanding of chemicals than biologicals—and some experts felt naively—prioritized "standoff" field detectors as their first procurement "wish" and physical protection as their secondary concern. DoD also focused on vaccines and diagnostics, for what we called the Soviet "dirty dozen" favorite pathogens, and threat and response education for healthcare providers. The US Army Medical Research Institute of Infectious Diseases (USAMRIID) was the core medical lab; its budget was approximately $22M in 1998, the year I retired as its commander. Most of the nonmedical research and development (sensors and protective clothing, respectively) was conducted at Edgewood Arsenal, MD, and the army's Natick Laboratories in Massachusetts. The entire S&T budget for biodefense within the DoD was $137M in 1997.[2]

[2] Compare that to Operation Warp Speed which was provided ca. $18 B—with a 'B'—for COVID vaccines during the Trump administration alone.

6 Post Cold War Threats

In January of 1991, as the former Soviet Union was imploding, the first team of the USA and UK biodefense scientists and diplomats visited several Biopreparat laboratories in Russia. A huge offensive program had apparently ended, but there was concern about what had happened to all the weapons experts. In December of 1991, a similar group of Russians, mostly from the Biopreparat BW industrial plants, visited Dugway Proving Ground, USAMRIID, and Pine Bluff, Arkansas, the latter a production site which had supported the now defunct US offensive program. The Soviet team returned home as Russians, with the USSR having dissolved in their absence. On December, 25, 1991, Gorbachev resigned, and Yeltsin took charge, now as the president of the Russian Federation. We, the hosting party at USAMRIID, saw the group depart for home with somber faces; they had come as Soviets and were returning as Russians. During that visit, Kanatzhan Alibekov, deputy director of Biopreparat, asked one of the USG hosts for a business card. He would later contact that individual and ask for help to defect to the USA.

In 1992 after serving as a member of the Soviet delegation, Ken Alibek did defect to the USA, like Pasechnik to the UK in 1989, bringing with him even more information about the massive offensive Soviet program. The publication of his autobiography, *Biohazard*, in 1999 would give the world the first comprehensive sense of the enormous scale of the secret program [16]. His estimate was that 30–40 thousand scientists and engineers had been employed, conducting relevant research, development, and testing of biological weapons in the former Soviet Union.

In 1992, as we breathed a sigh of relief at the demise of the Soviet BW program, the National Academies of Science Institute of Medicine (now the National Academy of Medicine) released the iconic and still relevant Lederberg-Shope report on *Emerging Infections* [17] reminding us that natural disease threats would continue and might intensify in the coming years. It was not a time to relax regarding microbial threats. The report underscored that microbes are subject to "genetic change and evolution." As a result, they might infect new hosts, have altered susceptibility to antimicrobials, and catch a ride with humans and animals into new geographic ranges. At the same time, the report reminded us that the human population was constantly changing, with respect to host immunity, travel, and behavior. Changes in climate and animal host ranges would also impact relationships of humans and animals with microbes. Sound familiar? While we had recently experienced great success with smallpox eradication, the authors reminded us that we should not forget that infectious disease would remain a major cause of illness and death worldwide. They went on to make still-relevant recommendations regarding the importance of preparation to be able to recognize diseases and intervene and carry out research, training, vaccine and drug development, vector control, public education, and behavioral change. We would do well to measure our current preparedness against that Lederberg-Shope timeless yardstick. Thirty years later, infectious diseases just keep coming, as underscored again in late 2022 by Anthony Fauci in a review paper published at the end of his 40 years chasing emerging diseases [18].

The Trilateral Agreement of 1992 was established between the USA and UK on one side and Russia on the other. The purpose was to attempt to assure that the Russian BW program no longer existed. Two US/UK teams visited four sites belonging to the Biopreparat: Pokrov, and Berdsk in October 1993, with Amb. Ed Lacey as the delegation head and Omutninsk and Obolinsk with Amb. Don Mahley.

After the Lacey and Mahley-led visits, relationships remained frosty between the US/UK delegation and the Russians. Little trust was built in the process of the on-site inspections. A Russian team then visited several sites in the USA which they charged might house a current BW program and the Russians also visited Evans Medical Laboratories in the UK. I was responsible for *prepping* personnel at those three US sites before the Russians arrived. The trips to the USA and UK, by the Russian inspection teams, were to take advantage of an agreed symmetrical solution to an obviously asymmetrical problem. The Russians' choice of the Pfizer facility at Groton, CN, raised the ire of Pfizer leadership such that we had to ask VP Al Gore to intervene with the Pfizer president before the visit could occur. Following the trilateral visits and several meetings of the three parties, held in Moscow, London, and Washington DC, the "Trilats" died a silent death, with the Russians still accusing the USA of hiding an offensive program, this time in overseas laboratories.[3]

When the activities planned and agreed to under the Trilateral Agreement ground to a halt, Senators Sam Nunn and Richard Lugar had another plan already underway. The "arms control" approach used in the Trilats gave us some access, but really very little progress in improving our knowledge of their offensive program and enhancing legitimate opportunities for former weapons scientists. The Nunn-Lugar Cooperative Threat Reduction (CTR) Program of 1991 provided DoD, and later Department of State (DoS), funds to support collaborative research between "former weapons scientists" in the Former Soviet Union (FSU) and Western scientists. This was probably the first major application of "science diplomacy" to biodefense. As we can see today, it wasn't a permanent solution, but it did have the effect of redirecting highly skilled weapons scientists to research and development in public health, food production, pharmaceuticals, and other areas that helped the Russian people and helped revive their economy at a time of desperate need. During these times, people in the Moscow ornate Soviet era subway stations walked silently dressed in gray and brown. In those early years, it was not unusual to see grandmotherly women standing along the side of the road with a pair of old shoes or a loaf of homemade bread, hoping to sell them for a few rubles with which to buy food for their families. In the Biopreparat labs, scientists turned from growing bacteria for weapons to growing bacteria like Bifidobacter to produce yogurt, or to sell for food or in supplements. Now they were pleased to have their names added to the list of "former weapons scientists" so that they might submit a proposal to the CTR program for funding.

[3]As I'm writing this, the Russians have been broadcasting dis/misinformation suggesting that the USA has biological weapon programs ongoing in public health laboratories in the Ukraine…30 years later. Some things never change.

A mechanism was developed to allow the transfer of funds to Russian scientists directly without being taxed by the Russian government. The International Science and Technology Center (ISTC) was established in Moscow in 1994 between Canada, the USA, the European Union, Japan, Norway, and South Korea (funding Parties), as well as Russia, Armenia, Belarus, Georgia, Kazakhstan, the Kyrgyz Republic, and Tajikistan (recipient Parties). In addition to funding scientists, the ISTC supported workshops, hands-on training, and other events to bring scientists from the Former Soviet Union (FSU) and the West together. It is said that the ISTC program supported nearly 60 thousand weapons scientists in hundreds of laboratories across Russia and the Commonwealth of Independent States. In 2015, President Putin announced the ISTC was no longer welcome in Russia; this was after the funding parties involved had contributed approximately $1B to support and help redirect former Soviet weaponeers.

At approximately, the same time (early to mid-1990s) that the US and UK diplomats and scientists were working to understand what was going on in the Russian program, the world became aware of Saddam Hussein's offensive BW program in Iraq. While Iraqi scientists and engineers began working on biological weapons in the early 1980s, it was not until the first half of the decade of the 1990s that we understood the scale and details of the program. The Iraqi BW program was a "rounding error" in scale and threat potential, compared to the Soviet program.

The UN Special Commission on Iraq (UNSCOM) [19] was established to assure that Iraqi weapons activities were uncovered and destroyed. Reporting to Ambassador Rolf Ekeus in the period of '93–'95, I led the third, fifth, and twentieth BW international inspection teams under UNSCOM. UNSCOM was created with the adoption of UN Security Council Resolution (UNSCR) 687, to oversee Iraqi compliance with the BWC and the destruction of Iraqi chemical, biological and missile weapons facilities and materiel. There were approximately 20 countries represented in the commission. The UNSCOM conducted a series of invasive inspections of suspect Iraqi facilities between the two Gulf Wars. The core Iraqi BW facilities, at Al Hakam, Salman Pak and Al Manal, were a tiny fraction of the size of those industrial-scale facilities supporting the Soviet program. Demilitarization of the Iraqi laboratories by UN teams and redirection of human staff were completed by 1996 and the facilities were ultimately destroyed before or during the second Gulf War in 2003. It was those experiences, working closely with the weaponeer-scientists involved, that caused me later to develop the concept of scientists "caught in the middle" between their own government and the global authorities attempting to ferret out the truth. That same concept may find relevance in the COVID-19 pandemic [20].

7 Advances in Technology and Concern about Non-state Actors

The Internet was invented in the early 1980s and became the popular means of communication in the 1990s. In 1989, D.A. Henderson of smallpox eradication fame proposed an Internet-based system to monitor and report emerging disease by allowing essentially any interested scientist or practitioner to report a suspicious disease in near real time. DA's vision would eventually emerge as Program for Monitoring Emerging Diseases (ProMED) mail 5 years later. The ProMED was founded in 1994 by Steven Morse, Barbara Hatch-Rosenburg, and Jack Woodal. Its purpose is to promote (rapid) communication among scientists, physicians, veterinarians, epidemiologists, and public health professionals internationally. It wasn't called "One Health" back then, but it became an excellent example of the concept.

The Sarin attack by a doomsday religious cult, the Aum Shinrikyo, in 1995, would, in the USA at least, have an influence on the awareness of potential biological attacks. That attack with a chemical agent stimulated discussions regarding homeland preparedness more effectively than the uncovering of the Soviet and Iraqi BW programs. We, in the USA, in those pre-911 days, still believed big oceans protected us on the east and west, and friendly neighbors, north and south. From my position as commander of the army's medical biodefense laboratory, I watched as the US government clearly took notice that, if such a chemical attack could occur on a public square in Japan, a similar chemical or biological attack could occur in a US city. It was more of a wake-up call than one would have expected from this very different kind of attack.

I recall that we, at my level in DoD, first started collaborating with Department of Health and Human Services (HHS) and seriously thinking about protecting the homeland from biological threats as the government began preparing for the 1996 Atlanta Olympics. Before then, our mission was focused on protecting the force, not the public. Note that this was before the Department of Homeland Security (DHS) had been established. DoD was the only department focused on medical and non-medical biodefense.

Joshua Lederberg, Nobelist and former president of Rockefeller University, was very interested in biological weapons and supportive of defensive programs. In 1998 he was interviewed for a *New Yorker* article by Richard Preston. As part of a verbal response to one of Richard's questions, Lederberg stated, "There is no technical solution to the problem of biological weapons. It needs an ethical, human, and moral solution if it's going to happen at all. Don't ask me what the odds are for an ethical solution, but there is no other solution."<*Then Lederberg paused and said,*> "But would an ethical solution appeal to a sociopath?" In that statement, Lederberg described a reality that was true before his time, true when he said it and remains possibly even more relevant today. With the advent of ever more powerful tools to manipulate microbes, while still an important warning from a man of wisdom, it is becoming an understatement.

In 1999, a mysterious spate of dead birds in New York City demonstrated the value of thinking in terms of "one health," when the pattern of species dying and species not affected, contributed to an ultimate identification of the West Nile virus in the USA. Here, what we had long called an "African disease" became endemic and spread across North America quickly, killing many birds, some horses and sickening… even killing…humans. The normal enzootic cycle is passage from birds, serving as reservoirs, to mosquitos and back to birds. Both susceptible mosquitoes and susceptible birds (not all species of birds) are found across the USA, so we were extremely vulnerable when either an infected bird or mosquito appeared in Queens,

NY…possibly having traveled by ship. When birds and humans started dying, the CDC initially identified the outbreak agent as the St Louis Encephalitis virus. But Tracey McNmara, an astute veterinary clinician at the Bronx Zoo noticed that crows, jays, and ravens were dying, while chickens were not. These findings prompted the CDC to take a second look at the diagnostics and eventually change the diagnosis to West Nile; an epidemiologist's dream…or nightmare. WNV spread rapidly across the USA. Because the horse is the amplifying host, immunization of horses is a very effective control measure. It is now endemic in the USA, a disease that did not exist here before 1999. And more than 100 humans die of West Nile annually.

In 1999, Matthew Messelsen, who served in the Nixon White House at the time the USA ended its offensive biological warfare program, made the following statement on the misuse of technologies. "Every major technology…has been intensively exploited, not only for peaceful purposes but also for hostile ones. Must this also happen with biotechnology, certain to be a dominant technology of the [twenty first] century."[4] Meselson's thoughtful quote reminded me that humans had used bacteria and viruses as weapons against other humans centuries before the concept of bacteria or viruses was understood and before either was discovered. Now, he was suggesting we think about how our new biotechnology tools might change things going forward…and not just in a good way.

Also in 1999, Joe Rosen, the driving force behind and co-editor of this book, envisioned, organized, and chaired a meeting at Dartmouth College on the topic of biosecurity, in the context of future health systems development. At the time, not many individuals or governments were interested in biosecurity. The meeting was visionary and I remember the meeting very well even after more than 20 years. Following that meeting, Rosen has continued exposing—a word I use cautiously—his Dartmouth students to the thinking of the experts, some of whom you will meet in the following chapters. Their lectures unlocked a treasure trove of information for the students. I'm sure some of what they heard in the classroom will stay with them throughout their careers and impact the decisions they make as they serve.

Just 2 years later, 2001, we were faced with anthrax-spore-laced letters that would change the way the US government would think about intentional biological threats, at least for a few years. The fact that the now infamous "Anthrax Letters" were sent within a month after the World Trade Center attacks only magnified their impact. Five people died, many were given prophylactic antibiotics, and over $300 million was spent on cleanup and future prevention technologies. The budget for biodefense jumped from $137M in 1997,[5] all within the Department of Defense, to a total biodefense budget of *ca.* $6B in 2002 spread across numerous agencies, but particularly increasing support for the National Institute of Allergy and Infectious Disease (NIAID) within the Department of Health and Human Services (HHS). Funding for biodefense would be greatly increased for 3–5 years and then, when there had not been another attack, the concern, emphasis, and spending would begin

[4] Messelsen, M. Personal communication. Jan 1999 at the Harvard School of Public Health.

[5] The budget for "Science and Technology".

to shift to naturally occurring emerging infectious diseases. The NIAID would build approximately eight or ten BSL-3 and two BSL-4 high-containment laboratories on university campuses. Both BSL-4s and most of the BSL-3s are functioning today, but the recipients have learned how expensive their operations and maintenance can be. It was *a gift that has kept on asking*.

As the USA built high-containment laboratories, they became a status symbol worldwide. We went from two BSL-4s in the USA before 2001 to 13 operational or planned today. Globally, we went from approximately 12 to more than 69 today, in operation, under construction, or planned [21].

COVID-19 isn't our first experience with SARS coronavirus. In 2002, an outbreak of SARS was first discovered in China, like the current one, and spread to more than 25 countries around the world, infecting approximately 8000 people and causing more than 700 deaths. There were cases in North America, but the USA was spared. China was hit the hardest by SARS 2002 and responded in the next couple of years by connecting hospitals across its vast country with glass-fiber communications systems, upgrading their high-containment laboratories, researching basic bench biology to understand coronaviruses, and beginning to seek countermeasures [22].

8 The Biotechnology Revolution

The biotechnology revolution, to a great extent, is built on the foundation first laid by Watson, Crick, and contemporaries and began a steep climb around 1980 and has continued to contribute tools for understanding and manipulating microbes. The cost and time of gene synthesis, sequencing of DNA, and the process of determining protein structures have made dramatic progress, with cost dropping and the speed and efficiency of sequencing, increasingly reminiscent of Moore's Law. In 2003, Rob Carlson published the book, *Biology is Technology* [23], bringing new awareness of the rapidly growing economic role of technology in biology and the importance of stepping up to a coming global competition. In 2014, during a one-on-one conversation in a Beijing hotel, a senior Chinese Party bureaucrat told me, with a smile, "The Europeans won the industrial revolution; you [US] won the information revolution; we [China] will win the biotech revolution."

The concept of One-Health has always existed to some degree, long before it was named or defined. Shortly after the turn of the century (2003–2004), likely influenced by some of the events discussed above, the concept began to crystalize in the minds of a few thoughtful physicians and veterinarians [24]. The US CDC website states, "One Health is a collaborative, multisectoral, and transdisciplinary approach—working at the local, regional, national, and global levels—with the goal of achieving optimal health outcomes recognizing the interconnection between people, animals, plants, and their shared environment." Essentially, it says that "we're all in this together," so we had better accept that fact and work across the disciplines impacting humans, animals, plants, and the environment. Nowhere is this more true than for infectious diseases. The CDC site goes on to say, "Scientists estimate that more than six out of every ten known infectious diseases in people can be spread from animals, and three out of every four new or emerging infectious diseases in people come from animals." While "one-health" collaborations between and across scientific disciplines are common, as well as efficient and critical to human animal and environmental welfare, it can be difficult to force the "one-health" concept. In my experience, it works best when technical experts work together because each is best equipped to contribute to finding a place on the board for a different piece of the puzzle [25].

The International Genetically Engineered Machines (iGEM) [26] competition was started by a small group of MIT students in 2003–2004 as a summer project. The purpose was to promote the safe and open development of engineering tools for biology. iGEM has grown in the last 20 years to become a worldwide competition and has also added a competitive category for high school students. The tools of synthetic biology are used in silico and in vitro in student laboratories at MIT. iGEM provides a registry of Standard Biological Parts maintained on a database as well as in a freezer. Students use these parts, or add their own to the Registry, to build biological "machines." The competition has grown in both the number of teams entering and the global location from which they come. In 2021, there were 350 teams and more than 7000 individuals participating. The "Overgrad" team winner was from Marburg, Germany; the "Undergrad" team from Toulouse, France, and the High School team winner from China. From the beginning, iGEM programs have inculcated safety, security, and ethical use of the tools of biotechnology.

A pivotal *2003* NASEM report, *Biotechnology Research in an Age of Terrorism* would formally alert the scientific community of potential dangers of misuse of some of the new technologies. Affectionally called "the Fink Report," after the chair Gerald Fink. It laid out examples of, not what should NOT be done, but rather what should trigger significant extra consideration about safety of the proposed research. Examples include, [The work…] "…would demonstrate how to make a vaccine ineffective"; or "…would alter the host range of a pathogen."[6] The report also called for the establishment of an oversite body; it became the *National Science Advisory Board for Biosecurity* (NSABB) in 2004. The NSABB laid out a *Proposed Framework for the Oversight of Dual Use Life Sciences Research: Strategies for Minimizing the Potential Misuse of Research Information* [27], in 2007, initially focusing on work with replicating infectious agents. Before this 2007 document from the NSABB, the policy had been, to await the decision to publish results until the manuscript was submitted for publication. The 2007 guidelines suggested a go/no-go decision be made during hypothesis development; much earlier in the research process. The NSABB further clarified the guidelines for these research activities in 2016 [28].

In 2005, shortly after the NSABB was established, the journals *Nature* and *Science* were faced with the question of whether to publish the newly determined sequence of the H1N1 1918 flu virus [29]. Some suggested that, if we publish the data, a bad actor will surely resurrect the virus and release it to the world. We will

[6] The seven examples came to be called, "The Seven Deadly Sins."

be facilitating another pandemic. Others argued that there are more "good" molecular virologists than "bad" and the new sequence data would support medical countermeasures development, thereby helping get ahead of any malicious reintroduction of the virus. Still, others suggested the virus would not have the same impact on the global population the second time around. In the end, the NSABB voted to publish the work. This was the first significant task for the board and would inform the group's thinking when it faced similar but even more difficult questions about the publication of the recipe for making highly pathogenic avian influenza virus (H5N1) transmissible via the respiratory route 5 or 6 years later [30].

Between 2005 and 2006, during the presidency of George W. Bush, the USA developed a flurry of Pandemic Preparedness Plans and Strategies, focused primarily and appropriately on pandemic influenza. One of the plans was updated by the Trump administration in 2017 and all were "on the shelf" when the COVID-19 pandemic began. What was that Gen Eisenhower quote? "When preparing for battle, I have always found that plans are useless, but planning is indispensable." I believe the jury is still out regarding the utility of those, now 15–20 year old plans.

The headline in the Frederick (MD) News-Post on August 2, 2008, was "Anthrax Case Turns." Bruce Ivins, a microbiologist at the US Army Medical Research Institute of Infectious Diseases had committed suicide, after an intensive investigation and application of significant psychological pressure on Bruce and his family from the FBI. From law enforcement's perspective, the hunt was over. Circumstantial evidence in the case has been called "overwhelming" that Bruce was the perpetrator. It's said that Bruce had begun drinking heavily and was seeing a psychiatrist for the previous year or two. Yet, a NAS report later found that the FBI may not have had definitive evidence when they "found Bruce guilty" [31]. Before accusing Ivins, the FBI had fingered Steve Hatfill, a physician and virologist who served as a National Research Council fellow at USAMRIID in the late 1990s. Hatfill was under surveillance for several years, his home aggressively raided and his phone tapped. He was fired from his job with a major defense contractor. Hatfill eventually filed lawsuits against several journalists and publications, claiming defamation. He won suits against *Vanity Fair* and *Readers Digest* and was awarded multiple million dollars in damages [32].

Four months after Bruce's death, in 2008, a simple but significant observation was published in a report entitled "World at Risk," a report of the "WMD Commission." Regarding biological threats it said, "The United States should be less concerned that terrorists will become biologists and more concerned that biologists will become terrorists" [33]. One can't help but wonder how much the timing of Bruce's suicide impacted that statement, calling for us to turn inward to look for malevolents in our biology labs. We had been thinking about Al Qaida or some other terrorist organization as the "bad guys," not a friendly middle-aged scientist, at a DoD lab, who composed songs and poems for departing colleagues and had taught a young people's mass in his local church for years. Almost immediately—on 8 Aug—House of Representative members John Dingle and Bart Stupak wrote a letter to the President: "If these allegations are true the FBI has identified serious weaknesses in the security at one of our Nation's premier laboratories for the study of some of the most deadly pathogens in the world. Their allegations also raise equally troubling security concerns about the thousands of other scientists and technicians who work at hundreds of laboratories across our country with select biological agents such as anthrax." The urgency to turn inward was underscored.

In 2009, the National Academies of Science released a pivotal report supporting Cooperative-Threat-Reduction entitled, *Global Security Engagement: A New Model for Cooperative Threat Reduction* (later informally named "CTR2.0"). The report praised the Nunn-Lugar program, ongoing in Russia since 1994, and sought to help prepare the program for the future. The engagements had begun to become bureaucratic and, in some cases, micromanaged from Washington DC; in some ways, it had gotten away from true "science diplomacy" that had been its core. The report recommended that the program should be expanded geographically, that it become nimbler, and that it refocus on the value of building human relationships and transparency through science collaboration.

A *National Strategy for Countering Biological Threats* was released by the Obama administration as its second Presidential Decision Directive (PDD) in November of 2009. For decades, our policy approaches have been "prevention" for nuclear threats and "protection, response and recovery" for biological threats. This Obama PDD emphasized both prevention of biological threats and, supporting the then recently released CTR 2.0 NAS report, added for the first time at the White House level the role of international engagement as an important component of reducing threats from biological warfare and terrorism.

In Sept of 2011, Ron Fouchier of Erasmus University in Holland announced at a scientific meeting in Malta some "...very bad news..." that he had modified H5N1 and described the resulting virus saying it was..."as efficiently transmitted as seasonal virus." [Note: H5N1 bird flu virus typically causes serious disease or even death in humans, but fortunately is not easily transmitted between humans by the respiratory route.] The following month in *Science*, he stated that his viral product was "Probably one of the most dangerous viruses you can make." When he and Yoshi Kawaoka of the University of Wisconsin each submitted similar papers to *Science* and *Nature*, respectively, the journals asked the NSABB to evaluate them for "safe" publication. The NSABB asked for a few redactions, a two-month moratorium on related research and that the question of what could be published be brought to an international committee of experts. The committee, convened at WHO HQ in Geneva in February and, reviewing now-revised manuscripts, said that the studies were valuable, that they highlighted safety and security concerns but should continue and that redaction and limiting distribution were not feasible. The NSABB re-reviewed revised manuscripts and by a majority, but not unanimous, vote agreed that the versions reviewed at WHO be communicated fully. It was a good lesson, reminding us that in a free society, it's probably not possible to withhold the results of academic research.

In 2012, CRISPR Cas9, a unique and powerful genome editing tool became the hottest new and promising molecular technology in global laboratories and at the same time quickly gave us pause. This new tool would be used by Chinese scientists to modify the germ line DNA in tens of human embryos terminated at 14 days, and later, in 2018, CRISPR was used, again in China, in a pair of twin humans carried to term. It was a major step forward in gene editing technology and will remain so, in most cases for the "good" of mankind.

Ebola virus, which had caused outbreaks and epidemics in Africa on numerous occasions since 1976, emerged again in 2014, resulting this time in a significantly large epidemic and more than 10,000 deaths in West Africa. This was the first time that an Ebola virus lethal to humans was seen in North America. There were two deaths in the USA, from a total of 11 cases; nine contracted the disease in Africa and two in the USA while caring for imported patients. For years we had noted and even responded to Ebola outbreaks in Africa but, like West Nile, had come to expect Ebola to stay in Africa. We were again reminded that we are living in an ever "smaller" world…and microorganisms do not respect borders or geographic distance.

In 2014, just 2 years after the Ron Fouchier–Yoshi Kawaoka research manipulating the genome of H5N1 to make it transmissible in ferrets via the respiratory route, a team of scientists from the University of Maryland and the Virginia-Maryland Veterinary College produced a variant of H7N1 also making it transmissible, but without technology. They did it, simply, by passaging the virus ten times in ferrets. Furthermore, there was no loss of virulence in the resulting virus. Repeating those Fouchier–Kawaoka experiments would require significant expertise and laboratory infrastructure whereas the H7N1 work could be done in a garage.

In May 2014, a research laboratory at Dugway Proving Ground mistakenly shipped inadequately inactivated anthrax spores to diagnostic laboratories in nine states and a joint DoD/Republic of Korea army laboratory in South Korea. No one became ill and recipient laboratories were able to inactivate the samples safely. A similar but unrelated safety failure at the Centers for Disease Control was made public, in July 2014. As many as 75 people may have been exposed when anthrax spore preparations, thought to be inactivated, were taken outside a BSL-3 containment suite. Some employees were administered antibiotics as a precautionary measure. While no illnesses resulted, national political leadership took notice and stepped in.

In October of 2014, we learned of the "White House Pause." Now, with the Fouchier–Kawaoka H5N1 experiments in mind, the UMD passaging H7N1 study and two near-miss human errors at two USG labs, the White House's National Security Council (NSC) took action. In Oct 2014, the NSC called for a halt to influenza, MERS, and SARS research involving the manipulation of viral genomes, now called "gain of function" studies. The NIH announced that, while such studies are important… "because they help define the fundamental nature of human-pathogen interactions" they also, "entail biosafety and biosecurity risks which need to be understood better" [34]. The "pause" would be lifted on December 19, 2017, by the director of the NIH in response to the *Framework for Guiding Funding Decisions about Proposed Research Involving Enhanced Potential Pandemic Pathogens* from the NIH [35]. The framework described the responsibilities of the NIH and other funding agencies to assess risk carefully.

Later in 2014, just as we thought the "White House Pause" would make us safer from high-risk viral research conducted by humans in high-containment laboratories, nature sneaked up on us yet again. Zika virus infections have been seen in the past, but primarily in Latin America and other equatorial regions where *Aedes* sp. vector mosquitos are found. This time, the epidemic reached the continental USA: Florida and Texas. While the exact mode of transfer to the USA is unknown, there is speculation that weather conditions were ideal to produce a large outbreak in Brazil. Then, traveling sporting teams may have carried infected Brazilian mosquitos north to regions where Zika is typically not endemic. The disease results in skin rash or hemorrhage, but the risk to pregnant women is more dramatic. An alarming clinical outcome, which received the most press and thus awareness in the USA, was an increased incidence of microcephaly and neurological anomalies when the virus was passed from a pregnant woman to the fetus. So Zika, like the story of West Nile virus, is spread primarily by insects, but in this case, the human, not the horse, is the major amplifying host. Medical countermeasures are under development.

And in 2019, the pandemic of our age occurred. While this book was prepared, there were more than 100 million cases of COVID-19 and more than one million deaths in the USA. Worldwide there have been more than 700 million illnesses and nearly seven million deaths [36]. Thus, COVID-19 changed our lives; it changed our families, our cities, our states, our nation, and the people of the globe. Some of the change was caused by the virus and some by humans themselves in response to the virus. While none of us has seen the entire COVID-19 elephant, we've all seen "enough." I leave that topic for the authors of the chapters that follow in this volume.

Finally, never say "Finally!" [37]. In 2022, as we were digging our way out of the COVID-19 pandemic, cases of monkeypox, a cousin of eradicated smallpox, were discovered in humans. There was a broad uptick in monkeypox in humans in Africa 10 or 15 years after the global community stopped vaccinating for smallpox, demonstrating mousepox cross-protection by the vaccine used to eradicate smallpox. The recent discovery would not have been a surprise, had these cases occurred in Africa. The disease had become smolderingly endemic on the African continent, with a rare spillover to the USA in imported pet rats in June of 2003 [38]. In response to that one incident, the importation of certain rat species to the USA from Africa was halted and the small outbreak was controlled. Then we learned—while the "why" is not totally clear— that the current (2022) outbreak was almost entirely limited to men who have sex with men. A couple of vaccines and drugs, developed for smallpox prophylaxis and therapy, are used currently in early 2023 to greatly reduce the number of active monkeypox cases in the USA and other countries, limiting its impact. We appear to be winning that one for now.

This has been just a glimpse of the pandemic and biosecurity-related disease challenges of the past century. We have no reason to believe that we have seen the end of such diseases, even with the latest tools available, such as mRNA vaccines and artificial intelligence to help us screen candidate drugs more efficiently and communication capabilities to help monitor the course of an outbreak, potentially to educate the populace at risk and to respond geographically in a more timely way.

9 Conclusion

9.1 Leadership

Keeping aware will require leadership. Our record in selecting and supporting leaders responsible for protecting the population from these threats and risks and giving them full responsibility and authority to operate is far from perfect. For the system to work most effectively, there must be "one boss for all" up there "someplace." Nathan Myhrvold, former Chief Technical Officer at Microsoft, gave us an interesting means of thinking about "who's in charge"; who is responsible overall? He said, look for "the most senior person in the organization whose full-time job is dedicated to function x."[7] In our system, when dealing with major health issues impacting the nation, the most senior person is typically found in the National Security Council (NSC) at the White House. However, it has not always been clear who is responsible. We normally call that traditionally small NSC team something like the Office of Global Health Security (and Biodefense). Most recently we've called the individual in charge the Special Assistant to the President for Global Health Security and Biodefense. That sounds right, but it's important that everyone in the greater enterprise knows who's in charge.

It may be helpful to review a brief history of that office. In 1998 a Health and Security Office was established at the NSC because of the visionary thinking of President Clinton's National Security Advisor, Sandy Berger. When the George W. Bush administration began, the office was abolished, but shortly thereafter we experienced 911 and the "anthrax letters" and the NSC Office of Health and Security was reestablished. When President Obama arrived, the office was again abolished…until the onset of the "Ebola '14" epidemic when it was once again reestablished. Then, with Ebola behind us, the Trump administration again abolished the office. Today we have a Senior Director for Global Health Security and Biodefense/ Special Assistant to the President, within the National Security Council. Are you beginning to see a pattern? [39]

It is important to understand that there are actually "leaders" up and down the operational "chain of command." Anyone can be a leader; it just requires, among other things, taking responsibility for one's part of the mission, using any authority one might possess wisely and working from within to build trust and be trustworthy.

9.2 Unity of Command

The result of many of these trusted relationships up and down an organization can be called Unity of Command. Unity of Command means that everyone with responsibility knows to whom he or she is responsible. For example, in the military, everyone knows who to salute, up the chain to the president, the Commander in Chief. It's not possible to maintain this "one person-one boss" principle perfectly, but it's

[7] Myhrvold, Nathan.

possible to come close to a healthy organizational structure. These human systems work best only when there is a relationship of trust between the individuals comprising the organization. A supervisor hires or accepts an employee and must trust the employee to accomplish a mission or set of goals. The employee must trust the supervisor to have his or her best interests in mind. Because the relationship is one of trust, the employee will not be micromanaged but will be given freedom to operate and to make his or her maximum contribution to the organization's mission. This could be called a "Responsibity-Trust-Freedom Triad," with trust as the cornerstone.

9.3 Deep Technical Expertise

Sound Leadership alone isn't enough. Deep technical expertise is needed to do the work, to think in the laboratory and in the field. Such expertise fleshes out the foundation of a pandemic "insurance policy" for the nation, a system of systems that will allow us to "accommodate surprise." Both are critical as we prepare for the next pandemic. One key outcome, of sound leadership, unity of command and especially deep technical expertise, is mentoring the next generation. Mentoring is automatic in a healthy culture. A culture where, ideally, responsibility for the safety and productivity of the organization remain close to the bench rather than in a distant bureaucracy inside another beltway. Unfortunately, mentoring is often one of the first things to be lost in an unhealthy culture [40].

Because there can be so many unknowns in the business of protecting citizens from infectious diseases, scientists must be given freedom to operate and, where infectious diseases are the focus, they must operate internationally. A "deep technical expert" knows his or her "bug" inside out and, very importantly, has a "Rolodex" filled with the names of the best experts on that pathogenic microorganism globally, just an email or phone call away. Developing a wiring diagram and giving a leader responsibility and authority over pandemic preparedness might take months. Building the cadre of experts takes much more time: it is measured in decades. Such individuals and teams are very hard-won but easily lost at the whim of a beltway politician.

9.4 Responsible Use of Powerful Tools

The sometimes heated discussion surrounding Gain-of-Function (GoF) research will likely continue. There is, quite reasonably, some confusion in the populous, regarding the real meaning of gain-of-function. I've even heard GoF equated with "biowarfare" in social media. CRISPR-Cas-9, one of the tools that makes GoF possible, is clearly very powerful and has broad application. Like powerful tools in other fields including chemistry and physics, outcomes are always better in the hands of responsible humans. Like any tool, they are equally powerful in the hands of "careful" humans and "careless" humans alike. As we regulate, we must take care not to negatively impact our legitimate infectious disease enterprise [41]. We are in a global competition.

Most of the authors represented in this volume have been speakers at Dr. Rosen's course at Dartmouth on the *Future of Health Care*. They will address the science of coronaviruses and COVID-19; medical countermeasures; the criticality of communication by experts and authorities on human behavior; and domestic and international responses to this pandemic and preparation for future ones. We must always expect surprises, but much of what is addressed in the following chapters can help us meet future surprises. No doubt, however, success will require ongoing effort between outbreaks, epidemics and pandemics, not just another "plan" on the shelf.

References

1. https://library.csi.cuny.edu/misinformation (2023) Misinformation and disinformation: thinking critically about information sources
2. Danzig R (2011) Driving in the dark: ten propositions about prediction and national security, Oct 26 2011. CNAS. https://www.cnas.org/publications/reports/driving-in-the-dark-ten-propositions-about-prediction-and-national-security%C2%A0
3. Barry JM (2020) The great influenza. Penguin Books, New York
4. Henderson DA (2009) Smallpox: the death of a disease. Prometheus Books, New York
5. Reardon S (2014) NIH finds forgotten smallpox store. Nature. https://doi.org/10.1038/nature.2014.15526
6. Roeder P, Rich K (2009) The global effort to eradicate rinderpest. IFPRI Discussion Paper 00923. https://ebrary.ifpri.org/utils/getfile/collection/p15738coll2/id/29876/filename/29877.pdf
7. Larson HJ, Gakidou E, Murray CJ (2022) The vaccine-hesitant moment. N Engl J Med 2022(387):58–65. https://www.nejm.org/doi/full/10.1056/nejmra2106441
8. Carlson R (2010) Biology is technology. President and Fellows of Harvard College, p 72
9. Bartlett, JMS, Stirling, D (2003) A short history of polymerase chain reaction. Methods in molecular biology. PCR Protocols, vol 226, 2nd edn. Humana Press, Townsend
10. UNAIDS Global AIDS Update (2022). https://www.unaids.org/sites/default/files/media_asset/2022-global-aids-update-summary_en.pdf
11. Semple RB. Nixon Counts on conversion of military facilities. New York Times. October 19, 1971, p 28
12. The Biological Weapons Convention. https://www.un.org/disarmament/biological-weapons/
13. Guillemin J (1999) Anthrax: the investigation of a deadly outbreak. University of California Press, Berkeley/Los Angeles
14. Guillemin J (2005) Biological weapons. Columbia University Press, New York, p 134
15. Preston R (1994) The hot zone. Anchor Books, New York
16. Alibek K (1999) Biohazard. Random House, New York
17. Lederberg J, Shope F (1992) Emerging infections: microbial threats to health in the United States. NAS Institute of Medicine
18. Fauci AS (2022) It aint't over till it's over…but it's never over—emerging and reemerging infectious diseases. N Engl J Med 387(22):2009–2011
19. Trevan T (1999) Saddam's secrets: the hunt for Iraw's hidden weapons. Harper/Collins, New York
20. Franz D (2021) Puzzling out the Iraqi Biological Weapons Program, Bulletin of the Atomic Scientists. July 21 2021. https://thebulletin.org/premium/2021-07/puzzling-out-the-iraqi-biological-weapons-program/
21. GlobalBioLabs. https://www.globalbiolabs.org/

22. Malave A, Emalin A (2010) Severe acute respiratory syndrome: lessons for future pandemics. AMA J Ethics. Virtual Mentor 12(9):719–725. https://journalofethics.ama-assn.org/article/severe-acute-respiratory-syndrome-sars-lessons-future-pandemics/2010-09
23. Carlson R. Biology is technology
24. One-Health Website. https://onehealthinitiative.com/contact/
25. CDC One Health Website. https://www.cdc.gov/onehealth/index.html
26. iGEM. https://igem.org/
27. NSABB (2007) Responsible life sciences research for global health security: a guidance document. https://www.ncbi.nlm.nih.gov/books/NBK305027/
28. https://osp.od.nih.gov/wp-content/uploads/2016/06/NSABB_Final_Report_Recommendations_Evaluation_Oversight_Proposed_Gain_of_Function_Research.pdf
29. Basler CF et al (2001) Sequence of the 1918 pandemic influenza viruss nonstructural gene (NS) segment and characteriztionof recombinant viruses bearing the 1918 NS genes. Proc Natl Acad Sci U S A. https://www.ncbi.nlm.nih.gov/pmc/articles/PMC30210/pdf/pq002746.pdf
30. Herfst S, Fouchier RA (2012) Airborne transmission of influenza A/H5N1 virus between ferrets. Science 337(6099):1534–1541. https://doi.org/10.1126/science.1213362
31. https://nap.nationalacademies.org/catalog/13098/review-of-the-scientific-approaches-used-during-the-fbis-investigation-of-the-2001-anthrax-letters
32. Ryan J, Cook T. DOJ Settles Hatfill Suit for $5.8 Million. ABC News. 27 June 2008. https://abcnews.go.com/TheLaw/DOJ/story?id=5264759&page=1
33. https://books.google.com/books/about/World_at_Risk.html?id=JfeCMvORRxsC
34. Lopatto E (2014) Science Report, Oct 20, 2014. https://www.theverge.com/2014/10/20/7020249/white-house-halts-research-into-super-strains-of-infectious-diseases
35. https://www.nih.gov/about-nih/who-we-are/nih-director/statements/nih-lifts-funding-pause-gain-function-research
36. https://www.worldometers.info/
37. Facui (2022) NEJM
38. https://www.cdc.gov/media/pressrel/r030702.htm
39. Bernard K. The White House Signals that Bioterrorism and Disease Don't Matter…again, Washington Post OpEd, 22 May 2018
40. Franz DR (2019) Facilities, equipment and procedures: an historic glimpse at high-containment laboratory safety and security. J Biosaf Biosecur. https://doi.org/10.1016/j.jobb.2018.12.002
41. Franz DR, Le Duc JW (2021) Technology advances, high-risk research and a safe way for ward. mBio 12:e02373–e02321. https://doi.org/10.1128/mBio.02373-21

Further Reading

Gronvall GK (2016) Synthetic biology: safety, security and promise. Health Security Press, Baltimore

Henderson DA (2009) Smallpox: the death of a disease. Prometheus Books, New York

Kahn LH (2009) Who's in charge: leadership during epidemics, bioterror attacks and other public health crises. Praeger Security International, Santa Barbara

Lele A, Roy K (2020) COVID-19: analysing the threat. Pentagon Press LLP, New Delhi

Lentzos F (2016) Biological threats in the 21st century: the politics, people, science and historical roots. Imperial College Press, London

Osterholm MT, Olshaker M (2017) Deadliest enemy: our war against killer germs. Little, Brown Spark, New York

History of Pandemics

Amy (Shu Ting) Liang and Ashley (Lin Ting) Liang

1 Introduction

COVID-19 has drastically transformed our lives and has lasted much longer than we had anticipated. Some countries were able to control the spread better than others. What could we have done differently to have minimized the 14,520 COVID-19-related deaths and what can we do differently in the future? An understanding of past pandemics and how they were contained can help us come to a better understanding.

2 Cholera Pandemics

Cholera is an acute gastrointestinal tract disease caused by *Vibrio cholerae*, a waterborne pathogen. The bacteria builds up in the small intestine and produces the cholera toxin. This toxin causes rapid and massive loss of bodily fluids, which leads to dehydration, hypovolemic shock, and death. It is often spread through contaminated water. There have been seven cholera pandemics so far, with the seventh still ongoing. The first six originated in India and were all the O1 classical biotype strain, and the current one originated in Indonesia and was caused by the El Tor biotype strain. The persistence of the virus is largely related to global issues of unsafe drinking water, insufficient sanitation, crowded housing, and lack of efficient sewage systems. Cholera affects people of all ages, but more recent pandemics mostly affect

A. (S.T). Liang (✉)
Dartmouth Medical School, Harvard Business School, Boston, MA, USA
e-mail: aliang@mba2023.hbs.edu

A. (L.T). Liang
Dartmouth College, Hanover, NH, USA

J. M. Rosen, R. R. Colwell (eds.), *The COVID-19 Pandemic*,
https://doi.org/10.1007/978-3-031-62772-9_2

developing countries. A brief history of each of the seven pandemic outbreaks is described in the following sections.

2.1 First World Cholera Pandemic (1817–1823)

The first cholera pandemic originated in August 1817 in Jessore, which is present-day Kolkata in India, and affected almost every country in Asia. Civil surgeons reported many incidences of severe gastrointestinal disease among patients and traced it back to contaminated rice. They also reported thousands of people collapsing and dying along with attacks of vomiting and diarrhea. The disease spread rapidly across the country, arriving in Sri Lanka. Overland, the disease traveled to Burma and Thailand. Seabound, it spread to Sumatra, Java, the Philippines, China, Japan, and other Southeast Asian countries. Cholera also spread to countries further west. It spread to Oman through infected British troops and was then carried along the east coast of Africa to Zanzibar. Cholera traveled northwest to Baghdad and Russia as well, but the severe winter in 1832 prevented the spread of cholera to Europe. The death toll is not known, but approximately hundreds of thousands of individuals died in India, and another 100,000 died in Java. By 1823, however, cholera largely disappeared.

2.2 Second Pandemic: 1829–1835

The second pandemic, also originating in India, was the first to reach Europe and the Americas. It reached Moscow and St. Petersburg by 1830. From there, cholera spread to Finland and Poland. It then reached countries like Germany and the UK. In 1832, cholera traveled with European immigrants to Canada and the United States. By 1833, it reached Mexico and Cuba. In England, approximately 22,000 died from 1831 to 1833 and another 52,000 died in England and Wales from 1833 to 1835.

2.3 Third Pandemic: 1846–1860 (Other Researches Believe Might Have Started as Early as 1837 and Lasted Until 1863)

The third pandemic is considered the deadliest with large estimated death tolls in Asia, Europe, North America, and Africa. It killed more than one million people in Russia, and around 23,000 people died in Britain in just one year (1854). In 1837, cholera once again resurged in the lower Bengal region of India and spread to Kaul. In 1840, another outbreak began in Bengal. Troops transported this disease to British Strait settlements and to China, where it then spread to the Philippines, Canton, Burma, and Central Asia to Bukhara in 1844. In 1845, cholera was present in Afghanistan as well as Punjab, Karachi, and Delhi in India. By 1846, it spread rapidly across the Arabian coast, affecting Iran, Iraq, and Baghdad as well. In 1847, cholera reached Russia. By the end of the 1840s, the pandemic was present all over Europe, the United States, Canada, Mexico, and North Africa. Throughout the 1950s, many outbreaks occurred across those continents. During this pandemic, John Snow succeeded in identifying contaminated water as the primary transmitter of the disease. He unfortunately died in 1858 during this pandemic.

2.4 Fourth Pandemic: 1863–1875; Fifth Pandemic: 1881–1896; Sixth Pandemic: 1899–1923

The fourth and fifth pandemics were overall less severe than the previous pandemics. Great Britain and the United States did not suffer as much due to improved water supplies and new quarantine measures. The sixth pandemic largely didn't affect Western Europe and North America. However, it was still destructive in India, Russia, the Middle East, and northern Africa. It killed more than one million people in India between 1918 and 1919.

2.5 Seventh Pandemic: 1961–Now

The current cholera pandemic began in Indonesia in 1961, differing from the rest which originated in India. The current pandemic also differs in that the *V. cholerae* strain is the El Tor biotype instead of the O1 classical biotype. Throughout the rest of the twentieth century, the pandemic mostly affected countries in South America and Africa. More recent epidemics include Zimbabwe in 2008, Haiti in 2010, Sierra Leon in 2012, and Yemen in 2017.

3 1918 Spanish Flu

The 1918 Spanish Flu was caused by an H1N1 influenza A virus postulated to be of avian origin [2]. The 1918 Spanish Flu lasted from 1918 to 1920 and consisted of four waves. The first wave lasted approximately from February 15, 1918, to June 1, 1918; the second lasted approximately from August 1, 1918, to December 2, 1918; the third lasted approximately from December 3, 1918, to April 30, 1919; and the fourth wave lasted approximately from December 1, 1919, to April 30, 1920 [3]. It infected about 500 million people, roughly a third of the world's population at that time, and resulted in the deaths of 50 million, including 675,000 Americans [2]. The first public news of the epidemic appeared in Madrid on May 22, 1918, in Madrid's ABC newspaper; hence, it became known as the Spanish Flu [4]. However, there is no definite evidence of origination, and most epidemiologists and virologists believe that the virus originated in either the United States or France [4]. A week later on May 28, 1918, King Alfonso XIII, the Prime Minister, and some cabinet members became ill [4]. As the influenza spread, basic services such as the postal service, telegraph services, and some banks were forced to temporarily close operations [4].

4 H1N1 2009 Swine Flu

The H1N1 swine flu was first detected in the United States in April of 2009 and was responsible for somewhere between 151,700 and 575,400 deaths worldwide during that first year. The virus originated in central Mexico, where it then spread quickly to the United States and then worldwide. The virus has been traced back to pigs in central Mexico. The parent virus, researchers say, had been present for a decade before a strain transferable to humans was generated. H1N1 is a result of quadruple reassortment between swine, birds, and humans. Symptoms of the swine flu include fever, cough, sore throat, runny or stuffy nose, body aches, headache, chills, and fatigue. Some other common symptoms include vomiting and diarrhea. Interestingly, an estimated 80% of H1N1-related deaths occurred in people younger than 65, differing from typical seasonal influenza where 70–90% of deaths occur in people 65 years and older. The Center of Disease Control and Prevention (CDC) highly recommends taking the H1N1 vaccine. Individuals over 10 years of age needed one dose and individuals under 10 years needed two. This was because people under 10 did not have an adequate amount of antibodies after only one dose. The CDC also approved the use of oseltamivir, zanamivir, peramivir, and baloxavir for the treatment of H1N1.

5 SARS

SARS began in November of 2002 when an unknown pneumonia emerged in southeastern Guangdong, China. In February 2003, the virus quickly spread to Hong Kong, Singapore, Vietnam, northern China, and Canada. There were an estimated 8098 cases of SARS worldwide and 774 deaths. SARS, as an airborne virus, can be spread through small droplets of saliva as well as indirectly via surfaces that have been touched by people infected with the virus. It was the first severe and easily transmissible new disease to emerge in the twenty-first century, prompting international efforts to isolate the virus and isolate infected and exposed patients. Most cases of SARS are composed of previously healthy adults aged 25–70. The first symptom was usually a high fever, sometimes with chills, as well as headache, malaise, and muscle pain. After a few days to a week, a dry, nonproductive cough or dyspnea begins to occur, sometimes accompanied by or progressing to hypoxemia. White blood cell and platelet counts also often decrease. There was no cure or vaccine for SARS.

6 MERS

Middle East Respiratory Syndrome (MERS) originated in 2012 in Saudi Arabia from dromedary camels. Symptoms included fever, coughing, and difficulty breathing. The suspected reservoir for the disease is bats. From 2012 until February 2022, there were 2585 laboratory-confirmed cases, 27 countries affected, and 890 known disease-related deaths. People aged 50–59 are at highest risk for the disease.

The largest MERS outbreak outside of the Arabian Peninsula was in South Korea in May–June of 2015. A South Korean businessman traveling back to his home country was treated in three hospitals over the course of eight days, resulting in 29 infections. A key element in preventing further spread was using temporary tents outside hospitals in South Korea to triage patients and isolate MERS from non-MERS patients.

Key learnings from this pandemic are that correct diagnosis and rapid quarantine prevented larger outbreaks, despite having no vaccinations and no treatment.

7 Comparison Between COVID-19 and 1918 Influenza

COVID-19 is often compared to the 1918 Influenza, so it is helpful to understand some similarities and differences. First, the patient population differs. While the 1918 Influenza killed a disproportionate number of 25–40 year olds, COVID-19 kills more 65+ with comorbidities [2, 5]. In particular, the mortality rate for the influenza rose to 8–10% for younger people compared to a 2.5% overall mortality, whereas the mortality rate for the 25- to 40-year-old age range is a mere 0.2% in contrast to the 2.4% overall mortality rate [2, 5]. Twent-five- to 40-year olds accounted for 40% of deaths from the 1918 Influenza, whereas those in the 18- to 44-year old range account for only 3.9% of deaths from COVID-19 [2, 5]. More countries were spared in the 1918 pandemic, whereas only the smaller Pacific Islands (Soloman Islands and Vanuata) remained COVID-19-free [2, 6]. The mortality rate for pregnant women with the Spanish Flu was 23–37% and 26% of those who survived lost their child, whereas the mortality rate of pregnant women with COVID-19 is unknown [2, 7]. The Spanish Flu resulted in acute illness in 25–30% of the world population, with over 50 million deaths, whereas COVID-19 has infected nearly 55 million to date, with 1.3 million deaths [2, 5]. In the United States alone, COVID-19 cases are at over 11 million as of November 16, 2020, which is nearly a 40% increase from the month prior [5].

Second, the two diseases kill via different mechanisms. While those with the Influenza died of secondary bacterial pneumonia, those with COVID-19 died from an overactive immune response that resulted in multiple organ failure [2, 8]. ARDS can develop in both cases [2, 8]. As a complication from the Influenza, ARDS had a 100% fatality rate compared to a 53.4% mortality rate as a complication from COVID-19 [2, 9].

The projected economic impact of COVID-19 on the US economy is a $5.76–$6.17 trillion decrease in GDP, based on Fitch Ratings and the US GDP according to the World Bank. The economic data during the 1918 pandemic is scarce, but it was noted that Mexico suffered a $9 billion loss [2].

Diagnoses, treatments, and vaccines were delayed in both cases. States developed different COVID-19 diagnostic tests since the initial one by CDC could not be confirmed. Currently, there are no FDA-approved COVID-19 treatments, but antivirals like Remdesivir, antibody, and IL-33 blockers are currently under investigation. Vaccines are also in development. In 1918, bleeding was initially used as treatment since such minimal progress had been made against pneumonia that even renowned William Osler still recommended it to relieve symptoms [2]. In 1917, Dr. Rufus Cole, Dr. Oswald Avery, and Dr. Alphonse Dochez, with help from six other Rockefeller researchers developed and tested a vaccination to prevent pneumonia caused by Types I, II, and III pneumococci. In March 1918, this vaccine was given to 12,000 troops on Long Island, with no vaccinated soldier developing pneumonia from those strains. In contrast, 101 out of 19,000 soldiers serving as controls, developed pneumonia from those strains [2].

Yet, since neither an influenza vaccine nor antibiotics to treat associated secondary bacterial infections were available, worldwide containment efforts relied heavily on isolation and quarantine similar to the current efforts against COVID-19 [2].

In terms of duration and origination, there is controversy over the origination of both viruses, and both consist of multiple waves. The 1918 Influenza lasted 25 months and may have originated in Spain, France, or the United States with no definite evidence of origination [3, 4]. The first wave lasted approximately from February 15, 1918, to June 1, 1918, and the fourth and final wave lasted approximately from December 1, 1919, to April 30, 1920 [3]. COVID-19 originated in Wuhan, China, on December 31, 2019, with controversy over whether it originated in a wet market or at the Wuhan Institute of Virology. Unlike in 1918, DNA sequencing of COVID-19 can predict whether infected individuals will be symptomatic or asymptomatic, based on a single base change (11083G > T) [10].

8 What We Have Learned for the Future

Table 1 highlights the key differences between COVID-19 and the 1918 influenza.

Table 1 Summary of major differences: COVID-19 versus 1918 influenza

	COVID-19	1918 influenza
Viral etiology	SARS-CoV-2	H1N1 influenza A virus
Mortality rate	2.40%	2.50%
Number of deaths	1.4 million	50 million
Highest risk population	65+ with comorbidities	25–40 year olds
Cause of death	Overactive immune system leading to end organ failure	Secondary bacterial infection
Place of origin	Wuhan (either in a wet market or Wuhan Institute of Virology)	Haskell County, Kansas
Virus type	Coronavirus	Orthomyxoviridae
Economic impact	\$5.76T–\$6.17T decrease in GDP	With minimal economic data, Mexico had a \$9B loss

9 Conclusion

Through these pandemics, we have learned the importance of having measures in place to improve the effectiveness of future responses. Particularly, better coordination, response logistics, communication, and treatment plans are essential for an improved outcome of future outbreaks.

First, coordination is key. Global coordination is critical to ensuring that all nations achieve rapid containment, which in the case of COVID-19, only certain nations achieved. In addition to global coordination, local coordination particularly with communities is necessary for education and disease surveillance. To do so, the Department of State, foreign governments, and nongovernmental sectors must work together to communicate regarding public health policies. Additionally, a platform or network for rapid information sharing can accelerate this process.

Second, proper response logistics are essential. Sufficient testing, personal protective equipment, and immediate implementation of strict control measures are the best practices for a strong response. To attain that, it is important to budget money for such responses, have optimized supply chains, and emergency plans with trained workers that can be quickly sent to areas of outbreak. Additionally, we learned the importance of training physicians including those in other specialties, nurses, and local healthcare workers, particularly by experienced healthcare responders in countries that have already encountered the infection. This sharing of best practices and learnings from mistakes will improve the effectiveness of responses by other nations.

Third, communicating with and educating the public is an essential part of an effective response. This includes timely and accurate public health messages to encourage physical distancing, face coverings, and stay-at-home measures. Social media movements have also helped with communicating to a wider audience. Most importantly, it is crucial to declare a state of emergency early on.

Finally, treatment and vaccination are key as well. This requires a scientific understanding, which involves investment in infectious disease and epidemiology research. The emergence of telemedicine and traveling nurses are alleviating the burden of pandemics on the healthcare system and with the improvement and investment in healthcare technology; this strain on the healthcare system should continue to decrease.

References

1. Centers for Disease Control and Prevention. 2009 H1N1 Flu ('Swine Flu') and You. Centers for Disease Control and Prevention, 10 Feb 2010. https://www.cdc.gov/h1n1flu/qa.htm
2. Centers for Disease Control and Prevention. 2009 H1N1 Pandemic (H1N1PDM09 Virus). Centers for Disease Control and Prevention, 11 June 2019. https://www.cdc.gov/flu/pandemic-resources/2009-h1n1-pandemic.html
3. Centers for Disease Control and Prevention. Origin of 2009 H1N1 Flu (Swine Flu): questions and answers. Centers for Disease Control and Prevention, 25 Nov 2009. https://www.cdc.gov/h1n1flu/information_h1n1_virus_qa.htm

4. World Health Organization. Regional Office for the Eastern Mediterranean (2022) MERS situation update: February 2022. World Health Organization. Regional Office for the Eastern Mediterranean. https://apps.who.int/iris/handle/10665/352631. License: CC BY-NC-SA 3.0 IGO
5. World Health Organization. Severe Acute Respiratory Syndrome (SARS). World Health Organization. https://www.who.int/health-topics/severe-acute-respiratory-syndrome#tab=tab_3
6. The Mount Sinai Hospital / Mount Sinai School of Medicine. 2009 swine flu pandemic originated in Mexico, researchers discover. ScienceDaily. ScienceDaily, 27 June 2016. www.sciencedaily.com/releases/2016/06/160627160935.htm
7. Centers for Disease Control and Prevention. Vaccine against 2009 H1N1 Influenza Virus. Centers for Disease Control and Prevention. https://www.cdc.gov/h1n1flu/vaccination/public/vaccination_qa_pub.htm
8. Piret J, Boivin G (2021) Pandemics throughout history. Front Microbiol 11:631736. https://doi.org/10.3389/fmicb.2020.631736
9. McGrew RE (1960) The first cholera epidemic and social history. Bull Hist Med 34(1):61–73. JSTOR, http://www.jstor.org/stable/44446659. Accessed 28 Oct 2022
10. Liang ST et al (2021) COVID-19: a comparison to the 1918 influenza and how we can defeat it. Postgrad Med J 97(1147):273–274. https://doi.org/10.1136/postgradmedj-2020-139070

Further Reading

Liang ST et al (2021) COVID-19: a comparison to the 1918 influenza and how we can defeat it. Postgrad Med J 97(1147):273–274. https://doi.org/10.1136/postgradmedj-2020-139070

Piret J, Boivin G (2021) Pandemics throughout history. Front Microbiol 11:631736. https://doi.org/10.3389/fmicb.2020.631736

Rewar S et al (2015) Treatment and prevention of pandemic H1N1 influenza. Ann Glob Health 81(5):645–653. https://doi.org/10.1016/j.aogh.2015.08.014

Use of Robotics in a Pandemic

Alexandra R. Jamieson, Pavan Nuthi, and Muthu B. J. Wijesundara

1 Introduction

The recent COVID-19 outbreak (SARS-CoV-2) has demonstrated how a rapidly progressing pandemic can overwhelm healthcare systems across the globe. With the onset of a pandemic, disease spreads rapidly, resulting in a massive influx of patients that overrun the operation limits of healthcare facilities, cause equipment and human resource shortages, and severely disrupt the workflow of hospital operations. In 2021, surges of coronavirus infection caused inpatient and intensive care unit (ICU) occupancy to surpass 90% for many of the responding hospitals in the United States, with some hospitals going beyond 100% capacity [24]. Depletion of critical care resources, such as ICU beds and mechanical ventilators, led to a sharp increase in the death toll during surges of COVID-19 cases [49, 51]. Healthcare workers experienced the burden of unmanageable workloads and long shifts due to the sheer number of patients, as well as direct exposure to infected patients since personal protective equipment (PPE) was scarce. Additionally, healthcare workers faced detrimental psychological effects due to social isolation, strained family relations, and the death of their colleagues [11, 27]. Healthcare systems have also taken on a heavy burden due to the severity of patient conditions, increased expenditures, and loss of revenue [28]. Overall, the COVID-19 pandemic brought to light the unpreparedness of global healthcare structures to handle a pandemic and how that has directly affected the quality of patient care, the well-being of healthcare workers, and the stability of healthcare systems.

While the experience of a pandemic in modern times has forced healthcare systems to learn how to better manage their resources, that alone cannot make up for a shortage of healthcare workers who are primary providers of the specialized care

A. R. Jamieson · P. Nuthi · M. B. J. Wijesundara (✉)
The University of Texas at Arlington Research Institute, Fort Worth, TX, USA
e-mail: muthuw@uta.edu

J. M. Rosen, R. R. Colwell (eds.), *The COVID-19 Pandemic*,
https://doi.org/10.1007/978-3-031-62772-9_3

that saves lives. In an effort to support healthcare workers on the front lines, scientists and engineers have been working diligently to create advanced technological solutions. Looking at the multitude of tasks that comprise a typical hospital workflow path during a pandemic, it is clear that many could be automated or assisted using solutions such as robotic devices, sensors, imaging, and artificial intelligence (AI) algorithms [33]. This chapter will focus on how robotic technology can be applied to provide innovative solutions to change the structure of healthcare and improve the response to future pandemics by reducing worker-to-patient contact, speeding up processes, and compensating for personnel shortages.

Figure 1 shows the nominal workflow of a hospital during a pandemic, namely, a logical map of services to support patient care, with consideration of patient infection status and severity. With this flowchart as guidance, each stage in the workflow will be chronologically examined to identify tasks that create challenges during a pandemic. By recognizing the bottlenecks in hospital workflow during patient influxes, the need for automation and technological assistance can then be determined.

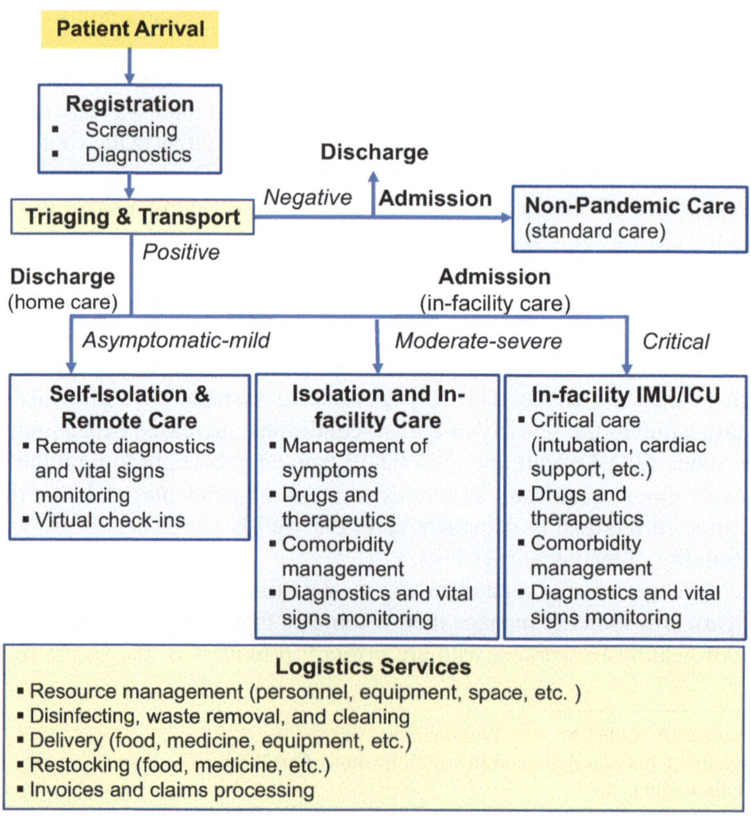

Fig. 1 Generalized workflow in a hospital during a pandemic

Starting with registration, patients arrive at a hospital where personnel assist with check-in, collect screening questionnaires, and run preliminary diagnostics. Since there is technically no limit to how many patients may enter the waitlist for hospital admission, this area can easily become congested as the incoming number of patients grows rapidly. As witnessed by the recent COVID-19 pandemic, shortages in staff resulted in delayed patient care and overcrowding of the registration area with infected individuals. Those factors caused major concern over the conditions of patients worsening while waiting, as well as the high risk of cross-infection. Resolving this blockage in the workflow path is critical to reducing overcrowding by identifying the severity of patients' conditions so they can quickly move to triage and transportation.

At the triage and transport stage, nurses and doctors, along with resource management workers, determine where patients will be transported, depending on the severity of their condition and availability of resources. This process can be severely hampered by delays in clinical decision-making due to low staff-to-patient ratio, inadequacy of resource allocation protocols, and shortage of available resources. Additionally, transporting patients across the facility can be time consuming and requires assistance from multiple staff due to the nature of large hospital landscapes and manual transportation devices. Addressing these shortcomings is critical to be able to streamline triaging and transporting to save patients' lives and reduce infectious spread within the hospital.

As outlined in Fig. 1, patients may be triaged into three possible groups: self-isolated home care, nonpandemic-related care facility, or isolated in-facility care including IMU (intermediate care) and ICU (intensive care). Patients who self-isolate at home will benefit the most from hospitals that can provide telehealth through remote diagnostics and vital signs monitoring, as well as virtual check-ins with a health professional to manage symptoms. For in-facility care and IMU/ICU treatment, around-the-clock management is needed to monitor patients' vital signs, perform diagnostics, deliver drugs and food, and provide treatment as necessary to manage the illness and comorbidities. Currently, all those patient care tasks are heavily human-centered and need highly specialized trained personnel and equipment. The limitations of this approach were evident during the COVID-19 pandemic, as there was a severe shortage of respiratory therapists and clinical personnel who could perform intubation and airway management, leading to higher death tolls. Current patient care protocols have also led to cross-infections due to long exposure times and/or close contact with infected individuals, resulting in further clinical personnel shortages. PPE is essential to protect workers and enable continuing patient care, yet the COVID-19 pandemic demonstrated that this equipment could hinder worker performance of procedures due to excessive sweating, fatigue and headaches, pressure on the skin, and suffocation [1]. Finding solutions that can reduce physical interactions between patients and healthcare workers is needed to better handle future pandemic situations.

In addition to personnel, inpatient care is heavily dependent on two other important aspects. *(1) Availability of equipment, facilities, and supplies*: As evident during the first wave of COVID-19 faced by New York hospitals, one of the main factors that contributed to inferior care and a higher death toll was the unavailability of required resources including acute care beds, ICU capacity, and ventilators. Shortages in PPE, such as N95 respirators and isolation gowns, hampered the ability to provide consistent and necessary care. Although the shortage of equipment, facilities, and supplies was identified to be caused by surges in the patient population and supply chain issues, those circumstances can be anticipated. Management systems must be advanced to improve resource allocation and sharing, not only within an individual hospital, but also among a group of collaborating hospitals, states, and the federal government. *(2) Management of logistics operation*: Across the entire hospital landscape, logistics support is an essential component to maintaining smooth workflow. Timely allocation of essential resources (equipment and personnel) at various stages of the workflow is key to maximizing hospital throughput. This includes scheduling clinical staff to perform routine patient care, deliver medicine, and handle emergency procedures, all in an efficient and timely manner. Apart from clinical staff allocation, other essential services, such as cleaning and waste management, delivering food to patients, room preparations, and transportation all should be coordinated. Current manual approaches to logistics planning and resource allocation are riddled with shortcomings, ranging from being human resource intensive to being too slow in a fast-evolving situation like a pandemic. Solutions are needed to assist in completing low-complexity tasks and optimizing decision-making for better management.

In summary, the recent COVID-19 outbreak revealed inadequacies in current healthcare systems across the globe to manage a pandemic situation [24]. Challenges arose due to insufficient resources, overwhelmed logistics management, overworked healthcare workers, and high infection transmission to staff. Many investigations identified robotics and automation as a solution to combat these common shortcomings. Technology-driven solutions that span from AI-based information sharing to decision-making algorithms to sensorized robotic platforms can be adapted to the needs of a hospital environment, and some existing systems have already taken major positive steps to support hospital operations in pandemic situations. In order for robotic and autonomous technologies to begin to have a real impact at the site of action, research, development, and implementation strategies need to be established that consider all aspects of healthcare delivery to be ready for future pandemics.

2 Robotics in Healthcare During a Pandemic

Currently, robotics and automation application to healthcare is a continuously evolving area of development and, at the same time, subject to changing regulations. Many systems are already well established and currently being used by the healthcare industry, including surgical robots and robotic intravenous compounding systems [41, 45]. The advancement of sensor technologies, wearable devices, medical imaging, computer vision, machine learning (ML), and data analytics will accelerate and diversify the role of robotics in healthcare. Learning from the COVID-19 pandemic, it is evident that the role of robotics may need to change significantly from traditional care settings to accommodate pandemic care during surges of incoming patients with unique needs. Therefore, it is imperative to pinpoint critical areas of need and identify how robotics and automation can complement human-centered care for patients, maintain efficient clinical workflow, reduce the spread of infection among patients and healthcare workers, and provide logistics support. As the healthcare industry has its own sophistication and complexities, close collaboration among researchers, the robotic industry, and healthcare systems is critical to developing and implementing robotic systems and auxiliary technologies to serve critical needs during a pandemic.

Many research reports have identified a range of potential roles for robotics and automation in hospital operations during a pandemic [38, 59]. These include disinfecting and cleaning facilities; handling and delivering drugs, food, and waste; restocking supplies; transporting patients; monitoring patients' conditions; operating medical equipment; and supporting medical interventions. However, robots require many auxiliary technologies and supporting infrastructure to operate in complex hospital environments. It is expected that the transition to robot implementation in hospitals will take time. A step-wise approach may be best where, for initial implementation, researchers and engineers first identify how to develop robotic systems that can adapt to existing hospital infrastructure and equipment with minimal modification. Collecting data and recognizing patterns of human behavior in a hospital should be a priority in order to train robots for seamless interaction and coexistence with humans. In regard to infrastructure, modifications may include adding dedicated robotic lanes with path identifiers for the navigation of delivery and transport robots. In some instances, modifications may be needed to the robot, as well as to the hospital infrastructure. For instance, disinfecting and cleaning robots with built-in sensors can be augmented by additional sensors in the facility for efficient and safe operation around people. Patient monitoring and medical equipment operating robotics will require novel interfaces for the robot to interact with existing medical equipment. New operation protocols must be in place for robots to co-exist with humans and complement each other to allow smooth and efficient hospital operations. Better information technology (IT) infrastructure will facilitate communication and operation of robotics within hospitals.

Some hospitals implemented robotic systems during the COVID-19 pandemic. Although the use of robots was limited, the evidence gathered and expert analyses showed that the incorporation of robotics and automation technologies to aid healthcare delivery in a pandemic situation is practical and has tremendous benefits [59]. Those studies suggest the potential for a future where robots equipped with sensors and data processing can assist hospital staff in patient screening and preliminary diagnostics to accelerate triaging and treatment planning. Vital signs monitoring and routine diagnostic tests can be performed by robotic arms and sensing modalities, reducing the need for frequent clinical rounds and allowing clinicians to focus on other critical care needs. Mobile robotic platforms with navigation and manipulation capabilities can be used for logistics support and delivery tasks. These types of robotic systems would be able to significantly reduce labor demand and disease transmission. As these robots can keep accurate records of deliveries and inventories, resource management and scheduling could be automated to support logistics and workflow. All these observations suggest that implementing robotic systems to become a commonplace standard can support high-level hospital operation and patient care.

Apart from healthcare delivery, robotic assistance has many other benefits, particularly for pandemic scenarios. Robots do not need PPE, are unlikely to spread pathogens, and can work around the clock. They also eliminate the need for training if they are preprogrammed to perform certain functions and can be easily replicated, unlike human counterparts requiring training for each new hire. It is also possible for robotics systems to move from one hospital location to another based on patient surge and demand. Thus, there is no need for accommodation or high pay rates, as seen in healthcare worker relocation during COVID-19. The current status, potential roles, and upward momentum of robotic technologies suggest that focused development of specialized robotic technologies for healthcare will transform how governments and healthcare systems respond and handle future pandemic scenarios.

The roles of robotics in hospital environments can vary widely both in complexity and involvement, spanning from passively monitoring patients to actively supporting complex tasks. The sophistication necessary for robotics to be effective in healthcare depends on many factors [55, 58] including a robot's ability to directly engage and interact with humans, coexist with a chaotic human-centric environment, and perform tasks collaboratively. All those factors need to be well understood, and stakeholders should work together to continue the advancement of robotics and auxiliary technologies to find practical solutions. From a robot design perspective, the level of human-robotic interaction (HRI) is an essential aspect that needs critical evaluation based on the designated functions of the robot to ensure its effectiveness and safety. Figure 2 shows a simplified classification of robot types demonstrating how their degree of HRI is dependent on the complexity of the task being performed [55, 58]. Wearables and autonomous robots would be well suited for tasks of very low uncertainty and minimal level of HRI, such as monitoring vital signs, delivering supplies, and sanitizing. A collaborative robotic partner has a moderate level of HRI, where autonomy may be built into their system for some tasks, but overall coordination between healthcare workers and the robotic systems is necessary. Collaborative robots can assist clinicians in performing meticulous and arduous tasks and even use AI to guide in data analysis and medical treatment planning, resulting in improved productivity, reduced errors, and less worker fatigue. These robots operate with the direction and command of clinicians, as the tasks they are involved in typically have some level of risk requiring expert attention. Teleoperated robots may be the only solution required for supporting complex tasks where risk is high, such as surgeries and airway management procedures like intubation. For those, the robot must be directly controlled by humans with specialized knowledge and skill to ensure the safe execution of the procedures.

By understanding the applications of different robot types (Fig. 2), as well as identifying the many bottlenecks in hospital workflow during a pandemic, stakeholders can determine how current systems should be deployed and how new systems are designed to benefit hospital operations. Hospital operations can be

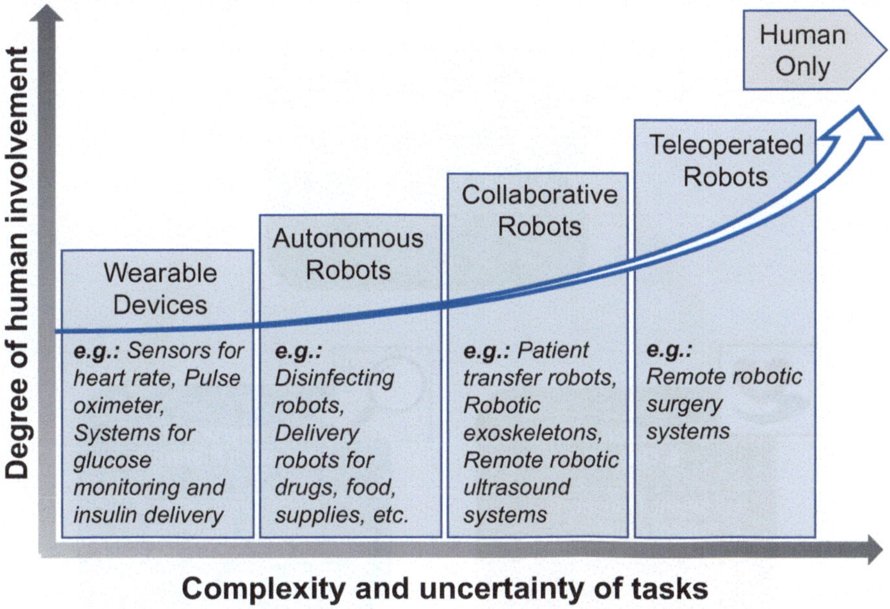

Fig. 2 Robot types and their degrees of task complexity and human involvement

categorized into three main aspects: data gathering, data interpretation and action planning, and execution of care and operations, as shown in Fig. 3. Each comprises many tasks that have the potential to be automated or assisted by robots. Data gathering can leverage various robot platforms and auxiliary technologies for patient care such as screening, diagnostics, and monitoring, as well as logistics operations, such as tracking deliveries and keeping inventories. Data interpretation and action planning involve data parsing and decision-making, where AI-based automation algorithms can be utilized to improve hospital workflow by optimizing resource allocation and assisting clinical and managerial decision-making. Execution of care and operations refers to executing various operations in the healthcare system: performing medical procedures; transporting patients; delivering food, therapeutics, and other supplies; and maintaining facility operations. The data-driven nature of all three allows continuous optimization of care processes and resource usage applicable to future pandemic situations and also better outcomes through the use of robotics and AI-based technologies.

2.1 Data Gathering

Data gathering in a healthcare setting can be broken down into several main areas: patient data, resource tracking, and pandemic-related data. Currently, there are many technology-based solutions, such as wearable vital signs sensors, electronic health records, and bar code scanning, which are used in healthcare settings.

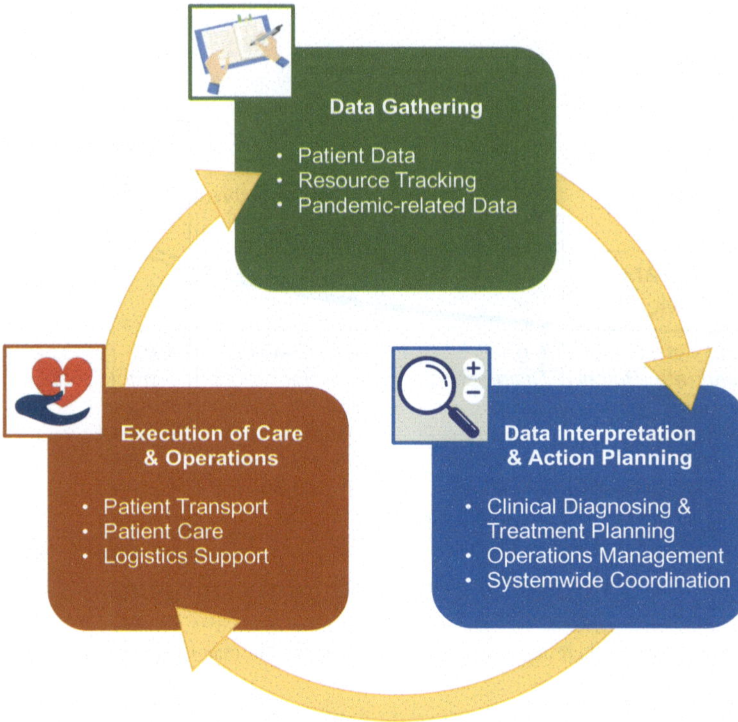

Fig. 3 Hospital operation aspects and their tasks

However, most data gathering and entering practices are currently human-centered and unsuitable in pandemic scenarios. By analyzing each area of data gathering and the methods used, the potential uses of robotics and intelligent systems for solving challenges and improving the efficiency of patient care and operations can be identified.

2.1.1 Patient Data Gathering

Patient-related data gathering encompasses screening, diagnostic tests, and monitoring. The screening process involves a range of tasks from obtaining registration paperwork to taking basic vitals, height, and weight. Diagnostic assessments are performed during admission as well as throughout patient care to determine treatments and evaluate recovery. In-facility patients are typically monitored for changes in vitals and conditions and receive regular in-person check-ins, while at-home care patients may be monitored remotely. In a pandemic with a surging patient population, there are many drawbacks to heavily human-centered data gathering that include high demand for human resources, increased person-to-person contact, exhausting of PPE, as well as proclivity to delays and errors. A close examination of all those data-gathering processes shows that many tasks are monotonous and repetitive. Therefore, the automation of these processes using robots and smart technologies is possible and would have many benefits, including lowering clinician workload and reducing cross-infections.

Patient Screening Registration and screening are low-effort tasks, yet hospital staff members have to perform those duties repeatedly as patients enter the hospital. This process exposes them to infected patients and occupies a large portion of their time, as well as causing delays in patient care due to increased workload with a high patient population in a pandemic. During COVID-19, a few robotic and intelligent systems have been used for patient screening in many parts of the world, and those implementations have clearly demonstrated the potential and benefit of such systems in pandemic care. For instance, the temi (by temi USA Inc.) is an AI-powered screening robot that is used in healthcare facilities throughout Southeast Asia. This robot was designed with a thermometer to screen for temperature, an autonomous navigation system, and a sink for hand washing. A similar robot, the AIMBOT (by UBTECH Robotics Inc.), is equipped with infrared (IR) and visible spectrum binocular cameras to measure patients' temperature, recognize if a mask is worn, and identify patients using facial recognition [59]. The Mitra Robot (by Invento Research Inc.) has strong capabilities as a receptionist robot with AI-driven intelligent communication. This robot also allows experts to guide patients remotely and stores patient data securely in an encrypted cloud.

While these robots show the potential for robotic-aided screening, the ideal capabilities of screening systems could be envisioned as the following: (1) Entering patients are greeted by a robot that can take their temperature and vitals specific to the pandemic and guide them to the appropriate waiting area. (2) The robot scans the patient's face for identification, adds them to the hospital's virtual health screening queue, and handles obtaining the person's admission paperwork. (3) When a person reaches the front of the virtual queue, the robot can call for the patient, identify them, and guide them to an area where additional diagnostics can be taken. (4) The robot can place appropriate sensors and/or guide the person through the health screening while receiving data from sensors wirelessly and storing them on an encrypted cloud. Once data are collected, the robot can either send the data to health professionals or AI-based algorithms for analysis before triaging. Current state-of-the-art market-ready robots certainly have the capability to be programmed to automate the full registration and screening process, yet hospitals have implemented only the most basic elements of the process thus far.

Diagnostics Automating diagnostic protocols can be considered a more complex task than screening due to added safety risks and uncertainties as well as increased HRI. While some sample collection processes can be simple such as urine collection or nasal swabbing, others may require professionals skilled in using specialized equipment like imaging machines (X-ray, CT-computed tomography scan, ultrasound, etc.) or performing minimally invasive procedures like biopsies, blood drawing, and endoscopies. Most of these procedures require close contact between the patient and the healthcare worker. Ideal solutions for a pandemic can be twofold: tele-operated robots where workers can remotely perform procedures or semi-autonomous robots where workers create the plan for the procedure and the robot executes it under supervision.

In cases where diagnostic procedure risk is low to the patient, introducing robots is relatively easy, and full automation can be envisioned. The most commonly used sample collection process during the recent COVID-19 pandemic was nasal and oral swabbing to test for SARS-CoV-2. Although swabbing is now commonly performed by patients themselves, concerns are on the rise of whether or not the proper technique is being used [29]. False-negative tests in a hospital environment can lead to errors in triaging, incorrect treatment, and a false sense of security for the patients and their caretakers. With the need for accurate, consistent, and rapid testing in hospitals, automated oral and nasal swabbing robots were developed. CAREEBO LLR S1 (by Lifeline Robotics) is one such example capable of performing oral swab sample collection by utilizing AI-based computer vision to identify anatomical markers to control the robotic arm for proper swabbing. Technology for other diagnostic procedures has also been reported. For example, there is current development of an automated blood drawing robotic device which can use near-IR and ultrasound imaging to locate veins, not requiring skin contact and automatically collecting blood samples into vials [16]. As these machines will soon be implemented in hospital settings, they will set a precedent for future robots with similar levels of automation for diagnostics.

For more invasive diagnostic procedures, robots would be better suited to be tele-operated or collaborative to provide clinicians greater control over the execution and safety of the procedure while reducing person-to-person contact. At present, many such robotic systems are being used, and they are known to increase the accuracy and consistency of procedures using machine vision and precise robot manipulators and end-effectors [53]. For instance, iSR'obot™ Mona Lisa (by Biobot Surgical Pte Ltd) is one of the latest in robot diagnostic technology that can perform a biopsy either by teleoperation or in an automated methodology with input from an expert clinician. It can automatically target the region of interest using magnetic resonance imaging (MRI) and ultrasound, and the physician can design a custom biopsy plan for the robot to execute. Procedures like endoscopies present greater challenges, as novel robotic arms must be designed to navigate through narrow and curving body pathways. Robots with articulated backbones have been proposed, developed, and studied to assist endoscopic procedures [18]. In a pandemic, these endoscopic robots could be an assistive tool to provide faster and more accurate diagnostics. For instance, during the recent COVID-19 pandemic, repeated sampling of distal lung tissue from patients on mechanical ventilation was critical. This highlighted the urgency for developing robotic-aided bronchoscopy, as the conventional process could obstruct ventilation, had limited dexterity, and required skilled operators. Efforts to develop such a tool were reported, which resulted in a robotic bronchoscope with high dexterity and small enough size to avoid interfering with patients on mechanical ventilation [40].

Medical imaging is a significant diagnostic tool used in every hospital. Although the equipment is noninvasive, it still requires operation by a skilled worker. To reduce contact between workers and potentially infected patients, imaging equipment must be advanced to a level where they can be highly automated, as well as capable of teleoperation. This type of system was demonstrated for point-of-care ultrasound scanning (POCUS), a primary determinative method for diagnosing and staging the severity of COVID-induced pulmonary diseases. This diagnostic process requires close contact of radiologists and ultrasound technicians with patients [4]. An autonomous POCUS robotic system has been developed which utilizes algorithms and sensors to optimize probe positioning for completing torso scans. The simulation results suggest the feasibility of this system for performing ultrasound scans remotely and autonomously and could potentially be translated to other forms of internal scanning diagnostics [4]. These systems would be significant during a pandemic, as they can assist medical personnel who are not as trained or experienced to perform imaging diagnostics.

Monitoring Monitoring patients at all stages of care is necessary for healthcare systems to manage and prioritize resources based on the severity of patients' conditions. In a rapidly evolving pandemic, illness progression is largely unknown and patient surges are unpredictable. Therefore, monitoring should be focused on all patient groups to streamline care operations, which includes patients waiting to be screened, at-home patients, and admitted patients.

Monitoring patients who are waiting to be admitted is critical and could benefit from mobile robotics able to perform rounds of monitoring continuously. Waiting rooms during a pandemic fill up quickly causing overcrowding and longer wait times. The conditions of these patients can deteriorate while waiting, leading to life-threatening situations. To handle this issue, mobile robots could be equipped with appropriate sensors to perform periodic checks on vitals, communicate with patients to assess their condition, and inform medical staff of necessary action if critical changes are detected. Dedicated robots for monitoring patients in waiting areas have yet to be developed, but there are many mobile robotic platforms currently available that could be modified for this purpose by integrating required end-effectors and sensor technologies [10, 30].

In a pandemic, mild cases are likely to be sent home to avoid overcrowding in the hospital. Monitoring these patients will help clinicians guide them through self-care and allow interventions as necessary. As was seen during the COVID-19 pandemic, telemedicine became common practice, where patients could be monitored by their doctors by a video call. In addition, remote patients could be sent home with wearable sensors that monitor their condition or use take-home vitals measurement devices that automatically transmit data to a hospital's central monitoring system. In Romania, a combination of a mobile app and a health monitoring device completed a successful pilot run with a few home care providers [50]. The system enables remote monitoring of multiple patients simultaneously for vitals such as temperature, respiratory rate, and blood oxygen levels.

For admitted patients, monitoring conditions and response to treatment is standard, and there are many monitoring equipment and wearable sensors that continuously track patient data (ECG-electrocardiogram, Vital Signs Monitor, Pulse Oximeter, etc.). However, staff must periodically enter the patient's room to check the data on equipment monitors or auxiliary software. This type of in-room checking during a pandemic can create significant bottlenecks in hospital workflow and be detrimental to patient care. With the latest developments in wireless sensor networks, hospital monitoring equipment could be retrofitted to transmit patient data in real time to a cloud computing application that monitors vital signs and data from all patients continuously. The COVID-19 pandemic initiated an investigation of a "smart hospital" vital signs monitoring system to determine how a centralized communication system could improve hospital efficiency during staff shortages [7]. These types of systems can enable hospital staff to observe the condition of all patients remotely and reduce the need for room visits. As an alternative solution to advanced wearable monitoring equipment, staff and patients could also rely on robotic nurses to perform room check tasks that human staff routinely do. A multifunctional sensory network has been proposed to work with an assembly of robotic nurses to collect information from patients and the environment and transmit the data to doctors and nurses. Additionally, robotic nurses would allow remote patient and clinician interaction [48].

2.1.2 Resource Tracking

Maximizing resources is critical during a pandemic so that staff is allocated efficiently where needed, supplies are delivered and restocked in a timely manner, and facilities are prepared and used efficiently. The data necessary for these operations are typically entered into management systems by humans, and very little automated technology is used at present. Those data include the availability of resources such as workers, hospital beds, equipment, PPE, food, drugs, and other supplies, as well as facility operations such as sanitization, maintenance, and room availability. To optimize hospital operations, there must be a continuous flow of incoming data on all key resources. This can be achieved by utilizing data collection devices such as sensors, cameras with machine vision, scanning systems, and robots equipped with any of those tools and algorithms.

For human staff and mobile robot systems, an automated scheduling network would help to optimize schedules for performing patient care, delivery tasks, and facility maintenance. Automated schedulers will require incoming data, such as availability and location of staff and robots in real time, which could be collected using a Hospital Real-time Location System (HRTLS) [23]. To achieve this kind of network, data on staff location in the hospital can be automatically tracked using wearable sensors such as radio-frequency identification (RFID) tags, wireless local area network (WLAN) nodes, Bluetooth®, and IR beacons. Hospitals could also be fitted with advanced cameras that can track staff through facial recognition or IR. Similarly, robots can be tracked not only for their location but also for the task they are performing. Novel management systems could also interpret descriptive data on hospital staff and robots so that their skills and specialized knowledge can be allocated by software to areas based on patient and facility needs.

In addition, data on hospital rooms and laboratories can be directed to the central management system. Current technology and approaches require human involvement to report data on facility status. All rooms in the hospital can have a data input terminal where clinicians and maintenance staff can input data about room status, i.e., if a room is occupied by a patient, if the room is clean or needs cleaning and/or restocked with supplies, if a patient needs an item delivered, etc. However, with the increased use of robots for basic functions in facility operations, some data can be gathered automatically by robots as they can record their completed tasks and the status of rooms and laboratories.

To monitor supplies, many hospitals use barcodes to track location, usage, and stock. This method can lead to errors, as scanning is typically a manual process performed by humans. Dedicated scanner robots can perform this repetitive task and reduce error. Other technologies use printable RFID tags or barcodes on medical equipment, PPE, consumables, medicine, and beds, which can be scanned automatically by gateway readers at hospital checkpoints. An existing management system called SmartX HUB (by SmartX Technology Inc.) can track hospital supplies using these techniques. A Command, Control, and Communications (C3) system was developed for a hospital in Singapore to monitor patients, staff, equipment, and inventory, supporting action planning using Internet Protocol cameras for image recognition of people, rooms, and items [31]. Employment of these advanced monitoring and communication systems would reduce time and resource usage, lower healthcare worker stress and work overload, improve patient care, and optimize hospital workflow.

2.1.3 Pandemic Data

Unique to a pandemic, communication of data and rapidly changing information regarding the disease is important to maintain a stable healthcare system. Hospitals must stay up to date with the latest information and protocols, which should come from a reliable, well-reviewed, and centralized source to avoid discrepancy and confusion. In Singapore, an AI toolkit called Bot MD (by 5 Health Inc) was developed, providing hospitals with the latest information on COVID-19 by gathering data from a variety of evidence-based sources [59]. Bot MD was successful in connecting local Singapore hospitals and nearby Asia-Pacific countries, demonstrating the usefulness of such systems. Similar healthcare information tools should continue to be developed and implemented to improve resource allocation and management among hospitals, as well as local, state, and federal agencies.

2.2 Data Interpretation and Action Planning

To attain efficient hospital operations and quality patient care in a pandemic, both clinical and operational data must be analyzed and interpreted in a timely manner to make clinical decisions, manage workflow, and operate the facility to its full capacity. Even during regular operations, many clinical and administrative/support staff are challenged by the amount of data to be handled and interpreted. In a pandemic, surges in the patient population will multiply the amount of data to be handled, which will overwhelm an already overworked staff. As a solution to these challenges in data interpretation and action planning, experts suggest using artificial intelligence (AI) algorithms to improve efficiency. Branches of AI such as machine learning (ML) and neural networks can be utilized to quickly analyze and interpret large amounts of data to identify patterns and predict outcomes. When the algorithms are combined with robotics and automation platforms, many operational and clinical duties can be automated or supported, allowing clinicians to attend to critical care.

To synthesize reliable and accurate algorithms for data analysis and interpretation within the scope of healthcare, large sets of preliminary data must first be available. For example, algorithms that diagnose and make decisions need a diverse and information-dense data set of patient conditions and health information related to the targeted disease. Similarly, a large set of operation and logistics data are needed to train algorithms targeted for the operational support of hospitals. Data analysis and interpretation can be performed using both offline and online ML methods which have their own strengths and weaknesses. Algorithms formulated through offline learning need to be explicitly trained by a preliminary data set prior to deployment. Online learning methods start with an algorithm similar to the offline technique; however, they are able to continue updating algorithms as new data points are added to the data set. As more data are fed to the online algorithm, the effectiveness and accuracy of the system's prediction and decision-making ability will increase. In a rapidly changing pandemic, algorithms must evolve and tailor to new incoming data.

Automation algorithms leverage the ability of computing elements to follow a decision tree algorithmically, which enables the reduction of labor and supported decision-making abilities. The healthcare industry has always been more conservative in its approach to automation due to the high stakes. However, it is moving toward integrating advanced computing power into many sectors that have been historically time and human labor intensive. For instance, AI algorithms are increasingly being utilized for the development of drugs, vaccines, and medical devices to reduce development time as well as improve efficacy and safety [6, 47]. The ability of these algorithms to utilize large sets of data in less time was dramatically evident by the discovery of a working vaccine for the pandemic virus SARS-CoV-2 in less than a year. Advancement of analytical, predictive, and decision-making algorithms for clinical practice and hospital operation is also feasible and should be pursued. Such AI algorithms are envisioned to analyze and interpret data for recommending action plans to humans or computing action plans for controlling robots and other automated systems.

Figure 4 shows the three main areas that will benefit from supporting and automating algorithms during a pandemic and general hospital operations. For algorithmically led operations, patient data could be used as input for creating treatment plans or diagnosing illness, while hospital resource data could be used to create

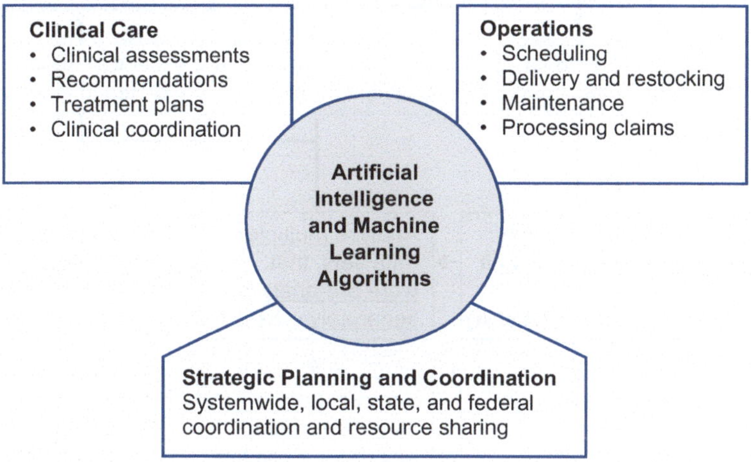

Fig. 4 Data-driven aspects of the hospital that would benefit from AI and ML

management plans for optimizing human resources, robotic utilities, as well as the use of supplies, equipment, space, and finances. In the future, hospitals could have a central algorithm that utilizes all data sources simultaneously to manage human resources optimally and control multifunctional robotic platforms to perform autonomous and collaborative tasks. Developing networks that utilize common algorithms and extract from the same pools of data, ranging from local to global scales, is another essential step to reaching the full potential AI has to offer in optimizing healthcare. Once fully developed and connected through a network, these algorithms can improve outcomes for patients and reduce burdens on healthcare workers, hospitals, and local and global healthcare systems.

2.2.1 Clinical Diagnosis and Treatment Planning

Quality of patient care depends on clinical assessment, recommendation, treatment plan, and clinical coordination. All aspects involve a large amount of data, including medical history, diagnostic test results, and expert observations. These data are generally processed manually, which is labor-intensive and error-prone. In most cases, medical data are fragmented and underutilized, leading to delays in care, medical errors, and subsequently increased morbidity and mortality [5]. A pandemic can further exacerbate difficulties in handling clinical data and making appropriate care decisions due to unpredictability and rapidly evolving information. In the early stages of a pandemic, the effects of illness chronic conditions are unknown, variability is unavoidable due to demographics, and treatment options are scarce and have yet to undergo long-term analysis. Clinicians will benefit greatly from systems that not only gather data efficiently but also utilize AI algorithms to analyze that data and provide potential diagnoses for triaging, clinical decision-making, and suggesting treatment plans.

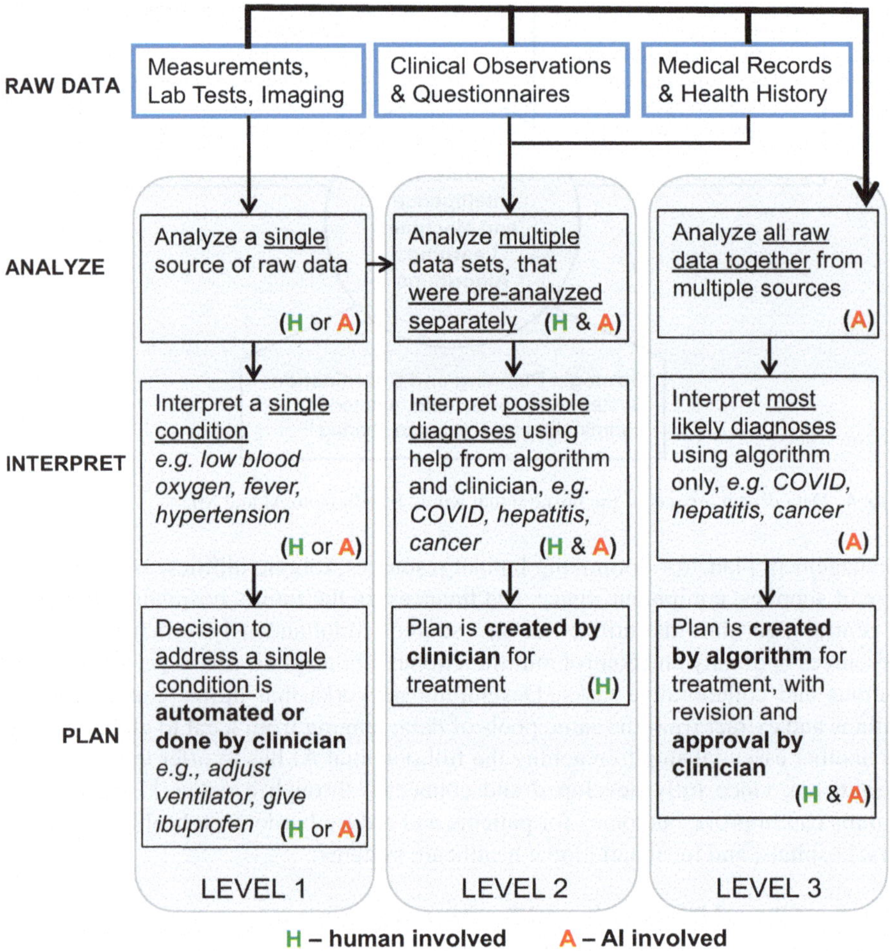

Fig. 5 Path of raw patient data from analysis and interpretation to the planning stage

Figure 5 shows the potential path of raw patient data passing through analysis and interpretation to the final stage of planning and decision-making. The path displayed considers both conventional methods (human-involved processes) and AI-driven methods, which is the focus of this section. To simplify the idea of AI utility in clinical care action planning, three levels are defined which correspond to the degree of complexity and human involvement in the data parsing path. Starting from the top row of Fig. 5, raw data in a hospital environment are collected from multiple sources. As described in the previous section, most of these data can come from automated data-gathering devices, while some data are collected by humans. Regardless of the source, three levels of data analysis, interpretation, and action planning can be envisioned. *Level 1—single source*: Analyze and interpret data from a single source to determine and address one condition. *Level 2—multisource*:

Using pre-analyzed data from individual sources (from level 1), review those analyses together for a more holistic interpretation. Possible diagnoses are determined by collaborative investigation between clinicians and AI. The clinician determines the patient's treatment plan. *Level 3—multisource with high autonomy*: Analyze data from all sources simultaneously, interpret a diagnosis, and create a treatment plan, all using AI. The clinician can review the AI result and modify it as needed before executing the plan.

Level 1—Single Source Medical professionals can determine how to help a patient by interpreting data from a single source to address a particular condition, such as fever, low blood oxygen, or hypertension. At this level, single data sources can be simple or complex. For example, analyzing body temperature data can help discern if a patient has a fever, then action can be taken to treat the fever. Alternatively, a single source of data from more complex diagnostic tools like X-ray, MRI, or lab results can also reveal a patient's condition. For instance, an X-ray scan may be the only diagnostic test needed for a radiologist to identify that the patient has a broken femur. At this level, analysis can be managed by humans, but AI could reduce the workload of the clinician and enhance the accuracy of diagnosis. Level 1 is used in many cases of patient data analysis, such as wearable sensors or diagnostic tools that can measure data for a single test and return an output that provides some information about the patient's condition. There is huge potential for advanced algorithm-enabled analysis and interpretation as today's healthcare industry is computerized.

For a data source that does not require complex interpretation, some automated systems have already been implemented, such as a pacemaker which can maintain ventricular pacing using automated algorithms. More recently, ML (machine learning) methods are being explored to enable remote monitoring of pacemaker data to identify abnormalities, notify the cardiologist, and allow for updating the pacemaker parameters [9]. Another area of automation with a single data source is artificial ventilation, which has come to the forefront during the COVID-19 crisis [35].

Recognition of patterns from images and detection of trends from medical tests have been traditionally performed by experts in the field. New ML-enabled algorithms are increasing the efficiency of clinical experts and sometimes outperforming them on average for analyzing complex data, such as ultrasound and MRI. For instance, recent developments in ML have made possible an online learning algorithm to interpret mammograms and detect breast cancer, improving the efficiency of radiologists in time, sensitivity, and specificity of detection [46]. Similarly, a recent study with 1500 patient cases of pulmonary function tests found that an ML algorithm outperformed independent interpretations of pulmonologists [56]. In summary, ML algorithms enable comprehension of complex patterns yet are flexible enough to change with historical data, all while reducing the chances of human error. Such studies make a strong case for ML-enabled decision support systems which can analyze data from single sources better, identify patterns hard for humans to comprehend, and increase the efficiency of human experts.

Although data interpretation algorithms, including ML methods, enable improved single data source diagnostics, a holistic view of human health is needed to diagnose and detect most complex diseases which manifest as a combination of simple conditions. For example, COVID-19 infection can manifest as a low-grade fever, cough, headache, and fatigue in the initial stages. Looking at multiple diagnostic tests to assess these conditions would provide appropriate care for the illness. Evaluating multiple data sources and types creates greater challenges for clinicians, but AI algorithms envisioned for Level 2 would be able to handle complex data to enable clinicians to make informed decisions with better outcomes for the patient population.

Level 2—Multisource Currently, in Level 2, most of the analysis and interpretation are done by clinicians using data from various sources that include previously analyzed data, as well as information from observation of patients and their medical history. Previously analyzed data can come from both human and AI interpretations. Clinicians, or clinicians assisted by AI, will use all data to make an informed diagnosis and make an appropriate clinical decision. As shown in Fig. 5, treatment and action planning are done solely by the clinician at this level. Collation of data from multiple sources is necessary for determining some specific diseases, as many illnesses share some common symptoms. Recent development of algorithms has shown the capability of using multiple sources for clinical diagnostics without the need for human involvement. For instance, an algorithm has been developed to utilize data from both X-ray and CT scans to identify COVID-19-induced pneumonia [37]. Such algorithms would have been valuable diagnostic tools for doctors in the early days of the COVID-19 pandemic, especially when PCR (polymerase chain reaction) tests were scarce. Screening algorithms with corroboration from multiple sources of data can be used as a first line of defense in pandemics.

At Level 2, the human and AI combination is anticipated to improve diagnostics, thereby reducing the workload of clinicians. While the aim is for a more holistic diagnosis and treatment planning by reviewing many analyses taken together, there is still room for improvement since the data used are initially analyzed separately, which can result in major deficiencies due to underutilization of all available data. To avoid this problem, more sophisticated AI algorithms can be developed to take data from all sources from the very beginning and throughout care and thereby enable diagnostic interpretation in an all-inclusive and comprehensive manner. The high-level data interpretation would also enable the development of ML algorithms that can be trained to create treatment plans, further reducing the workload on clinicians and improving treatment effectiveness.

Level 3—Multisource with High Autonomy Level 3 is envisioned more for the future, as it may take years of training algorithms to parse through accurately and interpret multiple raw data types collectively. At this level, AI would determine accurate diagnoses and create treatment plans, where the only human involvement would be for the clinician to review the action plan and modify it as needed. In a pandemic, this type of algorithm would allow identifying patterns at patient and group levels. Group-level data may enable algorithms to improve diagnostic procedures, identify better treatment options, and predict recovery trajectories, where all these aspects are important in a rapidly evolving pandemic to optimize healthcare utility. Although Level 3 sounds far-reaching, researchers are already developing preliminary versions of such algorithms. Computer-based algorithms called Clinical Decision Support Systems (CDSS) can analyze data in electronic health records to assist in making clinical decisions. Current CDSS use ML techniques to process data and provide key indicators and markers for clinicians to make decisions using evidence-based methods [54]. Level 3 algorithms are considered an evolution of CDSS where decision-making is done completely algorithmically, with little supervision. Such systems will drive healthcare in new directions by offloading the decision-making burden for humans and enhancing patient care during a pandemic.

At the highest automated level of decision-making, algorithms may be able to control some aspects of executing treatment or action plans by commanding robots or systems to automatically deliver care as needed. As patient care involves higher levels of risk, there are concerns about allowing technology to control precarious situations. Thus, easing the healthcare industry and society into accepting AI-driven data interpretation and decision-making systems will require rigorous training, testing, and safety verification. To do this, researchers should begin by focusing on systems at lower levels of data interpretation and automation and aim to involve humans in the loop. The use of ML and other data-intensive tools for clinical care also raises issues of ethics and consent. The importance of discussing such issues cannot be understated, and discussion is needed to involve decision-makers in developing such algorithms in an ethical and responsible way.

2.2.2 Operations Management

Efficient hospital operation is key to quality patient care. Hospital operations can be described as the management of four areas: human resources, equipment and supplies, facility, and finances. Currently, most hospitals rely on enterprise resource planning (ERP) solutions to streamline operations such as monitoring the current state of resources, as well as identifying areas with risk or in need of improvement. Although these systems provide decent support under normal operating conditions, they are not designed to provide planning and organizing that is sufficient during a pandemic due to rapid and unpredictable changes in demand.

Typically, managerial roles are assigned to make schedules, create action plans, keep inventory, and manage finances. An automated approach to resource management for the hospital could streamline those operations and optimally allocate resources. Automation methods driven by real-time data lead to efficient functioning of the entire hospital. For instance, a centralized system could contain information on the entire hospital staff, and also continuously monitor the number of personnel present in the building and tasks they perform. An automated data interpretation and decision-making algorithm can easily identify what is needed in real time and change the scheduling of staff accordingly. Similarly, automated inventory management technologies track supplies and equipment usage. Thus, the inventory management system can prioritize and plan logistics operations including equipment allocation, making delivery rounds, and maintaining stocks of materials and supplies. IDENTIPLATFORM™ (by INDENTI Medical) is one such AI-based inventory management system that automates data entry and communicates with ERP solutions while identifying patterns of use to replenish supplies accordingly. Timely cleaning, maintenance, and preparation of various patient rooms and laboratory spaces are a critical part of operation and these aspects of scheduling the custodial staff can be automated and tracked in a central resource management system. Overall, these types of AI-based resource management systems driven by real-time data will improve the effective utilization of resources and logistics operations. The improvements will reduce the burden on the healthcare system and improve patient care, which is critical during a pandemic.

Billing department staff in hospitals are overwhelmed by claims and invoices that need to be processed, even during normal operations. A pandemic exacerbates the issue, with surges of incoming patients creating an increased number of claims to process. AI-based claims management enables the identification of incorrect claims, freeing administrative staff to focus on cases requiring review. A uniform digital format developed to parse invoices, enabling automatic processing of claims and invoices, will further reduce workload and errors.

2.2.3 Coordination and Resource Management among Hospitals

Hospitals function as independent units, providing clinical care to their patient population during normal operations. However, the shortcomings of decentralized hospital operations became apparent during the COVID-19 pandemic. Coordination and communication among various local entities are necessary to minimize overcrowding of individual hospitals and improve resource sharing. A central command structure with an ML-driven resource management system can generate usage statistics in real time and also enable the coordination of multiple hospitals according to their surplus and demand needs. Data collected during the COVID-19 pandemic indicated the usefulness of coordination and spurred interest in developing solutions to facilitate collaboration among private, public, community, and federal hospital systems.

If statistical data on infection propagation in a pandemic are centralized by local, state, and federal governments, then models can be developed to predict infection rates and surges in a geographic area in real time. Such models are needed to enable authorities and hospital systems to plan more efficiently the allocation of resources since published infection rates are lagging indicators in a rapidly evolving pandemic. Some preliminary algorithms have shown the effectiveness of using sensor fusion and distribution estimation to drive the models, which could estimate COVID-19 infections in real time [32]. ML models were also used to predict COVID-19 infection using epidemiology data sets through offline learning [42]. Such algorithms allow foreseeing localized outbreaks and taking anticipatory measures to reduce the burden on local healthcare systems. A country-wide network of resource management systems informed by predictive disease propagation models could enable efficient use of resources through location-based sharing.

2.3 Execution of Care and Operations

Implementation of medical care involves many factors, including triaging and transporting patients, routine and emergency patient care, and logistics support to maintain care and manage hospital operations. Staff shortages and infection among healthcare workers have caused delays in critical care, delivery of drugs and essentials, and maintaining facility sanitation during the recent COVID-19 pandemic. To alleviate this, robotic technologies were proposed, as a potential solution, to perform a variety of patient care and logistics tasks typically human dependent. Most of the robots performed low-risk tasks, such as disinfecting and supply delivery. This limited use of robotic and autonomous systems demonstrated successfully their potential to assist in a healthcare environment [52, 59]. However, more technological development is required, considering the large number of tasks and varying degrees of HRI involved in patient care and logistics support operations.

2.3.1 Patient Transport

Patient transport within the hospital can generally be grouped into two categories: (1) taking patients triaged and admitted to designated units and rooms and (2) moving patients from hospital rooms to other facilities such as an operation theater and testing facilities. In general, patient transport, using chairs and beds, is time consuming and typically requires multiple workers. The process is a resource drain, a significant problem when resources are needed for critical care during a pandemic. The staff involved in the transport are at risk of cross-infection, as they cannot maintain social distancing requirements of a pandemic. Moreover, patient transport is recognized as a high occupational risk, notably physical injury to nurses and caregivers [19]. Power-drive assist technologies are used to reduce the physical injury risk of the individuals who transport patients [34]. However, it cannot address resource constraints and cross-infections.

Transport with self-driving robotic navigation is a potential solution to eliminate or partially reduce human resource requirements for moving patients. There are two possible approaches for implementing autonomous navigation with beds and transport chairs. The first approach is retrofitting self-driving robotic hardware to interface with existing beds and transport chairs. A research report shows the feasibility of attaching a motorized robotic bed mover to the bottom of a manual hospital bed [25]. A driverless wheelchair concept has also been proposed whereby requests from the hospital management system on pick-up and drop-off locations allow autonomous navigation through the hospital [8]. However, retrofitting beds and transport chairs would be a costly and onerous project. The second approach is to develop a robotic system that can adaptively attach and detach itself to many types of existing transport equipment in the hospital and has self-driving capability. This approach can be considered more attractive as it would require only the development of a single system that can operate multiple transport equipment types. Some early-stage feasibility trials of such a system have been reported [39]. Various technologies currently aid autonomous indoor navigation, such as image capture and processing, invisible paint on floors, and light-based distance measuring techniques.

These technologies, with advanced sensing methods and control algorithms, will ease the transformation of industrial indoor robots to assist in moving hospital beds and transport chairs.

2.3.2 Patient Care

A variety of tasks comprise executing care of patients admitted to a hospital. The majority of patients require routine care, while others may require intensive care. Routine patient care can be simplified into managing two areas: treatment and well-being. Treatment includes any action that is a direct part of a patient's treatment plan, where clinicians must deliver drugs, wound care, rehabilitation, and/or diagnostics at prescribed frequencies and duration. Management of patient well-being includes regularly checking in with patients to provide assistance, deliver food and water, ensure comfort, monitor condition, and provide advice and support. For intensive care, patients may require sudden critical intervention to deal with worsening conditions or they may arrive at the hospital in a critical state. Those patients will require round-the-clock care, monitoring, and treatment by specialized staff trained for high-risk interventions such as surgery, as well as setting up equipment, e.g., dialysis, ventilator, and feeding tube. With such a wide array of tasks to deliver patient care, maintaining efficiency, accuracy, and timeliness can be extremely difficult under pressure of a pandemic, when staff numbers are low and patient numbers are high. Nearly all listed tasks require healthcare workers to enter the patient's room and even have direct physical contact, which places them at high risk of cross-infection. Robots can be developed to support staff when patient volume is high by offloading tasks involved in drug delivery, rehabilitation and wound care, room visits, and intensive care.

Drug Delivery Many aspects must be considered when designing a robot for drug delivery. Patients can be in a variety of states that may shift during their hospital stay which can affect the drug delivery process, such as physical conditions not safe to use with the medicine (change in temperature, blood pressure, heart rate, etc.). In a traditional scenario, nurses will be able to communicate with the patient to assess their condition. To consider these nuances, robots could be built with sensors that measure body parameters before delivering medicine or receive data from monitoring sensors attached to the patient's body. They can also have teleoperation capability so that clinicians can communicate with patients through a virtual screen, to assess the patient's condition, provide instruction, and answer questions. With respect to actual drug delivery, robots can be equipped with robotic arms and end-effectors controlled by feedback data from sensors or image processing. There are many modes of drug delivery such as oral, inhalation, injection, and skin absorption. Some modes would benefit from an extended arm that could hand oral medicines to the patient, while others would require advanced manipulators, such as identifying the correct anatomical location on the arm to give injections automatically [16, 43]. These robots would need to be capable of accurately identifying the correct patient by utilizing advanced facial recognition software or RFID readers, as well as confirming that the patient takes the medicine. Other design considerations include ensuring the robot carrying the medicine is secured, prescriptions are accurate, and delivery is safe [15]. An intermediary concept may be the logical next step, where self-driving robots retrieve and transport medicine within the hospital, but medical personnel are involved and take drugs to hand to the patient. This strategy is being formulated, e.g., combining the Information Communication Technology platform and autonomous robot vehicles to optimize their travel [20]. While this concept is not ideal for reducing patient-to-worker contact, it can help optimize workflow and serve as a stepping-stone for robotic advancement in drug delivery.

Rehabilitation and Wound Care While areas of treatment like rehabilitation and wound care are less relevant to the COVID-19 pandemic, the nature of future pandemics is unpredictable and may demand robotic development in those areas. Continuing to develop robotics in all areas of patient treatment will ensure that foundational technology will be prepared for adoption in a pandemic.

To touch on some developments in rehabilitation, robotic wearables, such as exoskeletons, can be used to assist patients in learning to regain or maintain motor function without a physician needing to be present during the session. Exoskeletons would be very helpful in healthcare facilities for diseases caused by viruses that affect motor function, such as poliovirus, an epidemic which occurred in the early twentieth century [44]. Physicians would benefit from these devices as they could capture data on patient progress, develop rehabilitation plans using AI, and control sessions remotely.

A more complex aspect of treatment is wound care, which is overlooked in robotics. Wound treatment is a common task in hospitals and efforts have been made to reduce the necessity of frequent wound treatment, freeing clinician time and decreasing person-to-person contact. Robot suturing systems are being explored that are capable of closing wounds, using articulated arms and proximal force sensing [14]. Wound dressing technologies reduce the frequency of dressing changes, such as PICO™ (by Smith & Nephew plc). If these wound dressing technologies were designed to have a unified application, robots could be automated to apply dressings and be integrated into mobile teleoperated platforms and enable contactless wound care. Those technologies would be a valuable asset during a pandemic, notably smallpox or monkeypox virus. Robot systems that can evaluate wounds and aid in treatment planning would provide even greater assistance in a pandemic by reducing clinician involvement further. For example, a novel robot-driven wound recording system is being developed to quickly and accurately assess wounds using a 3D scanner and "next-based view" algorithm to create a digital reconstruction of a wound and assist clinicians in providing better wound care remotely [21]. Overall, progress in robotics and autonomous systems development for rehabilitation and wound care will help healthcare systems provide care regardless of patient influxes during a pandemic.

Room Visits There are many instances where clinical staff must visit patients, e.g., routine hospital rounds, responding to alarms, managing patient well-being, and delivering food and medicine. Automated and collaborative robot systems could support nurses on the floor by doing mundane and routine tasks, as well as provide company to patients. Some of these concepts have already been simulated to determine the impact of collaborative robots with end-effectors in a COVID-19 in-patient care environment [22]. The most common, mundane tasks requiring entry into a COVID-19 patient's room were to turn off false alarms and call buttons, as well as other activities involving knobs and buttons. The research showed that using robots to perform those tasks would greatly reduce the number of unnecessary room entries and significantly cut down PPE use. During a pandemic, the mental health of patients is a concern, especially when they are isolated from family and friends. Robots with human-like cognition have been explored to provide assistance and friendly interaction with patients [36]. A unique benefit of human-like robots is that they can assist the mental well-being of patients and reduce healthcare workers mental fatigue during a pandemic. Ultimately, more complex systems could be developed to combine social AI and the performance of simple nursing tasks to create an ideal nurse robot. Moxi (by Diligent Robotics Inc.) is one such example of a mobile nurse robot with built-in social intelligence and the capability to aid nurses in tasks such as running prescriptions, delivering laboratory samples, and bringing food to patients. Moxi robots have been deployed in various hospitals across the United States, and the response of the nursing staff has been overwhelmingly positive.

Intensive Care Many robotic systems are used to aid in intensive care and operating theaters, where high HRI occurs. These systems are mostly teleoperated robots since patient intervention is high risk and requires expert skill and supervision. They have been found to improve the stability, control, and accuracy of maneuvers during invasive patient treatment.

Tracheal intubation is a common procedure that is done before many types of surgeries and for emergency mechanical ventilation. During a pandemic caused by an airborne transmitting pathogen, clinicians are extremely vulnerable to infection due to the nature of the intubation procedure. Successful intubation requires high skill and sufficient time, both in high demand during a pandemic when staff are spread thin. Research on robot-assisted tracheal intubation is ongoing, with a teleoperated system recently reported to show potential for improving patient safety and procedure accuracy [13].

To date, the most widely used interventional care robots are for surgeries. Some systems are as simple as tool holders, while others are advanced enough to determine surgical plans and be teleoperated by surgeons [26]. Many tool-holding robots also have the capability to be programmed to assist surgeries by performing tasks such as needle insertion, cutting small incisions, or camera-handling. For instance, the AutoLap™ (by Medical Surgery Technologies Ltd.) uses image analysis to position a laparoscope. The most prevalent teleoperated system used in operating theaters today is the da Vinci by Intuitive Surgical Inc. It is fully controlled by the surgeon and comprises a multiarm robotic manipulator for handling tools, a high magnification viewing port, and tremor-filtration for stable operation. There are not nearly as many semi-autonomous or collaborative surgical robots, due to regulatory barriers and requirements for rigorous safety testing. The ROBODOC® Surgical System (by CUREXO Inc.) is one such advanced robot capable of formulating surgical plans using CT images and autonomous bone drilling, with the approval and supervision of a clinician. Surgical robots not only reduce the risk of cross-infection but they can also reduce postoperative time in the hospital, freeing up space for incoming patients during a pandemic [52]. Although surgical robots are the most developed robotic systems in healthcare, further development is needed to add capabilities that address overlooked procedures.

2.3.3 Logistics Support

To maintain a high level of patient care, supporting staff must perform a set of tasks, namely, sanitization, facility and room upkeep, food and medicine preparation and delivery, equipment delivery, and supplies restocking. It was apparent during the height of the COVID-19 pandemic that hospital personnel were stretched thin, leading to challenges in logistics operations. While these jobs are a critical component of hospital function, they are also low-uncertainty tasks that can be automated by creating robots with self-navigation and specially designed manipulators or tools. For instance, disinfection measures are critical to prevent an outbreak within a hospital and many autonomous mobile robotic platforms have been devised to perform ultraviolet-C light-based sanitization to kill ca. 100% of infective agents on surfaces [52]. However, further robotic development is needed to include additional tasks such as changing bed sheets and trash cans, removing food and other waste, and wiping down surfaces. These require robots with specially designed manipulators, sensors programmed to identify unique objects, and significant effort to train algorithms to perform such tasks in a dynamic environment. With respect to logistics support, self-navigating mobile robotic systems are already in use that transport items like food, medicine, bed sheets, trash, PPE, and more. Staff simply load the robot system with items to be transferred and verbally command it to go to the point of delivery. One such robot is the TUG (by ST Engineering, Aethon Inc.), used in many hospitals around the globe and proven to reduce the workload of staff and improve hospital workflow. For instance, the Alle-Kiski Medical Center in Philadelphia reported utilizing three TUG robots in their facility which resulted in a 20-minute decrease in turnaround time for laboratory results [12].

Autonomous delivery robots can be controlled using a command center algorithm that optimizes their paths so that items are delivered quickly and in a prioritized manner. Additionally, if two or more robots work collaboratively to complement each other's functions, more tasks can be fully automated to reduce human resource demand. For instance, manipulator robots can help with picking up trash and other materials and loading them into a delivery robot. Delivery robots have demonstrated value in a clinical setting. However, further development is needed on manipulation robotics and their collaborative performance with other robots to improve the efficiency of healthcare system operations and logistics. Also to be considered is a multifunctional robot by integrating capability for multiple tasks, such as cleaning, manipulation, and delivery, into a single platform.

Inventory management of supplies, equipment, and drugs is an important aspect of efficient workflow in a hospital. As mentioned in the Sect. 2.2 of this chapter, AI management software can play an important role in streamlining inventory data for better visibility and management of supplies. However, hospitals typically rely on staff to manually update data entry systems for procured stock, location of items and equipment, used and expired items or medicine, and billing. A study reported that nurses spend nearly 45 minutes of their total 8-hour shift sorting out operational failures [57]. As more robotic implementation is envisioned for hospitals in the future, most data entry tasks can be automated. For instance, robots can be connected to an online inventory management system so that when they are moving drugs and supplies, they can inform the system using RFID or barcode scanning if an item is picked up, en route, or delivered. That will automatically update the inventory, and the central management system can communicate with staff or robots to perform additional functions, such as restocking, ordering, and scheduling. An autonomous robot with an AI-based management system has been implemented in hospital pharmacy operations (XR2 by Omnicell Inc.) and has shown the feasibility of such systems to reduce human workload and improve efficiency. However, pharmacy management is only a small part of hospital operation and much more work is needed to achieve automated robotic-aided inventory management for the entire hospital.

3 Improvements and Investments

Intelligent robotic systems and AI have significant potential to support patient care and hospital operations at a time when the healthcare industry faces unprecedented challenges, notably during a pandemic. Extensive use of surgical and service robots has proven the applicability and feasibility of robotic assistance in hospitals so the outlook for robotics in healthcare is promising. However, the widespread implementation of robotic systems and AI in hospital environments has many barriers, not the least of which are technological hurdles, societal acceptance, organizational structure, and regulatory and legal processes [17]. Although technology is the most fundamental, any development and implementation must consider all mentioned factors together. Collaboration of researchers, engineers, healthcare workers, patients, and administrators is vital in designing and developing new robotic systems and testing, evaluating, and integrating systems into the hospital environment.

3.1 Technology Challenges

Transforming robotic concepts into the healthcare environment requires interdisciplinary teams and extensive resources. System designers should have comprehensive knowledge of the functions, level of HRI, and operating environment to design the targeted systems. Both functional and safety aspects of robotics must be addressed, testing protocols established, and vigorous in-lab and field evaluation of systems and components be conducted to achieve sufficient reliability, efficiency, and safety.

Robots are widely used in industry for materials handling, processing operations, assembly, and inspection. Over the years, surgical robots have penetrated the hospital environment and performed surgeries. A common feature of industrial and surgical robots is that they operate in controlled environments. However, robotic systems that support hospital operations and patient care during a pandemic encounter significant challenges because the hospital environment is dynamic and complex with a high level of uncertainty. Additionally, hospitals and supporting equipment are designed for human-centered operations and not for robots. Transforming robotic technologies to unstructured environments and operations requires significant technological advancement. The size, shape, and capabilities of robotic systems must match the operating environment and functional requirements. To support the full spectrum of functions, some systems may need to be fully autonomous, while others can be semi-autonomous or operated by a human. By considering all requirements and examining the current state-of-the-art robotic technologies, it is apparent that many advancements are needed in hardware, data processing, communication, and controls. Hardware components such as actuators and end-effectors, as well as sensors and electronics, must be examined to identify current status and development needs as they together enable the physical operation of the robotic system. Advanced data processing and interpretation, e.g., pattern recognition and ML, are needed when the robot is operating in unstructured dynamic environments where multiple robotic systems and humans work together. Communication infrastructure that allows robot-to-robot and human-to-robot data sharing is also vital to properly manage functions and aid each other in sharing responsibilities.

From the hardware point of view, state-of-the-art actuators and end-effectors that are currently being used in industrial, service, and surgical robots will be examined to identify uses and improvements needed for hospital applications. Currently, many available actuators give the flexibility to design robotic systems with application-specific precision, accuracy, torque, speed, size, and power. Therefore, not much development is required in actuator technologies. However, a significant advancement in end-effectors is required as most of the current end-effectors are designed only for targeted applications. Hospital operation and patient care involve handling a highly diverse set of objects with different sizes, shapes, and fragility. For instance, the robot may require operating or adjusting an existing medical device, such as a ventilator, which needs a specific end-effector based on the user interface type: touch, push button, dial, or combination of one or more types. In some cases, the end-effector may need to handle a pill bottle, glass of water, or bedding. Some instances may arise where they need to reach difficult locations. The current robotic market has no diverse set of end-effectors to match these unique end-application needs. Furthermore, research is needed to incorporate human factors into end-effector designs when HRI is involved. As hospitals, equipment, and materials and supplies are built based on anthropocentric design, end-effectors that can manipulate and handle objects and interfaces with dexterity and maneuverability similar to human capabilities must be explored. Both conventional rigid robotics, emerging soft robotics, and their combinations (hybrid) must be investigated. From a system point of view, two approaches should be considered, a robot with a designated end-effector for well-defined tasks or a multipurpose robot that can "hot swap" end-effectors from a bank of onboard end-effector options to perform a given task.

An important component of hardware is the sensors, which are the main components of robot feedback systems that provide environmental and operational input to the system. Robots may require multiple sensors and sensing modalities to navigate efficiently, interact with physical objects, and monitor patients and equipment. Demands for safety and efficiency in complex, dynamic, and unpredictable hospital environments require sensors to identify objects accurately, estimate movement (both robots and humans), and apply known forces. The array of sensor modalities required includes machine vision (camera, LiDAR-light detection and ranging, IR), tactile (force, contact), inertial sensing (accelerometer, gyroscope, magnetometer), and range estimation (laser, ultrasound) [2]. As all these sensors are readily available, no sensor development is required. However, investigations should be conducted to determine the optimal number of sensors, sensor modalities, and sensor integration schemes for the robotic platform. Required functions, operation constraints, and operating environments must be considered when examining the sensor requirements for the given robotic platform. In most operational scenarios, data from a few different sensing modalities may need to work in a complementary fashion to achieve accuracy, precision, repeatability, and safety. For example, inertial measurement, vision, and range estimation sensors working together can enhance navigation and manipulation capabilities. If a scenario arises where one sensor, such as the vision sensor, may be obscured, the range estimation and inertial measurement sensors should work to avoid collisions. When multiple sensors and different sensing modalities are needed, the next challenge is the integration of those sensors into a robotic platform and acquiring signals from them. The current approach is to use electronic circuitries custom designed for each platform. Circuit designs, manufacturing, and debugging require considerable effort and time, which can hinder rapid development and increase the cost of robotic systems. The requirement for custom circuit designs can be reduced if standard circuits are developed for sensor integration, similar to the motherboard and RAM (random access memory) in computers. Focused research to develop those standardized circuitry platforms will accelerate the development and advancement of an array of robotic platforms for healthcare.

When using multiple sensors and different sensor modalities, processing data for suitable inputs for decision-making and control is an important task. The data capture and interpretation are reasonably straightforward for many sensors, such as force and inertial sensors. Image processing and pattern recognition using AI have advanced in recent years in autonomous navigation and facial recognition. However, major improvements are needed to achieve the high-fidelity data needed for a robot to operate and navigate in healthcare settings. For example, robots need to identify humans with attached medical equipment and limbs with missing appendages or heavy wound dressings. They would also need to navigate a hospital environment with high uncertainties, such as hazardous materials and unknown objects. Data fusion is an aspect of data processing where data from multiple sensors will be collectively used to improve the accuracy and reliability of operation while reducing uncertainty and complexity. Since hospital robots have multiple sensing modalities and use many sensors, a more efficient data fusion algorithm needs to be developed for real-time data interpretation, decision-making, and operation of the robotic system. Overall, there needs to be a holistic approach to establishing sensor configurations, pattern recognition algorithms, and sensor fusion techniques to provide efficient robotic systems for the hospital environment.

Interconnectivity and communication of robotic systems, humans, and data are mandatory for the success of robotic-aided care and hospital operations during a pandemic. Healthcare systems and hospitals must have a highly reliable network and cloud computing infrastructure. Timely data sharing is necessary not only for robotic systems to operate but also to make clinical decisions, perform predictive analysis, and prioritize clinical and administrative tasks. From a robotic perspective, knowledge of the location of the robots, their capabilities, and completed tasks will allow robots to collaborate efficiently and humans to assign tasks as needed. For example, if there is a spill that needs to be cleaned up, a central command system should optimally assign the cleaning robot to the nearest location to handle the task. Also, communication systems are critical for robots to have situational awareness, such as hospital codes, so that the robot can respond appropriately and avoid interfering with emergency medical procedures. A higher number of robots in the hospital system increases instability and vulnerability to outside cyber-attacks and data breaches. Communication systems are needed with established security protocols and robustness to protect patient data, prevent hacking of robot control, and avoid the breakdown of the whole system.

3.2 Societal Acceptance

Many factors influence the acceptance of robotics in healthcare. From a healthcare worker's perspective, perceived usefulness, attitude toward technology, and ease of use significantly affect willingness to adopt robotic systems to support duties. Additionally, trust, privacy, and ethical concerns impose additional barriers to accepting robotics for healthcare delivery [3].

Robotics used in the recent COVID-19 pandemic demonstrated utility in hospital operation and patient care, undoubtedly changed the perception of robots, and underlined the need for their widespread use during a pandemic [59]. Recent surveys show a positive attitude among healthcare workers toward using robots in the hospital environment [38]. These surveys, along with other reports, suggest that clinicians' experience and knowledge regarding the use of robotic platforms are low, and ease of use is the main concern of healthcare workers. Robotic designers must focus on developing easy-to-operate interfaces, as well as providing foolproof operating instructions and training protocols so that the user can easily understand the usage and limits of the robotic system. Participatory design approaches that utilize user observations, interviews, and benchmark tests need to be considered to achieve user-centered designs.

Trust, privacy, and ethical concerns are key issues when any technology is used in healthcare. Both trust and privacy concerns are directly related to the safety of the systems. Safety has two aspects: physical harm and data breach. Like any technology, robotic systems can break down and fail, potentially leading to physical harm to humans and destruction of equipment. Therefore, developers must design and investigate robotic systems as fail-safe. Privacy concerns mainly arise from data breaches. To assure the privacy of patients, robust data security protocols and monitoring systems must be in place. From a public perspective, most ethical concerns arise when robots directly interact with the patient during care and other interventions with little or no human involvement. It is known that human touch and direct interaction support the mental well-being of patients and, in some cases, aid in speedy recovery. Having robots that can assist with direct communication between patients and caregivers or family members would ease most concerns. It should also be noted that in a pandemic, telepresence robots may improve personal interactions where caregivers are limited and family members cannot directly communicate with the patient.

3.3 Regulatory, Financial, and Legal Processes

In responding to alleviate human resource needs during the COVID-19 crisis, rapid development and deployment of various robotic platforms were in use throughout the world [59]. In many cases, the deployment may have bypassed or received a waiver from regulatory, financial, and legal entities to alleviate the burden on the healthcare system and patients. Looking to future scenarios, however, requires setting a framework for classifications, standards, and guidelines for robotics in healthcare. That framework will provide guidelines for all stakeholders, from robotic and software developers to care providers, on regulatory requirements, reimbursements, and legal processes for new technologies. A few international organizations have already begun efforts to develop standardized testing protocols and device requirements for the safety of medical robots with varying degrees of autonomy [26].

Regulatory approval is a significant challenge to the implementation of technology in hospitals. Healthcare robots consist of a wide array of platforms with varying functionalities. Some of these systems (e.g., diagnostic and monitoring) can be considered medical devices, while others (e.g., disinfecting, delivery, and other operations) may fall into a general use device category. To expand robotic deployment for healthcare, the establishment of standard regulatory procedures and protocols for healthcare robotics must take priority. Those procedures and protocols will enable quick adaptation of existing technologies to manage pandemic crises more efficiently and accelerate the development of new technologies. Reimbursement codes using robots in patient care are another major consideration for the healthcare industry. As many of the applications are new, it is essential to create reimbursement strategies and financial guidelines to inform healthcare providers and insurers to cover the cost of the acquisition and operation of healthcare robotics. Liability is a decisive factor since it can hinder the acceptance and adoption of robotic systems by the healthcare system. There is always a chance that a robotic system can malfunction or be incorrectly used, causing harm to patients, staff, and visitors, as well as damage to other equipment. If such a situation arises, there must be a legal framework to address the incident appropriately.

3.4 Organizational Structures

Responding to future pandemics depends on how well governmental and global organizations prepare for such scenarios. Nevertheless, individual hospital systems and their level of preparedness are crucial since they are the ultimate care providers. Therefore, it is necessary to evaluate and change the organizational aspects of current hospital systems, including operation and infrastructure, to ease the introduction and implementation of robotics in hospitals to support pandemic care [3, 17].

Discovering how robotics should integrate and function across various departments is key to efficient operation. It must be determined whether robotic operation is centralized throughout the hospital or decentralized, allowing autonomy to each major department, such as emergency care, acute care, and pharmacy, to manage their own systems. In addition to management structure, a healthcare professional should know how to respond to emergencies and unpredictable situations when a robotic system is in place. Studies need to be conducted to understand the need for and to develop new protocols for such situations. Overall, the introduction of robotic systems to healthcare will be expensive. Some costs include robotic hardware, expansion of existing building infrastructure to accommodate robotic systems, training clinical personnel to use and coexist with robots, and technical staff to maintain robots and communication systems. Most healthcare systems and individual hospitals worldwide will not have sufficient funds for development, acquisition, and appropriate infrastructure to integrate robots into their systems. Governments and world health organizations should consider building model hospitals and using existing healthcare systems to define infrastructure needs, operational aspects, and response plans, to ease the adaptation of robotics and AI technologies to individual hospitals and healthcare systems, reducing the financial burdens.

References

1. Agarwal A, Agarwal S, Motiani P (2020) Difficulties encountered while using PPE kits and how to overcome them: an Indian perspective. Cureus 12(11):e11652
2. Alatise MB, Hancke GP (2020) A review on challenges of autonomous mobile robot and sensor fusion methods. IEEE Access 8:39830–39846
3. Alla S, Pazos P (2019) Healthcare robotics: key factors that impact robot adoption in healthcare. Institute of Industrial and Systems Engineers, Orlando
4. Al-Zogbi L, Singh V, Teixeira B, Ahuja A, Bagherzadeh PS, Kapoor A et al (2021) Autonomous robotic point-of-care ultrasound imaging for monitoring of COVID-19–induced pulmonary diseases. Frontiers in Robotics and AI, 8: 645756
5. Anton N, Hornbeck T, Modlin S, Haque MM, Crites M, Yu D (2021) Identifying factors that nurses consider in the decision-making process related to patient care during the COVID-19 pandemic. PLoS One 16(7):e0254077
6. Arshadi AK, Webb J, Salem M, Cruz E, Calad-Thomson S, Ghadirian N et al (2020) Artificial intelligence for COVID-19 drug discovery and vaccine development. Front Artif Intell 3:65
7. Atta RM (2022) Cost-effective vital signs monitoring system for COVID-19 patients in smart hospital. Heal Technol 12(1):239–253
8. Baltazar AR, Petry MR, Silva MF, Moreira AP (2021) Autonomous wheelchair for patient's transportation on healthcare institutions. SN Appl Sci 3(3):1–13
9. Banta AR (2021) Algorithms toward a next generation pacemaker. Rice University
10. Bartosiak M, Bonelli G, Maffioli LS, Palaoro U, Dentali F, Poggialini G et al (2022) Advanced robotics as a support in healthcare organizational response: a COVID-19 pandemic case study. Healthc Manage Forum 35(1):8404704211042467
11. Billings J, Ching BCF, Gkofa V, Greene T, Bloomfield M (2021) Experiences of frontline healthcare workers and their views about support during COVID-19 and previous pandemics: a systematic review and qualitative meta-synthesis. BMC Health Serv Res 21(1):1–17
12. Bloss R (2011) Mobile hospital robots cure numerous logistic needs. Ind Robot: Int J 38:567
13. Boehler Q, Gage DS, Hofmann P, Gehring A, Chautems C, Spahn DR et al (2020) REALITI: a robotic endoscope automated via laryngeal imaging for tracheal intubation. IEEE Trans Med Robot Bionics 2(2):157–164
14. Cao L, Li X, Phan PT, Tiong AMH, Kaan HL, Liu J et al (2020) Sewing up the wounds: a robotic suturing system for flexible endoscopy. IEEE Robot Autom Mag 27(3):45–54
15. Capezio F, Mastrogiovanni F, Scalmato A, Sgorbissa A, Vernazza P, Vernazza T, Zaccaria R (2011) Mobile robots in hospital environments: an installation case study. In ECMR: 61–68
16. Chen AI, Balter ML, Maguire TJ, Yarmush ML (2020) Deep learning robotic guidance for autonomous vascular access. Nat Mach Intell 2(2):104–115
17. Cresswell K, Cunningham-Burley S, Sheikh A (2018) Health care robotics: qualitative exploration of key challenges and future directions. J Med Internet Res 20(7):e10410
18. de Carvalho Visconti TA, Otoch JP, de Almeida Artifon EL (2020) Robotic endoscopy. A review of the literature. Acta Cir Bras 35:e202000206
19. Dhelika R, Hadi AF, Yusuf PA (2021) Development of a motorized hospital bed with swerve drive modules for holonomic mobility. Appl Sci 11(23):11356
20. Fanti MP, Mangini AM, Roccotelli M, Silvestri B (2020) Hospital drugs distribution with autonomous robot vehicles. In 2020 IEEE 16th International Conference on Automation Science and Engineering (CASE): 1025–1030
21. Filko D, Marijanović D, Nyarko EK (2021) Automatic robot-driven 3D reconstruction system for chronic wounds. Sensors 21(24):8308
22. Freeman WD, Sanghavi DK, Sarab MS, Kindred MS, Dieck EM, Brown SM et al (2021) Robotics in simulated COVID-19 patient room for health care worker effector tasks: preliminary, feasibility experiments. Mayo Clin Proc Innov Qual Outcomes 5(1):161–170
23. Gholamhosseini L, Sadoughi F, Safaei A (2019) Hospital real-time location system (A practical approach in healthcare): a narrative review article. Iran J Public Health 48(4):593

24. Grimm CA (2021) Hospitals reported that the COVID-19 pandemic has significantly strained health care delivery. US Department of Health and Human Services, Washington
25. Guo Z, Xiao X, Yu H (2018) Design and evaluation of a motorized robotic bed mover with omnidirectional mobility for patient transportation. IEEE J Biomed Health Inform 22(6):1775–1785
26. Haidegger T (2019) Autonomy for surgical robots: concepts and paradigms. IEEE Trans Med Robot Bionics 1(2):65–76
27. Hall H (2020) The effect of the COVID-19 pandemic on healthcare workers' mental health. Jaapa 33(7):45–48
28. Hall K (2022) Financial effects of COVID-19: hospital outlook for the remainder of 2021. E-book KaufmanHall.com.
29. Higgins TS, Wu AW, Ting JY (2020) SARS-CoV-2 nasopharyngeal swab testing—false-negative results from a pervasive anatomical misconception. JAMA Otolaryngol Head Neck Surg 146(11):993–994
30. Huang H-W, Chen J, Chai PR, Ehmke C, Rupp P, Dadabhoy FZ et al (2022) Mobile robotic platform for contactless vital sign monitoring. Cyborg Bionic Syst 2022:9780497
31. Synapxe (2022) COVID-19 innovation: how the command, control and communications (C3 system) supports a public hospital. https://www.synapxe.sg/healthtech/covid-19/command-control-and-communications-c3
32. Jahja M, Chin A, Tibshirani RJ (2022) Real-time estimation of Covid-19 infections: deconvolution and sensor fusion. Stat Sci 37(2):207–228
33. Khan A, Anwar Y (2020) Robots in healthcare: a survey. In: Arai K, Kapoor S (eds) Advances in computer vision. CVC 2019. Advances in intelligent systems and computing, vol 944. Springer, Cham
34. Kotowski SE, Davis KG, Marras WS (2022) Patient handling through moving of the beds and stretchers. Int J Ind Ergon 87:103252
35. Kumar V, Kumar R, Kumar M, Wander GS, Gupta V, Sahani A (2022) Recent advances in low-cost, portable automated resuscitator systems to fight COVID-19. Heal Technol 12:181–191
36. Lee OE, Davis B (2020) Adapting 'sunshine,' a socially assistive chat robot for older adults with cognitive impairment: a pilot study. J Gerontol Soc Work 63(6–7):696–698. https://doi.org/10.1080/01634372.2020.1789256
37. Maghdid HS, Asaad AT, Ghafoor KZ, Sadiq AS, Mirjalili S, Khan MK (2021) Diagnosing COVID-19 pneumonia from X-ray and CT images using deep learning and transfer learning algorithms. In SPIE Multimodal image exploitation and learning, 11734: 99–110
38. Marín SDS, Gomez-Vargas D, Céspedes N, Múnera M, Roberti F, Barria P et al (2021) Expectations and perceptions of healthcare professionals for robot deployment in hospital environments during the COVID-19 pandemic. Front Robot AI 8:612746
39. Mitchell R (2020) Self-driving hospital beds. Electropages. https://www.electropages.com/blog/2020/07/self-driving-hospital-beds
40. Mitros Z, Thamo B, Bergeles C, Da Cruz L, Dhaliwal K, Khadem M (2021) Design and modelling of a continuum robot for distal lung sampling in mechanically ventilated patients in critical care. Front Robot AI 8:611866
41. Morrell ALG, Morrell-Junior AC, Morrell AG, Mendes J, Freitas M, Tustumi F, Morrell A (2021) The history of robotic surgery and its evolution: when illusion becomes reality. Rev Col Bras Cir 48:e20202798
42. Muhammad LJ, Algehyne EA, Usman SS, Ahmad A, Chakraborty C, Mohammed IA (2021) Supervised machine learning models for prediction of COVID-19 infection using epidemiology dataset. SN Comput Sci 2(1):1–13
43. Narayanan KL, Kumaran NM, Rajakumar G, Arshadh H, Dinesh I, Caleb V (2019) Design and fabrication of medicine delivery robots for hospitals. SSRN Electron J. https://doi.org/10.2139/ssrn.3432156
44. Neumann DA (2004) Polio: its impact on the people of the United States and the emerging profession of physical therapy. J Orthop Sports Phys Ther 34(8):479–492. https://doi.org/10.2519/jospt.2004.0301

45. Nurgat AZ, Alaboura D, Aljaber R, Mustafa O, Lawrence M, Mominah M et al (2021) Real-world experience of a standalone robotic device for batch compounding of small-volume parenteral preparations. Hosp Pharm 56(1):32–46
46. Pacilè S, Lopez J, Chone P, Bertinotti T, Grouin JM, Fillard P (2020) Improving breast cancer detection accuracy of mammography with the concurrent use of an artificial intelligence tool. Radiol Artif Intell 2(6):e190208
47. Pesapane F, Volonté C, Codari M, Sardanelli F (2018) Artificial intelligence as a medical device in radiology: ethical and regulatory issues in Europe and the United States. Insights Imaging 9(5):745–753
48. Peshko I, Pawluczyk R, Wick D (2013) Synergistic sensory platform: robotic nurse. J Low Power Electron Appl 3(2):114–158
49. Ranney ML, Griffeth V, Jha AK (2020) Critical supply shortages—the need for ventilators and personal protective equipment during the Covid-19 pandemic. N Engl J Med 382(18):e41
50. Săndulescu V, Puşcoci S, Petre M, Dumitrache M, Bota V, Gîrlea A (2021) mHealth application for remote health monitoring useful during the COVID 19 pandemic. https://doi.org/10.1109/MeMeA52024.2021.9478717
51. Sen-Crowe B, Sutherland M, McKenney M, Elkbuli A (2021) A closer look into global hospital beds capacity and resource shortages during the COVID-19 pandemic. J Surg Res 260:56–63
52. Shen Y, Guo D, Long F, Mateos LA, Ding H, Xiu Z et al (2020) Robots under COVID-19 pandemic: a comprehensive survey. IEEE Access 9:1590–1615
53. Siepel FJ, Maris B, Welleweerd MK, Groenhuis V, Fiorini P, Stramigioli S (2021) Needle and biopsy robots: a review. Curr Robot Rep 2(1):73–84
54. Sutton RT, Pincock D, Baumgart DC, Sadowski DC, Fedorak RN, Kroeker KI (2020) An overview of clinical decision support systems: benefits, risks, and strategies for success. NPJ Digit Med 3(1):1–10
55. Tavakoli M, Carriere J, Torabi A (2020) Robotics, smart wearable technologies, and autonomous intelligent systems for healthcare during the COVID-19 pandemic: an analysis of the state of the art and future vision. Adv Intell Syst 2(7):2000071
56. Topalovic M, Das N, Burgel P-R, Daenen M, Derom E, Haenebalcke C et al (2019) Artificial intelligence outperforms pulmonologists in the interpretation of pulmonary function tests. Eur Respir J 53(4):1801660
57. Tucker AL (2004) The impact of operational failures on hospital nurses and their patients. J Oper Manag 22(2):151–169
58. Yang G-Z, Nelson BJ, Murphy RR, Choset H, Christensen H, Collins HS et al (2020) Combating COVID-19—the role of robotics in managing public health and infectious diseases. Sci Robot 5:eabb5589. American Association for the Advancement of Science
59. Zhao Z, Ma Y, Mushtaq A, Rajper AMA, Shehab M, Heybourne A et al (2022) Applications of robotics, artificial intelligence, and digital technologies during COVID-19: a review. Disaster Med Public Health Prep 16:1634–1644

Environmental Detection and Monitoring of SARS-CoV-2 for COVID-19 Risk Prediction

Rita R. Colwell, Kyle D. Brumfield, Moiz Usmani, Anwar Huq, and Antarpreet S. Jutla

1 Introduction

Changing climate, along with increased urbanization, rapid long-distance travel, and population growth, has resulted in a global increased risk of disease outbreaks. At the end of 2019, reports of an unknown and life-threatening respiratory illness emerged in Wuhan, China, which rapidly circulated throughout the world. On March 11, 2020, it was obvious that a potential global outbreak was underway, and the World Health Organization [1, 2] declared the disease, determined as being caused by a coronavirus, COVID-19, a pandemic and named the causative agent the Severe Acute Respiratory Syndrome Coronavirus 2 (SARS-CoV-2).

SARS-CoV-2, an enveloped, positive-sense, single-stranded RNA-encoded virus, was subjected to genomic analysis and its sequence was quickly determined and the information circulated to laboratories around the world [3]. Because the virus was novel, its effect proved devastating for society and the economy, especially for public health in regions of the world where public health systems, notably in the United States, were unprepared for as massive an epidemic, as COVID-19 proved to be. Globally, by September 12, 2022, ca. 605 million confirmed cases of COVID-19 had been recorded and ca. 6.5 million had died [4]. Tracing sources of

R. R. Colwell (✉) · K. D. Brumfield
University of Maryland College Park, University of Maryland Institute for Advanced Computer Studies, College Park, MD, USA
e-mail: rcolwell@umd.edu

M. Usmani · A. S. Jutla
University of Florida, Herbert Wertheim College of Engineering, Engineering School of Sustainable Infrastructure and Environment, Gainesville, FL, USA

A. Huq
University of Maryland College Park, Maryland Pathogen Research Institute, College Park, MD, USA

J. M. Rosen, R. R. Colwell (eds.), *The COVID-19 Pandemic*,
https://doi.org/10.1007/978-3-031-62772-9_4

the virus and tracking its transmission were formidable tasks in the early months of the pandemic. Global transmission of the virus was rapid and within six months after the first reports in Wuhan, China, more than 216 nations, areas, or territories worldwide were affected [1]. Importantly, SARS-CoV-2 RNA proved detectible by molecular methods, namely, reverse transcription-quantitative polymerase chain reaction (RT-qPCR) and RNA sequencing, and was found in various patient samples, e.g., respiratory nasopharyngeal and oropharyngeal swabs [5, 6] and serum and tissues [7, 8]. Wu et al. (2020) [9] reported that SARS-CoV-2 RNA was excreted in feces and the virus was soon reported to be detectable in wastewater [10–14] and sludge [15, 16]. In addition to being a causative agent of respiratory illness, SARS-CoV-2 was shown to infect other organs, including the gastrointestinal tract, namely, glandular epithelial cells [7, 17], and was observed to be linked to gastrointestinal symptoms, namely, diarrhea, nausea, vomiting, and abdominal pain [2, 8]. Interestingly, Wang et al. in 2005 concentrated and detected SARS-CoV-2 RNA in sewage from hospitals in China and were able to detect the virus from presymptomatic individuals between one and five days before the onset of clinical symptoms [8, 18, 19], as well as from individuals with mild symptoms [8] and those who were seven to 11 days post-symptomatic infection [8, 9, 20].

The amount of SARS-CoV-2 shed in the stool was found to be highly variable, according to symptoms, and individuals with gastrointestinal symptoms typically showed higher fecal concentrations of the virus [21, 22]. The persistence of SARS-CoV-2 in fecal samples was reported early in the epidemic by Wang et al. (2005) [19]. Initially, there were conflicting reports suggesting SARS-CoV-2 was not detectable in feces, even when the patient carried a high viral load [21, 23]. However, early reports of the virus in fecal samples were verified and very quickly numerous studies monitoring wastewater for SARS-CoV-2 were carried out [24], and a worldwide focus was on the wastewater treatment plants that served ca. 2.1 billion people in ca. 105,600 treatment plants around the world [25]. The need to have a means of gauging public health risk led to the adoption of techniques developed during poliomyelitis epidemics decades earlier to track trends of the COVID-19 virus, a process now recognized as wastewater-based epidemiology (WBE) [26].

The premise of WBE is that the dynamics of the disease in the community can be measured rapidly, providing benefits over clinical testing alone. Specifically, WBE can account for both symptomatic and asymptomatic individuals shedding the virus. Furthermore, unlike individual testing, the issues of capacity, compliance, strategy of testing, and reporting of results do not affect the outcome. Since WBE metrics can be used to estimate incidence dynamics over time, a comparison of wastewater and case number incidence dynamics revealed estimates based on wastewater data tracked the timing of an infection peak better than the estimates derived from confirmed cases, namely, due to a lag in the time and resources required to aggregate clinical data and share the results. However, case confirmations were concluded to be a better estimate of the decline in infections [3]. In summary, WBE, which includes wastewater from households and nonhousehold sources such as rain and industrial use, is recognized as an additional tool that can be used to correlate the presence of risk-causing agents, including those of both biological and chemical origin, with public health [27].

The momentum of WBE proved compelling. Results of wastewater surveillance in Australia [13], France [28], Canada [29, 30], Italy [10, 31], Netherlands [11], Germany [32], Japan [33], United Arab Emirates [34], Spain [10], and the United States [35, 36] were soon reported. The spate of wastewater studies was organized and published by Weidhaas et al. (2021) based on their extensive analysis of the correlation of SARS-CoV-2 RNA in wastewater with COVID-19 disease [2]. An extensive systematic review of the application of wastewater surveillance to infer COVID-19 transmission was conducted by Shah et al. (2022), which assessed the performance of wastewater surveillance, with respect to its being an early warning system for community transmission of COVID-19 [37]. A total of 763 studies were examined, of which 92 were distributed across 34 countries between January 2020 and May 2021 and populations of 321 to 11,400,000 inhabitants [37]. The data obtained were informative in that wastewater signals were found to precede confirmed cases by up to 63 days, with 13 of the studies reporting positive samples before the first cases had been reported in the community. This provided early signs of transmission and allowed faster public health response. An important benefit of WBE for SARS-CoV-2 RNA proved to be that environmental surveillance, compared with declared COVID-19 cases, at the level of municipality, showed shedding of the virus even before the first cases were reported. Study results reported by Randazzo et al. (2020) showed wastewater surveillance to be indeed an effective strategy, namely, early indication of infection in a specific population, hence, a useful tool to assist decision-makers, for example, when to implement or lift a coronavirus lockdown [10].

Before the first cases of COVID-19 were occurring in the United States, a fortuitous decision had been made by the California State Water Resources Control Board, to develop statewide regulations for direct potable reuse [38]. This was accomplished by establishing an optimized standard operating protocol to characterize effectively the concentration of human pathogens in raw wastewater. Samples were collected over a 14-month monitoring campaign that began in December 2019, before the COVID-19 pandemic had begun, and ran until January 2021. The analysis included a range of enteric pathogens, namely *Giardia* and *Cryptosporidium* (protozoa pathogens) but also viruses and bacteria known to cause acute gastrointestinal infections. Because of the timing of the study, it was possible to determine from the results that the incidence of enteric pathogens neither increased nor decreased during the COVID-19 pandemic, i.e., no systemic change was noted with respect to acute gastrointestinal infections relative to COVID-19.

Another finding of WBE was from surveillance of aircraft wastewater, providing predictive value for COVID-19. A study was carried out on repatriation flights to Australia during the COVID-19 pandemic [39]. Passengers tested negative for SARS-CoV-2 before a flight, as did the wastewater prior to the flight, yet subsequently, the wastewater tested positive after arrival. The passengers were quarantined for 14 days post-arrival, and during quarantine, clinical testing identified cases of COVID-19. The study clearly showed aircraft wastewater surveillance for

SARS-CoV-2 provided an effective tool to determine viral transmission during aircraft transport of asymptomatic individuals. It is also worth noting a similar phenomenon has been observed through the surveillance of wastewater on commercial cruise ships [39].

A detailed study in Massachusetts, USA, of wastewater samples collected from March 2020 to May 2021, using RT-qPCR showed SARS-CoV-2 gene fragment copy numbers and COVID-19 clinical cases and deaths varied over time [40]. The investigation also concluded that wastewater data preceded clinically reported cases in the first wave of the pandemic but not as a leading indicator in the second wave, probably related to enhanced testing that allowed timely detection and reporting associated with the second wave of infection. Nevertheless, routine surveillance of community wastewater even post-pandemic allows surveillance that can be valuable to public health, especially for tracking virus variants, considering the rapid evolution observed for SARS-CoV-2 mutants. For example, a five-week warning of a COVID-19 peak prior to an Omicron variant surge in Detroit, Michigan, was determined using wastewater surveillance, further illustrating WBE as a public health tool when continually applied [41]. Similarly, studies undertaken in Belgium showed wastewater surveillance provided an early warning of a rise in infections and, importantly, of the circulation of specific variants in given geographical areas of Leuven, Belgium. Subsequently, shifts in dominant SARS-CoV-2 variants, namely, Alpha, Delta, and Omicron (BA.1 through BA.5), were detected by wastewater surveillance [4].

Composite 24-h wastewater samples, collected weekly between September 2021 and July 2022 from the Padua, Italy, wastewater treatment plant, using the ORF1b-nsp14 gene assay, revealed three major peaks in January, April, and July 2022 [42]. Total hospitalizations were found to follow similar trends, reflecting that of the viral detection rates in wastewater samples. By December 2021, Omicron BA.1 began replacing the Delta variant in the samples collected and the shift toward Omicron BA.2 was detected between February and April 2022 [42].

During the five years that have transpired since the first cases in 2019, that is, since the first indication of a COVID-19 pandemic and the spate of reports on various aspects of wastewater surveillance, there is general acceptance of WBE as a useful complement to clinical testing and a means of managing COVID-19. Assessment and improvement of the methodology and application have quickly followed. Clearly, a significant benefit of WBE is that it is unbiased, capturing data on an entire population rather than a subset of individuals tested in a clinic or those who carry out over-the-counter self-tests. To this point, Safford et al. (2022) investigated the relationship between wastewater and clinical data at a more granular spatial scale. Since clinical test results are often geographically aggregated to protect the privacy of testing individuals, they sought to develop and test a framework that probabilistically disaggregated the clinical testing data to improve comparison with wastewater data, using sampling sites from different parts of a community watershed. A conclusion of the study of Safford et al. (2022) was that, indeed, a framework could be developed to provide a reasonable agreement between wastewater and clinical data at highly granular spatial scales, again underscoring the public health value of WBE [43].

Evaluation of wastewater surveillance for SARS-CoV-2 was carried out on a countrywide scale in Norway from June 2022 to March 2023 [44]. The Norwegian Institute of Public Health piloted national wastewater surveillance for SARS-CoV-2 and variants, using criteria of sensitivity, specificity, timeliness, usefulness, representativeness, simplicity, flexibility, stability, and communication as a means of performance. Also carried out was a cross-correlation analysis to determine if the system gave early warning if a wave of new infections was imminent [44]. The results were favorable as most of the objectives of the Norwegian study were reached, notably in providing early warning one to two weeks prior to a peak in reported cases. Furthermore, the system was found to be useful for monitoring infections, notably at the population level, and proved helpful when individual testing rates were determined to be low. With the findings, a strong recommendation for standardization of WBE was made.

On the heels of the Norwegian study, a review of the analytical procedures and epidemiological modeling for WBE was published [45]. Analytical methods for sampling and viral detection from RNA quantification published by several investigators [46–48] were included in the review by Ciannella et al. (2023). The need for standardization, normalization of quantitative information (standard units for viral loads, for example), and analytical accuracy was highlighted and discussed extensively [45]. The findings reported include the necessity to standardize analytical procedures if results for different locations were to be compared. The lack of standardization in the various reports to date most likely accounts for the different analytical sensitivities in the analyses.

Subsequent studies of WBE have begun to focus on surveillance techniques and method standardization. Concentration, sample pre-treatment, and storage conditions for detection of SARS-CoV-2 RNA in wastewater were addressed by Cutrupi et al. (2023), whose findings showed initial sample concentration is critical, due to the low level of viral titers present, relatively, compared to the abundance of other microorganisms in the complex wastewater matrix. Two methods were tested, a biphasic separation system with polyethylene glycol (PEG)-dextran and PEG/NaCl precipitation [49]. The PEG/NaCl was concluded to perform better than biphasic separation, mainly because it allowed a more sensitive identification of the presence of the virus and detection of higher viral titer. Storage of samples for up to three weeks at 4.3 +/− 0.2 °C was found not to be detrimental. Importantly, pasteurization pre-treatment was deemed important as it increased operator safety while retaining the identification of viral concentration [49].

By the end of July 2023, the COVID-19 pandemic had grown significantly, with ca. 769 million individuals infected and ca. 6.9 million deaths globally [50]. The disease had now moved to endemicity. Unfortunately, many countries reduced their vigilance, just when the emergence of a variety of highly transmissible variants posed a renewed threat. The WHO defined at least five variants with increased transmissibility and risk, including Alpha (B.1.1.7), Beta (B.1.315), Gamma (P.1), Delta (B.1.617.2), and Omicron (B.1.1.529). Omicron variants having a large number of mutations had emerged and were identified, which proved significant as mutations occurring at primer binding sites reduced the sensitivity of traditional RT-qPCR assays, a phenomenon commonly referred to as primer dropout. Xu et al. (2023) refined methods for the detection of Omicron variants, to provide a more sensitive assay increasing the probability of detecting the virus in wastewater, hence providing early warning signals. A case study of Omicron variants was conducted by Xu et al. (2023), employing the modified detection method and assays having higher success rates with more accurate detection [51].

An exhaustive analysis of 524 wastewater sites across England, with sampling done at least twice a week, from November 2021 to February 2022, was carried out by Brunner et al. (2023), who employed amplicon-based sequencing to estimate SARS-CoV-2 variant frequencies. They were able to reconstruct the spread of the Omicron variant across England in close detail, showing alignment with the epidemiological estimates from individual test data. The temporal and spatial spread of Omicron within London was mapped, and the transition between Omicron subvariants BA.1 and BA.2 in February 2022 was tracked at both regional and national levels. The fast-paced dynamics of the SARS-CoV-2 variant frequencies clearly correlated with individual testing data, concluding WBE was effective, reliable, and affordable [52].

The Omicron BA.1 sub-lineage prevailed globally at the beginning of 2022 and was subsequently replaced by BA.2 in many countries. Naim et al. (2023) used RT-qPCR assays to detect SARS-CoV-2 and its variants in wastewater, observing the emergence of Omicron BA.1 replacing the Delta variant [53]. Furthermore, the investigators optimized prediction modeling of SARS-CoV-2 transmission. Successful demonstration of the dynamics of SARS-CoV-2 lineage spread and evolution in South Africa was achieved using surveillance from sentinel wastewater treatment plants in urban areas collected from April 2021 to the end of the Omicron BA.1 wave in January 2022.

By the end of July 2023, the BA.2.86 variant had been detected in nearly every country and on August 17, 2023, was classified as a variant to be monitored based on the number of mutations in the spike gene (Rasmussen et al., 2023). Wastewater samples collected in Bangkok, Thailand, between June and July of 2023 revealed the BA.2.86 variant circulating in Thailand with many mutations, particularly in the spike protein [54].

With ca. 10–30% of individuals experiencing long-term COVID-19 [55], surveillance clearly remains necessary, with detection methods for SARS-CoV-2 mutants a lingering critical aspect of the pandemic.

It can now be considered that surveillance of SARS-CoV-2 in wastewater is an accepted public health activity and a model to follow the evolution of endemic viruses. WBE has, in essence, become a key tool supporting clinical surveillance [56]. It is recognized that clinical testing, although fundamental to surveillance, is limited to the generalization of the data [50, 56], with a primary consideration that clinical testing, in general, detects symptomatic cases and there is a lack of uniformity in clinical testing, both temporally and geographically. WBE, in contrast, samples wastewater from an entire community, providing valuable data on transmission dynamics and accomplishing it less expensively. More than 70 countries now monitor and publicly report geographical and temporal trends of SARS-CoV-2 levels in wastewater [56, 57].

As WBE continues to mature as a public health tool, new techniques are being applied, including machine learning [58]. A regression model was developed to estimate the number of active COVID-19 infections across the Kent Ridge campus of the National University of Singapore. Improvements in the model were subsequently described, noting that viral shedding dynamics improved the reliability of the various models, emergence of new variants, reinfections, and vaccination rates [58]. Kanchan et al. (2024) used machine learning to predict deaths from COVID-19 based on concentrations of SARS-CoV-2 RNA detected in wastewater [59]. Wastewater surveillance data, combined with hospitalizations and death data, were used to generate machine learning-based artificial neural network models to predict future COVID-19 hospital admissions and death, essentially a more sophisticated early warning system for medical response teams and healthcare policymakers.

Distributed fixed lag modeling and artificial neural networks employed by Kanchan et al. (2024) assessed COVID-19 viral load in untreated wastewater to predict changes in hospital admissions in Idaho, USA, and deaths in Ada County, Idaho. The time lag between peaks in wastewater viral counts and COVID-19 hospitalizations and also deaths was 14 and 23 days, respectively [59]. The conclusion is that predictive artificial neural network modeling can combine wastewater surveillance data with hospitalizations and death data to generate machine learning-based models predicting future COVID-19 hospital admissions and deaths.

With WBE methods and applications having been amply demonstrated, an ethical review process was lacking. Bowes et al. (2023) addressed this deficiency by establishing an interdisciplinary workshop focused on developing a framework for a structured ethical review [60]. A consensus approach was taken to create the framework, having accepted the conclusion that wastewater-based testing to monitor enteric and respiratory pathogens, i.e., poliovirus and coronaviruses, can be successfully employed [26, 61], notably that wastewater testing captures symptomatic, asymptomatic, and pre-symptomatic carriers of infectious pathogens, regardless of individual access to healthcare [62]. Clearly, wastewater testing has been able to provide early warning of pathogen pressure in a community and both detect and track circulating and new genomic variants [63–65]. Nevertheless, the conclusion of this first step toward ethical review is that managing the technology in the future should safeguard the well-being of those under surveillance and maintain robust ethical guidelines to protect intended applications [60].

For the future, the value of WBE carries far beyond COVID-19 and the causative agent, SARS-CoV-2. When the utilization of wastewater analysis is coupled with advanced molecular methods, it offers a valuable public health tool. Real-time monitoring of infectious disease agents, as shown by the extensive results obtained during the COVID-19 pandemic, is now both possible and feasible. The microbiome analysis of wastewater was accomplished by Brumfield et al. (2022) using DNA metagenomics, RNA metatranscriptomics, and targeted SARS-CoV-2 sequencing. The microbiome analysis, including both DNA and RNA analyses, provided significant insight with respect to human health, namely, the detection not only of the pathogens but also the virulence/antibiotic resistance genes they carried and insights into functional gene profiles of the microorganisms. Furthermore, information that can be gleaned from wastewater analysis includes the presence of bacterial, fungal, viral, and parasitic infectious agents circulating in a community, associations among the microbial agents, and, when monitored persistently, a notice of rise, and fall in cadence and transmission within the community. In effect, microbiome analysis of wastewater provides a larger source of highly detailed information as well as an early warning of communicable diseases in timely alert to public health authorities and community physicians, local hospitals, and medical centers [66].

Climate-related incidence of pathogens can also be determined by WBE and can drive its application to the measurement of the health of communities globally. Key to the global application is standardization of methods and centralization of data. Comparison of data from region to region requires normalization and compliance, minimum standards, and the integration of human health surveillance geographically. Understanding the community structure of microorganisms beyond individual pathogens and including environmental parameters will make a global public health system possible.

A consensus study report issued by the National Academies of Science, Engineering, and Medicine (2023) addressed the potential for a national system to monitor the presence and spread of SARS-CoV-2 and its variants and called for movement toward a system providing real-time data on endemic and newly emerging microbial threats to public health, in effect a national wastewater surveillance system providing critical, ongoing data for public health decisions. The report outlined a framework for a national WBE program for the United States and the benefits, key characteristics, and candidate pathogens for monitoring and strategies to increase the public health impact of national wastewater surveillance [67]. The Centers for Disease Control and Prevention (CDC) has already established a National Wastewater Surveillance System [68] to track SARS-CoV-2 activity. However, until standardization of methods and data gathering and interpretation is achieved, the value of wastewater analysis as a public health tool will not be complete.

On a global scale geographically, outbreaks of communicable diseases have been shown to be influenced by many factors, notably environmental. With climate change increasingly evident, its effect on human health, especially infectious disease incidence and transmission would benefit from the application of lessons learned from dealing with global pandemics, such as influenza and cholera. Studies of climate factors influencing the incidence and spread of cholera have employed satellite monitoring of environmental factors driving the disease [69]. Predictive time series analysis linking cholera with terrestrial water storage [70], analysis of hydroclimatic processes [71, 72], and high-resolution land surface temperature data [73] provided the foundation for a satellite-based cholera pandemic prediction system. Assessment of the risk of cholera was successfully employed for Yemen [66] and a predictive intelligence for cholera in Ukraine (2022), Malawi (2023), and Gaza (2024) [74, 75]. Adapting from the cholera forecasting system, a heuristic score-based mathematical set of rules was established for COVID-19 as the core methodology, taking into account factors specific to COVID-19, namely, ambient air temperature, dew point temperature, population density, and other variables, a predictive model for estimating risk of COVID-19 transmission was developed [66, 76].

The combination of wastewater microbiome analysis and satellite sensor-based prediction of COVID-19, in the early stage of development, offers an extraordinarily valuable public health opportunity for future precision public health with global application, including locations lacking adequate resources. Such an effort was reported by Usmani et al. (2023), where authors identified a combined set of sociological and environmental parameters through careful integration of understanding of SARS-CoV-2 transmission via aerosols and wastewater surveillance of SARS-CoV-2. It was suggested that an accurate prediction of emerging infectious respiratory diseases such as COVID-19 poses significant challenges due to the novelty of the virus and an incomplete understanding of its transmission pathways within a human population. Hence, the authors advocated for the development of decision-making frameworks that incorporate predictive intelligence systems rather than aiming for absolute predictability. Previous research introduced a multifaceted climate-sociological hypothesis regarding the dissemination of COVID-19 among humans, suggesting its seasonal pattern in varying environmental conditions and its dependence on shifts in human behavior throughout the year. Further multi-model calibration and validation will be needed to be achieved on a geographically distinct and discrete temporal basis, which may identify pathways for a particular pathogen behavior in the human population (endemic or epidemic state). Lessons learned from cholera and COVID-19 suggest future pandemics will necessitate innovative mitigation and intervention strategies that synthesize interdisciplinary knowledge from fields such as microbiology, sociology, climate science, earth sciences, and public health, operating at a global scale.

Acknowledgments RRC, KDB, and co-authors are supported by the National Science Foundation (OCE1839171 and CCF1918749), the National Institute of Environmental Health Sciences, the National Institutes of Health (R01ES030317A), and the National Aeronautics and Space Administration (80NSSC20K0814 and 80NSSC22K1044).

References

1. Coronavirus Disease (COVID-19) Pandemic (2020) World Health Organization (WHO)
2. Weidhaas J, Aanderud ZT, Roper DK, Van Derslice J, Gaddis EB, Ostermiller J et al (2021) Correlation of SARS-CoV-2 RNA in wastewater with COVID-19 disease burden in sewersheds. Sci Total Environ 775:145790
3. Fernandez-Cassi X, Scheidegger A, Bänziger C, Cariti F, Tuñas Corzon A, Ganesanandamoorthy P et al (2021) Wastewater monitoring outperforms case numbers as a tool to track COVID-19 incidence dynamics when test positivity rates are high. Water Res 200:117252
4. Rector A, Bloemen M, Thijssen M, Delang L, Raymenants J, Thibaut J et al (2023) Monitoring of SARS-CoV-2 concentration and circulation of variants of concern in wastewater of Leuven, Belgium. J Med Virol 95(2):e28587

5. Azzi L, Maurino V, Baj A, Dani M, d'Aiuto A, Fasano M et al (2021) Diagnostic salivary tests for SARS-CoV-2. J Dent Res 100(2):115–123
6. Azzi L, Carcano G, Gianfagna F, Grossi P, Gasperina DD, Genoni A et al (2020) Saliva is a reliable tool to detect SARS-CoV-2. J Infect 81(1):e45–e50
7. Xiao F, Tang M, Zheng X, Liu Y, Li X, Shan H (2020) Evidence for gastrointestinal infection of SARS-CoV-2. Gastroenterology 158(6):1831–1833.e3
8. Wang W, Xu Y, Gao R, Lu R, Han K, Wu G et al (2020) Detection of SARS-CoV-2 in different types of clinical specimens. JAMA 323(18):1843–1844
9. Wu F, Zhang J, Xiao A, Gu X, Lee WL, Armas F et al (2020) SARS-CoV-2 titers in wastewater are higher than expected from clinically confirmed cases. mSystems 5(4):e00614–e00620
10. Randazzo W, Truchado P, Cuevas-Ferrando E, Simón P, Allende A, Sánchez G (2020) SARS-CoV-2 RNA in wastewater anticipated COVID-19 occurrence in a low prevalence area. Water Res 181:115942
11. Medema G, Heijnen L, Elsinga G, Italiaander R, Brouwer A (2020) Presence of SARS-coronavirus-2 RNA in sewage and correlation with reported COVID-19 prevalence in the early stage of the epidemic in The Netherlands. Environ Sci Technol Lett 7(7):511–516
12. Ahmed W, Bivins A, Simpson SL, Bertsch PM, Ehret J, Hosegood I et al (2022) Wastewater surveillance demonstrates high predictive value for COVID-19 infection on board repatriation flights to Australia. Environ Int 158:106938
13. Ahmed W, Angel N, Edson J, Bibby K, Bivins A, O'Brien JW et al (2020) First confirmed detection of SARS-CoV-2 in untreated wastewater in Australia: a proof of concept for the wastewater surveillance of COVID-19 in the community. Sci Total Environ 728:138764
14. Ahmed W, Bertsch PM, Bivins A, Bibby K, Farkas K, Gathercole A et al (2020) Comparison of virus concentration methods for the RT-qPCR-based recovery of murine hepatitis virus, a surrogate for SARS-CoV-2 from untreated wastewater. Sci Total Environ 739:139960
15. Peccia J, Zulli A, Brackney DE, Grubaugh ND, Kaplan EH, Casanovas-Massana A et al (2020) Measurement of SARS-CoV-2 RNA in wastewater tracks community infection dynamics. Nat Biotechnol 38(10):1164–1167
16. Graham KE, Loeb SK, Wolfe MK, Catoe D, Sinnott-Armstrong N, Kim S et al (2021) SARS-CoV-2 RNA in wastewater settled solids is associated with COVID-19 cases in a large urban Sewershed. Environ Sci Technol 55(1):488–498
17. Hamming I, Timens W, Bulthuis M, Lely A, Navis G, van Goor H (2004) Tissue distribution of ACE2 protein, the functional receptor for SARS coronavirus. A first step in understanding SARS pathogenesis. J Pathol 203(2):631–637
18. Tang A, Tong ZD, Wang HL, Dai YX, Li KF, Liu JN et al (2024) Detection of novel coronavirus by RT-PCR in stool specimen from asymptomatic child. China—Volume 26, Number 6—June 2020—Emerging Infectious Diseases Journal—CDC. [Cited 2024 Apr 1]; Available from: https://wwwnc.cdc.gov/eid/article/26/6/20-0301_article
19. Wang XW, Li J, Guo T, Zhen B, Kong Q, Yi B et al (2005) Concentration and detection of SARS coronavirus in sewage from Xiao Tang Shan hospital and the 309th Hospital of the Chinese People's liberation Army. Water Sci Technol 52(8):213–221
20. Chen Y, Chen L, Deng Q, Zhang G, Wu K, Ni L et al (2020) The presence of SARS-CoV-2 RNA in the feces of COVID-19 patients. J Med Virol 92(7):833–840
21. Wölfel R, Corman VM, Guggemos W, Seilmaier M, Zange S, Müller MA et al (2020) Virological assessment of hospitalized patients with COVID-2019. Nature 581(7809):465–469
22. Cheung KS, Hung IFN, Chan PPY, Lung KC, Tso E, Liu R et al (2020) Gastrointestinal manifestations of SARS-CoV-2 infection and virus load in fecal samples from a Hong Kong cohort: systematic review and meta-analysis. Gastroenterology 159(1):81–95
23. Peng L, Liu J, Xu W, Luo Q, Chen D, Lei Z et al (2020) SARS-CoV-2 can be detected in urine, blood, anal swabs, and oropharyngeal swabs specimens. J Med Virol 92(9):1676–1680
24. Sims N, Kasprzyk-Hordern B (2020) Future perspectives of wastewater-based epidemiology: monitoring infectious disease spread and resistance to the community level. Environ Int 1(139):105689

25. Hart OE, Halden RU (2020) Computational analysis of SARS-CoV-2/COVID-19 surveillance by wastewater-based epidemiology locally and globally: feasibility, economy, opportunities and challenges. Sci Total Environ 730:138875
26. Lago PM, Gary HE Jr, Pérez LS, Cáceres V, Olivera JB, Puentes RP et al (2003) Poliovirus detection in wastewater and stools following an immunization campaign in Havana, Cuba. Int J Epidemiol 32(5):772–777
27. Prevost B, Lucas FS, Goncalves A, Richard F, Moulin L, Wurtzer S (2015) Large scale survey of enteric viruses in river and waste water underlines the health status of the local population. Environ Int 79:42–50
28. Wurtzer S, Marechal V, Mouchel JM, Maday Y, Teyssou R, Richard E et al (2020) Evaluation of lockdown impact on SARS-CoV-2 dynamics through viral genome quantification in Paris wastewaters [Internet]. medRxiv. [Cited 2024 Mar 27]. p. 2020.04.12.20062679. Available from: https://www.medrxiv.org/content/10.1101/2020.04.12.20062679v2
29. D'Aoust PM, Graber TE, Mercier E, Montpetit D, Alexandrov I, Neault N et al (2021) Catching a resurgence: increase in SARS-CoV-2 viral RNA identified in wastewater 48 h before COVID-19 clinical tests and 96 h before hospitalizations. Sci Total Environ 770:145319
30. D'Aoust PM, Mercier E, Montpetit D, Jia JJ, Alexandrov I, Neault N et al (2021) Quantitative analysis of SARS-CoV-2 RNA from wastewater solids in communities with low COVID-19 incidence and prevalence. Water Res 188:116560
31. La Rosa G, Iaconelli M, Mancini P, Bonanno Ferraro G, Veneri C, Bonadonna L et al (2020) First detection of SARS-CoV-2 in untreated wastewaters in Italy. Sci Total Environ 736:139652
32. Westhaus S, Weber FA, Schiwy S, Linnemann V, Brinkmann M, Widera M et al (2021) Detection of SARS-CoV-2 in raw and treated wastewater in Germany—suitability for COVID-19 surveillance and potential transmission risks. Sci Total Environ 751:141750
33. Hata A, Honda R (2020) Potential sensitivity of wastewater monitoring for SARS-CoV-2: comparison with norovirus cases. Environ Sci Technol 54(11):6451–6452
34. Hasan SW, Ibrahim Y, Daou M, Kannout H, Jan N, Lopes A et al (2021) Detection and quantification of SARS-CoV-2 RNA in wastewater and treated effluents: surveillance of COVID-19 epidemic in The United Arab Emirates. Sci Total Environ 764:142929
35. Nemudryi A, Nemudraia A, Wiegand T, Surya K, Buyukyoruk M, Cicha C et al (2020) Temporal detection and phylogenetic assessment of SARS-CoV-2 in municipal wastewater. Cell Rep Med 1(6):100098
36. Wu F, Xiao A, Zhang J, Moniz K, Endo N, Armas F et al (2021) Wastewater surveillance of SARS-CoV-2 across 40 U.S. states from February to June 2020. Water Res 202:117400
37. Shah S, Gwee SXW, Ng JQX, Lau N, Koh J, Pang J (2022) Wastewater surveillance to infer COVID-19 transmission: a systematic review. Sci Total Environ 804:150060
38. Regulating Direct Potable Reuse in California | California State Water Resources Control Board [Internet]. [Cited 2024 Mar 26]. Available from: https://www.waterboards.ca.gov/drinking_water/certlic/drinkingwater/direct_potable_reuse.html
39. Ahmed W, Bertsch PM, Angel N, Bibby K, Bivins A, Dierens L et al (2020) Detection of SARS-CoV-2 RNA in commercial passenger aircraft and cruise ship wastewater: a surveillance tool for assessing the presence of COVID-19 infected travellers. J Travel Med 27(5):taaa116
40. Xiao A, Wu F, Bushman M, Zhang J, Imakaev M, Chai PR et al (2022) Metrics to relate COVID-19 wastewater data to clinical testing dynamics. Water Res 212:118070
41. Zhao L, Zou Y, Li Y, Miyani B, Spooner M, Gentry Z et al (2022) Five-week warning of COVID-19 peaks prior to the Omicron surge in Detroit, Michigan using wastewater surveillance. Sci Total Environ 844:157040
42. Baldovin T, Amoruso I, Fonzo M, Bertoncello C, Groppi V, Pitter G et al (2023) Trends in SARS-CoV-2 clinically confirmed cases and viral load in wastewater: a critical alignment for Padua city (NE Italy). Heliyon 9(10):e20571
43. Safford H, Zuniga-Montanez RE, Kim M, Wu X, Wei L, Sharpnack J et al (2022) Wastewater-based epidemiology for COVID-19: handling qPCR nondetects and comparing spatially granular wastewater and clinical data trends. ACS EST Water 2(11):2114–2124

44. Amato E, Hyllestad S, Heradstveit P, Langlete P, Moen LV, Rohringer A et al (2023) Evaluation of the pilot wastewater surveillance for SARS-CoV-2 in Norway, June 2022—March 2023. BMC Public Health 23(1):1714
45. Ciannella S, González-Fernández C, Gomez-Pastora J (2023) Recent progress on wastewater-based epidemiology for COVID-19 surveillance: a systematic review of analytical procedures and epidemiological modeling. Sci Total Environ 878:162953
46. Carducci A, Federigi I, Liu D, Thompson JR, Verani M (2020) Making waves: coronavirus detection, presence and persistence in the water environment: state of the art and knowledge needs for public health. Water Res 179:115907
47. Kabdaşlı I, Tünay O (2021) Concentration techniques tailored for the detection of SARS-CoV-2 genetic material in domestic wastewater and treatment plant sludge: a review. J Environ Chem Eng 9(5):106296
48. Kitajima M, Ahmed W, Bibby K, Carducci A, Gerba CP, Hamilton KA et al (2020) SARS-CoV-2 in wastewater: state of the knowledge and research needs. Sci Total Environ 739:139076
49. Cutrupi F, Rossi M, Cadonna M, Poznanski E, Manara S, Postinghel M et al (2023) Evaluation of concentration procedures, sample pre-treatment, and storage condition for the detection of SARS-CoV-2 in wastewater. Environ Sci Pollut Res Int 30(48):106660–106670
50. Environmental surveillance for SARS-COV-2 to complement public health surveillance—Interim Guidance [Internet]. [Cited 2024 Mar 14]. Available from: https://www.who.int/publications-detail-redirect/WHO-HEP-ECH-WSH-2022.1
51. Xu X, Deng Y, Ding J, Shi X, Zheng X, Wang D et al (2023) Refining detection methods for emerging SARS-CoV-2 mutants in wastewater: a case study on the omicron variants. Sci Total Environ 904:166215
52. Brunner FS, Payne A, Cairns E, Airey G, Gregory R, Pickwell ND et al (2023) Utility of wastewater genomic surveillance compared to clinical surveillance to track the spread of the SARS-CoV-2 omicron variant across England. Water Res 247:120804
53. Liddor Naim M, Fu Y, Shagan M, Bar-Or I, Marks R, Sun Q et al (2023) The rise and fall of omicron BA.1 variant as seen in wastewater supports epidemiological model predictions. Viruses 15(9):1862
54. Wannigama DL, Amarasiri M, Phattharapornjaroen P, Hurst C, Modchang C, Chadsuthi S et al (2023) Tracing the new SARS-CoV-2 variant BA.2.86 in the community through wastewater surveillance in Bangkok, Thailand. Lancet Infect Dis 23(11):e464–e466
55. Yoo SM, Liu TC, Motwani Y, Sim MS, Viswanathan N, Samras N et al (2022) Factors associated with post-acute sequelae of SARS-CoV-2 (PASC) after diagnosis of symptomatic COVID-19 in the inpatient and outpatient setting in a diverse cohort. J Gen Intern Med 37(8):1988–1995
56. Yousif M, Rachida S, Taukobong S, Ndlovu N, Iwu-Jaja C, Howard W et al (2023) SARS-CoV-2 genomic surveillance in wastewater as a model for monitoring evolution of endemic viruses. Nat Commun 14(1):6325
57. How Covid-19 Spurred Governments to Snoop on Sewage. The Economist [Internet]. [Cited 2024 Mar 27]. Available from: https://www.economist.com/international/2022/09/07/how-covid-19-spurred-governments-to-snoop-on-sewage
58. Senaratna KYK, Bhatia S, Giek GS, Lim CMB, Gangesh GR, Peng LC et al (2024) Estimating COVID-19 cases on a university campus based on wastewater surveillance using machine learning regression models. Sci Total Environ 906:167709
59. Kanchan S, Ogden E, Kesheri M, Skinner A, Miliken E, Lyman D et al (2024) COVID-19 hospitalizations and deaths predicted by SARS-CoV-2 levels in Boise, Idaho wastewater. Sci Total Environ 907:167742
60. Bowes DA, Darling A, Driver EM, Kaya D, Maal-Bared R, Lee LM et al (2023) Structured ethical review for wastewater-based testing [Internet]. medRxiv. [Cited 2024 Mar 27]. p. 2023.06.12.23291231. Available from: https://www.medrxiv.org/content/10.1101/2023.06.12.23291231v1

61. Ivanova OE, Yarmolskaya MS, Eremeeva TP, Babkina GM, Baykova OY, Akhmadishina LV et al (2019) Environmental surveillance for poliovirus and other enteroviruses: long-term experience in Moscow, Russian Federation, 2004–2017. Viruses 11(5):424
62. Chavarria-Miró G, Anfruns-Estrada E, Martínez-Velázquez A, Vázquez-Portero M, Guix S, Paraira M et al (2021) Time evolution of severe acute respiratory syndrome coronavirus 2 (SARS-CoV-2) in wastewater during the first pandemic wave of COVID-19 in the metropolitan area of Barcelona, Spain. Appl Environ Microbiol 87(7):e02750–e02720
63. Crits-Christoph A, Kantor RS, Olm MR, Whitney ON, Al-Shayeb B, Lou YC et al (2021) Genome sequencing of sewage detects regionally prevalent SARS-CoV-2 variants. MBio 12(1). https://doi.org/10.1128/mbio.02703-20
64. Lin X, Glier M, Kuchinski K, Ross-Van Mierlo T, McVea D, Tyson JR et al (2021) Assessing multiplex tiling PCR sequencing approaches for detecting genomic variants of SARS-CoV-2 in municipal wastewater. mSystems 6(5). https://doi.org/10.1128/msystems.01068-21
65. Sutton M, Radniecki TS, Kaya D, Alegre D, Geniza M, Girard AM et al (2022) Detection of SARS-CoV-2 B.1.351 (Beta) variant through wastewater surveillance before case detection in a community, Oregon, USA. Emerg Infect Dis—CDC 28(6) [Cited 2024 Mar 14]; Available from: https://wwwnc.cdc.gov/eid/article/28/6/21-1821_article
66. Brumfield KD, Leddy M, Usmani M, Cotruvo JA, Tien CT, Dorsey S et al (2022) Microbiome analysis for wastewater surveillance during COVID-19. MBio 13(4):e0059122
67. Wastewater-based Disease Surveillance for Public Health Action [Internet] (2023) Washington, D.C National Academies of Science, Engineering, and Medicine, 152. Available from: https://nap.nationalacademies.org/catalog/26767/wastewater-based-disease-surveillance-for-public-health-action
68. Anthes E. A new report outlines a vision for national wastewater surveillance. The New York Times [Internet] 2023 Jan 20 [cited 2024 Mar 27]. Available from: https://www.nytimes.com/2023/01/19/health/wastewater-surveillance-disease.html
69. Lobitz B, Beck L, Huq A, Wood B, Fuchs G, Faruque ASG et al (2000) Climate and infectious disease: use of remote sensing for detection of Vibrio cholerae by indirect measurement. Proc Natl Acad Sci 97(4):1438–1443
70. Jutla A, Akanda A, Unnikrishnan A, Huq A, Colwell R (2015) Predictive time series analysis linking Bengal cholera with terrestrial water storage measured from gravity recovery and climate experiment sensors. Am J Trop Med Hyg 93(6):1179–1186
71. Jutla A, Aldaach H, Billian H, Akanda A, Huq A, Colwell R (2015) Satellite based assessment of hydroclimatic conditions related to cholera in Zimbabwe. PLoS One 10(9):e0137828
72. Jutla A, Whitcombe E, Hasan N, Haley B, Akanda A, Huq A et al (2013) Environmental factors influencing epidemic cholera. Am J Trop Med Hyg 89(3):597–607
73. Khan R, Aldaach H, McDonald C, Alam M, Huq A, Gao Y et al (2019) Estimating cholera risk from an exploratory analysis of its association with satellite-derived land surface temperatures. Int J Remote Sens 40(13):4898–4909
74. Usmani M, Brumfield KD, Magers BM, Huq A, Barciela R, Nguyen TH et al (2022) Predictive intelligence for cholera in Ukraine? GeoHealth 6(9):e2022GH000681
75. Jutla A, Usmani M, Brumfield KD, Singh K, McBean F, Potter A et al (2023) Anticipatory decision-making for cholera in Malawi. MBio 14(6):e0052923
76. Usmani M, Brumfield KD, Magers B, Zhou A, Oh C, Mao Y et al (2024) Building environmental and sociological predictive intelligence to understand the seasonal threat of SARS-CoV-2 in human populations. Am J Trop Med Hyg. 110(3):518–528

mRNA Vaccines: Immunology and Present and Future Potential

Omar F. Khan, Julien Couture-Senécal, Yan Ming Anson Lau, Ana-Maria Oproescu, Janice Pang, Grayson Tilstra, and Brianna A. Walz

1 Introduction

Messenger ribonucleic acid (mRNA) is a molecule that cells use to encode instructions to produce proteins. Vaccines present unique proteins from pathogens, called antigens, to the immune system to develop protective immunity. If the vaccinated individual encounters the antigen in the future, such as during infection, the immune system will clear the pathogen associated with the antigen. When used as a vaccination tool, mRNA molecules encode the instructions to create antigens. These antigens are then presented to the immune system. In this way, cells of the individual manufacture antigens, which streamline the vaccination process.

2 How the Immune System Creates Protection Through Antibodies and T Cells

Adaptive immune system is one of the most important systems in the body. It is responsible for fighting off foreign pathogens that enter the body. The main cell types involved in this process are B cells and T cells. B cells are most known for their ability to randomly create a repertoire of antibodies to help fight infection. T cells are known for their helping abilities (CD4+ T cells) and their cytotoxicity (CD8+ T cells). Although very different, both B and T cells originate from the same precursor, the lymphoid progenitor originating from the bone marrow [1].

O. F. Khan (✉)
Institute of Biomedical Engineering, University of Toronto, Toronto, Canada

Department of Immunology, University of Toronto, Toronto, Canada
e-mail: dr.khan@utoronto.ca

J. Couture-Senécal · Y. M. A. Lau · A.-M. Oproescu · J. Pang · G. Tilstra · B. A. Walz
Institute of Biomedical Engineering, University of Toronto, Toronto, Canada

J. M. Rosen, R. R. Colwell (eds.), *The COVID-19 Pandemic*,
https://doi.org/10.1007/978-3-031-62772-9_5

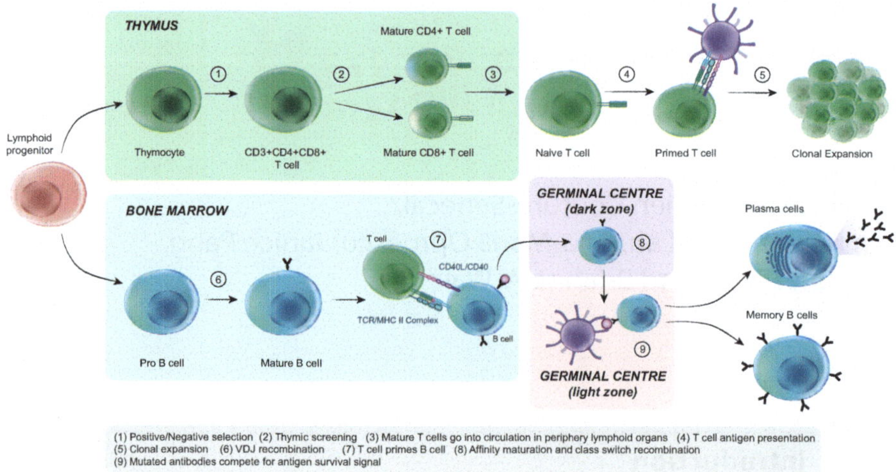

Fig. 1 Overview of adaptive immunity. A lymphoid progenitor gives rise to thymocytes. These cells feature receptors used to identify an antigen, which is a unique molecule that drives an immune response. These receptors are randomly generated. The T cells are screened to eliminate the ones that react to the body's own tissues (self-reactive) (1). Thymic selection results in CD8+ T cells, which kill cells infected with pathogens, and CD4+ T cells, which help B cells produce antibodies (2). Once the matured T cells (3) encounter and recognize their corresponding foreign antigen (4), they become primed and expand their numbers through cloning (5) to increase their ability to find their antigens. Similarly, B cells randomly create their receptors through a process called V(D)J recombination, with self-reactive B cells eliminated (6). If the B cells encounter their foreign (non-self) antigen in the spleen, and the same foreign antigen is recognized by a CD4+ T cell, the B cell becomes primed (7). The B cell travels to the germinal center dark zone to begin their refinement through the processes known as affinity maturation and class switching to create antibodies (8). Further antibody refinement occurs in the light zone of the spleen, where antibodies are randomly mutated in a process called somatic hypermutation that further enhances their affinity to the antigen (9). This process results in two types of specialized B cells. The first is a cell that travels to the bone marrow that will secrete low levels of antibodies into the blood. The second is a memory B cell that remains in the spleen and clones itself upon reinfection

Certain progenitor cells will leave the bone marrow and travel into the thymus. In the thymus, the progenitor cells receive signals, committing them to the T-cell lineage (Fig. 1(1)) [2]. The cells enter an intense period of proliferation in the thymus and can now be referred to as thymocytes [3]. The thymocytes at this point are referred to as double-negative cells, meaning they do not express the CD4 or the CD8 receptors. During the double-negative stage of the thymocytes, the T-cell receptor (TCR) begins rearrangement. The TCR is composed of an alpha and a beta chain. Both are rearranged individually and randomly. Cells with unsuccessful rearrangements will die at this point, and cells with successfully rearranged receptors undergo proliferation [4].

Now, the CD4 and CD8 receptors will both be expressed on the thymocyte, making them double-positive cells. They will next undergo thymic screening. First, positive selection begins on the double-positive thymocyte T-cell receptors. This process will determine if the T cell will express a CD4 or a CD8 receptor [5]. Depending on which type of MHC molecule the T-cell receptor binds to, the cell will become single positive. If the immature T cell binds to MHC I, then it will become a CD8+ T cell, and if it binds to MHC II, then it will become a CD4+ T cell (Fig. 1(2)). Once the cells are single positive, they are matured [6].

Before leaving the thymus, the T cells must undergo negative selection. This process will eliminate any self-reactive thymocytes. This process is facilitated by epithelial cells in the thymus and bone-marrow-derived macrophages or dendritic cells [7]. Once the thymic screening is complete, the T cell will undergo intense proliferation before entering the periphery (Fig. 1(3)). These T cells are now self-tolerant and MHC-restricted.

The mature but naïve T cells are now in circulation. When an antigen is encountered, an antigen-presenting cell will present a peptide on its MHC complex [8]. The T-cell CD4/CD8 receptor will recognize the peptide-loaded MHC molecule and become activated (Fig. 1(4)). The CD4-expressing T cells will become active helper T cells and will release an abundance of cytokines. When CD8+ T cells become active, they will also release cytokines and will be able to cytotoxically kill infected cells. Both CD4+ and CD8+ T cells will clonally expand once they are activated (Fig. 1(5)) [9].

The B cell uses an immunoglobin molecule to probe antigens. In simplified terms, an immunoglobin bound to the B cell's surface is a B-cell receptor (BCR), while the secreted form is an antibody. The bone marrow provides the essential signals to advance B-cell development. Through a series of signaling events and induction of transcription factors, the common lymphoid progenitor becomes a pro-B cell [10]. The next process is the rearrangement of the BCRs (which ultimately become antibodies when secreted). Antibodies are immunoglobin molecules that consist of a heavy and a light chain that need to be rearranged based on germline configuration. The pro-B cell begins to rearrange the heavy chain first. Once complete, the pro-B cell becomes a pre-B cell. From here the light chain rearranges, and the immature B cell is formed [11]; this process is called V(D)J recombination. This B cell will express only an IgM antibody, which is the earliest version of an antibody without additional optimization. Importantly, each B cell can produce only one specific antibody.

Before leaving the bone marrow, the B cell is tested for self-reactivity. The B cell is tested against self-molecules. If the IgM recognizes these self-molecules, they will die by apoptosis or become anergic [12]. If the B cell does not recognize self, then it will begin to express IgD and migrate to the peripheral lymphoid tissue. This is now considered a mature B cell (Fig. 1(6)). The cell continues to be monitored when in the periphery for a short amount of time for self-reactivity [13].

Each antibody/BCR contains a complementarity-determining region (CDR) that binds to antigens. This region is different for each population of antibodies [14]. When the B cell encounters an antigen that its immunoglobin can recognize, it will internalize it via the BCR, degrade it, and display it on the surface with an MHC II molecule. When a helper T cell recognizes this MHC:antigen complex, it will give signals to the B cell to aid with survival and proliferation (Fig. 1(7)). At this point, the B cell will enter the germinal center [15]. This interaction happens at the border of the T cell and B cell zone in the spleen [16].

Once this interaction occurs, the B cells form a germinal center. The purpose of the germinal center is to refine the antibody. This occurs through two processes: somatic hypermutation and class switch recombination (Fig. 1(8)) [17]. Somatic hypermutation is a process in which the variable region of the antibody is altered to find the best possible fit for the antigen. The specificity and affinity of the antibody are altered by small mutations during this process. Only the most superior antibodies will be selected from this process by testing their affinity with antigen-presenting cells (APCs) (Fig. 1(9)) [18]. In addition to somatic hypermutation, class switch recombination occurs in the germinal centers. This process involves gene rearrangement of the antibody to create different classes of antibodies [19]. Different classes of antibodies are needed for different locations and processes in the body. For example, IgG antibodies typically circulate in the blood, while IgA antibodies are found at mucosal surfaces.

After the antibodies have undergone the germinal center reaction, the B cells will become either plasma cells or memory B cells. Plasma cells immediately secrete antibodies into the extracellular environment. Plasma cells are the most important cell type in humoral immunity. Memory B cells express the mutated and class-switched immunoglobin. Memory B cells remain in circulation for years. If the same antigen re-enters the body in the future, the memory B cell will quickly recognize it and begins to mount a much faster adaptive immune response than the first infection [20].

3 What Happens with mRNA Vaccination

The goal of vaccination is the induction of protective immunity against disease-causing infectious pathogens, such as microorganisms, bacteria, viruses, and fungi. Most vaccines are designed to target natural defense mechanisms and activate the immune system in a manner similar to natural infections [21]. An effective protective immune response involves the innate and adaptive immune systems working sequentially together to identify invading pathogens and initiate the most effective defense response [21].

Contrasting Acquired Immunity from Natural Infection to mRNA Vaccination

In natural infection, innate immunity is the first line of defense against pathogens. Following pathogen invasion, the innate immune system must detect, contain, and eliminate the threat immediately [21]. Pathogens are detected through molecular-sensing surveillance mechanisms known as pattern recognition receptors (PRRs), which are expressed either on the surface of cells or in intracellular departments [22]. For example, toll-like receptors (TLRs) are PRRs that recognize pathogen-associated molecular patterns (PAMPs) shared by pathogens [22]. These receptors sense danger signals and activate and augment proinflammatory gene expression to facilitate host defense capabilities [22]. Ultimately, pathogen elimination in innate immunity is accomplished through phagocytosis of pathogens and subsequent intracellular destruction within intracellular vesicles that contain digestive enzymes [23]. However, the innate immune system only has a limited number of receptor molecules available to mount such protection. Innate defense mechanisms lack specificity and are not well equipped to induce a immunological memory response [24]. This is especially insufficient for vaccine-related protective immune responses that critically depend on immune memory responses. Thus, the adaptive immune system picks up the response so the body can acquire immunity against the specific pathogen, thereby limiting or preventing reinfection in the future [24]. The link between the innate and adaptive systems is the presentation of antigens. In natural infection, these antigens come from the pathogen.

Cells regularly take samples from their internal contents, load them onto MHC I molecules, and present them on their exterior surface [25]. CD8+ T cells probe the loaded MHC I with their TCRs, which only recognize foreign material [25]. If there is recognition, the CD8+ T cell triggers the death of the infected cell to limit the spread of the pathogen [25]. They also replicate themselves to expand their surveillance [26].

Some foreign material from the pathogen may not be located inside cells (intracellular); instead, it might be outside the cell (extracellular). Also, some intracellular pathogens may cause the death of their host cell, which releases them into the extracellular space [27]. Material from the extracellular space is processed by APCs, presented on their MHC II molecules, and shown to B cells [28]. If the BCR recognizes a foreign material, they undergo affinity maturation in the lymph node to create antibodies [29]. APCs can be macrophages or immature dendritic cells, which can phagocytose pathogens and foreign material [30]. Following the ingestion of antigens, APCs migrate to the draining lymph nodes [30]. As such, APCs from innate immunity act as messengers to precisely define the level of danger and convey such information to secondary lymphoid organs relevant to activating an adaptive immune response.

As shown in Fig. 2, the major difference between natural infection and vaccination is the source of the foreign material (antigen). In natural infection, the source is the pathogen and the trauma it causes as it consumes body tissue to replicate itself. Combating infection is a time- and energy-intensive process for the immune system. In contrast, vaccination is a short cut that provides the antigen without the trauma of infection. Vaccination can result in the antigen entering cells, which leads to presentation on MHC I and interaction with CD8+ T cells to kill infected cells. Antigens in the extracellular space can be processed by APCs for presentation on MHC II and interaction with B cells to create antibodies. Critically, while the source of the antigen differs between natural infection and vaccination, all of the processes after antigen presentation are common. Thus, vaccination activates the adaptive immune response by efficiently priming existing naïve T cells and B cells.

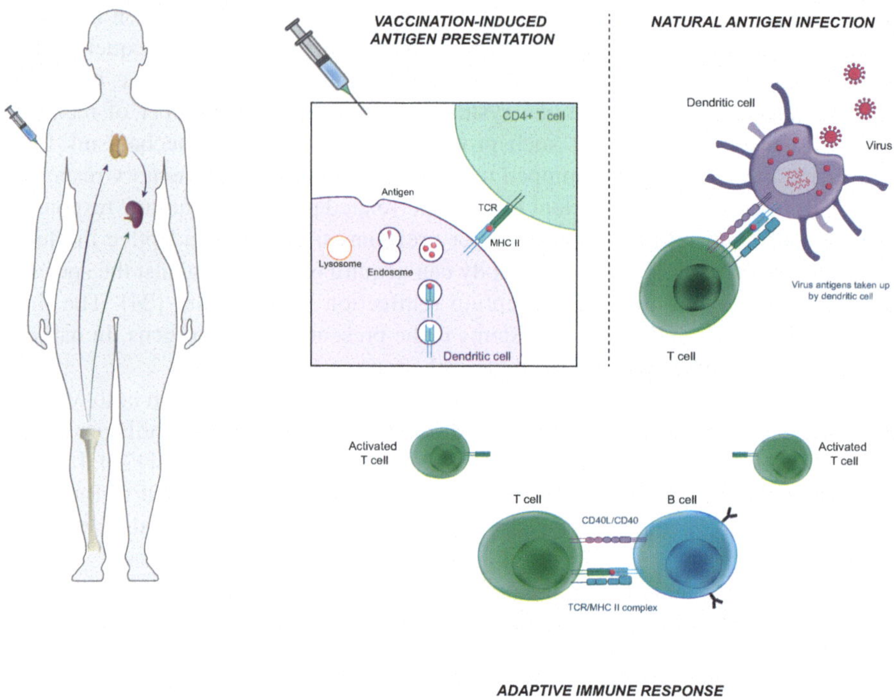

Fig. 2 Contrasting immunity produced by natural infection and vaccination. (**a**) Antigens are unique fragments of a pathogen. The immune system develops protective T cells and antibodies against the pathogens via these antigens. Prior to immunization, T cells are generated in the bone marrow, travel to the thymus to screen out autoreactive cells, and can enter circulation or a lymph organ such as the spleen. Similarly, early B cells originate from the bone marrow and travel to the spleen via the blood. The antigens that are supplied through vaccination ultimately make their way to lymph organs. T and B cells probe the antigen to see if there is recognition. If there is recognition, the body will develop protective immunity. (**b**) In natural infection, the pathogen causes trauma as it uses the body to replicate itself. The antigen is sourced directly from the infecting pathogen. In contrast, vaccination provides the antigen without the trauma of infection. Despite a lack of infection, the immune system is still able to utilize the antigen to develop protective T cells and antibodies. Antigens, despite their source, are all processed in the same way to develop protective immunity

4 mRNA Vaccines

The earliest forms of vaccination involved the use of partially or fully inactive pathogens, which were injected into the patient. Because of inactivation, the immune system could easily attack and process the pathogen to develop acquired immunity. However, not all pathogens are amenable to inactivation. In contrast, subunit vaccines do not use the entire pathogen. Instead, pathogen-derived antigens are injected into patients. These antigens can also be paired with an adjuvant. Adjuvants create different levels of irritation that trigger the innate immune response to help convey artificial danger signals to secondary lymphoid organs and activate the adaptive immune response.

Messenger ribonucleic acids (mRNAs) are information-carrying molecules that encode instructions the cells need to produce proteins. When used as a tool for vaccination, mRNA encodes instructions to produce the antigen (Fig. 3). The mRNA is delivered to some of the patient's cells, which produce the antigen via the ribosome. This contrasts to subunit protein vaccines, where antigens are produced separately, such as in a bioreactor. An advantage to producing antigens from mRNA is speed. While subunit vaccines must be manufactured, purified, and paired with the correct adjuvant, the sequence of mRNA can be quickly changed and deployed. Additionally, some antigens can be difficult to manufacture and store in a form that can be readily administered to patients, such as highly insoluble proteins. Such manufacturing, processing, and storage issues do not exist with mRNA vaccines because the cells produce the antigens. However, the use of mRNA as a vaccination tool can be severely impacted by poor cellular delivery; the mRNA must be effectively delivered to a portion of cells in the patient with high enough efficiency so that enough antigen can be produced to drive an immune response.

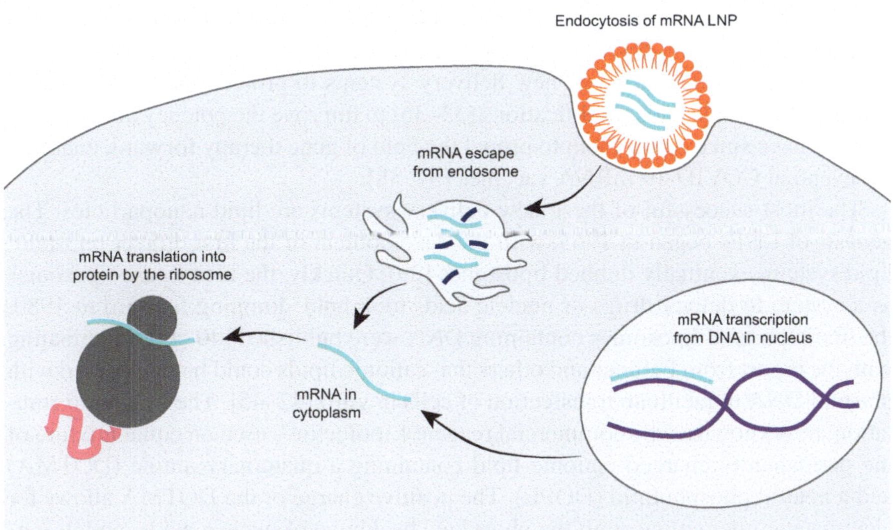

Fig. 3 Deoxyribonucleic acid (DNA) in the nucleus of cells encodes the information needed to produce the cell's proteins. To convert the information in DNA into proteins, it must first be transcribed into messenger ribonucleic acid (mRNA). The mRNA leaves the nucleus and associates with a ribosome. The ribosome translates the mRNA into protein. To create a vaccine, an mRNA is created that encodes an antigen. This antigen-encoding mRNA is packaged into a lipid nanoparticle (LNP), which enters the cell via endocytosis. Once in the endosome, the nanoparticle releases the mRNA, which enters the cytoplasm. In the cytoplasm, the mRNA is translated into the protein antigen through the same ribosomal process as the mRNA created from the cell's DNA

5 Development of mRNA Vaccines

Historically, the clinical translation of RNA-based drugs has been hampered by low delivery efficiency to target cells. RNA delivery is challenging because RNA is a large and negatively charged molecule that cannot penetrate the cell membrane and is readily degraded by endogenous enzymes. Over the past 35 years, these challenges have been steadily overcome through the development of lipid nanoparticles (LNPs) that can safely and efficiently deliver RNA to cells.

While it feels as if gene therapy, and specifically mRNA technology, has risen to the forefront of medicine in the past few years, its central ideas can be traced back decades. In 1972, Friedmann and Roblin, spurred by their colleagues' results, published a high-profile paper in *Science* advocating for a significant push in gene therapy research with regard to genetic diseases. However, it wasn't until 1990 that the first report of the successful use of in vitro transcription (IVT) in animals was published, when mRNA encoding for reporter genes was shown to produce detectable levels of protein [31]. This initial study was a major milestone, but still it was held back by rudimentary RNA transcripts that were delivered directly through intramuscular injection, and thus were relatively inefficient at reaching cells. Despite a follow-up study that showed a physiological response to vasopressin-encoding mRNA in rats, there were nagging issues of high innate immunogenicity, significant degradation of the free transcripts by RNases, and poor cell transfection that prevented widespread adoption [32]. Clearly, improvements were required. Over the intervening decades, a combination of new delivery systems to protect nucleic acid cargo and chemical nucleoside modifications [33–36] to improve the potency and stability of RNA have since come about to propel the field of gene therapy forward, enabling the eventual COVID-19 mRNA vaccines [37, 38].

The most successful of these new delivery systems are lipid nanoparticles. The genesis of LNPs began in 1965, with the development of the first bilayer phospholipid system, eventually dubbed liposomes [39]. Quickly, the idea to use liposomes as a system to deliver drugs or nucleic acids took hold. Jumping forward to 1980, the first reports of liposomes containing DNA were published [40, 41], culminating with the report from Felgner and others that cationic lipids could be complexed with plasmid DNA to facilitate transfection of cells in vitro [42–45]. The Felgner formulation, now known as the commercial reagent Lipofectin™, used an equal mixture of the permanently charged cationic lipid containing a quaternary amine (DOTMA) and a neutral phospholipid (DOPE). The positive charge of the DOTMA allows for electrostatic interactions with the phosphate backbone of nucleic acids, and its carbon tails promoted assembly into a particle. What followed was a burst of activity to develop new cationic lipids to deliver nucleic acids. However, it soon became clear that cationic particles caused toxicity and were not suitable for all applications [46, 47]. In 1994, Bailey and Cullis found that they could replace the cationic lipid with an "ionizable" analog, DODAP, that contained a tertiary amine and formed particles that would remain neutral at physiological pH and gain a positive charge in acidic conditions [48]. However, the formulation reported by Bailey and Cullis suffered from instability, and particles tended to aggregate over time. A further improvement reported in two 2001 studies by Maurer et al. and Semple et al. showed that the addition of a lipid-anchored polyethylene glycol (PEG) stabilized particles at lower diameters [49] and increased circulation half-life of LNPs [50]. Furthermore, Semple et al. showed that the introduction of cholesterol and DSPC significantly improved encapsulation of nucleic acids and stability of the particles. With the floodgates of ionizable lipid development opened, thousands of new structures have been synthesized and tested [51–61]. Two key discoveries during this period were the ionizable lipid DLinDMA in 2005, which improved the efficacy of RNAi in vivo, and its direct descendent, DLin-MC3-DMA, which further improved potency and tolerability. DLin-MC3-DMA would eventually be used in the landmark gene therapy Onpattro [62, 63]—the first LNP-based siRNA therapeutic to be approved by the FDA and confirmation of the potential of LNPs and nucleic acids as a whole.

During the COVID-19 pandemic, LNP and mRNA technology enabled rapid development of the first vaccines against SARS-CoV-2: BNT162b2 [38] and mRNA-1273 [37]. These two vaccines are remarkably similar, with the primary difference being the ionizable lipids used in their formulations—ALC-0315 for BNT162b2 and SM-102 for mRNA-1273. The development of LNPs for vaccine application predates the urgency of the pandemic, beginning in 2018 with the first report of SM-102 by Sabnis et al. (referred to as Lipid 8) [64]. In a follow-up study by Hassett et al. in 2019 [65], a similar series of lipids were tested for intramuscular vaccines, and SM-102 was found to produce high antibody titers while remaining well tolerated. These initial reports justified the use of SM-102 in clinical trials for mRNA vaccines. ALC-0315, the ionizable lipid in BNT162b2, displays similar structural motifs to SM-102, but its development has not been documented in the scientific literature. ALC-0315 has similar mRNA delivery efficacy to SM-102 but a longer degradation half-life in vivo. A brief chronological list of notable LNPs are shown in Fig. 4.

And thus, we arrive at the modern incarnation of LNPs: a four-component system comprising an ionizable lipid, cholesterol, a phospholipid, and a lipid-PEG, all of which are key for the encapsulation and delivery of mRNA (Fig. 5). Removed from their historical context, it is worth describing in detail what the exact function of each of these components is, and why they are included in LNP formulations. Altogether, the four-component LNP serves to encapsulate, protect, and deliver mRNA to cells with high efficacy and was a key technology that enabled rapid development and disbursement of COVID-19 vaccines.

6 mRNA Vaccine Mechanism of Action

To achieve a robust immune response, many subunit and inactivated virus vaccines require the use of adjuvants. In contrast to traditional flu vaccines, COVID-19 mRNA vaccines do not require an exogenous adjuvant to induce a strong adaptive

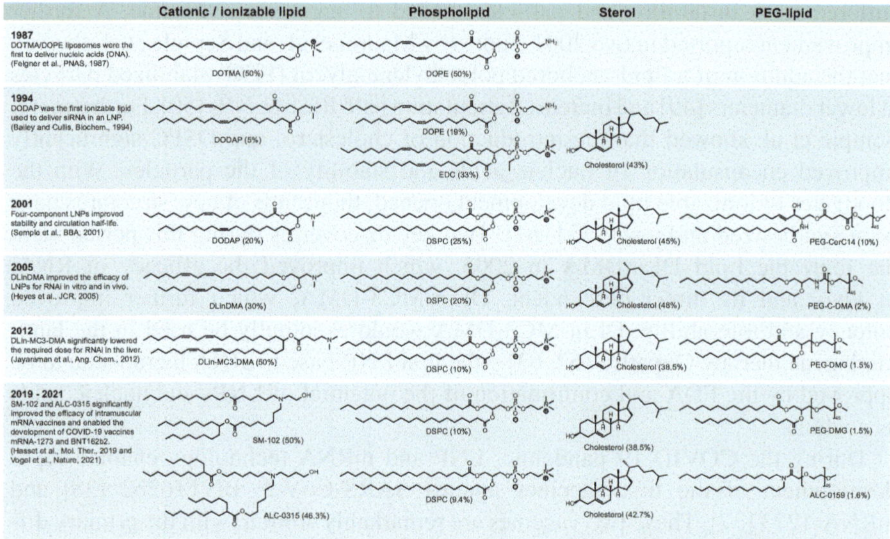

Fig. 4 Development timeline of lipids and LNPs leading up to the two COVID-19 mRNA-LNP vaccines, mRNA-1273 and BNT162b2. Molar ratios of lipid components in each formulation are given in brackets

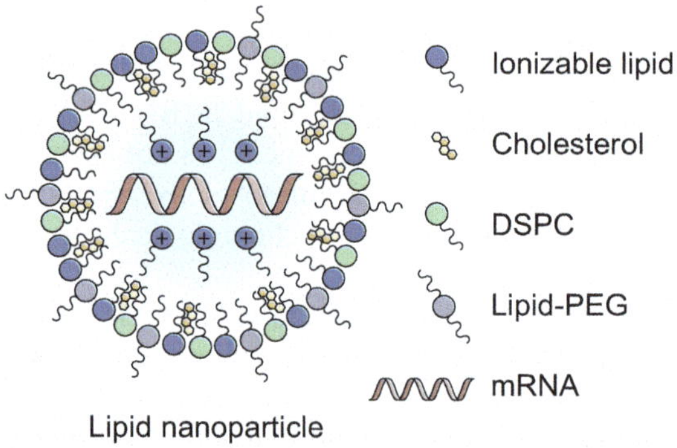

Fig. 5 General representation of lipid nanoparticles (LNPs) used in mRNA vaccines. Common components of an mRNA LNP vaccine include an ionizable lipid to help bind mRNA, cholesterol to provide rigidity, and distearoylphosphatidylcholine (DSPC) to enhance LNP entry into the cell and mRNA release into the cytoplasm. Lipid-anchored poly(ethylene glycol) (lipid-PEG) helps the LNPs remain stable in solution

immune response [66]. Moreover, exogenous mRNA is a highly immunogenic molecule that can trigger the innate immune system. Since improvements in mRNA chemistry and manufacturing significantly reduced mRNA-mediated innate immune signaling [36, 67–69], the role of LNPs (and their individual components) as immune agonists has become increasingly clear and provides insight into their unique mechanisms of action.

6.1 Ionizable Lipid

Ionizable lipids are responsible for binding mRNA during formulation, and they act as the main driver of mRNA delivery to the cytosol of a cell. Across the wide spectrum of different structures that have been tested as ionizable lipids, there are a few common features that have been found to improve delivery:

(1) The pK_a of the ionizable tertiary amine is key. This parameter prominently affects the apparent pK_a of formulated LNPs and mRNA expression after intramuscular LNP administration [70]. For potent intramuscular vaccines, Hassett et al. [65] reported that an apparent LNP pK_a between 6.6 and 6.9 is ideal. Particles formulated with SM-102 fall within this range, although particles formulated with ALC-0315 have much lower pK_a values of approximately 6.0–6.1 [71]. This discrepancy can be reconciled by considering the next two important factors of ionizable lipid design.
(2) The lipid should include hydrocarbon tails that promote self-assembly into an LNP during formulation and increase its ability to fuse with cell membranes. Both SM-102 and ALC-0315 include multiple tails ranging between 6 and 12 carbons in length.
(3) The lipid should exhibit a general "cone" shape, formed by the hydrocarbon tails, that tapers to a point at the tertiary amine headgroup. This shape helps disrupt membranes by inducing transformation from their normal lamellar phase to a hexagonal phase [72]. SM-102 and ALC-0315 achieve this by including branched tails that promote hexagonal packing. In particular, SM-102 exhibits higher fusogenicity and endosomal escape efficiency than DLin-MC3-DMA, which has fewer tails and a more cylindrical shape overall [64].

The molecular structure of ionizable lipids has been shown to significantly impact mRNA translation efficiency and immune response to mRNA vaccines in preclinical studies [65]. Multiple studies have also shown that removing the ionizable lipid component from LNPs eliminates the adjuvant activity. For instance, mice coadministered with a protein antigen and empty LNPs lacking ionizable lipids did not produce antigen-specific antibodies [73, 74].

The mechanism through which ionizable lipids induce immune activation is still being actively investigated. So far, it has been established that ionizable lipids drive an acute inflammatory response, characterized by the production of inflammatory cytokines such as IL-6 and IL-1b, which leads to the induction of Tfh and germinal center B cells [73–76]. Consequently, concerns have emerged in the scientific community about their pro-inflammatory effects and potential link to side effects reported in human patients [77–80]. The other components within the LNP formulation, typically referred to as "helper lipids," each have a key role in nanoparticle formation, delivery, and vaccine performance.

6.2 Cholesterol

Cholesterol, with its characteristic fused-ring structure, is a structural excipient that helps stabilize the particle by reducing fluidity throughout the LNP and can also assist with fusogenicity [81, 82]. This stability is particularly important in serum to prevent premature leakage of nucleic acid cargo before cellular entry.

6.3 DSPC

Phospholipids, such as distearoylphosphatidylcholine (DSPC) which is used in mRNA-1273 and BNT162b2, are more polar than the ionizable lipid and tend to form a layer on the periphery of the particles [83]. The presence of phospholipids in an LNP increases its fusogenicity with cellular membranes to enable delivery to the cytosol.

DSPC is a substrate for phospholipases that activate inflammation through the production of arachidonic acid upon hydrolysis of the sn-2 ester bond [84]. DSPC has also been shown to activate B-cell subsets, which have been linked to accelerated blood clearance phenomena [85].

6.4 Lipid-PEG

The lipid-PEG anchors itself within the outer layer of the LNP and controls the size distribution of the particles. PEG also provides a large steric barrier around each particle that minimizes aggregation during prolonged storage. Once the LNP is in dilute conditions after administration, the lipid-PEG can dissociate from the particles, allowing serum proteins to opsonize the LNPs and form a "protein corona," which dictates cellular uptake and biodistribution [86].

It is worth noting that PEG is an allergen that has been linked to rare hypersensitivity reactions and anaphylaxis [87]; vaccines that include lipid-PEG, like mRNA-1273 and BNT162b2, are therefore not recommended for patients with PEG allergy. Studies have reported a significant increase in anti-PEG antibodies after mRNA vaccination, which correlates with higher systemic reactogenicity after multiple doses [88–90]. Anti-PEG antibodies also contribute to accelerated blood clearance of LNPs, which may inhibit the efficacy of mRNA vaccines upon repeated dosing [85].

6.5 Cryoprotectants

The two COVID-19 vaccines also leverage a cryoprotectant, sucrose. The cryoprotectant is not part of the LNP; instead, it is added to the liquid solution that suspends the LNPs after the LNPs are formulated. The cryoprotectant preserves the physiochemical properties of the particles when they are frozen for storage and distribution [91, 92].

7 Advanced Preclinical and Early Clinical Trial Work Pre-COVID-19

Prior to the outbreak of the COVID-19 pandemic, many preclinical studies and early clinical trials were underway to evaluate mRNA–LNPs for vaccines against cancers and infectious diseases. It is worth discussing the advantages that mRNA–LNP vaccines hold over protein subunit, inactivated or live virus vaccines, or DNA vaccines. Foremost, compared to DNA, mRNA does not integrate into the genome and thus does not pose a risk of insertional mutagenesis. mRNA is also degraded by normal cellular processes and only produces protein for a short time. Compared to protein or viral vaccines, mRNA–LNP vaccines can also be generated extremely rapidly once the sequence of a viral pathogen becomes available, as all that needs to change is the sequence of the mRNA that is included in the LNP. As of 2018, preclinical studies are underway of mRNA–LNP vaccines against influenza, HIV, zika virus, Ebola virus, respiratory syncytial virus, human cytomegalovirus, rabies, human papillomavirus, and melanoma, ovalbumin-expressing cancer, and colon cancer. Clinical trials for mRNA–LNPs against influenza, zika virus, HIV-1, and rabies had also been initiated [32].

So why do mRNA–LNPs work so well as vaccines? In simple terms, vaccines have two requirements: an antigen, to which you want the body to develop an immune response, and an adjuvant, an additive that invokes an initial immune response to train the body against the antigen. The key to the potency of mRNA–LNPs is that they both deliver antigen as an mRNA transcript and act as their own adjuvant, driving an immune response through the act of delivery. In 2018, Pardi et al. [93] published a mechanism for this potent immune activation. mRNA–LNPs are powerful activators of specific CD4+ cells known as T follicular helper cells (T_{fh}). In turn, these T_{fh} promote the increased formation of germinal center (GC) B cells, which mature into memory B cells and long-lived plasma cells that actually produce the high antibody titers [93]. Therefore, the quality of antibody response produced by a vaccine is directly related to its ability to activate T_{fh} cells—an area in which mRNA–LNPs excel.

8 Application of mRNA Vaccines to SARS-CoV-2

The COVID-19 pandemic prompted an urgent need for an effective vaccine to combat SARS-CoV-2. Sahin et al. [94] leveraged the lipid nanoparticle (LNP)-formulated mRNA vaccine technology to develop BNT162b1. This mRNA vaccine encoded for the receptor-binding domain (RBD) of SARS-CoV-2 spike protein [94]. The RBD antigen was fused to a T4 fibritin-derived "foldon" trimerization domain to enhance immunogenicity via multivalent display [94]. The RNA was optimized for high stability and translation efficiency through incorporating 1-methylpseudouridine to replace uridine [94]. This dampened innate immune sensing and increased mRNA translation in vivo [94]. To evaluate the efficacy of the vaccine, researchers measured antibody and T-cell responses following BNT162b1 vaccination of healthy adults aged 18–55 years [94]. Following two doses of 1–50 μg of BNT162b1, strong vaccine-induced antibody responses were detected, with RBD-specific neutralizing antibodies being 0.7-fold (1 μg dose) to 3.5-fold (50 μg dose) higher than in recovered individuals [94]. Majority of the participants mounted T helper (Th)1-skewed T-cell immune responses with strong RBD-specific CD4+ and CD8+ T-cell responses [94]. The robust RBD-specific immunological responses and favorable cytokine responses of the BNT162b1 mRNA vaccine highlight the remarkable capabilities of mRNA technology.

In a later study by Turner et al. [95], they investigated the dynamics of antibody-secreting plasmablasts (which is a precursor to the plasma cell) and germinal center B cells induced by BNT162b2, an mRNA–LNP vaccine encoding the spike protein of SARS-CoV-2. This study examined antigen-specific B-cell responses in peripheral blood and draining lymph nodes in 14 individuals who had two doses of BNT162b2 [95]. Circulating levels of IgG- and IgA-secreting plasmablasts targeting spike protein peaked 1 week after the second immunization, and levels became undetectable 3 weeks later [95]. High frequencies of S-binding germinal center B cells and plasmablasts were observed in draining lymph nodes for at least 12 weeks after booster immunization [95]. The S-binding monoclonal antibodies originating from germinal center B cells were found to mainly target the receptor-binding domain of the S protein [95]. In comparison, fewer clones bound to the N-terminal domain or to epitopes shared with the S proteins of human betacoronaviruses [95]. The cross-reactive B-cell clones had increased somatic hypermutation compared to B-cell clones that recognized only SARS-CoV-2 protein, indicating memory B cell origin [95]. Collectively, this study found that SARS-CoV-2 mRNA-based vaccination of humans induces persistent germinal center B-cell response, which is critical in generating robust humoral immunity.

9 Clinical Evidence of COVID-19 mRNA Vaccine Efficacy

Human data demonstrated the ability of mRNA vaccines to engage components of the adaptive immune response to generate acquired immunity. Using a combination of fine-needle aspirates and analysis of blood samples, direct evidence of antigen-specific CD4+ and CD8+ T-cell responses was observed.

Germinal centers (GC) are structures in the lymphatic system that enable B cells to acquire affinity-enhancing somatic hypermutations (SHM) [96, 97]. The surviving clones in these structures differentiate into memory B cells (MBCs) and long-lived bone marrow plasma cells (BMPCs) [98]. With SARS-CoV-2 mRNA vaccination, there is a persistent GC response for at least 6 months in humans [95]. However, the fate of responding GC B cells and the functional consequences of the lasting GC response require further elucidation. In a study by Kim et al. [99], they detected SARS-CoV-2 spike protein-specific MBCs in individuals who received two doses of BNT162b2 6 months prior. They found that spike-specific IgG-secreting BMPCs were detected in nine of eleven individuals [99]. In addition, they sequenced the B-cell receptors and tracked the expression of corresponding monoclonal antibodies to track the evolution of spike-specific B-cell clones [99]. Results indicated that early blood-spike-specific plasmablasts exhibited lowest SHM frequencies. This contrasted with a 3.5-fold increase observed in spike-specific GC B cells within 6 months following vaccination [99]. Notably, spike-specific MBCs and BMPCs amassed high levels of SHM, which parallels enhanced anti-spike antibody avidity in blood and improved affinity, as well as the neutralization capacity of BMPC-derived monoclonal antibodies [99]. Thus, persistence of the GC reaction induced by SARS-CoV-2 mRNA vaccination in humans results in affinity-matured long-term antibody responses that can effectively neutralize the virus.

SARS-CoV-2 mRNA vaccines can lead to protection from severe disease as early as 10 days following prime vaccination [100]. Vaccine-induced CD8+ T cells are likely the main mediators of such protection during early stages [101]. Oberhardt et al. [102] sought to uncover details of CD8+ T-cell induction in comparison to natural infection and in association with other branches of vaccine-induced immunity. They demonstrated a stable and fully functional CD8+ T-cell response mobilized one week following prime vaccination with BNT162b2, when circulating CD4+ T cells and neutralizing antibodies are still at a low level [102]. Although the functional capacity or memory precursor T-cell pool was unaffected, highly differentiated effector CD8+ T cells underwent expansion, following boost vaccination [102]. Relative to natural infection, vaccine-induced early memory T cells had similar functional capacities but had differing subset distribution [102]. These results suggest that CD8+ T cells are important effector cells that are expanded during early protection following prime vaccination [102]. This process precedes maturation of other effector branches of vaccine-induced immunity and can be stably maintained following boost vaccination [102].

The development of nucleoside-modified mRNA-formulated lipid nanoparticles encoding SARS-CoV-2 spike glycoprotein was investigated in BNT162b2. This stabilized the protein's prefusion conformation and led to a 95% efficacy in preventing COVID-19 [100]. Sahin et al. extended previous studies' phase I/II trial reports [103] by studying the immune response induced by BNT162b2 prime-boost vaccination from an additional phase I/II trial in healthy adults (aged 18–55 years) [104]. At one week after vaccination boost, SARS-CoV-2 serum geometric mean 50% neutralizing titers were up to 3.3-fold relative to samples from individuals who had recovered from COVID-19. The resulting sera from BNT162b2 neutralized 22 pseudoviruses with the spike glycoprotein of different SARS-CoV-2 variants. In addition, most participants showed strong responses of IFNγ+ or IL-2+ CD8+ and CD4+ T helper type 1 cells that were detectable throughout a nine-week period subsequent to the boost [104]. Epitope-specific CD8+ T cells of early differentiated effector memory phenotype made up 0.02–2.92% of all circulating CD8+ T cells [104]. These cells were also detectable at 0.01–0.28% eight weeks later [104]. This study revealed that BNT162b2 is not only well-tolerated but also able to elicit a strong adaptive humoral and poly-specific cellular immune response against conserved epitopes of a broad range of SARS-CoV-2 variants [104].

Cases of myocarditis, inflammation of the heart muscle, and pericarditis, inflammation of the thin tissue surrounding the heart, the pericardium, have been associated with the two COVID-19 mRNA vaccines. Incidence of myocarditis following vaccination is rare for most vaccines, with the exception of smallpox vaccines in adults, and reliance of patients seeking healthcare makes it difficult to determine the true rate of incidence for mRNA COVID-19 vaccines [105]. Notably, no cases of myocarditis were reported in clinical trials of mRNA vaccines, and adverse cardiac events were reported in less than 0.1% of trial participants, with no difference noted between recipients of vaccine and recipients of placebo [106, 107]. US Centers for Disease Control and Prevention estimates a rate of 12.6 cases of myocarditis per million second doses of an mRNA vaccine [108].

As of June 2021, approximately 296 million doses of mRNA vaccines had been administered in the United States, and 1226 cases of myocarditis had been reported, primarily in young males (median age of 26 years) who had received multiple doses of BNT162b2 or mRNA-1273. A total of 309 of these cases resulted in hospitalization and none in death. After reviewing available evidence on benefit–risk balance of the mRNA vaccines, the Advisory Committee on Immunization Practices (ACIP) determined that the benefits of vaccination outweighed the risks of myocarditis in all populations [109]. In a study focused on members of the US military published in June 2021, a total of 23 cases of myocarditis had been reported after more than 2.8 million doses of mRNA vaccines had been administered. Similar to the general population, most cases occurred in young men (median age of 25 years) who received multiple doses of BNT162b2 (7 cases) or mRNA-1273 (16 cases) [110]. Other studies of myocarditis following COVID-19 vaccination have been reported in the literature and occur predominantly after a second dose of an mRNA vaccine [111–116].

10 Heterogenous Responses to Vaccines

10.1 Antigen Drift, Recall, and Cross-Reactivity

A common question posed by the public is why do some people get sick despite being vaccinated? Answers to this question are usually antigen drift and antibody cross-reactivity. If there is sufficient change in the pathogen (antigen), a person's antibodies and plasma cells may not have enough affinity for efficient recall. In short, the pathogen is treated as if new. Most viruses, including COVID-19, are prone to mutations that lead to genetic drift [117]. This allows the virus to escape recognition by the immune system. This needs to be taken into consideration during vaccine design [118]. As viruses evolve, existing antibodies against an older version of the antigen may not have high enough affinity to stop the infection; however, they may have enough affinity to achieve sufficient binding to prevent severe disease [119]. Protection against infection and prevention of severe disease are different, but both are necessary for long-term protection [120].

Over time, antibody titers normally wane, leaving a minimal level circulating in the blood. If the pathogen is encountered in the future after the titer wanes, the plasma cell will mass produce antibodies again (recall). However, if the antibody titers have waned and antigen drift occurs, recall may not be as effective. If there is insufficient recognition, the immune system may treat the pathogen as a novel infection and restart the acquired immunity process. However, if antibody titers do not wane and there is antigen drift, it still may be able to neutralize some of the pathogens. Having more of the lower affinity antibodies immediately available in the blood can help compensate for the lack of recognition and recall [121]. This is one rationale for boosting people with the same vaccine more frequently during a pandemic, despite the increasing difference between vaccine antigen and circulating antigen. Boosts initiate an artificial recall to increase immediately the amount of lower affinity antibodies to assist in overwhelming evolved, less recognized pathogen. In this way, the higher quantity helps compensate for lower affinity.

The heterogeneity in protection among individuals is also influenced by the antibody generation process itself. Antibodies are unique due to the random process involved in their generation and affinity maturation; thus, some individuals randomly generate antibodies that serendipitously have a high affinity for the evolved pathogen (cross-reactive antibodies). In such cases, the individual's antibodies will have targeted a region unchanged on the antigen. This is the case for individuals with antibodies against an older beta coronavirus that were cross-reactive with the spike protein on SARS-CoV-2 [122]. Similar variability in T-cell responses is also possible because of the random processes used in TCR development.

Generally, these affinity-related issues and the randomness of antibody production and TCR creation are reasons why clinical trials measure vaccine efficacy in multiple ways [119, 120]. The first measure is protection against infection. Infection prevention can occur when high affinity antibodies immediately neutralize the pathogen. A second measure is protection against severe disease, where antibodies may not have high enough affinity to sterilize the infection immediately; however, the antibodies can still impede progress of the pathogen. In such cases, patients can still become infected, but severity of the disease is lower because there is some degree of protection. Thus, when COVID-19 underwent significant antigen drift, antibodies generated by the original vaccine could not prevent outright infection but instead prevented unchecked virus replication and spread.

10.2 Haplotypes, Genetic Variability, and the Impact on Vaccine Performance

Single nucleotide polymorphisms (SNPs) are DNA sequence variations found among individuals resulting from the substitution of one nucleotide base for its counterpart (e.g., C to T or A to G), producing a pair of distinct polymorphisms, or "alleles." Many SNPs are correlated with each other, and teasing apart their individual effect on phenotype is difficult. In contrast, haplotypes are groups of alleles at different locations along the same chromosome inherited as a unit. Fundamentally, the difference between haplotypes and individual genotypes caused by SNPs is that alleles of haplotypes are assigned to a single chromosome. This means that each individual will have two haplotypes for a given stretch of genome— representing their maternal and paternal chromosomes [123]. Therefore, for each protein-coding gene, an individual can produce at most two distinct proteins corresponding to their haplotypes. The folding, kinetics, stability, and other properties of these proteins depend on interactions of a specific combination of amino acids. In turn, haplotypes may alter these combinations and interactions of amino acids, making them directly relevant to biological function [124].

SNPs and haplotypes illustrate the immense genetic variety present in humans and are useful for categorizing large populations into smaller subsets with common genotypic features [125, 126]. Of particular interest to the mRNA COVID-19 vaccines are the haplotypes of the major histocompatibility complex (MHC). The MHC is one of the most gene-dense and polymorphic stretches of the human genome, with thousands of known alleles, and is widely associated with autoimmune and infectious diseases. MHC class I and class II are sets of cell-surface proteins tasked with coordinating the innate and adaptive immune systems through the presentation of peptide antigens.

MHC molecules are loaded with peptide fragments intracellularly. Through a variety of processes, a protein will be degraded into peptides, and this is what will eventually be presented on the outside of the cell on the MHC molecules [127]. The enzymatic machinery that processes the antigens into peptides can vary among individuals. Furthermore, each MHC molecule will have different affinities for certain peptides. MHC molecules have a peptide-binding cleft, which interacts with the amino acids of the peptide. The affinity of this binding interaction is determined by the polymorphic residues lining the cleft and their interaction with the peptide [128]. Lower affinities mean that certain antigen-derived peptides are not stably bound to MHC, and thus not reliably displayed to the immune system. Ultimately, these differences in affinity can result in heterogeneity of antigen presentation across haplotypes, where lower affinity individuals will be unable to reliably display the antigen and drive a robust immune response.

Many associations of MHC with autoimmune conditions have been linked to individual alleles of MHC class I and II, although fewer associations have been found with infectious diseases [129]. However, one study by Mudd et al. [130] investigated T_{FH}) receptors after vaccination with the BNT162b2 vaccine and found that a common human leukocyte antigen (HLA) allele, a region of the human genome that includes the MHC, was immunodominant. This finding suggests that haplotypes should be considered when selecting antigens for mRNA vaccines to ensure that they are not biased toward specific populations.

In practice, these variabilities mean that it is crucial to design clinical trials that include a diverse patient population to capture a range of genotypes. However, evidence shows that White and non-Hispanic individuals are commonly overrepresented in clinical trials [131], and in particular, the clinical trials for mRNA-1273 and BNT162b2 have been criticized for lack of diversity [132]. This lack of diversity considerations in clinical trial design, in combination with the disproportionate effect that the COVID-19 pandemic has had on communities of color [133, 134], highlights the need to improve the relationship between marginalized communities and the scientific community [135].

11 Mixing mRNA Vaccines with Other Vaccine Types

11.1 Heterologous Boost Safety

Homologous boosting is when a follow-up immunization uses the same vaccine. In contrast, a heterologous boost is when a follow-up immunization uses a different type of vaccine. During the pandemic, some regions had access to two major types of vaccines: mRNA–LNPs from Moderna and BioNTech/Pfizer, and adenoviruses modified to express the COVID-19 spike protein (a viral vector vaccine) from AstraZeneca. In response to the early intermittent supply chain shortage of mRNA vaccines and the occurrence of rare adverse events after vaccination with the AstraZeneca ChAdOx1-S nCov-19 viral vector vaccine, a heterologous prime-boost regimen consisting of ChAdOx1 nCov-19 followed by Pfizer-BioNtech BNT162b2 was proposed. The inception of heterologous immunization stemmed from evidence suggesting that younger women and older individuals were more suspectable to severe side effects of ChAdOx1-S nCov-19, such as thrombotic thrombocytopenia syndrome or myocarditis [136–138]. This was potentially caused by the viral nature of ChAdOx1 nCov-19, where the adenovirus vector technology was central for the expression of the SARS-CoV2 spike protein as an antigen. To ensure adequate vaccination of the population, several countries including Germany implemented the heterologous vaccination strategy [139].

A study by Hillus et al. [140] assessed the safety and efficacy of heterologous prime-boost vaccination regimens compared to homologous immunization of BNT162b2 or ChAdOx1 nCov-19. Participants first received a prime vaccination with either BNT162b2 or ChAdOx1 nCov-19. After 3 weeks, BNT162b2-primed participants received a homologous boost vaccination of BNT162b2. Participants who were primed with ChAdOx1 nCov-19 received either a homologous or a heterologous boost vaccination of ChAdOx1 nCov-19 or BMT162b2, respectively, after 10–12 weeks. Heterologous ChAdOx1 nCov-19-BNT162b2 immunization had a higher frequency of local reactions such as pain and tenderness at the injection site; however, the frequency of systemic reactions such as fatigue, myalgia, headaches, feverishness, and chills were lower than in other regimens [140]. This suggested that the heterologous ChAdOx1 nCov-19-BNT162b2 regime did not produce severe adverse effects.

11.2 Heterologous Boost Efficacy and Immunological Outcomes

Recipients of ChAdOx1 nCov-19-BNT162b2 had the highest circulating anti-SARS-CoV-2 RBD, anti-SARS-CoV-2 spike protein S1 domain (S1) IgG levels, and anti-S1 IgA compared to homologous immunization regimens [140]. In addition to antibody levels, the avidity of antibodies, specifically the anti-S1 IgG, was also highest in ChAdOx1 nCov-19-BNT162b2 recipients [140]. The neutralizing capacity of serum antibodies of ChAdOx1 nCov-19-BNT162b2 recipients in inhibiting ACE2-RBD binding against the original and variant (alpha, beta, and gamma) pseudoviruses was highest compared to other immunization regimens [140, 141]. Moreover, differences in T-cell responses were also observed with mixed vaccination regimes. Heterologous ChAdOx1 nCov-19-BNT162b2 immunization exhibited significantly higher levels of T-cell reactivity as measured by the amount of interferon-γ (IFN-γ) secreted, thus suggesting greater T-cell activation [140, 141]. The heterologous prime-boost regimen also significantly increased the frequency of circulating antigen-specific CD4+ and CD8+ T cells and their reactivity toward peptide pools covering the whole spike protein antigen, suggesting a robust and diverse T-cell immunity [141, 142]. Interestingly, a large subgroup of CD8+ T cells are IFN-γ^+ and tumor necrosis factor-α^+ (TNF-α), suggesting the heterologous prime-boost regimen elicited a more robust T-cell response [141]. Lastly, although insignificant, the frequency of CD19$^+$ CD20$^+$ B cells that carried membrane-bound immunoglobulin (B-cell receptor) for the spike protein was elevated compared to that in the homologous prime-boost regimen [141]. This potentially suggests that a heterologous prime-boost regimen can enhance B-cell responses.

12 Outlook for mRNA Vaccines

12.1 Multivalent Vaccines for COVID-19

The emergence of SARS-CoV-2 variants of concerns and risks of decreased neutralization generates a need for assessing booster doses and variant-specific vaccines [143]. Specifically, in an interim analysis by Choi et al. [144], they evaluated the safety and immunogenicity of a single booster dose of mRNA-1273 or variant-modified mRNAs, such as multivalent mRNA-1273.211, for use in humans. Results showed that immediately prior to boost dose, neutralizing antibodies against wild-type SARS-CoV-2 were decreased relative to peak titers against the same strain measured one month after the primary series [144]. Neutralization titers for beta, gamma, and delta variants were either low or undetectable [144]. Notably, following booster vaccination with the multivalent mRNA-1273.211, neutralization titers significantly increased against the wild-type SARS-CoV-2, and epsilon and iota SARS-CoV-2 variants relative to peak titers against wild-type one month after the primary series [144]. In addition, neutralization titers for the beta, gamma, and delta variants were comparable to peak titers measured after the primary vaccine series against wild-type SARS-CoV-2 [144]. Thus, these results showcased the potential for multivalent vaccines to offer broader protection that will be valuable for the development of new mRNA vaccines.

12.2 mRNA Vaccines Beyond COVID-19

By reaching a new milestone in clinical efficacy and receiving approval for emergency use against COVID-19, mRNA vaccines will likely be central to the rapid development of mRNA-based vaccines against a plethora of infectious diseases. For instance, mRNA vaccines can be used for influenza vaccine development. A key challenge for influenza vaccine development is that certain strains of influenza virus appear late in the season after manufacturers have already begun producing the vaccine virus [145]. Thus, an mRNA-based platform that reduces vaccine production timeline will be valuable in enabling more data collection to take place before strain selection. The mRNA payload within this platform may also offer generic manufacturing processes, whereby one only needs to switch mRNA construct designs while maintaining the same delivery system [146]. Thus, this substantially reduces the resources required for vaccine manufacturing for seasonal vaccines. Alternatively, multivalent influenza vaccines are also being explored to provide more broad-spectrum protection against this rapidly evolving virus [147]. In one example, prototype seasonal influenza vaccines consist of three or four hemagglutinin (HA) antigens from different viral subtypes. Additionally, neuraminidase (NA), a tetrameric membrane protein found in influenza virus, is known to be linked to protective immunity against severe viral disease [147]. These formulations generated strong functional antibody and cellular responses in nonhuman primates and were protective against viral challenges in mice [147].

mRNA vaccines can potentially be used to treat infection with chikungunya virus (CHIKV), which is causing substantial morbidity worldwide due to its spread via mosquitoes [148]. This infection can lead to acute illness characterized by fever, rash, and arthralgia [148]. CHIKV infection can also result in chronic arthritis or death [148]. Since it remains to be a challenge to implement antibody therapy for the treatment of CHIKV infection, Kose et al. explored the development of lipid-encapsulated mRNA encoding a potent neutralizing human monoclonal antibody to treat CHIKV infection [149]. Treatment of mice with LNPs expressing the CHIKV monoclonal antibody led to protection from arthritis, musculoskeletal tissue infection, and death [149]. In addition, studies in macaques demonstrated a dose–response effect following one dose of mRNA, and this response was maintained after the second dose [149]. Together, the preclinical data suggest the use of an mRNA vaccine platform to protect against CHIKV infection.

mRNA cancer vaccines have been in development for many years and are continuing their progression toward clinical use. For cancer, the aim is to neutralize tumor-promoting inflammation to modify the tumor microenvironment that regulates T-cell activity and broadens the T-cell repertoire. Oberli et al. developed an LNP formulation as a vaccine to treat aggressive forms of melanoma, where their formulation incorporated ionizable lipid, phospholipid, cholesterol, lipid-PEG, and an additive for the delivery of mRNA vaccines [150]. When mice were injected with melanoma cells and treated with the LNPs, they showed slower tumor growth and shrinkage and higher levels of CD8+ T cells up to 7 days after treatment [150]. This study was a proof-of-concept that mRNA delivered via LNPs may be efficacious for treating cancers.

Other types of ribonucleic acids are also being explored for vaccination. Self-amplifying ribonucleic acids (saRNA) are long mRNA molecules that encode two items. The first is a polymerase, and the second is an antigen. When the ribosome translates the saRNA, it makes both the polymerase and the antigen. The polymerase then makes copies of the saRNA, which the ribosome translates into more polymerases and antigens, thereby creating self-amplification. These saRNA molecules were used to develop preclinical vaccines against Ebola, a deadly virus that causes hemorrhagic fever, and *Toxoplasma gondii*, an incurable intracellular parasite. Both vaccines were 100% protective in mice that were exposed to lethal levels of Ebola or *T. gondii* [151].

Trans-amplifying ribonucleic acid (taRNA) systems are also being studied. Similar to saRNA, the taRNA system encodes a polymerase and an antigen. However, unlike saRNA, each component is encoded in separate mRNA molecules. While saRNA is a very large molecule, each mRNA in the taRNA system is shorter, which may improve manufacturability. In mouse preclinical studies, this technology mounted a protective immune response against infection with live influenza virus [152].

A potential obstacle for the development of mRNA vaccines against nonviral pathogens is post-translational modification. When the ribosome in a cell makes a protein (translation), the proteins can also be modified post-translation to become fully functional. The repertoire of post-translational modifications a human cell performs can be different from other forms of life, such as bacteria [153]. Viruses are a special case because of the use of the infected person's cells to make more virus copies. Thus, the proteins that the virus needs to create more copies of itself, including post-translational modifications, are fully compatible with human protein production pathways. Accordingly, when an mRNA that encodes a viral antigen is used as a vaccine, the antigen can be correctly produced. In contrast, some parasites, such as extracellular bacteria or flatworms, do not use human cells to replicate themselves, and thus do not rely on human cells for protein translation or post-translational modifications. As a result, some of their post-translational modifications are exotic and cannot be performed by human cells. Thus, if one of their antigens are encoded as an mRNA and used in a vaccine, the resulting protein may not be correctly processed and so will appear different. Thus, when expanding mRNA vaccines beyond viruses, careful consideration must be given to the type of antigen that is being encoded in the mRNA and the final form of that antigen. Ultimately, the immune system must be shown an antigen that has enough fidelity to the pathogen's protein to ensure immune system recall after exposure post-vaccination.

13 Conclusion

As a vaccination tool, mRNA has been proven to be safe, efficacious, adaptable, and capable of rapidly responding to ever-changing pathogens. When it comes to creating protective vaccines against viruses, these types of vaccines have shown outstanding utility in the clinic and across the world. With the successful debut of this technology during the COVID-19 pandemic, we can expect new efforts to adapt it for pathogens and diseases that are not viral in nature, such as cancers, bacterial infections, intracellular parasitic protozoans, and perhaps even multicellular organisms like immunity-evading flatworms.

References

1. Yatim KM, Lakkis FG (2015) A brief journey through the immune system. CJASN 10:1274–1281. https://doi.org/10.2215/CJN.10031014
2. Rothenberg EV (2012) Transcriptional drivers of the T-cell lineage program. Curr Opin Immunol 24:132–138. https://doi.org/10.1016/j.coi.2011.12.012
3. Borowski C, Martin C, Gounari F, Haughn L, Aifantis I, Grassi F, von Boehmer H (2002) On the brink of becoming a T cell. Curr Opin Immunol 14:200–206. https://doi.org/10.1016/S0952-7915(02)00322-9
4. Pang SS, Berry R, Chen Z, Kjer-Nielsen L, Perugini MA, King GF, Wang C, Chew SH, La Gruta NL, Williams NK, Beddoe T, Tiganis T, Cowieson NP, Godfrey DI, Purcell AW, Wilce MCJ, McCluskey J, Rossjohn J (2010) The structural basis for autonomous dimerization of the pre-T-cell antigen receptor. Nature 467:844–848. https://doi.org/10.1038/nature09448
5. Merkenschlager M, Graf D, Lovatt M, Bommhardt U, Zamoyska R, Fisher AG (1997) How many thymocytes audition for selection? J Exp Med 186:1149–1158. https://doi.org/10.1084/jem.186.7.1149
6. Singer A, Adoro S, Park J-H (2008) Lineage fate and intense debate: myths, models and mechanisms of CD4- versus CD8-lineage choice. Nat Rev Immunol 8:788–801. https://doi.org/10.1038/nri2416
7. Kishimoto H, Sprent J (1997) Negative selection in the thymus includes semimature T cells. J Exp Med 185:263–272. https://doi.org/10.1084/jem.185.2.263
8. Germain RN (1994) MHC-dependent antigen processing and peptide presentation: providing ligands for T lymphocyte activation. Cell 76:287–299. https://doi.org/10.1016/0092-8674(94)90336-0
9. Smith-Garvin JE, Koretzky GA, Jordan MS (2009) T cell activation. Annu Rev Immunol 27:591–619. https://doi.org/10.1146/annurev.immunol.021908.132706
10. Busslinger M (2004) Transcriptional control of early B cell development. Annu Rev Immunol 22:55–79. https://doi.org/10.1146/annurev.immunol.22.012703.104807
11. Geier JK, Schlissel MS (2006) Pre-BCR signals and the control of Ig gene rearrangements. Semin Immunol 18:31–39. https://doi.org/10.1016/j.smim.2005.11.001
12. Cornall RJ, Goodnow CC, Cyster JG (1995) The regulation of self-reactive B cells. Curr Opin Immunol 7:804–811. https://doi.org/10.1016/0952-7915(95)80052-2
13. Goodnow CC, Crosbie J, Jorgensen H, Brink RA, Basten A (1989) Induction of self-tolerance in mature peripheral B lymphocytes. Nature 342:385–391. https://doi.org/10.1038/342385a0
14. Padlan EA (1994) Anatomy of the antibody molecule. Mol Immunol 31:169–217. https://doi.org/10.1016/0161-5890(94)90001-9
15. Crotty S (2014) T follicular helper cell differentiation, function, and roles in disease. Immunity 41:529–542. https://doi.org/10.1016/j.immuni.2014.10.004
16. Okada T, Cyster JG (2006) B cell migration and interactions in the early phase of antibody responses. Curr Opin Immunol 18:278–285. https://doi.org/10.1016/j.coi.2006.02.005
17. Allen CDC, Okada T, Cyster JG (2007) Germinal-center organization and cellular dynamics. Immunity 27:190–202. https://doi.org/10.1016/j.immuni.2007.07.009
18. Li Z, Woo CJ, Iglesias-Ussel MD, Ronai D, Scharff MD (2004) The generation of antibody diversity through somatic hypermutation and class switch recombination. Genes Dev 18:1–11. https://doi.org/10.1101/gad.1161904
19. Matthews AJ, Zheng S, DiMenna LJ, Chaudhuri J (2014) Regulation of immunoglobulin class-switch recombination. In: Advances in immunology, vol 122. Elsevier, pp 1–57. https://doi.org/10.1016/B978-0-12-800267-4.00001-8

20. Shlomchik MJ, Weisel F (2012) Germinal center selection and the development of memory B and plasma cells: germinal center differentiation and selection. Immunol Rev 247:52–63. https://doi.org/10.1111/j.1600-065X.2012.01124.x
21. Zepp F (2016) Principles of vaccination. In: Thomas S (ed) Vaccine design: methods and protocols: volume 1: Vaccines for human diseases. Springer, New York, pp 57–84. https://doi.org/10.1007/978-1-4939-3387-7_3
22. Kumar H, Kawai T, Akira S (2011) Pathogen recognition by the innate immune system. Int Rev Immunol 30:16–34. https://doi.org/10.3109/08830185.2010.529976
23. Kawai T, Akira S (2009) The roles of TLRs, RLRs and NLRs in pathogen recognition. Int Immunol 21:317–337. https://doi.org/10.1093/intimm/dxp017
24. Turvey SE, Broide DH (2010) Innate immunity. J Allergy Clin Immunol 125:S24–S32. https://doi.org/10.1016/j.jaci.2009.07.016
25. Neefjes J, Jongsma MLM, Paul P, Bakke O (2011) Towards a systems understanding of MHC class I and MHC class II antigen presentation. Nat Rev Immunol 11:823–836. https://doi.org/10.1038/nri3084
26. Albanese M, Tagawa T, Bouvet M, Maliqi L, Lutter D, Hoser J, Hastreiter M, Hayes M, Sugden B, Martin L, Moosmann A, Hammerschmidt W (2016) Epstein–Barr virus microRNAs reduce immune surveillance by virus-specific CD8+ T cells. Proc Natl Acad Sci 113:E6467–E6475. https://doi.org/10.1073/pnas.1605884113
27. Labbé K, Saleh M (2008) Cell death in the host response to infection. Cell Death Differ 15:1339–1349. https://doi.org/10.1038/cdd.2008.91
28. Jørgensen KW, Buus S, Nielsen M (2010) Structural properties of MHC class II ligands, implications for the prediction of MHC class II epitopes. PLOS One 5:e15877. https://doi.org/10.1371/journal.pone.0015877
29. Chen H, Zhang Y, Ye AY, Du Z, Xu M, Lee C-S, Hwang JK, Kyritsis N, Ba Z, Neuberg D, Littman DR, Alt FW (2020) BCR selection and affinity maturation in Peyer's patch germinal centres. Nature 582:421–425. https://doi.org/10.1038/s41586-020-2262-4
30. Hampton HR, Chtanova T (2019) Lymphatic migration of immune cells. Front Immunol 10. https://www.frontiersin.org/articles/10.3389/fimmu.2019.01168. Accessed 3 Nov 2022
31. Wolff JA, Malone RW, Williams P, Chong W, Acsadi G, Jani A, Felgner PL (1990) Direct gene transfer into mouse muscle in vivo. Science 247:1465–1468. https://doi.org/10.1126/science.1690918
32. Pardi N, Hogan MJ, Porter FW, Weissman D (2018) mRNA vaccines—a new era in vaccinology. Nat Rev Drug Discov 17:261–279. https://doi.org/10.1038/nrd.2017.243
33. Sahin U, Karikó K, Türeci Ö (2014) mRNA-based therapeutics—developing a new class of drugs. Nat Rev Drug Discov 13:759–780. https://doi.org/10.1038/nrd4278
34. Chaudhary N, Weissman D, Whitehead KA (2021) mRNA vaccines for infectious diseases: principles, delivery and clinical translation. Nat Rev Drug Discov 20:817–838. https://doi.org/10.1038/s41573-021-00283-5
35. Weissman D (2015) mRNA transcript therapy. Expert Rev Vaccines 14:265–281. https://doi.org/10.1586/14760584.2015.973859
36. Karikó K, Muramatsu H, Welsh FA, Ludwig J, Kato H, Akira S, Weissman D (2008) Incorporation of pseudouridine into mRNA yields superior nonimmunogenic vector with increased translational capacity and biological stability. Mol Ther 16:1833–1840. https://doi.org/10.1038/mt.2008.200
37. Corbett KS, Flynn B, Foulds KE, Francica JR, Boyoglu-Barnum S, Werner AP, Flach B, O'Connell S, Bock KW, Minai M, Nagata BM, Andersen H, Martinez DR, Noe AT, Douek N, Donaldson MM, Nji NN, Alvarado GS, Edwards DK, Flebbe DR, Lamb E, Doria-Rose NA, Lin BC, Louder MK, O'Dell S, Schmidt SD, Phung E, Chang LA, Yap C, Todd J-PM, Pessaint L, Van Ry A, Browne S, Greenhouse J, Putman-Taylor T, Strasbaugh A, Campbell T-A, Cook A, Dodson A, Steingrebe K, Shi W, Zhang Y, Abiona OM, Wang L, Pegu A, Yang ES, Leung K, Zhou T, Teng I-T, Widge A, Gordon I, Novik L, Gillespie RA, Loomis RJ, Moliva JI, Stewart-Jones G, Himansu S, Kong W-P, Nason MC, Morabito KM, Ruckwardt TJ, Ledgerwood JE, Gaudinski MR, Kwong PD, Mascola JR, Carfi A, Lewis MG, Baric

RS, McDermott A, Moore IN, Sullivan NJ, Roederer M, Seder RA, Graham BS (2020) Evaluation of the mRNA-1273 vaccine against SARS-CoV-2 in nonhuman primates. N Engl J Med 383:1544–1555. https://doi.org/10.1056/NEJMoa2024671
38. Vogel AB, Kanevsky I, Che Y, Swanson KA, Muik A, Vormehr M, Kranz LM, Walzer KC, Hein S, Güler A, Loschko J, Maddur MS, Ota-Setlik A, Tompkins K, Cole J, Lui BG, Ziegenhals T, Plaschke A, Eisel D, Dany SC, Fesser S, Erbar S, Bates F, Schneider D, Jesionek B, Sänger B, Wallisch A-K, Feuchter Y, Junginger H, Krumm SA, Heinen AP, Adams-Quack P, Schlereth J, Schille S, Kröner C, de la Caridad Güimil Garcia R, Hiller T, Fischer L, Sellers RS, Choudhary S, Gonzalez O, Vascotto F, Gutman MR, Fontenot JA, Hall-Ursone S, Brasky K, Griffor MC, Han S, Su AAH, Lees JA, Nedoma NL, Mashalidis EH, Sahasrabudhe PV, Tan CY, Pavliakova D, Singh G, Fontes-Garfias C, Pride M, Scully IL, Ciolino T, Obregon J, Gazi M, Carrion R, Alfson KJ, Kalina WV, Kaushal D, Shi P-Y, Klamp T, Rosenbaum C, Kuhn AN, Türeci Ö, Dormitzer PR, Jansen KU, Sahin U (2021) BNT162b vaccines protect rhesus macaques from SARS-CoV-2. Nature 592:283–289. https://doi.org/10.1038/s41586-021-03275-y
39. Bangham AD, Standish MM, Watkins JC (1965) Diffusion of univalent ions across the lamellae of swollen phospholipids. J Mol Biol 13:238-IN27. https://doi.org/10.1016/S0022-2836(65)80093-6
40. Fraley R, Straubinger RM, Rule G, Springer EL, Papahadjopoulos D (1981) Liposome-mediated delivery of deoxyribonucleic acid to cells: enhanced efficiency of delivery by changes in lipid composition and incubation conditions. Biochemistry 20:6978–6987. https://doi.org/10.1021/bi00527a031
41. Fraley R, Subramani S, Berg P, Papahadjopoulos D (1980) Introduction of liposome-encapsulated SV40 DNA into cells. J Biol Chem 255:10431–10435
42. Felgner PL, Gadek TR, Holm M, Roman R, Chan HW, Wenz M, Northrop JP, Ringold GM, Danielsen M (1987) Lipofection: a highly efficient, lipid-mediated DNA-transfection procedure. PNAS 84:7413–7417. https://doi.org/10.1073/pnas.84.21.7413
43. Mizuguchi H, Nakagawa T, Nakanishi M, Imazu S, Nakagawa S, Mayumi T (1996) Efficient gene transfer into mammalian cells using Fusogenic liposome. Biochem Biophys Res Commun 218:402–407. https://doi.org/10.1006/bbrc.1996.0070
44. Felgner JH, Kumar R, Sridhar CN, Wheeler CJ, Tsai YJ, Border R, Ramsey P, Martin M, Felgner PL (1994) Enhanced gene delivery and mechanism studies with a novel series of cationic lipid formulations. J Biol Chem 269:2550–2561
45. Mok KW, Cullis PR (1997) Structural and fusogenic properties of cationic liposomes in the presence of plasmid DNA. Biophys J 73:2534–2545. https://doi.org/10.1016/S0006-3495(97)78282-1
46. Lappalainen K, Jääskeläinen I, Syrjänen K, Urtti A, Syrjänen S (1994) Comparison of cell proliferation and toxicity assays using two cationic liposomes. Pharm Res 11:1127–1131. https://doi.org/10.1023/A:1018932714745
47. Filion MC, Phillips NC (1997) Toxicity and immunomodulatory activity of liposomal vectors formulated with cationic lipids toward immune effector cells, Biochimica et Biophysica Acta (BBA). Biomembranes 1329:345–356. https://doi.org/10.1016/S0005-2736(97)00126-0
48. Bailey AL, Cullis PR (1994) Modulation of membrane fusion by asymmetric transbilayer distributions of amino lipids. Biochemistry 33:12573–12580. https://doi.org/10.1021/bi00208a007
49. Maurer N, Wong KF, Stark H, Louie L, McIntosh D, Wong T, Scherrer P, Semple SC, Cullis PR (2001) Spontaneous entrapment of polynucleotides upon electrostatic interaction with ethanol-destabilized cationic liposomes. Biophys J 80:2310–2326. https://doi.org/10.1016/S0006-3495(01)76202-9
50. Semple SC, Klimuk SK, Harasym TO, Dos Santos N, Ansell SM, Wong KF, Maurer N, Stark H, Cullis PR, Hope MJ, Scherrer P (2001) Efficient encapsulation of antisense oligonucleotides in lipid vesicles using ionizable aminolipids: formation of novel small multilamellar vesicle structures, Biochimica et Biophysica Acta (BBA). Biomembranes 1510:152–166. https://doi.org/10.1016/S0005-2736(00)00343-6

51. Han X, Zhang H, Butowska K, Swingle KL, Alameh M-G, Weissman D, Mitchell MJ (2021) An ionizable lipid toolbox for RNA delivery. Nat Commun 12:1–6. https://doi.org/10.1038/s41467-021-27493-0
52. Hou X, Zaks T, Langer R, Dong Y (2021) Lipid nanoparticles for mRNA delivery. Nat Rev Mater 6:1079–1094. https://doi.org/10.1038/s41578-021-00358-0
53. Akinc A, Zumbuehl A, Goldberg M, Leshchiner ES, Busini V, Hossain N, Bacallado SA, Nguyen DN, Fuller J, Alvarez R, Borodovsky A, Borland T, Constien R, de Fougerolles A, Dorkin JR, Narayanannair Jayaprakash K, Jayaraman M, John M, Koteliansky V, Manoharan M, Nechev L, Qin J, Racie T, Raitcheva D, Rajeev KG, Sah DWY, Soutschek J, Toudjarska I, Vornlocher H-P, Zimmermann TS, Langer R, Anderson DG (2008) A combinatorial library of lipid-like materials for delivery of RNAi therapeutics. Nat Biotechnol 26:561–569. https://doi.org/10.1038/nbt1402
54. Semple SC, Akinc A, Chen J, Sandhu AP, Mui BL, Cho CK, Sah DWY, Stebbing D, Crosley EJ, Yaworski E, Hafez IM, Dorkin JR, Qin J, Lam K, Rajeev KG, Wong KF, Jeffs LB, Nechev L, Eisenhardt ML, Jayaraman M, Kazem M, Maier MA, Srinivasulu M, Weinstein MJ, Chen Q, Alvarez R, Barros SA, De S, Klimuk SK, Borland T, Kosovrasti V, Cantley WL, Tam YK, Manoharan M, Ciufolini MA, Tracy MA, de Fougerolles A, Mac Lachlan I, Cullis PR, Madden TD, Hope MJ (2010) Rational design of cationic lipids for siRNA delivery. Nat Biotechnol 28:172–176. https://doi.org/10.1038/nbt.1602
55. Jayaraman M, Ansell SM, Mui BL, Tam YK, Chen J, Du X, Butler D, Eltepu L, Matsuda S, Narayanannair JK, Rajeev KG, Hafez IM, Akinc A, Maier MA, Tracy MA, Cullis PR, Madden TD, Manoharan M, Hope MJ (2012) Maximizing the potency of siRNA lipid nanoparticles for hepatic gene silencing in vivo. Angew Chem 124:8657–8661. https://doi.org/10.1002/ange.201203263
56. Akinc A, Goldberg M, Qin J, Dorkin JR, Gamba-Vitalo C, Maier M, Jayaprakash KN, Jayaraman M, Rajeev KG, Manoharan M, Koteliansky V, Röhl I, Leshchiner ES, Langer R, Anderson DG (2009) Development of Lipidoid–siRNA formulations for systemic delivery to the liver. Mol Ther 17:872–879. https://doi.org/10.1038/mt.2009.36
57. Akinc A, Querbes W, De S, Qin J, Frank-Kamenetsky M, Jayaprakash KN, Jayaraman M, Rajeev KG, Cantley WL, Dorkin JR, Butler JS, Qin L, Racie T, Sprague A, Fava E, Zeigerer A, Hope MJ, Zerial M, Sah DW, Fitzgerald K, Tracy MA, Manoharan M, Koteliansky V, de Fougerolles A, Maier MA (2010) Targeted delivery of RNAi therapeutics with endogenous and exogenous ligand-based mechanisms. Mol Ther 18:1357–1364. https://doi.org/10.1038/mt.2010.85
58. Whitehead KA, Dorkin JR, Vegas AJ, Chang PH, Veiseh O, Matthews J, Fenton OS, Zhang Y, Olejnik KT, Yesilyurt V, Chen D, Barros S, Klebanov B, Novobrantseva T, Langer R, Anderson DG (2014) Degradable lipid nanoparticles with predictable in vivo siRNA delivery activity. Nat Commun 5:4277. https://doi.org/10.1038/ncomms5277
59. Dahlman JE, Barnes C, Khan OF, Thiriot A, Jhunjunwala S, Shaw TE, Xing Y, Sager HB, Sahay G, Speciner L, Bader A, Bogorad RL, Yin H, Racie T, Dong Y, Jiang S, Seedorf D, Dave A, Singh Sandhu K, Webber MJ, Novobrantseva T, Ruda VM, Lytton-Jean AKR, Levins CG, Kalish B, Mudge DK, Perez M, Abezgauz L, Dutta P, Smith L, Charisse K, Kieran MW, Fitzgerald K, Nahrendorf M, Danino D, Tuder RM, von Andrian UH, Akinc A, Panigrahy D, Schroeder A, Koteliansky V, Langer R, Anderson DG (2014) In vivo endothelial siRNA delivery using polymeric nanoparticles with low molecular weight. Nature Nanotech 9:648–655. https://doi.org/10.1038/nnano.2014.84
60. Hajj KA, Ball RL, Deluty SB, Singh SR, Strelkova D, Knapp CM, Whitehead KA (2019) Branched-tail lipid nanoparticles potently deliver mRNA in vivo due to enhanced ionization at endosomal pH. Small 15:1805097. https://doi.org/10.1002/smll.201805097
61. Ball RL, Hajj KA, Vizelman J, Bajaj P, Whitehead KA (2018) Lipid nanoparticle formulations for enhanced co-delivery of siRNA and mRNA. Nano Lett 18:3814–3822. https://doi.org/10.1021/acs.nanolett.8b01101
62. Akinc A, Maier MA, Manoharan M, Fitzgerald K, Jayaraman M, Barros S, Ansell S, Du X, Hope MJ, Madden TD, Mui BL, Semple SC, Tam YK, Ciufolini M, Witzigmann D, Kulkarni

JA, van der Meel R, Cullis PR (2019) The Onpattro story and the clinical translation of nanomedicines containing nucleic acid-based drugs. Nat Nanotechnol 14:1084–1087. https://doi.org/10.1038/s41565-019-0591-y

63. Adams D, Gonzalez-Duarte A, O'Riordan WD, Yang C-C, Ueda M, Kristen AV, Tournev I, Schmidt HH, Coelho T, Berk JL, Lin K-P, Vita G, Attarian S, Planté-Bordeneuve V, Mezei MM, Campistol JM, Buades J, Brannagan TH, Kim BJ, Oh J, Parman Y, Sekijima Y, Hawkins PN, Solomon SD, Polydefkis M, Dyck PJ, Gandhi PJ, Goyal S, Chen J, Strahs AL, Nochur SV, Sweetser MT, Garg PP, Vaishnaw AK, Gollob JA, Suhr OB (2018) Patisiran, an RNAi therapeutic, for hereditary transthyretin amyloidosis. N Engl J Med 379:11–21. https://doi.org/10.1056/NEJMoa1716153
64. Sabnis S, Kumarasinghe ES, Salerno T, Mihai C, Ketova T, Senn JJ, Lynn A, Bulychev A, McFadyen I, Chan J, Almarsson Ö, Stanton MG, Benenato KE (2018) A novel amino lipid series for mRNA delivery: improved endosomal escape and sustained pharmacology and safety in non-human primates. Mol Ther 26:1509–1519. https://doi.org/10.1016/j.ymthe.2018.03.010
65. Hassett KJ, Benenato KE, Jacquinet E, Lee A, Woods A, Yuzhakov O, Himansu S, Deterling J, Geilich BM, Ketova T, Mihai C, Lynn A, McFadyen I, Moore MJ, Senn JJ, Stanton MG, Almarsson Ö, Ciaramella G, Brito LA (2019) Optimization of lipid nanoparticles for intramuscular administration of mRNA vaccines. Mol Ther Nucleic Acids 15:1–11. https://doi.org/10.1016/j.omtn.2019.01.013
66. Liang F, Lindgren G, Lin A, Thompson EA, Ols S, Röhss J, John S, Hassett K, Yuzhakov O, Bahl K, Brito LA, Salter H, Ciaramella G, Loré K (2017) Efficient targeting and activation of antigen-presenting cells in vivo after modified mRNA vaccine administration in Rhesus Macaques. Mol Ther 25:2635–2647. https://doi.org/10.1016/j.ymthe.2017.08.006
67. Karikó K, Buckstein M, Ni H, Weissman D (2005) Suppression of RNA recognition by toll-like receptors: the impact of nucleoside modification and the evolutionary origin of RNA. Immunity 23:165–175. https://doi.org/10.1016/j.immuni.2005.06.008
68. Anderson BR, Muramatsu H, Nallagatla SR, Bevilacqua PC, Sansing LH, Weissman D, Karikó K (2010) Incorporation of pseudouridine into mRNA enhances translation by diminishing PKR activation. Nucleic Acids Res 38:5884–5892. https://doi.org/10.1093/nar/gkq347
69. Nelson J, Sorensen EW, Mintri S, Rabideau AE, Zheng W, Besin G, Khatwani N, Su SV, Miracco EJ, Issa WJ, Hoge S, Stanton MG, Joyal JL (2020) Impact of mRNA chemistry and manufacturing process on innate immune activation. Sci Advances 6:eaaz6893. https://doi.org/10.1126/sciadv.aaz6893
70. Tilstra G, Couture-Senécal J, Lau YMA, Manning AM, Wong DSM, Janaeska WW, Wuraola TA, Pang J, Khan OF (2023) Iterative design of ionizable lipids for intramuscular mRNA delivery. J Am Chem Soc 145:2294–2304. https://doi.org/10.1021/jacs.2c10670
71. Ansell SM, Du X (2022) Novel lipids and lipid nanoparticle formulations for delivery of nucleic acids. US20220072155A1. https://patents.google.com/patent/US20220072155A1/en. Accessed 4 Oct 2022
72. Heyes J, Palmer L, Bremner K, MacLachlan I (2005) Cationic lipid saturation influences intracellular delivery of encapsulated nucleic acids. J Control Release 107:276–287. https://doi.org/10.1016/j.jconrel.2005.06.014
73. Ndeupen S, Qin Z, Jacobsen S, Bouteau A, Estanbouli H, Igyártó BZ (2021) The mRNA-LNP platform's lipid nanoparticle component used in preclinical vaccine studies is highly inflammatory. IScience 24:103479. https://doi.org/10.1016/j.isci.2021.103479
74. Alameh M-G, Tombácz I, Bettini E, Lederer K, Sittplangkoon C, Wilmore JR, Gaudette BT, Soliman OY, Pine M, Hicks P, Manzoni TB, Knox JJ, Johnson JL, Laczkó D, Muramatsu H, Davis B, Meng W, Rosenfeld AM, Strohmeier S, Lin PJC, Mui BL, Tam YK, Karikó K, Jacquet A, Krammer F, Bates P, Cancro MP, Weissman D, Luning Prak ET, Allman D, Locci M, Pardi N (2021) Lipid nanoparticles enhance the efficacy of mRNA and protein subunit vaccines by inducing robust T follicular helper cell and humoral responses. Immunity 54:2877–2892.e7. https://doi.org/10.1016/j.immuni.2021.11.001

75. Tahtinen S, Tong A-J, Himmels P, Oh J, Paler-Martinez A, Kim L, Wichner S, Oei Y, McCarron MJ, Freund EC, Amir ZA, de la Cruz CC, Haley B, Blanchette C, Schartner JM, Ye W, Yadav M, Sahin U, Delamarre L, Mellman I (2022) IL-1 and IL-1ra are key regulators of the inflammatory response to RNA vaccines. Nat Immunol 23:532–542. https://doi.org/10.1038/s41590-022-01160-y
76. Ndeupen S, Bouteau A, Herbst C, Qin Z, Jacobsen S, Powers NE, Hutchins Z, Kurup D, Diba LZ, Watson M, Ramage H, Igyártó BZ (2022) Langerhans cells and cDC1s play redundant roles in mRNA-LNP induced protective anti-influenza and anti-SARS-CoV-2 immune responses. PLoS Pathog 18:e1010255. https://doi.org/10.1371/journal.ppat.1010255
77. Igyártó BZ, Jacobsen S, Ndeupen S (2021) Future considerations for the mRNA-lipid nanoparticle vaccine platform. Curr Opin Virol 48:65–72. https://doi.org/10.1016/j.coviro.2021.03.008
78. Szebeni J, Storm G, Ljubimova JY, Castells M, Phillips EJ, Turjeman K, Barenholz Y, Crommelin DJA, Dobrovolskaia MA (2022) Applying lessons learned from nanomedicines to understand rare hypersensitivity reactions to mRNA-based SARS-CoV-2 vaccines. Nat Nanotechnol 17:337–346. https://doi.org/10.1038/s41565-022-01071-x
79. Moghimi SM, Simberg D (2022) Pro-inflammatory concerns with lipid nanoparticles. Mol Ther 30:2109–2110. https://doi.org/10.1016/j.ymthe.2022.04.011
80. Kobiyama K, Ishii KJ (2022) Making innate sense of mRNA vaccine adjuvanticity. Nat Immunol 23:474–476. https://doi.org/10.1038/s41590-022-01168-4
81. Yang S-T, Kreutzberger AJB, Lee J, Kiessling V, Tamm LK (2016) The role of cholesterol in membrane fusion. Chem Phys Lipids 199:136–143. https://doi.org/10.1016/j.chemphyslip.2016.05.003
82. Patel S, Ashwanikumar N, Robinson E, Xia Y, Mihai C, Griffith JP, Hou S, Esposito AA, Ketova T, Welsher K, Joyal JL, Almarsson Ö, Sahay G (2020) Naturally-occurring cholesterol analogues in lipid nanoparticles induce polymorphic shape and enhance intracellular delivery of mRNA. Nat Commun 11:983. https://doi.org/10.1038/s41467-020-14527-2
83. Kulkarni JA, Darjuan MM, Mercer JE, Chen S, van der Meel R, Thewalt JL, Tam YYC, Cullis PR (2018) On the formation and morphology of lipid nanoparticles containing ionizable cationic lipids and siRNA. ACS Nano 12:4787–4795. https://doi.org/10.1021/acsnano.8b01516
84. Pniewska E, Pawliczak R (2013) The involvement of phospholipases A2 in asthma and chronic obstructive pulmonary disease. Mediat Inflamm 2013:793505. https://doi.org/10.1155/2013/793505
85. Besin G, Milton J, Sabnis S, Howell R, Mihai C, Burke K, Benenato KE, Stanton M, Smith P, Senn J, Hoge S (2019) Accelerated blood clearance of lipid nanoparticles entails a biphasic humoral response of B-1 followed by B-2 lymphocytes to distinct antigenic moieties. Immuno Horizons 3:282–293. https://doi.org/10.4049/immunohorizons.1900029
86. Dilliard SA, Cheng Q, Siegwart DJ (2021) On the mechanism of tissue-specific mRNA delivery by selective organ targeting nanoparticles. PNAS 118. https://doi.org/10.1073/pnas.2109256118
87. Moghimi SM (2021) Allergic reactions and anaphylaxis to lipid nanoparticle-based COVID-19 vaccines. Mol Ther 29:898–900. https://doi.org/10.1016/j.ymthe.2021.01.030
88. Carreño JM, Singh G, Tcheou J, Srivastava K, Gleason C, Muramatsu H, Desai P, Aberg JA, Miller RL, P. Study Group, Pardi N, Simon V, Krammer F (2022) mRNA-1273 but not BNT162b2 induces antibodies against polyethylene glycol (PEG) contained in mRNA-based vaccine formulations. Vaccine 40:6114–6124. https://doi.org/10.1016/j.vaccine.2022.08.024
89. Ju Y, Lee WS, Pilkington EH, Kelly HG, Li S, Selva KJ, Wragg KM, Subbarao K, Nguyen THO, Rowntree LC, Allen LF, Bond K, Williamson DA, Truong NP, Plebanski M, Kedzierska K, Mahanty S, Chung AW, Caruso F, Wheatley AK, Juno JA, Kent SJ (2022) Anti-PEG antibodies boosted in humans by SARS-CoV-2 lipid nanoparticle mRNA vaccine. ACS Nano 16:11769–11780. https://doi.org/10.1021/acsnano.2c04543
90. Guerrini G, Gioria S, Sauer AV, Lucchesi S, Montagnani F, Pastore G, Ciabattini A, Medaglini D, Calzolai L (2022) Monitoring anti-PEG antibodies level upon repeated lipid nanoparticle-

based COVID-19 vaccine administration. Int J Mol Sci 23:8838. https://doi.org/10.3390/ijms23168838

91. Ball RL, Bajaj P, Whitehead KA (2016) Achieving long-term stability of lipid nanoparticles: examining the effect of pH, temperature, and lyophilization. Int J Nanomedicine 12:305–315. https://doi.org/10.2147/IJN.S123062
92. Zhao P, Hou X, Yan J, Du S, Xue Y, Li W, Xiang G, Dong Y (2020) Long-term storage of lipid-like nanoparticles for mRNA delivery. Bioact Mater 5:358–363. https://doi.org/10.1016/j.bioactmat.2020.03.001
93. Pardi N, Hogan MJ, Naradikian MS, Parkhouse K, Cain DW, Jones L, Moody MA, Verkerke HP, Myles A, Willis E, LaBranche CC, Montefiori DC, Lobby JL, Saunders KO, Liao H-X, Korber BT, Sutherland LL, Scearce RM, Hraber PT, Tombácz I, Muramatsu H, Ni H, Balikov DA, Li C, Mui BL, Tam YK, Krammer F, Karikó K, Polacino P, Eisenlohr LC, Madden TD, Hope MJ, Lewis MG, Lee KK, Hu S-L, Hensley SE, Cancro MP, Haynes BF, Weissman D (2018) Nucleoside-modified mRNA vaccines induce potent T follicular helper and germinal center B cell responses. J Exp Med 215:1571–1588. https://doi.org/10.1084/jem.20171450
94. Sahin U, Muik A, Derhovanessian E, Vogler I, Kranz LM, Vormehr M, Baum A, Pascal K, Quandt J, Maurus D, Brachtendorf S, Lörks V, Sikorski J, Hilker R, Becker D, Eller AK, Grützner J, Boesler C, Rosenbaum C, Kühnle MC, Luxemburger U, Kemmer-Brück A, Langer D, Bexon M, Bolte S, Karikó K, Palanche T, Fischer B, Schultz A, Shi PY, Fontes-Garfias C, Perez JL, Swanson KA, Loschko J, Scully IL, Cutler M, Kalina W, Kyratsous CA, Cooper D, Dormitzer PR, Jansen KU, Türeci Ö (2020) COVID-19 vaccine BNT162b1 elicits human antibody and TH1 T cell responses. Nature 586:594–599. https://doi.org/10.1038/s41586-020-2814-7
95. Turner JS, O'Halloran JA, Kalaidina E, Kim W, Schmitz AJ, Zhou JQ, Lei T, Thapa M, Chen RE, Case JB, Amanat F, Rauseo AM, Haile A, Xie X, Klebert MK, Suessen T, Middleton WD, Shi P-Y, Krammer F, Teefey SA, Diamond MS, Presti RM, Ellebedy AH (2021) SARS-CoV-2 mRNA vaccines induce persistent human germinal centre responses. Nature 596:109–113. https://doi.org/10.1038/s41586-021-03738-2
96. Victora GD, Nussenzweig MC (2012) Germinal centers. Annu Rev Immunol 30:429–457. https://doi.org/10.1146/annurev-immunol-020711-075032
97. Cyster JG, Allen CDC (2019) B cell responses: cell interaction dynamics and decisions. Cell 177:524–540. https://doi.org/10.1016/j.cell.2019.03.016
98. Radbruch A, Muehlinghaus G, Luger EO, Inamine A, Smith KGC, Dörner T, Hiepe F (2006) Competence and competition: the challenge of becoming a long-lived plasma cell. Nat Rev Immunol 6:741–750. https://doi.org/10.1038/nri1886
99. Kim W, Zhou JQ, Horvath SC, Schmitz AJ, Sturtz AJ, Lei T, Liu Z, Kalaidina E, Thapa M, Alsoussi WB, Haile A, Klebert MK, Suessen T, Parra-Rodriguez L, Mudd PA, Whelan SPJ, Middleton WD, Teefey SA, Pusic I, O'Halloran JA, Presti RM, Turner JS, Ellebedy AH (2022) Germinal centre-driven maturation of B cell response to mRNA vaccination. Nature 604:141–145. https://doi.org/10.1038/s41586-022-04527-1
100. Polack FP, Thomas SJ, Kitchin N, Absalon J, Gurtman A, Lockhart S, Perez JL, Pérez Marc G, Moreira ED, Zerbini C, Bailey R, Swanson KA, Roychoudhury S, Koury K, Li P, Kalina WV, Cooper D, Frenck RW, Hammitt LL, Türeci Ö, Nell H, Schaefer A, Ünal S, Tresnan DB, Mather S, Dormitzer PR, Şahin U, Jansen KU, Gruber WC (2020) Safety and efficacy of the BNT162b2 mRNA Covid-19 vaccine. N Engl J Med 383:2603–2615. https://doi.org/10.1056/nejmoa2034577
101. Kalimuddin S, Tham CYL, Qui M, de Alwis R, Sim JXY, Lim JME, Tan H-C, Syenina A, Zhang SL, Le Bert N, Tan AT, Leong YS, Yee JX, Ong EZ, Ooi EE, Bertoletti A, Low JG (2021) Early T cell and binding antibody responses are associated with COVID-19 RNA vaccine efficacy onset. Med 2:682–688.e4. https://doi.org/10.1016/j.medj.2021.04.003
102. Oberhardt V, Luxenburger H, Kemming J, Schulien I, Ciminski K, Giese S, Csernalabics B, Lang-Meli J, Janowska I, Staniek J, Wild K, Basho K, Marinescu MS, Fuchs J, Topfstedt F, Janda A, Sogukpinar O, Hilger H, Stete K, Emmerich F, Bengsch B, Waller CF, Rieg S, Sagar N, Boettler T, Zoldan K, Kochs G, Schwemmle M, Rizzi M, Thimme R, Neumann-

Haefelin C, Hofmann M (2021) Rapid and stable mobilization of CD8+ T cells by SARS-CoV-2 mRNA vaccine. Nature 597:268–273. https://doi.org/10.1038/s41586-021-03841-4
103. Walsh EE, Frenck RW, Falsey AR, Kitchin N, Absalon J, Gurtman A, Lockhart S, Neuzil K, Mulligan MJ, Bailey R, Swanson KA, Li P, Koury K, Kalina W, Cooper D, Fontes-Garfias C, Shi P-Y, Türeci Ö, Tompkins KR, Lyke KE, Raabe V, Dormitzer PR, Jansen KU, Şahin U, Gruber WC (2020) Safety and immunogenicity of two RNA-based Covid-19 vaccine candidates. N Engl J Med 383:2439–2450. https://doi.org/10.1056/NEJMoa2027906
104. Sahin U, Muik A, Vogler I, Derhovanessian E, Kranz LM, Vormehr M, Quandt J, Bidmon N, Ulges A, Baum A, Pascal KE, Maurus D, Brachtendorf S, Lörks V, Sikorski J, Koch P, Hilker R, Becker D, Eller A-K, Grützner J, Tonigold M, Boesler C, Rosenbaum C, Heesen L, Kühnle M-C, Poran A, Dong JZ, Luxemburger U, Kemmer-Brück A, Langer D, Bexon M, Bolte S, Palanche T, Schultz A, Baumann S, Mahiny AJ, Boros G, Reinholz J, Szabó GT, Karikó K, Shi P-Y, Fontes-Garfias C, Perez JL, Cutler M, Cooper D, Kyratsous CA, Dormitzer PR, Jansen KU, Türeci Ö (2021) BNT162b2 vaccine induces neutralizing antibodies and poly-specific T cells in humans. Nature 595:572–577. https://doi.org/10.1038/s41586-021-03653-6
105. Kuntz J, Crane B, Weinmann S, Naleway AL (2018) Myocarditis and pericarditis are rare following live viral vaccinations in adults. Vaccine 36:1524–1527. https://doi.org/10.1016/j.vaccine.2018.02.030
106. US Food and Drug Administration (n.d.) Pfizer-BioNTech COVID-19 vaccine (BNT162, PF-07302048): Vaccines and Related Biological Products Advisory Committee briefing document
107. US Food and Drug Administration (n.d.) Moderna MRNA-1273 sponsor briefing document for vaccines and related biological. Products Advisory Committee
108. Bozkurt B, Kamat I, Hotez PJ (2021) Myocarditis with COVID-19 mRNA vaccines. Circulation 144:471–484. https://doi.org/10.1161/CIRCULATIONAHA.121.056135
109. Gargano JW, Wallace M, Hadler SC, Langley G, Su JR, Oster ME, Broder KR, Gee J, Weintraub E, Shimabukuro T, Scobie HM, Moulia D, Markowitz LE, Wharton M, McNally VV, Romero JR, Talbot HK, Lee GM, Daley MF, Oliver SE (2021) Use of mRNA COVID-19 vaccine after reports of myocarditis among vaccine recipients: update from the advisory committee on immunization practices—United States, June 2021, morbidity and mortality weekly. Report 70:977–982. https://doi.org/10.15585/MMWR.MM7027E2
110. Ryan M, Montgomery J, Engler R, Hoffman D, McClenathan B, Collins L, Loran D, Hrncir D, Herring K, Platzer M, Adams N, Sanou A, Cooper LT Jr (2021) Myocarditis following immunization with mrna covid-19 vaccines in members of the us military. JAMA Cardiol 6:1202–1206. https://doi.org/10.1001/jamacardio.2021.2833
111. Kim HW, Jenista ER, Wendell DC, Azevedo CF, Campbell MJ, Darty SN, Parker MA, Kim RJ (2021) Patients with acute myocarditis following mRNA COVID-19 vaccination. JAMA Cardiol 6:1196–1201. https://doi.org/10.1001/jamacardio.2021.2828
112. Muthukumar A, Narasimhan M, Li Q-Z, Mahimainathan L, Hitto I, Fuda F, Batra K, Jiang X, Zhu C, Schoggins J, Cutrell JB, Croft CL, Khera A, Drazner MH, Grodin JL, Greenberg BM, Mammen PPA, Morrison SJ, de Lemos JA (2021) In-depth evaluation of a case of presumed myocarditis after the second dose of COVID-19 mRNA vaccine. Circulation 144:487–498. https://doi.org/10.1161/CIRCULATIONAHA.121.056038
113. Albert E, Aurigemma G, Saucedo J, Gerson DS (2021) Myocarditis following COVID-19 vaccination. Radiol Case Report 16:2142–2145. https://doi.org/10.1016/j.radcr.2021.05.033
114. D'Angelo T, Cattafi A, Carerj ML, Booz C, Ascenti G, Cicero G, Blandino A, Mazziotti S (2021) Myocarditis after SARS-CoV-2 vaccination: a vaccine-induced reaction? Can J Cardiol 37:1665–1667. https://doi.org/10.1016/j.cjca.2021.05.010
115. McLean K, Johnson TJ (2021) Myopericarditis in a previously healthy adolescent male following COVID-19 vaccination: a case report. Acad Emerg Med 28:918–921. https://doi.org/10.1111/acem.14322
116. Bautista García J, Peña Ortega P, Bonilla Fernández JA, Cárdenes León A, Ramírez Burgos L, Caballero Dorta E (2021) Acute myocarditis after administration of the BNT162b2 vac-

cine against COVID-19. Rev Esp Cardiol (Engl Ed) 74:812–814. https://doi.org/10.1016/j.rec.2021.04.005
117. Yewdell JW (2021) Antigenic drift: Understanding COVID-19. Immunity 54:2681–2687. https://doi.org/10.1016/j.immuni.2021.11.016
118. Koyama T, Weeraratne D, Snowdon JL, Parida L (2020) Emergence of drift variants that may affect COVID-19 vaccine development and antibody treatment. Pathogens 9:E324. https://doi.org/10.3390/pathogens9050324
119. Chemaitelly H, Yassine HM, Benslimane FM, Al Khatib HA, Tang P, Hasan MR, Malek JA, Coyle P, Ayoub HH, Al Kanaani Z, Al Kuwari E, Jeremijenko A, Kaleeckal AH, Latif AN, Shaik RM, Abdul Rahim HF, Nasrallah GK, Al Kuwari MG, Al Romaihi HE, Al-Thani MH, Al Khal A, Butt AA, Bertollini R, Abu-Raddad LJ (2021) mRNA-1273 COVID-19 vaccine effectiveness against the B.1.1.7 and B.1.351 variants and severe COVID-19 disease in Qatar. Nat Med 27:1614–1621. https://doi.org/10.1038/s41591-021-01446-y
120. Whitley RJ, Babiker A, Cooper LA, Ellenberg SS, Fix A, Griffin M, Joffe S, Kalil J, Levine MM, Makgoba MW, Tsiatis AA, Moore RH, Hamill RJ, Lipsitz L, Rosenberg ES, Faugno A, Bruno R, Gorman R, Hamilton H, Horwith G, Huynh C, Mytle N, Pavetto C, Tong X, Treanor J, El Sahly HM, Baden LR, Essink B, Doblecki-Lewis S, Martin JM, Anderson EJ, Campbell TB, Clark J, Jackson LA, Fichtenbaum CJ, Zervos M, Rankin B, Eder F, Feldman G, Kennelly C, Han-Conrad L, Levin M, Neuzil KM, Corey L, Gilbert P, Janes H, Follmann D, Marovich M, Polakowski L, Mascola JR, Ledgerwood JE, Graham BS, August A, Clouting H, Deng W, Han S, Leav B, Manzo D, Pajon R, Schödel F, Tomassini JE, Zhou H, Miller J (2021) Efficacy of the mRNA-1273 SARS-CoV-2 vaccine at completion of blinded phase. N Engl J Med 385:1774–1785. https://doi.org/10.1056/NEJMoa2113017
121. Guthmiller JJ, Utset HA, Wilson PC (2021) B cell responses against influenza viruses: short-lived humoral immunity against a life-long threat. Viruses 13. https://doi.org/10.3390/v13060965
122. Zhou P, Yuan M, Song G, Beutler N, Shaabani N, Huang D, He W, Zhu X, Callaghan S, Yong P, Anzanello F, Peng L, Ricketts J, Parren M, Garcia E, Rawlings SA, Smith DM, Nemazee D, Teijaro JR, Rogers TF, Wilson IA, Burton DR, Andrabi R (2022) A human antibody reveals a conserved site on beta-coronavirus spike proteins and confers protection against SARS-CoV-2 infection. Sci Transl Med 14:eabi9215. https://doi.org/10.1126/scitranslmed.abi9215
123. Crawford DC, Nickerson DA (2005) Definition and clinical importance of haplotypes. Annu Rev Med 56:303–320. https://doi.org/10.1146/annurev.med.56.082103.104540
124. Clark AG (2004) The role of haplotypes in candidate gene studies. Genet Epidemiol 27:321–333. https://doi.org/10.1002/gepi.20025
125. Auton A, Abecasis GR, Altshuler DM, Durbin RM, Abecasis GR, Bentley DR, Chakravarti A, Clark AG, Donnelly P, Eichler EE, Flicek P, Gabriel SB, Gibbs RA, Green ED, Hurles ME, Knoppers BM, Korbel JO, Lander ES, Lee C, Lehrach H, Mardis ER, Marth GT, McVean GA, Nickerson DA, Schmidt JP, Sherry ST, Wang J, Wilson RK, Gibbs RA, Boerwinkle E, Doddapaneni H, Han Y, Korchina V, Kovar C, Lee S, Muzny D, Reid JG, Zhu Y, Wang J, Chang Y, Feng Q, Fang X, Guo X, Jian M, Jiang H, Jin X, Lan T, Li G, Li J, Li Y, Liu S, Liu X, Lu Y, Ma X, Tang M, Wang B, Wang G, Wu H, Wu R, Xu X, Yin Y, Zhang D, Zhang W, Zhao J, Zhao M, Zheng X, Lander ES, Altshuler DM, Gabriel SB, Gupta N, Gharani N, Toji LH, Gerry NP, Resch AM, Flicek P, Barker J, Clarke L, Gil L, Hunt SE, Kelman G, Kulesha E, Leinonen R, McLaren WM, Radhakrishnan R, Roa A, Smirnov D, Smith RE, Streeter I, Thormann A, Toneva I, Vaughan B, Zheng-Bradley X, Bentley DR, Grocock R, Humphray S, James T, Kingsbury Z, Lehrach H, Sudbrak R, Albrecht MW, Amstislavskiy VS, Borodina TA, Lienhard M, Mertes F, Sultan M, Timmermann B, Yaspo M-L, Mardis ER, Wilson RK, Fulton L, Fulton R, Sherry ST, Ananiev V, Belaia Z, Beloslyudtsev D, Bouk N, Chen C, Church D, Cohen R, Cook C, Garner J, Hefferon T, Kimelman M, Liu C, Lopez J, Meric P, O'Sullivan C, Ostapchuk Y, Phan L, Ponomarov S, Schneider V, Shekhtman E, Sirotkin K, Slotta D, Zhang H, McVean GA, Durbin RM, Balasubramaniam S, Burton J, Danecek P, Keane TM, Kolb-Kokocinski A, McCarthy S, Stalker J, Quail M, Schmidt JP, Davies CJ, Gollub J, Webster T, Wong B, Zhan Y, Auton A, Campbell CL, Kong Y, Marcketta A,

Gibbs RA, Yu F, Antunes L, Bainbridge M, Muzny D, Sabo A, Huang Z, Wang J, Coin LJM, Fang L, Guo X, Jin X, Li G, Li Q, Li Y, Li Z, Lin H, Liu B, Luo R, Shao H, Xie Y, Ye C, Yu C, Zhang F, Zheng H, Zhu H, Alkan C, Dal E, Kahveci F, Marth GT, Garrison EP, Kural D, Lee W-P, Leong WF, Stromberg M, Ward AN, Wu J, Zhang M, Daly MJ, DePristo MA, Handsaker RE, Altshuler DM, Banks E, Bhatia G, del Angel G, Gabriel SB, Genovese G, Gupta N, Li H, Kashin S, Lander ES, McCarroll SA, Nemesh JC, Poplin RE, Yoon SC, Lihm J, Makarov V, Clark AG, Gottipati S, Keinan A, Rodriguez-Flores JL, Korbel JO, Rausch T, Fritz MH, Stütz AM, Flicek P, Beal K, Clarke L, Datta A, Herrero J, McLaren WM, Ritchie GRS, Smith RE, Zerbino D, Zheng-Bradley X, Sabeti PC, Shlyakhter I, Schaffner SF, Vitti J, Cooper DN, Ball EV, Stenson PD, Bentley DR, Barnes B, Bauer M, Cheetham RK, Cox A, Eberle M, Humphray S, Kahn S, Murray L, Peden J, Shaw R, Kenny EE, Batzer MA, Konkel MK, Walker JA, MacArthur DG, Lek M, Sudbrak R, Amstislavskiy VS, Herwig R, Mardis ER, Ding L, Koboldt DC, Larson D, Ye K, Gravel S, The 1000 Genomes Project Consortium, Corresponding authors, Steering committee, Production group, Baylor College of Medicine, BGI-Shenzhen, Broad Institute of MIT and Harvard, Coriell Institute for Medical Research, E.B.I. European Molecular Biology Laboratory, Illumina, Max Planck Institute for Molecular Genetics, McDonnell Genome Institute at Washington University, US National Institutes of Health, University of Oxford, Wellcome Trust Sanger Institute, Analysis group, Affymetrix, Albert Einstein College of Medicine, Bilkent University, Boston College, Cold Spring Harbor Laboratory, Cornell University, European Molecular Biology Laboratory, Harvard University, Human Gene Mutation Database, Icahn School of Medicine at Mount Sinai, Louisiana State University, Massachusetts General Hospital, McGill University, N. National Eye Institute (2015) A global reference for human genetic variation. Nature 526:68–74. https://doi.org/10.1038/nature15393

126. Hoehe MR, Church GM, Lehrach H, Kroslak T, Palczewski S, Nowick K, Schulz S, Suk E-K, Huebsch T (2014) Multiple haplotype-resolved genomes reveal population patterns of gene and protein diplotypes. Nat Commun 5:5569. https://doi.org/10.1038/ncomms6569
127. Garcia KC, Adams EJ (2005) How the T cell receptor sees antigen—a structural view. Cell 122:333–336. https://doi.org/10.1016/j.cell.2005.07.015
128. Fremont DH, Hendrickson WA, Marrack P, Kappler J (1996) Structures of an MHC class II molecule with covalently bound single peptides. Science 272:1001–1004. https://doi.org/10.1126/science.272.5264.1001
129. Trowsdale J, Knight JC (2013) Major histocompatibility complex genomics and human disease. Annu Rev Genom Hum Genet 14:301–323. https://doi.org/10.1146/annurev-genom-091212-153455
130. Mudd PA, Minervina AA, Pogorelyy MV, Turner JS, Kim W, Kalaidina E, Petersen J, Schmitz AJ, Lei T, Haile A, Kirk AM, Mettelman RC, Crawford JC, Nguyen THO, Rowntree LC, Rosati E, Richards KA, Sant AJ, Klebert MK, Suessen T, Middleton WD, Team SJTRCS, Wolf J, Teefey SA, O'Halloran JA, Presti RM, Kedzierska K, Rossjohn J, Thomas PG, Ellebedy AH (2022) SARS-CoV-2 mRNA vaccination elicits a robust and persistent T follicular helper cell response in humans. Cell 185:603–613.e15. https://doi.org/10.1016/j.cell.2021.12.026
131. Flores LE, Frontera WR, Andrasik MP, del Rio C, Mondríguez-González A, Price SA, Krantz EM, Pergam SA, Silver JK (2021) Assessment of the inclusion of racial/ethnic minority, female, and older individuals in vaccine clinical trials. JAMA Network Open 4:e2037640. https://doi.org/10.1001/jamanetworkopen.2020.37640
132. Khalil L, Leary M, Rouphael N, Ofotokun I, Rebolledo PA, Wiley Z (2022) Racial and ethnic diversity in SARS-CoV-2 vaccine clinical trials conducted in the United States. Vaccine 10. https://doi.org/10.3390/vaccines10020290
133. Sequist TD (2020) The disproportionate impact of covid-19 on communities of color, NEJM catalyst innovations in care delivery. https://catalyst.nejm.org/doi/full/10.1056/CAT.20.0370. Accessed 23 Oct 2022

134. Kantamneni N (2020) The impact of the COVID-19 pandemic on marginalized populations in the United States: a research agenda. J Vocat Behav 119:103439. https://doi.org/10.1016/j.jvb.2020.103439
135. Andrasik MP, Broder GB, Wallace SE, Chaturvedi R, Michael NL, Bock S, Beyrer C, Oseso L, Aina J, Lucas J, Wilson DR, Kublin JG, Mensah GA (2021) Increasing black, indigenous and people of color participation in clinical trials through community engagement and recruitment goal establishment. PLoS One 16:e0258858. https://doi.org/10.1371/journal.pone.0258858
136. Schultz NH, Sørvoll IH, Michelsen AE, Munthe LA, Lund-Johansen F, Ahlen MT, Wiedmann M, Aamodt A-H, Skattør TH, Tjønnfjord GE, Holme PA (2021) Thrombosis and thrombocytopenia after ChAdOx1 nCoV-19 vaccination. N Engl J Med 384:2124–2130. https://doi.org/10.1056/NEJMoa2104882
137. Greinacher A, Thiele T, Warkentin TE, Weisser K, Kyrle PA, Eichinger S (2021) Thrombotic thrombocytopenia after ChAdOx1 nCov-19 vaccination. N Engl J Med 384:2092–2101. https://doi.org/10.1056/NEJMoa2104840
138. Pottegård A, Lund LC, Karlstad Ø, Dahl J, Andersen M, Hallas J, Lidegaard Ø, Tapia G, Gulseth HL, Ruiz PL-D, Watle SV, Mikkelsen AP, Pedersen L, Sørensen HT, Thomsen RW, Hviid A (2021) Arterial events, venous thromboembolism, thrombocytopenia, and bleeding after vaccination with Oxford-AstraZeneca ChAdOx1-S in Denmark and Norway: population based cohort study. BMJ 373:n1114. https://doi.org/10.1136/bmj.n1114
139. RKI—Empfehlungen der STIKO—Mitteilung der STIKO zur COVID-19-Impfung: Impfabstand und heterologes Impfschema nach Erstimpfung mit Vaxzevria (1.7.2021), (n.d.). https://www.rki.de/DE/Content/Kommissionen/STIKO/Empfehlungen/PM_2021-07-01.html. Accessed 3 Nov 2022
140. Hillus D, Schwarz T, Tober-Lau P, Vanshylla K, Hastor H, Thibeault C, Jentzsch S, Helbig ET, Lippert LJ, Tscheak P, Schmidt ML, Riege J, Solarek A, von Kalle C, Dang-Heine C, Gruell H, Kopankiewicz P, Suttorp N, Drosten C, Bias H, Seybold J, Al-Rim B, Bardtke L, Beheim-Schwarzbach JI, Behn K, Bergfeld L, Bethke N, Bleicker T, Briesemeister D, Brumhard S, Conrad C, Dieckmann S, Frey D, Gabelich J-A, Georg P, Gläser U, Hasler L, Hetey A, Hiller AL, Horn A, Hülso C, Kegel L, Koch W, Krannich A, Kroneberg P, Lisy M, Mackeldanz P, Maeß B, Münn F, Olk N, Peiser C, Pohl K, Hermel A, Rönnefarth M, Rubisch C, Sanchez Rezza A, Schellenberger I, Schenkel V, Schlesinger J, Schmidt S, Schwanitz G, Sinnigen A-S, Stubbemann P, Tesch J, Treue D, Wendisch D, Zvorc S, Klein F, Kurth F, Corman VM, Sander LE, S.G. EICOV/COVIM (2021) Safety, reactogenicity, and immunogenicity of homologous and heterologous prime-boost immunisation with ChAdOx1 nCoV-19 and BNT162b2: a prospective cohort study. Lancet Respir Med 9:1255–1265. https://doi.org/10.1016/S2213-2600(21)00357-X
141. Barros-Martins J, Hammerschmidt SI, Cossmann A, Odak I, Stankov MV, Morillas Ramos G, Dopfer-Jablonka A, Heidemann A, Ritter C, Friedrichsen M, Schultze-Florey C, Ravens I, Willenzon S, Bubke A, Ristenpart J, Janssen A, Ssebyatika G, Bernhardt G, Münch J, Hoffmann M, Pöhlmann S, Krey T, Bošnjak B, Förster R, Behrens GMN (2021) Immune responses against SARS-CoV-2 variants after heterologous and homologous ChAdOx1 nCoV-19/BNT162b2 vaccination. Nat Med 27:1525–1529. https://doi.org/10.1038/s41591-021-01449-9
142. Maringer Y, Nelde A, Schroeder SM, Schuhmacher J, Hörber S, Peter A, Karbach J, Jäger E, Walz JS (2022) Durable spike-specific T-cell responses after different COVID-19 vaccination regimens are not further enhanced by booster vaccination. Sci Immunol 7:eadd 3899. https://doi.org/10.1126/sciimmunol.add3899
143. Wu K, Choi A, Koch M, Elbashir S, Ma L, Lee D, Woods A, Henry C, Palandjian C, Hill A, Jani H, Quinones J, Nunna N, O'Connell S, McDermott AB, Falcone S, Narayanan E, Colpitts T, Bennett H, Corbett KS, Seder R, Graham BS, Stewart-Jones GBE, Carfi A, Edwards DK (2021) Variant SARS-CoV-2 mRNA vaccines confer broad neutralization as primary or booster series in mice. Vaccine 39:7394–7400. https://doi.org/10.1016/j.vaccine.2021.11.001

144. Choi A, Koch M, Wu K, Chu L, Ma LZ, Hill A, Nunna N, Huang W, Oestreicher J, Colpitts T, Bennett H, Legault H, Paila Y, Nestorova B, Ding B, Montefiori D, Pajon R, Miller JM, Leav B, Carfi A, McPhee R, Edwards DK (2021) Safety and immunogenicity of SARS-CoV-2 variant mRNA vaccine boosters in healthy adults: an interim analysis. Nat Med 27:2025–2031. https://doi.org/10.1038/s41591-021-01527-y
145. Houser K, Subbarao K (2015) Influenza vaccines: challenges and solutions. Cell Host Microbe 17:295–300. https://doi.org/10.1016/j.chom.2015.02.012
146. Jackson NAC, Kester KE, Casimiro D, Gurunathan S, DeRosa F (2020) The promise of mRNA vaccines: a biotech and industrial perspective. Npj Vaccines 5:1–6. https://doi.org/10.1038/s41541-020-0159-8
147. Chivukula S, Plitnik T, Tibbitts T, Karve S, Dias A, Zhang D, Goldman R, Gopani H, Khanmohammed A, Sarode A, Cooper D, Yoon H, Kim Y, Yan Y, Mundle ST, Groppo R, Beauvais A, Zhang J, Anosova NG, Lai C, Li L, Ulinski G, Piepenhagen P, DiNapoli J, Kalnin KV, Landolfi V, Swearingen R, Fu T-M, DeRosa F, Casimiro D (2021) Development of multivalent mRNA vaccine candidates for seasonal or pandemic influenza. Npj Vaccines 6:1–15. https://doi.org/10.1038/s41541-021-00420-6
148. Schwartz O, Albert ML (2010) Biology and pathogenesis of chikungunya virus. Nat Rev Microbiol 8:491–500. https://doi.org/10.1038/nrmicro2368
149. K. N, F. Jm, S. G, B. R, T. Rn, de S. Ad, E. Sm, T. Ma, H.-N. E, C. G, H. S, D. Ms, C. Je (2019) A lipid-encapsulated mRNA encoding a potently neutralizing human monoclonal antibody protects against chikungunya infection. Sci Immunol 4. https://doi.org/10.1126/sciimmunol.aaw6647
150. Oberli MA, Reichmuth AM, Dorkin JR, Mitchell MJ, Fenton OS, Jaklenec A, Anderson DG, Langer R, Blankschtein D (2017) Lipid nanoparticle assisted mRNA delivery for potent cancer immunotherapy. Nano Lett 17:1326–1335. https://doi.org/10.1021/acs.nanolett.6b03329
151. Chahal JS, Khan OF, Cooper CL, McPartlan JS, Tsosie JK, Tilley LD, Sidik SM, Lourido S, Langer R, Bavari S, Ploegh HL, Anderson DG (2016) Dendrimer-RNA nanoparticles generate protective immunity against lethal Ebola, H1N1 influenza, and toxoplasma gondii challenges with a single dose. PNAS 113:E4133–E4142. https://doi.org/10.1073/pnas.1600299113
152. Beissert T, Perkovic M, Vogel A, Erbar S, Walzer KC, Hempel T, Brill S, Haefner E, Becker R, Türeci Ö, Sahin U (2020) A trans-amplifying RNA vaccine strategy for induction of potent protective immunity. Mol Ther 28:119–128. https://doi.org/10.1016/j.ymthe.2019.09.009
153. Walsh G, Jefferis R (2006) Post-translational modifications in the context of therapeutic proteins. Nat Biotechnol 24:1241–1252. https://doi.org/10.1038/nbt1252

Coronavirus Virology

Ethan J. Fritch and Timothy P. Sheahan

1 Introduction

Coronaviruses (CoVs) have emergence potential and pose a significant threat to global health, evidenced by the emergence of severe acute respiratory syndrome coronavirus (SARS-CoV) in 2003, the discovery of Middle East respiratory syndrome coronavirus (MERS-CoV) in 2012, and the emergence of SARS-CoV-2 in 2019. SARS-CoV-2 has caused over seven million deaths globally, including over one million in the USA in an ever-expanding pandemic characterized by the emergence of novel variants of concern that escape vaccines and immunotherapeutics. A significant percentage of those that survive acute COVID-19 (20–40%) may progress to a post-acute syndrome with neurological, renal, cardiac, and pulmonary complications and whose long-term disease outcomes remain uncertain. The CoV family is comprised of a large genetically diverse group of viruses infecting a wide range of mammalian and avian species, including multiple livestock and companion animal species. Spillover and novel CoV emergence are a concern not just for human health, as zoonotic CoV outbreaks and epidemics have also been reported in economically important livestock. Thus, for increased pandemic preparedness, a more holistic approach should be taken to understand CoV ecology, biology, and emergence potential.

E. J. Fritch
Department of Microbiology and Immunology, School of Medicine, University of North Carolina at Chapel Hill, Chapel Hill, NC, USA

T. P. Sheahan (✉)
Department of Microbiology and Immunology, School of Medicine, University of North Carolina at Chapel Hill, Chapel Hill, NC, USA

Department of Epidemiology, Gillings School of Public Health, University of North Carolina at Chapel Hill, Chapel Hill, NC, USA
e-mail: sheahan@email.unc.edu

J. M. Rosen, R. R. Colwell (eds.), *The COVID-19 Pandemic*,
https://doi.org/10.1007/978-3-031-62772-9_6

2 Coronavirus Virology

2.1 Coronavirus Phylogeny and Genome Structure

2.2 CoV Phylogeny

Coronaviruses (CoVs) are members of the subfamily *Orthocoronavirinae* of the virus family *Coronaviridae* in the order *Nidovirales*. CoVs gained their name following inspection of viral particles under electron microscopy, wherein the pleomorphic, roughly spherical particles of about 100–120 nM in diameter appear to have a crown or solar corona surrounding them. The viral particles contain a ~ 30 kb positive sense single-stranded RNA genome encoding at a minimum of five open reading frames (ORFs) and 20 genes. The International Committee on the Taxonomy of Viruses has divided the *Orthocoronavirinae* subfamily into four genera: *Alphacoronavirus*, *Betacoronavirus*, *Gammacoronavirus*, and *Deltacoronavirus*. Phylogenetic evidence supports bats as the viral origin for *Alpha* and *Betacoronavirus* genera and birds as the viral origin for *Gamma* and *Deltacoronavirus* genera [1].

The *Alphacoronavirus* genus contains species capable of infecting livestock and companion mammals and two species that cause mild respiratory disease in humans. Species of note from the *Alphacoronavirus* genus are porcine epidemic diarrhea virus (PEDV), canine coronavirus (CCoV), feline coronavirus (FCoV), and porcine transmissible gastroenteritis virus (TGEV), and two species that actively circulate in human populations are human coronavirus 229E (HCoV-229E) and human coronavirus NL63 (HCoV-NL63). While HCoV-229e and NL63 cause mild respiratory symptoms in humans, a recent surveillance study of hospitalized pneumonia patients in Malaysia implicated canine coronavirus HuPn-2018 (CCoV-HuPn-2018) as the causative agent [2]. This is the first evidence of a canine-feline CoV species capable of infecting humans and causing disease.

Thus far, the CoV genus that causes the greatest impact on human health is *Betacoronavirus*. This genus is divided into four lineages. Key members of lineage A or *Embecovirus* include bovine coronavirus (BCoV), human coronavirus OC43 (HCoV-OC43), mouse hepatitis virus (MHV), and human coronavirus HKU1 (HCoV-HKU1). Interestingly, it has been hypothesized that the recently emerged HCoV-OC43 was the causative agent of the 1889–1891 respiratory disease pandemic, which was originally thought to have been caused by influenza [3]. Lineage B or *Sarbecovirus* covers all severe acute respiratory syndrome (SARS)-related CoVs, including SARS-CoV-1, SARS-CoV-2, and Bat SARS-like coronavirus WIV1. Lineage C or *Merbecovirus* contains Middle East respiratory syndrome coronavirus (MERS-CoV) and related bat CoV HKU4 and HKU5. Lineage D or *Nobecovirus* contains CoV species that have been isolated from bats in sampling studies.

The final two genera, *Gammacoronavirus* and *Deltacoronavirus,* are not yet known to harbor major human pathogens. The *Gammacoronavirus* genus contains species that primarily affect wild and domestic bird populations, and the most studied member, infectious bronchitis virus (IBV), is a major avian livestock pathogen affecting poultry and egg production. The discovery of two marine mammal *Gammacoronavirus*, the beluga whale CoV SW1 and bottlenose dolphin CoV HKU22, demonstrates viruses in this genus are not limited to terrestrial avian hosts. *Deltacoronavirus* genus contains species largely found in wild birds and pigs. Porcine delta coronavirus (PDCoV) was first identified in 2012 in moribund neonatal piglets and remains an agricultural pathogen of concern. Recent zoonotic PDCoV infection and disease in Haitian children highlight the broad host range of this virus and its potential for continued emergence in the human population [4].

2.2.1 Genome Structure

CoV has a hallmark genome structure where the genes that encode the replication machinery reside in the first two-thirds of the genome, while structural and group-specific accessory proteins are encoded in the last third. The 5′ most open reading frame (ORF) ORF1a/b encodes conserved enzymes and cofactors that facilitate RNA synthesis. ORF1a/b is translated as a large polyprotein that is cleaved by viral proteases to form 16 discreet proteins that have a wide array of functions, including protease polyprotein processing, remodeling of the host cell, innate immune antagonism, RNA synthesis, proofreading, and capping of viral RNA genomes and transcripts. Following ORF1a/b are structural genes whose gene products comprise the structure of the physical virus particle. Unlike ORF1a/b, these genes are all encoded on individual ORFs in the same canonical order of spike (S), envelope (E), membrane (M), and nucleocapsid (N) (Fig. 1). The viral genomic RNA is packaged by the N protein into a helical-shaped ribonucleoprotein complex, which is encapsulated by a lipid envelope decorated with S, M, and E [5]. The E and M glycoproteins regulate particle maturation and release, while S is a 180/90 kDa glycoprotein trimer that mediates species specificity, cell binding, and entry, the main determinant of protective immunity, and provides virus particles their unique appearance in electron microscopy [6]. Overall, structural gene functions are conserved among CoVs even though these proteins, especially S, are genetically variable.

The group-specific accessory ORFs, interspersed among structural genes in the 3′ end of the genome (Fig. 1), have a diverse array of functions, many of which have been described to antagonize innate immunity. Although accessory ORF proteins have been described as playing key roles in pathogenesis, they are typically dispensable for replication, as evidenced by the fact that accessory ORF deletion viruses are replication-competent [7]. The origin of these ORFs is hypothesized to be from viral recombination events with host species and there is evidence of recombination with other viral families, resulting in the expression of accessory genes [8].

The two noncoding regions of the genome, 5′ and 3′ untranslated regions (UTRs), flank the coding region of the CoV genome. The 3' UTR is included on all genomic and subgenomic messenger RNAs (sgmRNA) produced during viral replication, while a portion of the 5′ UTR, termed the leader, is included on all sgmRNA. The 5′ and 3′ UTRs of CoVs are structured like eukaryotic host mRNAs, having a 5′ cap and a 3′ polyadenylated tail, which aid in the stability of viral RNA and evasion of

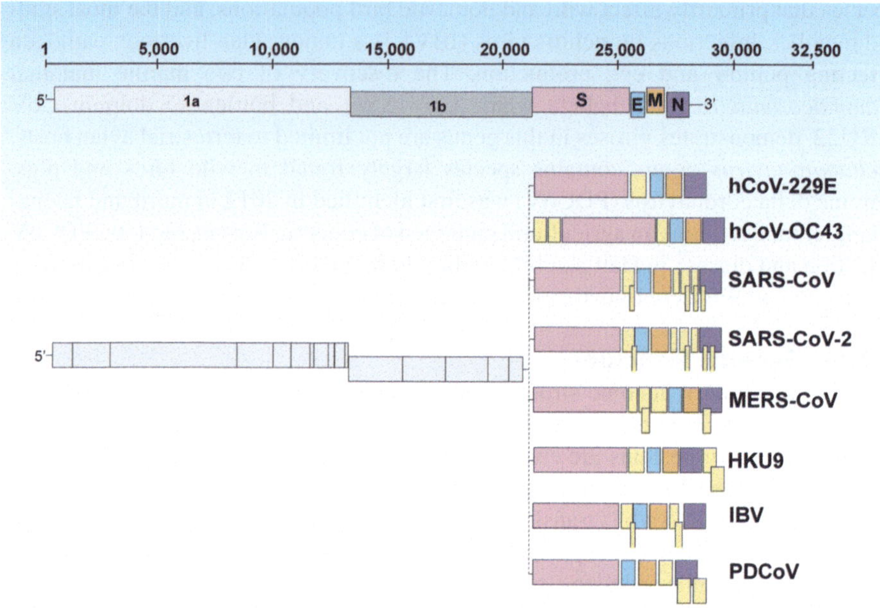

Fig. 1 The coronavirus genome and typical gene order. The gene order for representative members of the CoV family is shown. ORF1a/b, which encodes for the nonstructural proteins, is shown in gray. The accessory ORFs are shown in yellow. Spike (pink), envelope (E), membrane (M), and nucleocapsid (N) are shown

the innate immune system. The 5′ UTR is capped via viral machinery and contains up to five stem-loop (SL) structures, proposed to function in genome length and subgenomic length RNA synthesis. Three of the SL structures, SL1, SL2, and SL4, are conserved among alpha-, beta-, and gammacoronaviruses, while SL5 is conserved among alpha- and betacoronaviruses [9]. The leader transcriptional sequence (TRS-L) lies within the 5′ UTR, and it, along with the conserved body TRS (TRS-B), is necessary for the production of subgenomic messenger RNAs via a process called discontinuous transcription, where approximately 70 nucleotides of leader RNA from the 5′ end of the genome are joined to the 5′ end of each mRNA. The TRS of SARS-CoV, MERS-CoV, and SARS-CoV-2 are identical, ACGAAC; however, the TRS is not conserved among CoVs and varies depending on clade. The 3′ UTR is composed of a highly structured region followed by a poly-A tail. The 5′ most structure is the bulged stem loop (BSL), the base of which is capable of changing conformation to pair with the loop of the immediate 3′ pseudoknot (PK) hairpin, forming a pseudoknot structure. The different conformations of the BSL and PK have been hypothesized to form a molecular switch that regulates viral RNA synthesis. The loop formed by the pseudoknot has been identified as a binding site for Nsp8 and Nsp9, which is hypothesized to initiate the formation of the RTC on the 3′ UTR. Following the pseudoknot hairpin is a long-branched hairpin structure that contains the hypervariable region (HVR) of the 3′ UTR in a loop, the conserved stem-loop 2-like motif (s2m), and an octanucleotide sequence conserved among alpha- and betacoronaviruses. The s2m element has been proposed to act as a kissing complex to dimerize CoV genomes, while the function of the octanucleotide sequence remains unknown [10]. The 3′ most terminal structure is composed of a triple helix junction that folds back upon the flanking sequence of the pseudoknot hairpin, from which follows the poly-A tail [9].

2.3 Coronavirus Lifecycle

2.3.1 Entry, Uncoating, and Replication Initiation

Viral entry is mediated via the spike structural protein binding to host receptors on the outer leaflet of the cellular membrane, with specificity varying, depending on the structural composition of the receptor-binding domain of the protein (Fig. 2). After attachment, virions enter the cell via endocytosis, and fusion of the viral envelope and cellular membrane is mediated by the interaction of the S fusion peptide with host membranes, combining the two, resulting in deposition of the ribonucleoprotein complex into the cytosol of the host cell [11]. Exposure of the ribonucleoprotein complex to ATP is hypothesized to induce partial uncoating of the viral RNA, allowing the translation of viral proteins and initiation of viral replication [12].

In the cytosol, the positive sense CoV genome undergoes cap-dependent translation initiation and the 5′ most ORF1a is translated, encoding nonstructural proteins (Nsps) 1–11. However, when the translating ribosome reaches a slippery sequence and associated pseudoknot within the Nsp11 gene, it can experience a − 1

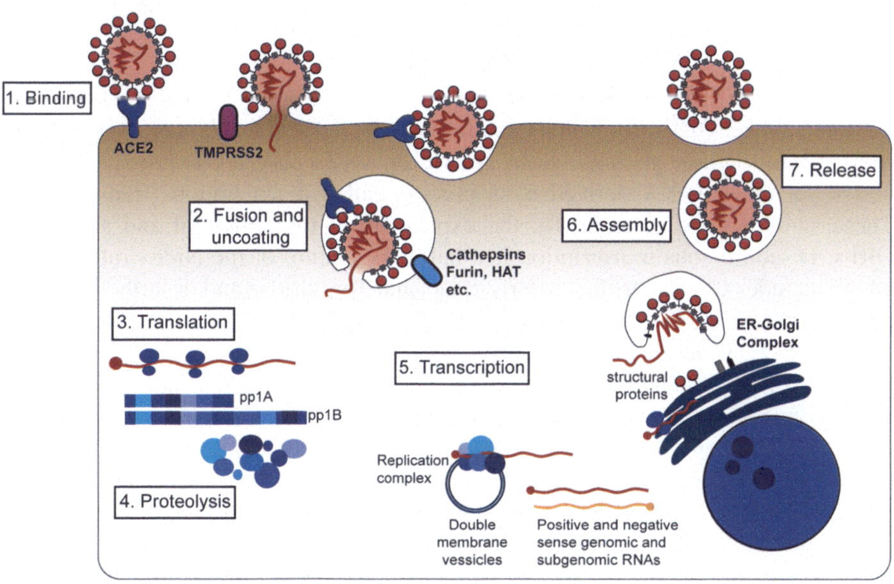

Fig. 2 The coronavirus life cycle. The coronavirus life cycle is shown using SARS-CoV-2 as an example. Key steps in the virus life cycle, divided into seven steps, are shown

frameshift, resulting in the translation of ORF1b, encoding Nsps 12–16, viral proteins necessary for RNA replication, transcription, proofreading, and capping. While the efficiency of the −1 frameshift varies depending on the CoV species, it typically falls between 15% and 30% and assists in regulating the amount of Nsp12–16% during viral replication, yielding an excess of Nsp1–11 as a result [13]. During the translation of ORF1a/b, viral proteases Nsp3 and Nsp5 cleave the polyprotein into individual Nsps. Nsp1 shuts down host translation of mRNA via binding to and blocking the entry channel of the 40S subunit of the ribosome, thus reducing translation of the innate immune system effectors and preserving amino acid resources for the production of viral proteins. Nsps 3/4/6 remodel the intracellular membranes creating double-membrane vesicles (DMVs), wherein the viral replication-transcription complex (RTC) is formed. The double-membrane vesicles shield viral double-stranded RNA intermediates from host innate immune intracellular sensors. Nsp15 is an endoribonuclease that degrades dsRNA, and it is proposed to antagonize innate immunity by targeting polyuridine RNAs, avoiding recognition of viral RNA intermediates by MDA5.

The original positive sense genome is copied into negative sense gRNA by a complex of Nsps 7/8/12, wherein Nsps 7/8 acts as a processivity factor for primer-dependent Nsp12 RdRp activity, and these three Nsps form the core of the RTC. During this process, Nsp13 acts as a superfamily I helicase, unwinding RNA duplexes, and Nsp9 is suspected to hold the unwound strands apart as a ssRNA binding protein. Nsp14 utilizes proofreading exoribonuclease (ExoN) activity to counteract the error-prone Nsp12, relying upon Nsp10 binding of Nsp14 as a cofactor to stabilize the ExoN domain. Viral RNAs are capped in four steps by Nsps 12/13/14/16, wherein Nsp16 activity is dependent on Nsp10. More specific activities of Nsps are described in further detail in the following sections.

2.3.2 Discontinuous Transcription

As noted above, ORF1a/b, which encodes viral replication machinery, is translated from the full-length genomic RNA. However, ORFs 2–9 are expressed from a nested set of sgmRNA derived by a process called discontinuous transcription. These sgmRNA are essential for the expression of the structural and accessory ORFs. Discontinuous transcription is a hallmark feature of the Nidovirales order, which includes Coronaviridae, Arteriviridae, and Toroviridae and describes the production of a nested (nido in Latin) set of subgenomic transcripts that all contain a common 5′ UTR, termed 5′ leader RNA, joined to the body sequences of each subgenomic mRNA, including 3′ co-terminal ends [14].

Discontinuous transcription occurs during negative-strand RNA synthesis from full-length genomic RNA and is mediated by the strategically located body TRS (TRS-B) sequences just upstream of each subgenomic gene ORF. During this process, the RTC synthesizes negative sense RNA in the 3′ to 5′ direction, beginning at the 3′ end of the genome, until it reaches a TRS-B at which time the RTC disassociates and reassociates with complementary leader transcription regulatory sequence (TRS-L) encoded near the 5′ end of the genome where the RTC continues transcription of the upstream 5′ UTR upstream at TRS-L creating a complete subgenomic negative-strand sgmRNA. Thus, the replication strategy gets its name "discontinuous transcription" because there are thousands of nucleotides between the TRS-B and TRS-L. With this strategy, all sgmRNAs share the same 5′ end containing the leader TRS-L. Ultimately, a nested set of negative sense sgmRNAs is synthesized for each subgenomic ORF due to read through of TRS-B, which is diminished in frequency moving from 3′ to 5′. As such, it would be predicted that the most abundant sgmRNA is nucleocapsid, the most 3′ subgenomic ORF, and the least abundant sgmRNA is spike, the most 5′ subgenomic ORF; however, the sgmRNA for E is commonly the least abundant sgmRNA produced, indicating that there may be an additional underlying mechanism important in guiding discontinuous transcription. It is hypothesized that RNA structures in the UTRs may interact with cis-acting elements in the coding RNA to help guide the joining of TRS-L and TRS-B [15]. Evidence for such interaction has been described for one sgmRNA of TGEV, to date; however, there has been little evidence suggesting conserved RNA secondary structures exist in the CoV coding regions that may guide this process. It has been demonstrated that 5′ UTR structure SL1/2/4 is necessary for discontinuous transcription; however, exact interactions of viral RNA and proteins, other than Nsp1 and nucleocapsid, with these elements remain unknown [9].

The nested set of negative sense sgmRNA then serves as templates for positive-strand sgmRNA synthesis from which proteins are produced. Positive-stranded sgmRNAs may appear to be polycistronic, but only the most '5 ORF is canonically translated, even though these sgmRNAs contain the sequence for all genes that reside between the translated ORF and the 3' UTR. For example, the spike gene sgmRNA also contains the sequence for spike, ORF 3a, E, M, 6, 7, 7a, 8, and N, yet only the spike ORF is canonically translated. There are examples of sgmRNA in many CoV species that function in bi- and tri-cistronic manners, translating multiple ORFs through leaky scanning [16]. The polycistronic nature of the sgmRNA creates a possibility for cap-independent translation initiation, but there has yet to be evidence of a CoV-encoded IRES or mechanism for such. The lack of such evidence for cap-independent translation is unsurprising since CoV appears to rely on cap-dependent translation initiation and encodes nonstructural genes for capping viral RNA.

Recent studies of SARS-CoV-2 and HCoV-229E have shown that the process of discontinuous transcription is not precise and limited to canonical TRSs; instead, a much larger library of sgmRNA is produced during viral transcription [17]. These studies proposed a unique hypothesis that sgmRNA may be acting not only as templates for a larger "undiscovered" viral proteome but may be involved in recombination events to improve viral fitness. In addition, it has been shown that it is possible to rewire the primary sequence of the TRS network of SARS-CoV and still isolate replication-competent virus, indicating plasticity within the transcription regulatory network of CoVs that may be advantageous for survival [18]. TRS remodeling not only attenuates the recovery of RNA recombinants but also attenuates virus pathogenesis, providing a strategy to develop recombination-resistant attenuated live virus vaccines.

2.3.3 Virion Assembly and Release

After synthesis of viral sgmRNAs and translational structural protein gene products, virus particle assembly takes place (Fig. 2). Virion assembly occurs at the interface of the endoplasmic reticulum and Golgi apparatus where spike, envelope, and membrane viral proteins are anchored via their transmembrane domains into the plasma membrane of vesicles. The viral genome is wound onto nucleocapsid proteins in a "beads on a string" process, and the RNA-nucleocapsid complex is packaged within the virion through budding into the endoplasmic reticulum Golgi intermediate complex, after which vesicles harboring nascent virus particles are trafficked to the cell membrane and released into extracellular space [6].

2.4 Viral Protein Function in Replication and Disease

2.4.1 Nonstructural Proteins

CoVs produce 16 Nsps in total, all encoded on ORF1a/b of the CoV genome, and each plays key roles during infection. The majority of the Nsps work to form the viral replication-transcription complex (RTC), which functions to produce, then cap, and proofread genomic and subgenomic RNA. The RTC resides within double-membrane vesicles (DMVs) created by the interaction of three transmembrane Nsps, which help hide viral RNA intermediates from innate immune recognition. Importantly, many Nsps are multifunctional not only facilitating viral RNA synthesis and polyprotein processing, but some also antagonize the host antiviral response to infection. Below, we summarize the functions of Nsps and reference reviews should additional information be desired.

Nsp1 controls host mRNA translation via binding to and blocking the entry channel of the 40S subunit of the ribosome, thus reducing the translation of innate immune system effectors and preserving amino acid resources for the production of viral proteins [19]. This translation inhibition function is circumvented by the SL1 structure in the 5′ UTR, which binds to Nsp1 and allows translation of viral genomic and sgmRNA. Nsp1s of CoVs have additionally been reported to induce host mRNA degradation and prevent nuclear export of host mRNA. Degradation of host mRNA occurs in a manner dependent on ribosomal binding, hypothesized to recruit endonucleases; however, the mechanism behind the degradation has still yet to be defined. Inhibition of nuclear export has been attributed to Nsp1 binding to nuclear pore proteins, suppressing translation of interferons whose promoters are activated following viral infection.

Nsp2 does not have a described function. However, cryo-EM, along with folding algorithms, has revealed a conserved zinc ion binding site. Deletion of Nsp2 in MHV resulted in replication-competent virus, indicating that it may be dispensable in some CoVs [20].

Nsp3 is the largest Nsp and contains multiple domains that carry out a variety of functions. The common domains that exist in all CoVs are two ubiquitin-like domains, an acidic hypervariable region, papain-like protease domains, macrodomain region, two transmembrane regions surrounding an ectodomain, and two domains of unknown function. Nsp3 is anchored in double-membrane vesicles and helps facilitate viral replication, transcription, and innate immune antagonism [21]. The most N-terminal domain of Nsp3 is the ubiquitin-like domain 1 (Ubl1), and the Ubl1-N interaction is thought to tether the viral genome to the RTC early in infection and has been shown to be essential in genome transcription. The binding of nucleocapsid to the TRS of the genomic RNA blocks Ubl1-N interaction; thus, it has been hypothesized that Ubl1 may act as a switch for replication and transcription by competing with the TRS to bind to nucleocapsid. CoVs can encode 1 or 2 papain-like protease (PL^{pro}) domains. The PL^{pro} cleaves the junctions between Nsp1/2, Nsp2/3, and Nsp3/4, releasing the first three Nsps from the polyprotein. However, if there are two PL^{pro} domains encoded by the CoV, there is specificity in which Nsp junctions are cleaved by each PL^{pro}.

CoV Nsp3s can encode up to three macrodomains, but only the first macrodomain, known as Mac1, is conserved among CoVs. Mac 1 can hydrolyze mono-ADP-ribose (MAR), and this deMARylation activity counteracts the antiviral activity of poly-ADP-ribose polymerases induced by interferon signaling. In MHV, mutations in Mac1 reduce deMARylation activity resulting in reduced lethality and pathogenesis in mice, further solidifying the role of Mac1 in suppressing innate immune signaling [22]. Following Mac1 is a non-conserved region that was first described in SARS-CoV-1, containing three subdomains: Macrodomain 2, Macrodomain 3, and a frataxin-like fold. This non-conserved region was originally called the SARS-unique domain (SUD); however, it was renamed to Mac2, Mac3, and the domain proceeding Ubl2 and PL2 ^{pro} (DPUP) after being found in other CoVs. The DPUP subdomain has been hypothesized to play a role in metal ion binding, and if deleted, it results in attenuated RNA synthesis.

The ubiquitin-like 2 domain follows the DPUP subdomain and is anchored to the $PL2^{pro}$ domain in MERS-CoV and SARS-CoV. The role of Ubl2 remains undetermined, as there are contradictory studies reporting its presence blocks innate immune signaling and others showing deletion has no effect. It has been speculated that Ubl2 may modulate the targeting of $PL2^{pro}$ to specific substrates of its deubiquitinating activity. The Nsp3 $PL2^{pro}$ domain directly follows the Ubl2 domain and is the conserved PL^{pro} among coronaviruses. CoVs with only one PL^{pro} can process the first three Nsp junctions using $PL2^{pro}$. However, $PL2^{pro}$ is also capable of deubiquitinating (DUB) and deISGylating activities through specific ubiquitin-binding sites. The ubiquitin-binding sites allow for $PL2^{pro}$ to hold onto and process polyubiquitin chains, thus combating the ubiquitin and ISG15 post-translational modifications that the host cell uses to combat viral infections. $PL2^{pro}$ has been implicated in antagonizing the innate immune response via targeting the IRF3 and NF-κB pathways. However, the direct target of $PL2^{pro}$ in these pathways remains disputed.

The two transmembrane domains flank the ectodomain and anchor Nsp3 into the endoplasmic reticulum. The ectodomain interacts with Nsp4 within the lumen, and the two proteins work in conjunction to form double-membrane vesicles. Other than the ectodomain and transmembrane domains, Nsp3 domains are cytosolic. A recent cryo-EM study has shown that hexamerization of Nsp3 allows the transmembrane domains to form a pore to allow the movement of viral proteins and RNA across double-membrane vesicles [23]. The final two domains of Nsp3 are known as the Y1 and CoV-Y domains, both of which have undescribed functions. The Y1 domain is conserved among all of the order *Nidovirales,* and the CoV-Y domain is conserved among all coronaviruses.

Nsp4 is a transmembrane protein that works in conjunction with Nsp3 and Nsp6 to remodel the endoplasmic reticulum of the host cell and form double-membrane vesicles. There are four transmembrane domains on Nsp4, and between the first two domains is a luminal loop that interacts with the Nsp3 ectodomain [24]. The Nsp3–4 luminal interaction pulls together the endoplasmic reticulum membranes, forming a double-membrane vesicle structure stabilized by Nsp6 [25]. The double-membrane vesicles are thought to hide viral double-stranded RNA intermediates from host innate immune sensors and to centralize the viral RNA and protein synthesis during replication.

Nsp5 is known as the 3C-like protease (3CL^{pro}) due to its similarity to the picornavirus 3C protease. However, it is also known as the major protease (M^{pro}) of the CoV and M^{pro} is responsible for cleaving the remaining junctions between Nsps 4 and 16 not cleaved by the PL^{pro} domains of Nsp3 [26]. It has been proposed that Nsp5 plays a role in antagonizing the innate immune response by cleaving key proteins in antiviral signaling cascades. However, evidence for such has been found only in PEDV and PDCoV [27, 28].

Nsp6 is the final protein that works in concert with Nsps3 and msps4 to help remodel the endoplasmic reticulum of the host cell for DMV formation. There are six transmembrane domains on Nsp6, winding the protein in and out of the cytoplasm and lumen of the endoplasmic reticulum, with a c-terminal cytoplasmic amphipathic helix [25]. Nsp6 homodimerizes and zippers the endoplasmic reticulum when expressed by itself. However, when working alongside Nsp3/4, the three Nsps form clusters of double-membrane vesicles with Nsp6 organizing the Nsp3/4 double-membrane structures and tethering lipid droplets to facilitate efficient replication [29].

Nsps 7 and 8 work together to form a primase complex during RNA replication and transcription. Nsp7 has been hypothesized to act as an RNA-binding protein and to stabilize the Nsp8 in the complex. Nsp8 has been described to function as a primase, creating short oligonucleotides for Nsp12 to extend into genomic and sgmRNA species [30]. There is additional evidence in HCoV-229E that Nsp8 is responsible for terminal adenylyltransferase activity and elongated poly-A-tails on newly transcribed viral RNAs [31]. Several complexes of Nsp7 and 8 have been solved, including a hexadecamer in SARS-CoV, a tetramer in SARS-CoV-2, and a heterotrimer in feline coronavirus (FCoV). When Nsp7/8 is expressed, along with Nsp12, it forms the core of the replication-transcription complex and is capable of producing nascent RNA strands. The actual number and positioning of Nsp7 and 8 proteins involved in the RTC are inconsistent, based on the crystal structures that have been solved and which Nsps are coexpressed to create the structure [32]. One such structure shows Nsp7 acting as a bridge to create a dimer RTC that would facilitate strand switching for the production of sgmRNA.

Nsp9 has been shown to act as a sequence nonspecific single-stranded DNA/RNA-binding protein [33]. The binding of Nsp9 to the genome has been hypothesized to stabilize newly transcribed RNA during replication and transcription or protect it from degradation. Structural modeling of SARS-CoV-2 Nsp9 in a complex with Nsp12 has been verified, in which Nsp9 binds and inhibits the capping functionality of the Nsp12 nucleotidyltransferase domain [34].

Nsp10 functions as a cofactor for the enzymatic activity of Nsps 14 and 16 [35, 36]. Binding interfaces between Nsp10/14 and Nsp10/16 slightly overlap, indicating that interaction with one excludes the other, but given its position in ORF1a prior to the −1 frameshift, Nsp10 is produced in excess of its ORF1b partners. Nsp10 contains two conserved zinc fingers predicted to allow binding to single- and double-stranded RNA and DNA. Disruption of zinc finger domains in MHV results in decreased polyprotein processing and mutant viruses defective in negative-strand RNA production.

Nsp11 is made up of the N terminus of Nsp12 prior to the −1 frameshift. Its function remains unknown and is a small intrinsically disordered protein, with approximately 13–23 amino acids in CoVs.

Nsp12 is most commonly known as the RNA-dependent RNA polymerase (RdRp) of the CoVs and is made up of two domains conserved across the *Nidovirales*. The N-terminal domain of Nsp12 is known as the nidovirus RdRp-associated nucleotidyltransferase domain (NiRAN) and the C-terminal domain is the RNA-dependent RNA polymerase (RdRp) [37]. Nsp12 performs the critical role of copying viral RNA and also performs the second step of the capping process [34]. The NiRAN domain of Nsp12 is capable of transferring nucleoside monophosphates to substrates. Evidence of the NiRAN activity transferring nucleoside monophosphate to Nsp9 has been demonstrated. However, other protein substrates remain to be described. NiRAN has also been shown to perform guanylyltransferase activity necessary for performing the second step in 5′ cap synthesis by transferring a guanine monophosphate to the 5′ diphosphosphate created in the first step and creating a cap −1 structure [34]. The RdRp domain of Nsp12 is capable of primer-dependent polymerase activity. Nsp12 complex formation with Nsp7/8 greatly increases RdRp processivity and RNA binding, indicating the latter acts as a primase to aid in CoV replication and transcription. Additional association with Nsp13 allows the unwinding of dsRNA intermediates created by Nsp12, while the association with Nsp14 provides a proofreading function to compensate for the high error rate of the CoV RdRp.

Nsp13 is well conserved among *Nidovirales* as a superfamily 1 helicase made up of five domains: an N-terminal zinc-binding domain, a stalk domain, domain 1B, and two C-terminal RecA-like domains. The zinc-binding domains and stalk domains are needed to preserve the helicase activity of Nsp13; however, their functions have not been described. Domain 1B helps to form the RNA-binding channel around the core RecA-like domains that carry out the primary helicase activity of Nsp13. As a superfamily 1 helicase, Nsp13 is capable of binding to ssRNA and unwinding dsRNA and DNA from 5′ to 3′ in an ATP-dependent manner, and recent evidence has also indicated that the helicase activity of Nsp13 allows for backtracking during RNA synthesis to enable excision of mismatched base pairs [38]. The helicase function of Nsp13 relies on nucleotide triphosphatase activity (NTPase) to provide energy for unwinding RNA duplexes and translocation down ssRNA. The same active site that provides NTPase activity also provides Nsp13 with RNA 5′-triphosphatase (RTPase) activity, allowing CoVs to carry out the first step in the capping process and create a cap −2 structure at the 5′ end of viral RNA [39]. Nsp13 has been shown to be a part of the RTC as its function is necessary to unwind dsRNA intermediates and initiate the capping process [34]. While the primary function of Nsp13 appears to assist in the production of nascent viral RNA, there is evidence that it also participates in inhibiting type I interferon production. SARS-CoV-2 Nsp13 is capable of binding to TANK binding kinase 1 and inducing its degradation. Nsp13 has also been shown to limit NF-κB phosphorylation and nuclear translocation.

Nsp14 has two distinct domains that provide essential functions: an N-terminal exoribonuclease (ExoN) domain and a C-terminal N7-methyltransferase domain. Nsp14 has also been implicated in antagonizing innate immune response to viral infection. Overexpression of SARS-CoV-2 Nsp14 results in the shutdown of host translation and blocks ISG induction, which depends on both enzymatic domain functions being intact [40]. The ExoN domain provides a unique proofreading mechanism for CoVs, allowing them to maintain larger RNA genomes than other Nidovirales, namely, arteriviruses, which lack an ExoN activity. ExoN processes double-stranded RNA 3′ to 5′, identifying mismatches in Watson-Crick base pairing at the 3′ terminus and excising the mismatch pair [36]. Nsp10 binding to SARS-CoV-1 ExoN stabilizes the domain and greatly increases enzymatic activity by 35-fold [36]. Mutations in the ExoN domain of Nsp14 result in viruses that are replication-competent, but have a 21-fold increase in mutation frequency and are severely attenuated in vivo. The N7-methyltransferase domain performs the third step in the production of the 5′ cap, methylation of the guanine at the N7 position, and production of a cap 0 structure [41]. The binding of Nsp10 to the ExoN domain does not affect the enzyme activity of the N7-methyltransferase domain.

Nsp15 contains three domains, and the C-terminal domain is the most well studied and is a uridine-specific endoribonuclease (EndoU), which is conserved across *Nidovirales* [42]. The catalytic function of EndoU requires oligomerization to stabilize the active site, whether it is with itself or other Nsps. However, its function is not necessary for viral replication. EndoU plays a key role in antagonizing the innate immune response by cleaving single- and double-stranded polyuridine RNAs to avoid the recognition of viral RNA intermediates by MDA5. There is evidence that Nsp15 interacts with Nsp12 and Nsp8; however, it is unclear whether this interaction is transient as there have yet to be any crystal structures solved identifying this interaction.

Nsp16 provides coronaviruses with 2’-O-methytransferase activity, creating a cap 1 structure following the enzymatic action of Nsps 12–14 [43]. The enzymatic activity of Nsp16 depends on complex formation with Nsp10 [35]. Ablation of the Nsp16 enzymatic function in MERS-CoV results in attenuated mutants sensitive to

MDA5 and IFIT recognition of viral RNAs, indicating the cap 1 structure created by Nsp16 is critical in CoV evasion of host innate immune sensing.

2.4.2 Structural Proteins

Coronaviruses create enveloped virions using four different structural proteins, with the exception of some species of *betacoronavirus* that also contain a fifth protein, hemagglutinin esterase (HE). The structural proteins sequester plasma membranes from the host cell, package coronavirus genomic RNA, and provide the means of transmission between host species. The common CoV structural proteins are spike (S), envelope (E), membrane (M), and nucleocapsid (N).

Spike glycoproteins contain receptor-binding and membrane fusion portions of the CoV virion and are the major determinant of CoV host species specificity [44]. CoV spike proteins are translated as one ORF approximately 1300 amino acids long and contain multiple domains. Following translation, the spike proteins are subsequently glycosylated and trafficked to the endoplasmic reticulum where they form a homotrimer that protrudes from the lipid envelope of newly formed virions. Spike proteins are cleaved into two subunits, S1 and S2, during viral packaging or after release from the host cell. Given its role in receptor binding and membrane fusion, the spike protein is the main antigenic target of vaccines and antibody therapies that aim to neutralize CoV virions. The S1 subunit, in the N-terminal portion of the protein spike trimer, harbors the receptor-binding domain (RBD) protein and is located in the most apical portion of the structure. S1 is made up of two domains; the N-terminal domain (NTD) and C-terminal domain (CTD), one or both of which function in receptor binding depending on the CoV. For example, NTD of MHV harbors the RBD binding to receptor carcinoembryonic antigen-related cell adhesion molecule 1 (CEACAM1), while the CTD of SARS-CoV-1/2 and MERS-CoV contains the RBDs binding to angiotensin-converting enzyme 2 (ACE2) and dipeptidyl peptidase 4 (DPP4), respectively. Studies have shown that the S1 domain can adopt multiple conformations and an open conformation is required in SARS-CoV-1/2 and MERS-CoV to bind their protein receptors. S2 is made up of three domains: the ectodomain, the transmembrane domain, and the cytoplasmic tail. The ectodomain of S2 forms the stalk of the spike and interacts with the S1 domain, coupling both by noncovalent bonds after proteolytic cleavage [45]. The membrane fusion components of the ectodomain are a fusion peptide and two heptad repeats, common to class 1 fusion proteins [11]. Studies indicate the fusion peptide is exposed after receptor binding of the S1 subunit, allowing cleavage and activation of the fusion peptide by host proteases, either on the surface of the cell or in endosomes. The activated fusion peptide subsequently interacts with host membranes and the heptad repeats are able to fold back on each other, collapsing the S2 subunit to draw the virion to the host membrane for fusion. The transmembrane domain follows the ectodomain and anchors the spike protein into the plasma membranes of the host cell or virions. The cytoplasmic tail is at the C terminus of the S2 subunit and plays roles in trafficking spike to the endoplasmic reticulum and interacting with the membrane protein to form virions. The cytoplasmic tail also may play a role in membrane fusion, as inhibition of palmitoylation of cysteine residues leads to decreased membrane fusion of virions. Plasticity in the spike leads to downstream changes in the recognition of proteins that may act as ligands for the RBD and changes in the range of host species, infectivity efficiency, and preexisting host immunity. Mutations occur rapidly in the CTD of S1, which has been hypothesized as caused by its more outward position on the spike protein, enduring higher selective pressure to escape immune recognition. Spike mutations, especially in the RBD, have been demonstrated best during the SARS-CoV-1 outbreaks and SARS-CoV-2 pandemic [46, 47]. The RBD-ACE2 interacting residues evolved quickly between variants during the SARS-CoV-2 pandemic, allowing the virus to escape antibody neutralization and increase viral transmissibility. In addition to RBD mutations, other mutations in S1 of spike, such as the D614G mutant that emerged early in the SARS-CoV-2 pandemic, can affect spike conformations of the RBD to increase infectivity.

Hemagglutinin esterase is a part of the viral envelope of some *betacoronaviruses*; however, its function remains poorly understood. The esterase activity of HE appears to be important in CoVs that use glycans for receptor binding, as it destroys receptors that may be used by the spike protein [48]. The lectin binding function of HE is not conserved, however, and recent studies have demonstrated that abrogation of lectin binding in BCoV leads to a decrease in viral output and mutations in the spike protein, indicating a potential balance between spike receptor affinity and HE activity.

Envelope proteins are the least abundant in the virion and the smallest of the structural proteins, yet contribute notably to viral release and fitness [49]. The E protein is produced in excess during viral replication and has been shown to act not only as a structural component in the virion but also as a viroporin and virulence factor [50]. The E protein contributes to the curvature of the virion envelope, facilitates trafficking, is predicted to allow the scission of virions from the host cell membranes, and helps modulate apoptotic pathways during infection. Deletion of E in MHV and SARS-CoV-1 results in replication-competent mutants, while TGEV mutants lacking E are propagation-deficient, indicating that not all CoV E utilizes the same mechanisms to facilitate viral maturation and release. The E protein comprises a short, N-terminal domain, transmembrane domain, and long, C-terminal domain. The orientation of the N- and C-terminal domains varies, depending on the CoV, and their cytoplasmic or lumen may also depend on where the E protein is localized within the cell. There is no conserved function for the N-terminal domain; however, it appears that it may contribute to Golgi localization in SARS-CoV-1. The C-terminal domain interacts with the CoV membrane protein, encodes a Golgi localization signal, and in some alpha- and betacoronaviruses contains a post-synaptic density protein 95/Drosophila disk large tumor suppressor/zonula occludens-1 protein (PDZ)-binding motif (PBM). The PBM of the SARS-CoV-1 E C-terminal domain has been shown to interact with a number of host cell proteins, most notably its interaction with syntenin triggers activation of mitogen-activated protein kinase (MAPK) signaling and downstream inflammatory cytokines that lead to acute lung injury in vivo. Deletion of the PBM or treatment of mice with a MAPK inhibitor attenuated SARS-CoV-1 pathogenesis; however, in vitro passaging of

PBM mutants resulted in compensatory mutations or reversion, indicating that the PBM is a key virulence factor for SARS-CoV-1 [51]. The transmembrane domain of E anchors the protein onto host membranes and the virion. The core of the transmembrane domain is made up of an amphipathic helix that allows E to oligomerize into a pentamer and act as a viroporin with ion channel activity. Viroporins contribute to multiple steps of the viral life cycle, and the viroporin activity of E has been shown to be cation-selective, unnecessary for replication, but enhances viral replication and pathogenesis in vivo.

Membrane proteins are the most abundant in CoV virions and act as a scaffold to form virus particles by interacting with themselves and the other CoV structural proteins. Expression of M with E or M with N proteins is sufficient to form virus-like particles, and M expression is necessary for viral replication. The M protein homodimerizes, can adopt two different conformations, and is made up of a short N-terminal ectodomain, three transmembrane domains, and a long C-terminal endodomain [52]. The transmembrane domains of M are responsible for dimerization and interaction with the E protein. Following translation, the M protein is trafficked to the Golgi and helps to facilitate virion production via the interaction of its C-terminal endodomain with S and N proteins. In E deletion mutants, there is duplication of a mutant M gene that is hypothesized to compensate for E virion assembly and viroporin activity. However, there is no evidence supporting viroporin activity for M proteins, even though the protein structure does resemble a viroporin [52]. The M protein has also been implicated in antagonizing the innate immune system through interaction with a number of host proteins involved in the double-stranded RNA sensing pathway.

Nucleocapsid proteins are the most abundant CoV protein produced during viral replication and facilitate genomic RNA packaging into virions [53]. While classified as a structural protein, there is evidence that the N protein participates in CoV replication and transcription [54, 55]. The N protein is heavily phosphorylated and made up of an N-terminal intrinsically disordered region (NTR), N-terminal structured domain (NTD), an intrinsically disordered linker domain (LKR), a C-terminal structured domain (CTD), and a C-terminal intrinsically disordered region (CTR). The NTR has been implicated in N interaction with stress granule component Ras-GTPase-activating protein SH3-domain-binding protein (G3BP). The NTD contains an RNA-binding domain that has been shown to bind to TRS and complementary TRS (cTRS) sequences to melt RNA duplexes, and this activity has also implicated the N protein as an RNA chaperone. The LKR domain contains a serine-arginine-rich region that is heavily phosphorylated, contains a nuclear localization signal, and participates in interaction with viral RNA and Nsp3. The flexibility of the LKR domain has been hypothesized to affect N protein multimerization and allow for differential protein-protein interaction and N activity, and the multiple functions of the LKR region are hypothesized to be controlled by its post-translational phosphorylation. The CTD of N allows for homodimerization, in some CoVs, has a nuclear export signal, and also has demonstrated RNA-binding activity. The CTR domain allows for tetramerization and has residues that interact with M for packaging within the virion. Most domains of N have demonstrated RNA-binding capabilities, further highlighted its key purpose in packaging, and facilitated viral replication and transcription through these interactions. Building upon the importance of N:RNA interaction, there have been many reports published throughout the SARS-CoV-2 pandemic regarding phase separation induced by N [56]. The studies suggest that the phase separation properties of N are key to inhibiting stress granule formation, forming ribonucleoprotein complexes with CoV gRNA, and interacting with M to package ribonucleoprotein complexes into virions. The phase separation activity of N has been shown to be controlled via post-translational phosphorylation of the LKR, which may direct activity to switch between N dimer or oligomer forms. Of note, CTD and CTR have been shown to be critical in the recognition of the packaging signal on the gRNA of MHV. However, for many coronaviruses, the exact packaging signal and N interaction residues remain to be defined by reverse genetics [6]. The N protein has also been implicated in antagonizing interferon signaling by interfering with TRIM25-mediated RIG-I signaling, TBK1-IRF3 interaction, and STAT phosphorylation. N is one of the most studied CoV proteins; however, given its domain structure, multimerization, variable post-translational modifications, and range of functions it may perform, its exact activity during CoV replication remains enigmatic.

2.4.3 Accessory ORFs

As previously mentioned, accessory ORFs are dispensable for viral replication and most commonly provide CoVs with functions to antagonize the host immune response to viral infection. Accessory ORFs are highly diverse between CoV species, and gaps in knowledge remain within the CoV field with respect to the function of many accessory ORFs in pathogenesis. Many of the proposed functions of accessory ORFs have been characterized using overexpression, and their functions still need to be evaluated using single ORF knockouts in infectious clone systems. Due to the diversity of accessory ORFs, we highlight functions described for the three highly pathogenic CoVs: SARS-CoV-1, MERS-CoV, and SARS-CoV-2.

MERS-CoV accessory ORFs have been widely implicated in antagonizing interferon (IFN) induction and signal transduction. ORF3 is similar to SARS-CoV-2 ORF3a and has demonstrated the ability to induce apoptosis, but it is degraded quickly in comparison. ORF4a is a double-stranded RNA-binding protein that suppresses activation of RIG-I and MDA5, the PKR antiviral stress response, and stress granule formation. ORF4b contains a phosphodiesterase domain that can inhibit RNase L activation, a nuclear localization signal capable of binding karyopherin 4 and inhibiting NF-κB nuclear transport, and has demonstrated suppression of IRF3 phosphorylation and nuclear translocation through interaction with IKKε and TBK1. ORF5 has been described as an IFN antagonist with no underlying mechanism behind its activity and has a structure similar to that of a viroporin. ORF8b suppresses RIG-I-like receptors and IFN expression through suppression of IKKε.

SARS-CoV-2 accessory ORF functions have been described to be like those of closely related SARS-CoV since many of the ORFs are conserved between the two viruses [57]. ORF3a acts as a viroporin, inducing caspase, inflammasome, and mitochondrial death pathway activation, and has also been described as reducing

IFNAR expression. ORF3b inhibits IFN production through a hypothesized interaction with MAVS, and SARS-CoV-2 ORF3b appears to be evolving to suppress IFN more efficiently. ORF6 inhibits the nuclear translocation of STAT and induces the degradation of STAT interacting proteins. ORF7a induces p38 MAPK and NF-κB activation and inhibits MHC-1 antigen presentation. ORF7b has been described to suppress type I and III IFNs and induce TNF-α-mediated apoptosis. SARS-CoV ORF8a and ORF8b mediate ubiquitin-dependent degradation of IRF3, both of which induce inflammasome activation, and ORF8a forms an ion channel. SARS-CoV-2 ORF8 acts as a histone mimic to disrupt chromatin modifications and epigenetic control of transcription during infection [58]. ORF9b limits IFN production through the degradation of MAVS downstream signaling components and interference with ubiquitination necessary for NF-κB activation.

2.5 Genetic Systems for the Study of CoV Biology

Reverse genetics has provided the virology field with the means to study viral gene function for decades by creating infectious clone systems in which genetic sequences can be manipulated to create mutant viruses (Fig. 3). While some positive-stranded RNA viruses are sufficiently short in length to be encoded into a single plasmid system, CoV and their long, structured RNA genomes pose a challenge for traditional reverse genetics. The field has employed the advances made in biotechnology to develop multiple infectious clone systems that have yielded key insights into the replication and pathogenesis of coronaviruses.

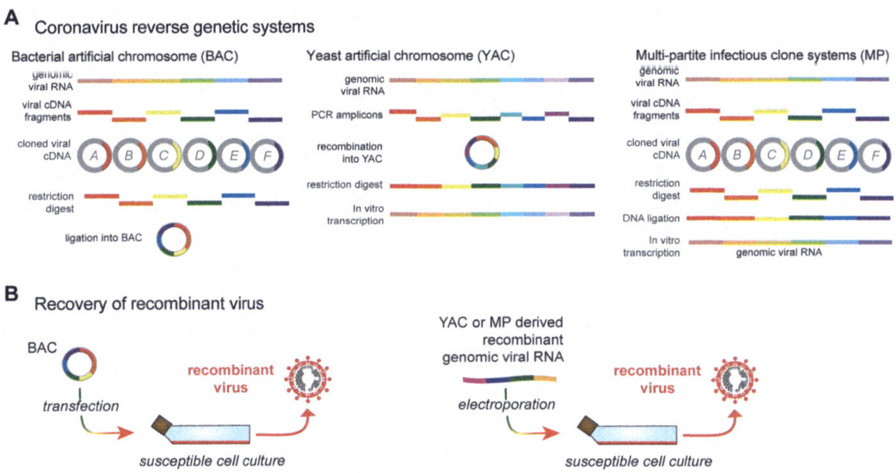

Fig. 3 Coronavirus reverse genetic systems. Several CoV reverse genetic systems have been developed. (**a**) Bacterial artificial chromosome (BAC), yeast artificial chromosome (YAC), and multi-partite (MP) systems and the steps to engineer recombinant viruses are shown. (**b**). Recovery of recombinant viruses in the BAC, YAC, and MP systems requires the transfection of DNA or RNA into cells that are susceptible to CoV infection

2.5.1 Bacterial Artificial Chromosome (BAC) Infectious Clone Systems

BAC systems allow a full-length cDNA copy of the CoV genome to be propagated in bacteria (Fig. 3). In order to produce the BAC, the viral genome must be segmented via preexisting or engineered restriction sites and amplified with plasmid vectors. These fragments are filled into a BAC backbone to create a full-length cDNA that is subsequently transformed into bacteria to produce additional copies. Recombinant virus launch is facilitated by transfection of the BAC cDNA into cells where viral genomic RNA is driven off a promoter that is functional in mammalian cells, such as CMV. The main advantage of the BAC system is the ability to propagate full-length cDNA in bacteria and create recombinant virus directly from DNA.

2.5.2 Yeast Artificial Chromosome (YAC) Infectious Clone Systems

YAC systems utilize transformation-associated recombination (TAR) in order to produce a YAC containing a full-length cDNA of the viral genome (Fig. 3). Viral genomic RNA is used as a template for targeted reverse transcription, employing site-specific primers to create overlapping regions within the viral genome and to a YAC backbone at the 5′ and 3′ ends. These cDNA fragments are PCR-amplified, pooled, and transformed into yeast where full-length cDNA is generated by recombination of overlapping regions. After propagation of the full-length YAC in yeast, full-length viral genomic RNA is then generated by in vitro transcription using purified YAC cDNA as template. Recombinant virus launch is facilitated by the transfection of viral genomic RNA into susceptible host cells. The main advantage of the YAC system is the ease of assembly of full-length viral cDNA since it does not require typical recombinant DNA ligation or the use of restriction sites.

2.5.3 Multipartite Infectious Clone Systems

In multipartite systems, the CoV genome is split among several plasmid vectors (Fig. 3), which are propagated in *Escherichia coli*. Directional assembly of full-length cDNA is facilitated by the ligation of DNA fragments containing unique restriction enzyme junctions on the 5′ and 3′ ends of each fragment. Full-length genomic cDNA serves as a template for in vitro transcription, and the subsequent viral RNA is electroporated into host cells for viral production. The main advantage of the multipartite system lies in the modular nature of the cDNA. If mutations in one particular gene are desired, then manipulation of only the plasmid containing this region is needed. Multipartite systems have facilitated the elucidation of several molecular determinants of CoV pathogenesis, including spike as a determinant of host range, ExoN as a mediator of replication fidelity, and various Nsp/accessory ORF proteins as innate immune antagonists.

2.6 Coronavirus Disease

2.6.1 Enzootic Coronaviruses

Enzootic coronaviruses infect a wide range of avian and mammalian species and are of concern due to their role as livestock or companion animal pathogens and the associated economic impact of these infections. A number of enzootic CoVs have been circulating for many years, with novel CoV emergence into livestock species observed as recently by 2012, with PDCoV. Enzootic CoVs typically cause either enteric illness or respiratory illness; however, some are capable of infecting the gut and lung systems, causing disease and bovine coronavirus (BCoV), for example. Porcine coronaviruses, PEDV, PDCoV, and TEGV cause enteric disease, resulting in high rates of mortality in infant piglets; thus, these CoVs are agricultural pathogens of concern. IBV and related gammacoronaviruses can infect a wide range of avian species, including chickens, causing respiratory illness fatal in chicks, reducing egg production, and spreading via feces. CCoV causes enteric illness, while canine respiratory CoV (CRCoV) causes respiratory illness, both spread in areas with high population density such as kennels. CRCoV, a betacoronavirus, was first detected in 2003 and encodes a HE gene like the human common cold-causing CoV OC43. FCoV is an interesting example of a CoV capable of causing enteric and systemic disease as it is hypothesized that FCoV is first transmitted as feline enteric coronavirus (FECV) and undergoes mutations in persistently infected populations to develop into a different biotype, feline infectious peritonitis virus (FIPV), capable of infecting monocytes and macrophages and spreading to the spleen, liver, and central nervous system, ultimately causing mortality. In addition to these commonly circulating enzootic CoVs, there are less common equine and camelid CoVs that have been detected by sampling livestock species. Many of these enzootic CoVs have shown an ability to infect human cells and ex vivo cultures, as well as demonstrating a wide host range, indicating we have just scratched the surface of the diversity of CoV species.

2.6.2 Human Coronaviruses (hCoV)

The four known hCoV, HCoV-229E, HCoV-OC43, HCoV-NL63, and HCoV-HKU1 (Fig. 4), are endemic in human populations and typically cause common cold-like illness in children and healthy adults but can cause severe respiratory disease in the aged and immunocompromised. All have global distributions and actively circulate in the human population, exhibiting seasonal epidemiological disease patterns and age-dependent disease severity, similar to influenza and rhinoviruses [59]. In most healthy individuals, symptoms include general malaise, dyspnea, headache, sneezing, nasal discharge, sore throat, croup, and less common fever and cough. However, in the elderly and immunocompromised, hCoV infections can result in pneumonia, bronchiolitis, febrile seizures, and even death. Phylogenetic studies of HCoV-OC43 and HCoV-229E spike receptor-binding domains suggest that CoVs may use adaptive evolution to evade host immune memory to remain endemic in the human population [60].

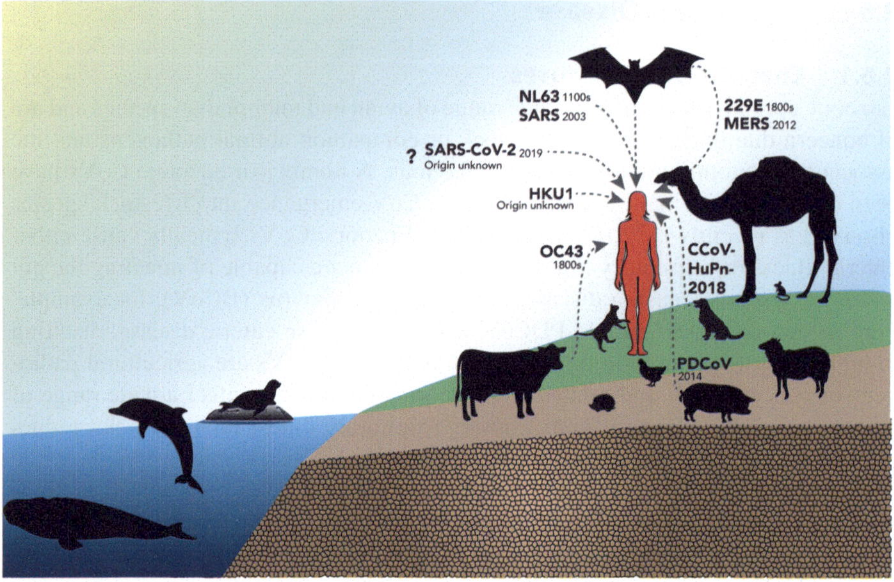

Fig. 4 The coronavirus family has emergence potential. CoV infects many different birds and mammals depicted in this figure as silhouettes. Through molecular clock analysis, almost all known human CoVs were originally believed to have emerged as zoonoses. For example, OC43 is closely related to bovine CoV and is thought to have spilled over into humans in the 1800s

2.6.3 Zoonotic CoV

Infections and emergence of three highly pathogenic CoVs over the past twenty years have demonstrated that novel CoV spillover events are not uncommon. Bats, other mammals, and birds are natural reservoirs for many viral families, including CoVs, and surveillance of these populations, especially bats in multiple geographic areas, has provided a wealth of CoV phylogeny. Additionally, reverse genetics studies utilizing full-length Bat-CoVs or recombinant chimeras encoding the spike glycoprotein have highlighted that numerous CoV species are capable of replicating efficiently in multiple host species and are poised for spillover into human populations. Previous studies demonstrated civets and raccoon dogs functioned as intermediate species for the 2003 SARS-CoV spillover into human populations, primarily occurring in open markets in southern China. In addition, dromedary camels were quickly identified as the most likely key intermediate species for the emergence of a bat MERS-CoV and its subsequent transmission into human hosts, implicating both exotic animal species and livestock in CoV spillover events. While the intermediate species of SARS-CoV-2 has yet to be determined, severally closely related SARS-CoV-2-like viruses have been identified in pangolins. However, these animals are rare and not maintained in large commercial operations for human consumption. Rather, illegal trade involving a few animals predominates so it remains unclear whether this species is a true intermediate host or a pass-through host, promoting the transmission of bat viruses into dense human environments. Theories regarding the belief that the Russian flu pandemic was caused by BCoV spillover to become OC43 further suggest the potential for livestock spillover events [3]. Zooanthroponosis has the potential to contribute to the recombination and generation of novel CoV species, given the evidence of transmission of SARS-CoV-2 to multiple mammalian species, including mink and deer. In addition, there is the potential for companion animals to be a source of zoonotic CoV spillover, based on recent evidence of a CCoV zoonotic infection [2].

2.6.4 Emergent CoVs

The 2003 SARS-CoV, 2012 MERS-CoV, and 2019 SARS-CoV-2 are highly pathogenic emerging coronaviruses that cause severe respiratory disease and mortality in human populations (Fig. 4). The mortality rate is variable among the emergent CoVs, with MERS-CoV (35%) being the most severe, followed by SARS-CoV (10%) and SARS-CoV-2 (1%). Interestingly, both SARS-CoV and SARS-CoV2 utilize the same receptor for entry, angiotensin-converting enzyme 2 (ACE2), yet their disease presentations and epidemiology are quite different, which remains a topic of scientific inquiry. Following infection, common symptoms of these three emergent CoVs include malaise, headache, fever, chills, myalgia, dyspnea, nonproductive cough, pneumonia, and diarrhea in some cases. The disease severity of SARS, MERS, and COVID-19 is age-dependent, as significantly higher mortality rates are recorded in elderly populations. Innate and adaptive immune dysregulation is more pronounced in elderly populations, indicating increased mortality of CoV disease in the latter age group may be a result of prolonged viremia and increased inflammation following clearance of viral replication. Of the three emergent CoVs, the course of SARS-CoV-2 disease has been studied in the greatest detail and is similar in many ways to both SARS-CoV and MERS-CoV-2 pathogenesis. In general, emergent CoV disease is typically biphasic where the first phase is driven by virus replication, while the second phase is mediated by immunopathological responses to virus-induced damage and late innate immunity recognition of viral infection. As highlighted in the previous sections, CoV structural, nonstructural, and accessory proteins are capable of antagonizing innate antiviral responses, allowing viral replication, cell death, and delayed IFN responses. Unchecked viral replication may lead to the accumulation of pathogen and damage-associated molecular patterns, causing macrophage and neutrophil infiltration and increased immune pathological disease in some cases. Increased immune cell infiltrates further increase inflammation that has been hypothesized to drive cytokine storms, with severely ill patients developing acute lung injury (ALI) that may progress to more severe disease, including end-stage lung disease and acute respiratory distress syndrome (ARDS).

ALI/ARDS is characterized as an acute inflammatory disorder associated with a deterioration of the alveolar epithelium and endothelium. Breakdown of these cellular barriers, release of pro-inflammatory cytokines by alveolar macrophages, excessive immune infiltration of neutrophils, and activation of fibroblasts are associated with initiation of the repair process. Unfortunately, these processes initiated for repair, when overexuberant, can drive increased tissue damage. The pathological features of ALI/ARDS include diffuse alveolar damage (DAD), alveolar septal thickening, and formation of hyaline membranes (HM). ALI/ARDS patients have increased respiration and decreased blood oxygen saturation, potentially leading to multi-organ failure and mortality, which has been the case with SARS, MERS, and COVID-19.

SARS-CoV-1 likely caused few asymptomatic infections during the global epidemic during 2002–2003 [61]. While the overall mortality rate of that epidemic was 10%, it increased with age where individuals of the elderly populations had a mortality rate of over 50% during the Hong Kong outbreak [61]. Comorbidities were more common in elderly populations. However, they did not account for the total increase compared to younger populations, indicating an increased complexity to the age-dependent increase in disease severity. Gastrointestinal illness was the most common extrapulmonary symptom of SARS-CoV-1 infection, followed by hepatic dysfunction and abnormal urinalysis. There were a few cases documenting seizures and detection of SARS-CoV-1 RNA in the cerebrospinal fluid. However, the mechanism behind SARS-CoV-1 invasion of the central nervous system remains to be understood.

SARS-CoV-1 utilizes ACE2 as a receptor for viral entry, which was of special interest given the protective role that ACE2 plays in acute lung injury and downregulation of the renin-angiotensin pathway. ACE2 is expressed on epithelial cells throughout the body, and SARS-CoV-1 viremia was documented in severe cases of SARS by quantification of viral RNA in extrapulmonary tissue. SARS-CoV-1 spike protein downregulates ACE2 expression, resulting in increased inflammation in the lung and worsening of SARS disease severity in a murine model. Increased inflammatory signatures have been documented in severe SARS cases, with patients having elevated serum levels of IFN-α/γ, IL-1, IL-6, IL-8, IL-12, CCL2, TGFβ, CXCL9, and CXCL10 with low levels of anti-inflammatory IL-10. SARS-CoV-1 abortive infection of hematopoietic cells demonstrated upregulation of many of these cytokines, indicating that non-replicative cellular infections may play important roles in immunopathogenic disease.

MERS-CoV is endemic in camels in the Middle East and has a human case fatality rate of 34%, with approximately 40% of cases resulting from nosocomial outbreaks [62]. Severe and fatal MERS-CoV cases occur more often in men, elderly, and immunocompromised populations and are often observed in those with comorbidities, including obesity, diabetes, cardiac disease, and lung disease. Mortality rates may be inflated due to lack of widespread surveillance, similar mild symptoms compared to other respiratory tract infections, and potential asymptomatic infections. Human-to-human spread of MERS-CoV is limited and appears to require close contact, with most transmission events occurring between household and healthcare contacts of symptomatic individuals. Acute kidney injury, gastrointestinal illness, and vomiting are the most common extrapulmonary symptoms.

Unlike the SARS-like CoVs, MERS-CoV uses CD26/DPP4 as a receptor for viral entry, which is expressed widely throughout the body and on hematopoietic cells. In the respiratory tract, DPP4 expression is lowest in the nasal mucosa, giving a possible explanation for why nasopharyngeal swabs of mild MERS can be

PCR-negative for MERS-CoV further suggesting that the lower levels of DPP4 expression in the upper respiratory tract may contribute to poor transmissibility between humans. DPP4 expression increases in the bronchioles, alveoli, and vascular endothelial cells. Interestingly, DPP4 expression is upregulated in the lungs of smokers and chronic obstructive pulmonary disorder (COPD) patients, which may contribute to increased susceptibility and comorbidity of MERS-CoV infection. Immune dysregulation has been described in MERS-CoV like that of SARS-CoV-1. Delayed increase in pro-inflammatory IFN, IL-1β, IL-6, IL-8, CCL5, and CXCL10 was detected in severe cases of MERS-CoV, correlating with increased levels of neutrophils and macrophages in the lung. MERS-CoV can replicate in macrophage and dendritic cells, unlike SARS-CoV-1, resulting in the induction of proinflammatory cytokines. Primary cell studies demonstrated that MERS-CoV infection of microvascular endothelial cells results in cytopathic effect, which may contribute to vascular leakage into the alveoli and the development of pneumonia.

SARS-CoV-2 remains a global pandemic and will most likely become endemic in the human population, causing annual seasonal outbreaks like currently endemic CoVs. The global mortality rate of COVID-19 is approximately 1%; however, transmission, disease severity, and mortality have fluctuated among variants throughout the pandemic. Elderly, immunocompromised, males, and individuals with comorbidities, such as cardiovascular disease, chronic lung disease, chronic kidney disease, diabetes, and obesity, incur the most severe symptoms and higher mortality. Asymptomatic SARS-CoV-2 infections are common, and most healthy individuals clear disease with mild respiratory tract symptoms that do not require hospitalization [63]. SARS-CoV-2 disease is typically characterized as biphasic, where the first phase is mediated by viral replication, primarily in the olfactory epithelium, ciliated airway epithelium, basal cells, and alveolar type 2 (AT2) cells in the alveoli, resulting in cell death, leading to edema and excessive damage to the gas exchange region of the lung. Anosmia following COVID-19 has been reported, even in mild cases, which may be a result of viral infection in the nasal passage and olfactory bulbs [64]. Gastrointestinal illness, nausea, and vomiting are common extrapulmonary symptoms of infection with neurological complications being rare. Severe cases of COVID-19 in males may lead to acute cardiac injury and failure, even in individuals without a history of heart disease.

SARS-CoV-2, like SARS-CoV-1, utilizes ACE2 as a receptor for viral entry. The spike glycoprotein of SARS-CoV-2 has a higher binding affinity to human ACE2 than SARS-CoV-1, partially explaining the increased transmissibility of SARS-CoV-2. As with SARS-CoV-1, SARS-CoV-2 is capable of replicating in airway and alveolar cells and abortive infection of hematopoietic cells. Overactivation of proinflammatory cytokines is common during severe cases of COVID-19, with many patients having elevated serum levels of IL-1β, IL-6, IL-8, IL-10, and TNF-α. Cytokine storm induces increased infiltration of macrophages, neutrophils, and T-lymphocytes, further damage to the alveoli, and vascular leakage. Activation of the complement system has been described in COVID-19 patients, and systemic activation was associated with respiratory failure. Additionally, coagulopathies have been reported during COVID-19 and patients with severe illness typically have elevated serum levels of D-dimer, thrombocytopenia, and low fibrinogen levels.

An ongoing consequence of the COVID-19 pandemic has been the prevalence of extended symptoms after recovery that have been termed long-haul COVID-19 or post-acute sequelae of COVID-19 (PASC). PASC symptoms have been documented in survivors of SARS and MERS as well. However, the higher number of COVID-19 cases has brought more attention to the burgeoning public health issue [65]. PASC includes a large number of symptoms across multiple organ systems that can last for months, following clearance of SARS-CoV-2 acute infection. Common symptoms are fatigue, malaise following physical activity, and cognitive dysfunction. Individuals that experienced asymptomatic acute infections can also develop PASC, and in general, aged individuals surviving severe disease may be more likely to develop long-term sequelae. Interestingly, women appear to be less likely to develop PASC than men. Most likely, multiple underlying mechanisms contribute to PASC development. For example, PASC respiratory sequelae may be a result of pulmonary fibrosis caused by the repair process, as noted above. Coagulation disorders described in COVID-19 cases are of special interest given that extended clotting and hypoxia may promote long-COVID disease phenotypes.

2.7 CoV Disease Models and Application in Vaccine and Therapeutic Development

2.7.1 Cell Lines

Cancer cell lines have long been used to study virus infection and replication. Many CoVs are capable of infecting and rapidly causing cytopathic effects in cell culture, but the susceptibility of cell lines to CoV infection is largely dependent on the expression of the target receptor for the CoV spike. Since cells can be transduced with vectors to knock-in and knock-out genes, they serve as important tools to understand how specific proteins can affect the viral life cycle. Engineering cell lines have been useful for studying the host range of CoV receptor binding, determinants of viral replication, and immune response to infections. The main drawback of cell lines is that they are a monoculture, which does not represent the diversity of cell populations present during infection of the host. Cell lines have aberrant cell cycles as well, which are essential for maintaining the lines and passaging, but also mean that they may not accurately recapitulate viral infections in vivo. Additionally, some cell lines are deficient in signaling pathways that may play important roles in responding to CoV infections, so special consideration must be taken in choosing the right cell line for the hypothesis in question.

2.7.2 Primary Cell Models

Primary cell culture models have grown in popularity in the study of CoVs since they are more representative of the diversity of cell types present in the airway during CoV infection while allowing easy manipulation and infection [66]. These models range from ex vivo systems in which cells are harvested from human lung tissue to pluripotent or embryonic stem cell systems that are specifically differentiated, both of which can form 2D or 3D cultures. Primary cells can mimic the large airways of the bronchi, small airways of the bronchioles, and alveoli, depending on the location from where they were harvested or how they are differentiated. The lung contains multiple cell types, including ionocytes, ciliated cells, club cells, basal cells, goblet cells, alveolar type 1 (AT1) cells, and alveolar type 2 (AT2) cells. Distribution of these cells changes from the proximal to the distal airway, so studying CoV replication and pathogenesis in the presence of specific cell populations allows for key insights into the infection dynamics of the human lung, without requiring invasive procedures.

Air/liquid interface cultures can be created from stem cells or patient samples to simulate the 2D dynamics of diverse lung cell populations during CoV infection. The cultures consist of cells grown on a transwell system so that the basolateral surface is submerged in media, and the apical surface is exposed to the air to mimic the air-liquid interface of the lung.

Organoids are 3D cultures that can be created from stem cells or patient samples to simulate lung tissue structures composed of diverse cell types during a CoV infection. The organoids can produce lumens and are able to produce mucus like the air-liquid interface cultures. The polarity of the cultures can be controlled to orient cells either inward or outward, depending on the need of the experiment, and culturing AT2 cells allows differentiation and production of alveolospheres to specifically simulate alveoli. Organoids can also be implanted into mice to study infection in the context of vascularization and immune cell infiltration.

An important attribute of ex vivo systems is their reliance on cells harvested during lung transplants or from nasal tissues. Cultures are created from an individual donor patient. This can be both a boon and a detriment to studies, depending on the reproducibility of the findings across multiple donors and the limited supply of cells from each patient. However, these cells may exhibit a more natural gene expression profile compared to pluripotent stem cell systems [66].

2.7.3 Animal Models. Animal Models

Animal models are critical to understand pathogenesis of infections in vivo and possible interventions that can reduce the disease burden. For agricultural pathogenic CoVs, viral infection and pathogenesis can be studied directly in the species of concern in controlled laboratory environments, so this section will focus on models for human CoV disease. Rodent species and nonhuman primates are the most widely used models for human CoV disease, but there has been work during the COVID-19 pandemic using ferrets as well. Each model comes with its own advantages and disadvantages to consider, but all have had specific studies done in which they proved most useful. Transgenic mice are commonly used to study CoV disease due to the ease in which it is possible to establish the model. Most human CoVs do not readily cause pathogenesis in mice due to differences in the human and mouse receptors to which the RBD of the CoV spike protein binds. To alleviate the mismatch between CoV RBD and mouse receptor binding, the murine ortholog can be "humanized" by gene replacement or through the introduction of point mutations, which facilitate virus binding and entry. These models also allow for the evaluation of pathogenic potential and zoonotic spillover for evaluating bat CoV isolates.

Transgenic cytokeratin-18 human angiotensin-converting enzyme 2 (K18-*hACE2*) mice were created following the SARS-CoV-1 pandemic to study the pathogenesis of SARS without mouse adaptation in vivo [67]. These mice express the human ACE2 gene on a K18 promoter so that SARS-CoV-1 can directly infect the mice and cause disease. SARS-CoV-2 also uses the ACE2 receptor, so these mice were quickly repurposed at the beginning of the COVID-19 pandemic to help preclinical studies of COVID-19. The hACE2 receptor is overexpressed in all epithelial cells in this model including those in the CNS, leading to extrapulmonary disease during SARS-CoV-1 and SARS-CoV-2 infections. Most notably, most K18-*hACE2* mice die from central nervous system complications following SARS-CoV-1 infection, which does not reflect human disease outcomes [67]. SARS-CoV-2 does not induce severe pathology in the brain of infected K18-*hACE2,* but does induce severe respiratory pathogenesis. This led to some investigators proposing K18-*hACE2* as a model to study the neuroinvasion seen following SARS-CoV-1 and SARS-CoV-2 infections.

Mouse-adapted virus models are created by infecting mice intranasally with CoVs, followed by serial passaging in mice to facilitate adaptive evolution. Resultant mouse-adapted viruses can be cloned using the reverse genetic systems noted above, facilitating the generation of recombinant virus and reverse genetic studies to identify the viral determinants of pathogenesis. The benefit of these systems is the ability of the mouse-adapted virus to cause disease in common laboratory strains of mice of all ages, allowing pathogenesis studies in outbred mouse lines and gene knockout mice without having to backcross to a transgenic line with the human CoV receptor. This is unfortunately not the case for MERS-CoV, as even mouse-adapted viruses still depend on transgenic DPP4 mice to establish infection and cause pathogenesis. An important caveat to mouse-adapted viruses is the mutations that occur to cause disease in mice may not fully reflect the pathogenesis demonstrated in humans and may interfere with the evaluation of antibody-based therapeutics and vaccines designed against the human variant virus.

While SARS-CoV-1 Urbani is capable of infecting and causing pathogenesis in aged mice, severe disease was not observed in young adult mice typically used for preclinical studies. Thus, a mouse-adapted model was created via passaging SARS-CoV-1 in 6-week-old BALB/c mice 15 times (MA15) to develop a strain capable of fatal disease outcome [68]. Six coding mutations were identified and subsequently cloned into an infectious clone to create the MA15 strain of SARS-CoV-1. The MA15 strain is capable of causing mortality in common laboratory strains of mice and has proven to be an invaluable tool for evaluating SARS and host determinants of pathogenesis.

During the COVID-19 pandemic, there grew a need to create a SARS-CoV-2 mouse-adapted model, given the limited supply of K18-*hACE2* and the desire to study the pathogenesis in preestablished knockout laboratory mouse strains. SARS-CoV-2, unlike SARS-CoV-1, did not replicate in mice, so a mouse-adapted (MA) SARS-CoV-2 was created by mutating residues Q498 and P499 of the S protein predicted to interfere with S binding to the mACE2. Upon infection, SARS-CoV-2 MA was cleared by young BALB/c mice, with slower clearance in aged mice and mild pathology in both age groups. Ten rounds of serial passaging of SARS-CoV-2 MA resulted in a highly pathogenic SARS-CoV-2 mouse-adapted strain (MA10) capable of severe respiratory disease, acute lung injury, and mortality [69]. MA10 causes less severe pathology in C57B6/J mice, indicating that disease outcomes may depend on the genetic background of the particular laboratory mice used.

Transgenic human dipeptidyl peptidase 4 (hDPP4) mice were generated to facilitate the study of MERS-CoV pathogenesis due to the inability of MERS-CoV to bind to the murine DPP4 receptor. Point mutations, knock-in of specific hDPP4 exons, or gene replacement strategies were used to create transgenic MERS-CoV models. As DPP4 plays a role in regulating T-cell function, these processes may be affected by species incompatibilities, which can vary, depending on the transgenic mouse strategy. For example, the two murine models that introduce the fewest changes to the mDPP4 receptor likely result in the most native DPP4 function and distribution. One example is a hDPP4 knock-in of human exons 10–12 important in MERS-CoV binding to the murine DPP4 receptor so that MERS-CoV can directly infect mice [70]. A second example includes a minimally modified DPP4 as residues 288 and 330, key binding sites for the MERS spike protein; however, passaging and adaptation of MERS-CoV were necessary to create a lethal disease model [71].

Humanized mouse models for CoV infection are created by either depleting the immune system of mice or using bone marrow-liver-thymus mice that are implanted with human lung tissue [72]. This allows for the lung implants on these mice to be directly infected with clinical CoV isolates and to study immune cell infiltration during CoV infection. These mouse models are typically more expensive to use and more labor-intensive, so they are not as advantageous in cost compared to standard laboratory strains. However, they provide a unique model in which to study CoV pathogenesis in the context of human immune cells.

Collaborative cross (CC) mice allow for the evaluation of how host genetic variation affects disease outcomes and pathogenesis [73]. The CC is large panel recombinant inbred lines that were created from eight founder strains, five classical laboratory strains, and three wild-derived strains, using a funnel breeding scheme. The defined genetics and diversity among the lines allow precise genetic mapping of loci that contribute to specific disease outcomes and host responses to infection. The use of the CC requires that the CoV of interest be able to readily infect mice. CC mice are more expensive and exhibit less docile behavior compared to laboratory strains of mice; however, they provide a unique opportunity to map how host genetics contributes to the variability in CoV disease outcomes. This is of special interest with SARS-CoV-2 given the range of symptoms that have been documented throughout the COVID-19 pandemic.

Golden Syrian hamster models have been employed for SARS-CoV-1 and SARS-CoV-2 due to their ability to be readily infected without viral adaptation [74]. SARS-CoV-1 replicates to high titers, but requires hamsters to be immunocompromised to develop severe morbidity and mortality. Hamsters infected with

SARS-CoV-2 develop inflammatory immune responses and severe pathology; however, there is no disease-induced mortality. Hamsters shed SARS-CoV-2 and can transmit the virus if cohoused with uninfected hamsters. Hamsters are easy to handle, but there are less abundant research resources available to detect hamster proteins and fewer knockout lines that are established, limiting their usage.

Ferrets, unlike mice and hamsters, can develop a cough following respiratory viral infections. For this reason, ferrets have historically been used to study transmission of influenza in vivo and have recently been used to study SARS-CoV-2 transmission [75]. Ferret ACE2 contains critical ACE2 SARS-CoV-2 binding residues, making the animals susceptible to infection and mild disease symptoms. Most importantly, simulations of direct and indirect contact have been able to reproduce viral transmission between ferrets, making them an essential model to study CoV transmission. Unfortunately, for the ferret model to be used, the receptor for CoV binding must have the residues for spike to interact or CoVs must be adapted to infect ferrets. Additionally, ferrets are not as easily handled and do not breed as quickly as smaller rodents.

Nonhuman primates (NHPs) are considered the most ideal animal to model CoV disease due to their physiological similarity and close phylogenetic relatedness to humans. Thus, NHPs have been employed for many viral families to evaluate vaccines and therapeutics, including all three highly pathogenic CoVs. MERS-CoV and SARS-CoV-1/2 readily infect many NHPs and cause respiratory pathology. However, depending on the NHP that is used, there may not be overt symptoms indicative of viral infection. Both rhesus macaques and common marmosets have been used to study SARS-CoV-1/2 and demonstrate overt clinical symptoms and lung pathology following MERS-CoV infection. SARS-CoV-1 symptoms appear to be viral strain-dependent in rhesus macaques. Infections of common marmosets display more severe symptoms including fever, pneumonia, and diarrhea. SARS-CoV-2 causes age-dependent disease in rhesus macaques and milder disease in common marmosets. Cynomolgus macaques are capable of being infected by both SARS-CoV-1/2, but only SARS-CoV-2 causes overt disease symptoms that are lesser than those observed in rhesus macaques, while SARS-CoV-1 causes diffuse alveolar damage. NHP models are by far the most challenging to use due to limitations on the number of animals, specific facilities required to conduct the research, and ethical questions regarding the importance of the studies and treatment of animals.

The power of the above animal models is poignantly exemplified by their having been used in the preclinical development of multiple approved vaccines and therapeutics, including the Moderna vaccine and the antiviral therapies remdesivir and molnupiravir, as well as an array of human monoclonal antibodies.

3 Conclusion

Several novel CoVs have emerged in the past 20 years (SARS1, MERS, SARS-2, SADS, and PEDV) with increased pathogenesis and increased range throughout the world. Thus, we will likely continue to experience outbreaks and epidemics of CoVs in the future, creating the necessity to continue to develop CoV-specific vaccines and therapeutics. The diversity of CoV host species, their unique replication machinery, the ability of viral proteins to antagonize innate immunity, and the high amount of recombination within their genomes have created three potential once-in-a-lifetime events within the past twenty years. How many more are to come, and will we be prepared to respond? This is a question that is impossible to answer but we must try.

References

1. Cui J, Li F, Shi ZL (2019) Origin and evolution of pathogenic coronaviruses. Nat Rev Microbiol 17(3):181–192. https://doi.org/10.1038/s41579-018-0118-9
2. Vlasova AN, Diaz A, Damtie D, Xiu L, Toh TH, Lee JS et al (2022) Novel canine coronavirus isolated from a hospitalized patient with pneumonia in East Malaysia. Clin Infect Dis 74(3):446–454. https://doi.org/10.1093/cid/ciab456
3. Brussow H, Brussow L (2021) Clinical evidence that the pandemic from 1889 to 1891 commonly called the Russian flu might have been an earlier coronavirus pandemic. Microb Biotechnol 14(5):1860–1870. https://doi.org/10.1111/1751-7915.13889
4. Lednicky JA, Tagliamonte MS, White SK, Elbadry MA, Alam MM, Stephenson CJ et al (2021) Independent infections of porcine deltacoronavirus among Haitian children. Nature 600(7887):133–137. https://doi.org/10.1038/s41586-021-04111-z
5. Barcena M, Oostergetel GT, Bartelink W, Faas FG, Verkleij A, Rottier PJ et al (2009) Cryo-electron tomography of mouse hepatitis virus: insights into the structure of the coronavirion. Proc Natl Acad Sci USA 106(2):582–587. https://doi.org/10.1073/pnas.0805270106
6. Masters PS (2019) Coronavirus genomic RNA packaging. Virology 537:198–207. https://doi.org/10.1016/j.virol.2019.08.031
7. Menachery VD, Mitchell HD, Cockrell AS, Gralinski LE, Yount BL Jr, Graham RL et al (2017) MERS-CoV accessory ORFs play key role for infection and pathogenesis. MBio 8(4). https://doi.org/10.1128/mBio.00665-17
8. Huang C, Liu WJ, Xu W, Jin T, Zhao Y, Song J et al (2016) A bat-derived putative cross-family recombinant coronavirus with a reovirus gene. PLoS Pathog 12(9):e1005883. https://doi.org/10.1371/journal.ppat.1005883
9. Yang D, Leibowitz JL (2015) The structure and functions of coronavirus genomic 3′ and 5′ ends. Virus Res 206:120–133. https://doi.org/10.1016/j.virusres.2015.02.025
10. Imperatore JA, Cunningham CL, Pellegrene KA, Brinson RG, Marino JP, Evanseck JD et al (2022) Highly conserved s2m element of SARS-CoV-2 dimerizes via a kissing complex and interacts with host miRNA-1307-3p. Nucleic Acids Res 50(2):1017–1032. https://doi.org/10.1093/nar/gkab1226
11. Bosch BJ, van der Zee R, de Haan CA, Rottier PJ (2003) The coronavirus spike protein is a class I virus fusion protein: structural and functional characterization of the fusion core complex. J Virol 77(16):8801–8811. https://doi.org/10.1128/jvi.77.16.8801-8811.2003
12. Dang M, Li Y, Song J (2021) ATP biphasically modulates LLPS of SARS-CoV-2 nucleocapsid protein and specifically binds its RNA-binding domain. Biochem Biophys Res Commun 541:50–55. https://doi.org/10.1016/j.bbrc.2021.01.018

13. Plant EP, Perez-Alvarado GC, Jacobs JL, Mukhopadhyay B, Hennig M, Dinman JD (2005) A three-stemmed mRNA pseudoknot in the SARS coronavirus frameshift signal. PLoS Biol 3(6):e172. https://doi.org/10.1371/journal.pbio.0030172
14. Sawicki SG, Sawicki DL (1998) A new model for coronavirus transcription. Adv Exp Med Biol 440:215–219. https://doi.org/10.1007/978-1-4615-5331-1_26
15. Sola I, Almazan F, Zuniga S, Enjuanes L (2015) Continuous and discontinuous RNA synthesis in coronaviruses. Annu Rev Virol 2(1):265–288. https://doi.org/10.1146/annurev-virology-100114-055218
16. Liu DX, Cavanagh D, Green P, Inglis SC (1991) A polycistronic mRNA specified by the coronavirus infectious bronchitis virus. Virology 184(2):531–544. https://doi.org/10.1016/0042-6822(91)90423-9
17. Kim D, Lee JY, Yang JS, Kim JW, Kim VN, Chang H (2020) The architecture of SARS-CoV-2 transcriptome. Cell 181(4):914–21 e10. https://doi.org/10.1016/j.cell.2020.04.011
18. Yount B, Roberts RS, Lindesmith L, Baric RS (2006) Rewiring the severe acute respiratory syndrome coronavirus (SARS-CoV) transcription circuit: engineering a recombination-resistant genome. Proc Natl Acad Sci USA 103(33):12546–12551. https://doi.org/10.1073/pnas.0605438103
19. Vora SM, Fontana P, Mao T, Leger V, Zhang Y, Fu TM et al (2022) Targeting stem-loop 1 of the SARS-CoV-2 5′ UTR to suppress viral translation and Nsp1 evasion. Proc Natl Acad Sci USA 119(9). https://doi.org/10.1073/pnas.2117198119
20. Graham RL, Sims AC, Brockway SM, Baric RS, Denison MR (2005) The nsp2 replicase proteins of murine hepatitis virus and severe acute respiratory syndrome coronavirus are dispensable for viral replication. J Virol 79(21):13399–13411. https://doi.org/10.1128/JVI.79.21.13399-13411.2005
21. Lei J, Kusov Y, Hilgenfeld R (2018) Nsp3 of coronaviruses: structures and functions of a large multi-domain protein. Antivir Res 149:58–74. https://doi.org/10.1016/j.antiviral.2017.11.001
22. Voth LS, O'Connor JJ, Kerr CM, Doerger E, Schwarting N, Sperstad P et al (2021) Unique mutations in the murine hepatitis virus macrodomain differentially attenuate virus replication, indicating multiple roles for the macrodomain in coronavirus replication. J Virol 95(15):e0076621. https://doi.org/10.1128/JVI.00766-21
23. Wolff G, Limpens R, Zevenhoven-Dobbe JC, Laugks U, Zheng S, de Jong AWM et al (2020) A molecular pore spans the double membrane of the coronavirus replication organelle. Science 369(6509):1395–1398. https://doi.org/10.1126/science.abd3629
24. Hagemeijer MC, Monastyrska I, Griffith J, van der Sluijs P, Voortman J, van Bergen en Henegouwen PM et al (2014) Membrane rearrangements mediated by coronavirus nonstructural proteins 3 and 4. Virology 458–459:125–135. https://doi.org/10.1016/j.virol.2014.04.027
25. Angelini MM, Akhlaghpour M, Neuman BW, Buchmeier MJ (2013) Severe acute respiratory syndrome coronavirus nonstructural proteins 3, 4, and 6 induce double-membrane vesicles. MBio 4(4). https://doi.org/10.1128/mBio.00524-13
26. Lu Y, Lu X, Denison MR (1995) Identification and characterization of a serine-like proteinase of the murine coronavirus MHV-A59. J Virol 69(6):3554–3559. https://doi.org/10.1128/JVI.69.6.3554-3559.1995
27. Wang D, Fang L, Shi Y, Zhang H, Gao L, Peng G et al (2016) Porcine epidemic diarrhea virus 3C-like protease regulates its interferon antagonism by cleaving NEMO. J Virol 90(4):2090–2101. https://doi.org/10.1128/JVI.02514-15
28. Zhu X, Wang D, Zhou J, Pan T, Chen J, Yang Y et al (2017) Porcine deltacoronavirus nsp5 antagonizes type I interferon signaling by cleaving STAT2. J Virol 91(10). https://doi.org/10.1128/JVI.00003-17
29. Ricciardi S, Guarino AM, Giaquinto L, Polishchuk EV, Santoro M, Di Tullio G et al (2022) The role of NSP6 in the biogenesis of the SARS-CoV-2 replication organelle. Nature 606(7915):761–768. https://doi.org/10.1038/s41586-022-04835-6
30. Imbert I, Guillemot JC, Bourhis JM, Bussetta C, Coutard B, Egloff MP et al (2006) A second, non-canonical RNA-dependent RNA polymerase in SARS coronavirus. EMBO J 25(20):4933–4942. https://doi.org/10.1038/sj.emboj.7601368

31. Tvarogova J, Madhugiri R, Bylapudi G, Ferguson LJ, Karl N, Ziebuhr J (2019) Identification and characterization of a human coronavirus 229E nonstructural protein 8-associated RNA 3′-terminal adenylyltransferase activity. J Virol 93(12). https://doi.org/10.1128/JVI.00291-19
32. Zhang C, Li L, He J, Chen C, Su D (2021) Nonstructural protein 7 and 8 complexes of SARS-CoV-2. Protein Sci 30(4):873–881. https://doi.org/10.1002/pro.4046
33. Egloff MP, Ferron F, Campanacci V, Longhi S, Rancurel C, Dutartre H et al (2004) The severe acute respiratory syndrome-coronavirus replicative protein nsp9 is a single-stranded RNA-binding subunit unique in the RNA virus world. Proc Natl Acad Sci USA 101(11):3792–3796. https://doi.org/10.1073/pnas.0307877101
34. Yan L, Ge J, Zheng L, Zhang Y, Gao Y, Wang T et al (2021) Cryo-EM structure of an extended SARS-CoV-2 replication and transcription complex reveals an intermediate state in cap synthesis. Cell 184(1):184–93 e10. https://doi.org/10.1016/j.cell.2020.11.016
35. Bouvet M, Debarnot C, Imbert I, Selisko B, Snijder EJ, Canard B et al (2010) In vitro reconstitution of SARS-coronavirus mRNA cap methylation. PLoS Pathog 6(4):e1000863. https://doi.org/10.1371/journal.ppat.1000863
36. Bouvet M, Imbert I, Subissi L, Gluais L, Canard B, Decroly E (2012) RNA 3′-end mismatch excision by the severe acute respiratory syndrome coronavirus nonstructural protein nsp10/nsp14 exoribonuclease complex. Proc Natl Acad Sci USA 109(24):9372–9377. https://doi.org/10.1073/pnas.1201130109
37. Lehmann KC, Gulyaeva A, Zevenhoven-Dobbe JC, Janssen GM, Ruben M, Overkleeft HS et al (2015) Discovery of an essential nucleotidylating activity associated with a newly delineated conserved domain in the RNA polymerase-containing protein of all nidoviruses. Nucleic Acids Res 43(17):8416–8434. https://doi.org/10.1093/nar/gkv838
38. Malone B, Chen J, Wang Q, Llewellyn E, Choi YJ, Olinares PDB et al (2021) Structural basis for backtracking by the SARS-CoV-2 replication-transcription complex. Proc Natl Acad Sci USA 118(19). https://doi.org/10.1073/pnas.2102516118
39. Ivanov KA, Thiel V, Dobbe JC, van der Meer Y, Snijder EJ, Ziebuhr J (2004) Multiple enzymatic activities associated with severe acute respiratory syndrome coronavirus helicase. J Virol 78(11):5619–5632. https://doi.org/10.1128/JVI.78.11.5619-5632.2004
40. Hsu JC, Laurent-Rolle M, Pawlak JB, Wilen CB, Cresswell P (2021) Translational shutdown and evasion of the innate immune response by SARS-CoV-2 NSP14 protein. Proc Natl Acad Sci USA 118(24). https://doi.org/10.1073/pnas.2101161118
41. Chen Y, Cai H, Pan J, Xiang N, Tien P, Ahola T et al (2009) Functional screen reveals SARS coronavirus nonstructural protein nsp14 as a novel cap N7 methyltransferase. Proc Natl Acad Sci USA 106(9):3484–3489. https://doi.org/10.1073/pnas.0808790106
42. Frazier MN, Riccio AA, Wilson IM, Copeland WC, Stanley RE (2022) Recent insights into the structure and function of coronavirus ribonucleases. FEBS Open Bio 12:1567. https://doi.org/10.1002/2211-5463.13414
43. Decroly E, Imbert I, Coutard B, Bouvet M, Selisko B, Alvarez K et al (2008) Coronavirus nonstructural protein 16 is a cap-0 binding enzyme possessing (nucleoside-2′O)-methyltransferase activity. J Virol 82(16):8071–8084. https://doi.org/10.1128/JVI.00407-08
44. Li F (2016) Structure, function, and evolution of coronavirus spike proteins. Annu Rev Virol. 3(1):237–261. https://doi.org/10.1146/annurev-virology-110615-042301
45. Li Z, Tomlinson AC, Wong AH, Zhou D, Desforges M, Talbot PJ et al (2019) The human coronavirus HCoV-229E S-protein structure and receptor binding. elife 8:8. https://doi.org/10.7554/eLife.51230
46. Chinese SMEC (2004) Molecular evolution of the SARS coronavirus during the course of the SARS epidemic in China. Science 303(5664):1666–1669. https://doi.org/10.1126/science.1092002
47. Li Y, Zheng P, Liu T, Shi C, Wang B, Xu Y et al (2022) Structural requirements and plasticity of receptor-binding domain in human coronavirus spike. Front Mol Biosci 9:930931. https://doi.org/10.3389/fmolb.2022.930931

48. Desforges M, Desjardins J, Zhang C, Talbot PJ (2013) The acetyl-esterase activity of the hemagglutinin-esterase protein of human coronavirus OC43 strongly enhances the production of infectious virus. J Virol 87(6):3097–3107. https://doi.org/10.1128/JVI.02699-12
49. Schoeman D, Fielding BC (2019) Coronavirus envelope protein: current knowledge. Virol J 16(1):69. https://doi.org/10.1186/s12985-019-1182-0
50. Westerbeck JW, Machamer CE (2015) A coronavirus E protein is present in two distinct pools with different effects on assembly and the secretory pathway. J Virol 89(18):9313–9323. https://doi.org/10.1128/JVI.01237-15
51. Jimenez-Guardeno JM, Regla-Nava JA, Nieto-Torres JL, DeDiego ML, Castano-Rodriguez C, Fernandez-Delgado R et al (2015) Identification of the mechanisms causing reversion to virulence in an attenuated SARS-CoV for the design of a genetically stable vaccine. PLoS Pathog 11(10):e1005215. https://doi.org/10.1371/journal.ppat.1005215
52. Zhang Z, Nomura N, Muramoto Y, Ekimoto T, Uemura T, Liu K et al (2022) Structure of SARS-CoV-2 membrane protein essential for virus assembly. Nat Commun 13(1):4399. https://doi.org/10.1038/s41467-022-32019-3
53. Chang CK, Hou MH, Chang CF, Hsiao CD, Huang TH (2014) The SARS coronavirus nucleocapsid protein—forms and functions. Antivir Res 103:39–50. https://doi.org/10.1016/j.antiviral.2013.12.009
54. Zuniga S, Cruz JL, Sola I, Mateos-Gomez PA, Palacio L, Enjuanes L (2010) Coronavirus nucleocapsid protein facilitates template switching and is required for efficient transcription. J Virol 84(4):2169–2175. https://doi.org/10.1128/JVI.02011-09
55. McBride R, van Zyl M, Fielding BC (2014) The coronavirus nucleocapsid is a multifunctional protein. Viruses 6(8):2991–3018. https://doi.org/10.3390/v6082991
56. Cascarina SM, Ross ED (2022) Phase separation by the SARS-CoV-2 nucleocapsid protein: consensus and open questions. J Biol Chem 298(3):101677. https://doi.org/10.1016/j.jbc.2022.101677
57. Fang P, Fang L, Zhang H, Xia S, Xiao S (2021) Functions of coronavirus accessory proteins: overview of the state of the art. Viruses 13(6). https://doi.org/10.3390/v13061139
58. Kee J, Thudium S, Renner DM, Glastad K, Palozola K, Zhang Z et al (2022) SARS-CoV-2 disrupts host epigenetic regulation via histone mimicry. Nature 610:381. https://doi.org/10.1038/s41586-022-05282-z
59. Su S, Wong G, Shi W, Liu J, Lai ACK, Zhou J et al (2016) Epidemiology, genetic recombination, and pathogenesis of coronaviruses. Trends Microbiol 24(6):490–502. https://doi.org/10.1016/j.tim.2016.03.003
60. Kistler KE, Bedford T (2021) Evidence for adaptive evolution in the receptor-binding domain of seasonal coronaviruses OC43 and 229e. elife 10:10. https://doi.org/10.7554/eLife.64509
61. Peiris JS, Guan Y, Yuen KY (2004) Severe acute respiratory syndrome. Nat Med 10(12 Suppl):S88–S97. https://doi.org/10.1038/nm1143
62. Azhar EI, Hui DSC, Memish ZA, Drosten C, Zumla A (2019) The Middle East Respiratory Syndrome (MERS). Infect Dis Clin N Am 33(4):891–905. https://doi.org/10.1016/j.idc.2019.08.001
63. Nishiura H, Kobayashi T, Miyama T, Suzuki A, Jung SM, Hayashi K et al (2020) Estimation of the asymptomatic ratio of novel coronavirus infections (COVID-19). Int J Infect Dis 94:154–155. https://doi.org/10.1016/j.ijid.2020.03.020
64. Najafloo R, Majidi J, Asghari A, Aleemardani M, Kamrava SK, Simorgh S et al (2021) Mechanism of anosmia caused by symptoms of COVID-19 and emerging treatments. ACS Chem Neurosci 12(20):3795–3805. https://doi.org/10.1021/acschemneuro.1c00477
65. O'Sullivan O (2021) Long-term sequelae following previous coronavirus epidemics. Clin Med (Lond) 21(1):e68–e70. https://doi.org/10.7861/clinmed.2020-0204
66. Edwards CE, Tata A, Baric RS (2022) Human lung organoids as a model for respiratory virus replication and countermeasure performance in human hosts. Transl Res 250:36. https://doi.org/10.1016/j.trsl.2022.07.002

67. McCray PB Jr, Pewe L, Wohlford-Lenane C, Hickey M, Manzel L, Shi L et al (2007) Lethal infection of K18-hACE2 mice infected with severe acute respiratory syndrome coronavirus. J Virol 81(2):813–821. https://doi.org/10.1128/JVI.02012-06
68. Roberts A, Deming D, Paddock CD, Cheng A, Yount B, Vogel L et al (2007) A mouse-adapted SARS-coronavirus causes disease and mortality in BALB/c mice. PLoS Pathog 3(1):e5. https://doi.org/10.1371/journal.ppat.0030005
69. Leist SR, Dinnon KH 3rd, Schafer A, Tse LV, Okuda K, Hou YJ et al (2020) A mouse-adapted SARS-CoV-2 induces acute lung injury and mortality in standard laboratory mice. Cell 183(4):1070–85 e12. https://doi.org/10.1016/j.cell.2020.09.050
70. Li K, Wohlford-Lenane CL, Channappanavar R, Park JE, Earnest JT, Bair TB et al (2017) Mouse-adapted MERS coronavirus causes lethal lung disease in human DPP4 knockin mice. Proc Natl Acad Sci USA 114(15):E3119–E3E28. https://doi.org/10.1073/pnas.1619109114
71. Douglas MG, Kocher JF, Scobey T, Baric RS, Cockrell AS (2018) Adaptive evolution influences the infectious dose of MERS-CoV necessary to achieve severe respiratory disease. Virology 517:98–107. https://doi.org/10.1016/j.virol.2017.12.006
72. Wahl A, De C, Abad Fernandez M, Lenarcic EM, Xu Y, Cockrell AS et al (2019) Precision mouse models with expanded tropism for human pathogens. Nat Biotechnol 37(10):1163–1173. https://doi.org/10.1038/s41587-019-0225-9
73. Threadgill DW, Miller DR, Churchill GA, de Villena FP (2011) The collaborative cross: a recombinant inbred mouse population for the systems genetic era. ILAR J 52(1):24–31. https://doi.org/10.1093/ilar.52.1.24
74. Roberts A, Vogel L, Guarner J, Hayes N, Murphy B, Zaki S et al (2005) Severe acute respiratory syndrome coronavirus infection of golden Syrian hamsters. J Virol 79(1):503–511. https://doi.org/10.1128/JVI.79.1.503-511.2005
75. Kim YI, Kim SG, Kim SM, Kim EH, Park SJ, Yu KM et al (2020) Infection and rapid transmission of SARS-CoV-2 in ferrets. Cell Host Microbe 27(5):704–9 e2. https://doi.org/10.1016/j.chom.2020.03.023

COVID-19 and the Role of Conspiracy Theories, Disinformation, and Fake News

Peter Katona

1 Introduction

> If we don't have the capacity to distinguish what's true from what's false then by definition the marketplace of ideas doesn't work, then by definition our democracy doesn't work. We are entering into an epistemological (how we know what we know) crisis.
>
> Barack Obama 2020

The Chinese describe the word "crisis" in two pictograms or characters. One is the symbol for danger and the other is for opportunity. As we are currently emersed in a vast information mismanagement crisis, information is all too often spun, slanted, or outright lied about. The "truth" has become more and more blurry. Subsequently, science, public health, and democracy take a toll as more and more people can't distinguish scientific or political truth from falsehood. Almost three-quarters of doctors say misinformation made it harder to treat patients for COVID-19 and made patient outcomes worse, according to a recent USA today poll." They also reported that, "nearly as many flagged problematic misinformation about weight loss, supplements, mental health and other vaccines" [20].

In the early stages of the COVID-19 pandemic in 2020, the World Health Organization (WHO) declared a global epidemic of *both* disease and misinformation when Director-General Tedros Adhanom Ghebreyesus stated that "we're fighting an infodemic" of information and a pandemic of disease. This gained great momentum with the US presidential election and the release of mRNA COVID-19 vaccines in late 2020.

By 2022, little had changed. Surgeon General Vivek H. Murthy called COVID-19 fake news an "urgent threat to public health." He specifically singled out technology companies that enabled misinformation "to poison our information environment with little accountability to their users" [21]. Both Surgeon General Murthy and

P. Katona (✉)
UCLA Schools of Medicine and Public Health, Los Angeles, CA, USA

J. M. Rosen, R. R. Colwell (eds.), *The COVID-19 Pandemic*,
https://doi.org/10.1007/978-3-031-62772-9_7

President Biden placed much of the blame on social media platforms, with President Biden going as far as to say "They're killing people," with the spread of misinformation. Physicians agree, especially regarding medical rumors [19, 20]. Some states have even tried punishing physicians for bad choices [13]. With this in mind we will delve into how we define the terms, how we got here, the sources and consequences, mistrust, and make suggestions on combatting this plague [29].

2 Definitions

Fake news is more complex than merely being untrue [22]. The perspective of the perpetrator and the receiver are both relevant. There are three main categories: First, there is news that is intentionally untrue. There may be a covert or overt motive, often with spin or exaggeration. Second, there may be news that is inadvertently untrue, including having a misconception, bad reporting, ambiguity, or being duped. Third, there is news that you simply don't like or don't want to hear, leading to denials, spin, and more untruths. Fake news stories can use technology, newscasts, and social media to look like "proper" news sites. Even an intentional omission of an important story falls under the same umbrella. An example was the initial day of the televised January 6 Committee hearings which was not covered by Fox News. They even took out commercials just in case a viewer might want to temporarily channel surf to more liberal media. Often, specific target demographics are singled out to reinforce a point over and over again until it just has to be true.

Misinformation is false information widely spread. Malinformation is truthful information published in a way so as to mislead and harm. Disinformation is false information spread deliberately with the intent to mislead or misorient. "Influence operations" involve coordinated actions by a state or substate actor or bot[1] done through means other than direct force to influence a specified target audience's attitudes, behaviors, or decisions. All these terms fall under the term "fake news."

A cookie is information that a website puts on or collects from a user's computer. Cookies store limited information from a Web browser session on a given website that can then be retrieved in the future. Some websites will ask if they can use cookies. Deep Fakes (video and/or voice) show people saying or doing things that appear very real but in fact are not. The close presidential election in Slovakia was affected by a deep fake, probably sent from Russia [30].

Those that perpetrate fake news either do so knowingly or unknowingly. For those who purposely spread false information, they may have a political, economic, religious, or scientific agenda. Some look for easy answers to complicated problems. Some spread misinformation with malice, some are ignorant or misled. And if it is repeated over and over again, it is more likely to be believed.

We also have to assume that there is no "absolute" truth, only degrees of it. Almost every news item or scientific thought can carry some degree of "spin" which

[1] a bot is a software application that runs automated tasks over the Internet, usually with the intent to imitate human activity on the Internet, including messaging, on a large scale.

is at the border of truth and fake news. An example of spin leading to fake news was when, within hours of the brutal attack on Paul Pelosi, House Speaker Pelosi's husband, right-wing activists and media figures began circulating groundless conspiracy claims—nearly all of them sinister and many homophobic. All were quickly withdrawn as spin is not as powerful as outright lying.

There is the excuse or counterargument that everybody lies and thus lacks at least some credibility. This takes us to a misleading false equivalence. News truthfulness can be ambiguous, where the line between truth and falsehood is blurred. Anyone can say or write anything, and you can't disprove even an absurd idea such as aliens had landed and controlled our lives.

Fake news can be intentionally untrue, where there is use of exaggeration or spin or outright lying to achieve a covert or overt motive. There might be news that is inadvertently untrue based on an uneducated misconception. It might also be merely unresearched or undesirable. While leaving out an important news item from a daily news report, for example, may not technically be fake news, it is intentionally misleading and thus falls under the umbrella of disinformation.

3 How Did We Get Here?

Fake news crisis periods are recurrent and cyclical and have had many causes. As Rand's excellent report "Truth Decay" [7] stated, a surge in American fake news in the last century occurred in 20- to 30-year cycles. In other words, it didn't start with COVID-19. These cycles were the following:

- 1880s–1890s with yellow journalism
- 1920s–1930s with jazz journalism
- 1960s–1970s with new journalism
- 2000s with easy weaponization of truth on many new and old platforms, especially cable TV and social media

Yellow journalism and yellow press, peaking in the late nineteenth century, are American terms for journalism and associated newspapers that present little or no legitimate, well-researched news while instead using eye-catching headlines for increased sales. Techniques may include exaggerations of news events, scandal-mongering, or sensationalism.

Jazz journalism is a sensationalist style of news that matched its era, the rebellious Roaring Twenties, a dramatic change from the somber news of World War I. It was a term applied to American sensational newspapers in the 1920s. Focused on entertainment, celebrities, sports, scandal, and crime, the style was a New York phenomenon, practiced primarily by three new tabloid-size daily newspapers in a fight for circulation.

New Journalism is a style of news writing and journalism, developed in the 1960s and 1970s, that used literary techniques unconventional at the time. It is characterized by a subjective perspective, a literary style reminiscent of long-form nonfiction. It combined journalistic research with the techniques of fiction writing in the reporting of stories about real-life events.

Rand based their analysis on several societal trends: the heightened disagreement about facts and analytical interpretations of facts and data; a blurring of the line between opinion and fact, an increase in the relative volume and influence of opinion and personal experience over fact; and diminished trust in formerly respected institutions as sources of factual information. Unlike prior periods, the current political polarization and weaponization of truth encompasses all of these, an alarming situation.

Historically, in Walter Cronkite's era as CBS anchor for 19 years (1962–1981), most information (or news) was limited to three channels, a few newspapers, and radio. Noteworthy is the fact that his tenure as the most trusted newsman in America coincided with new journalism, the American literary movement that pushed the boundaries of traditional journalism and nonfiction writing.

During this time, there were only three US TV channels, no cellphones with cameras, no social media, no cable news, and no extreme politically driven TV. The US, despite the Vietnam War, was also less divided than today. Limited information eventually evolved into too much information. Alvin Toffler's often-quoted phrase "information overload" quickly became popularized in his 1970 book *Future Shock* as more and more sources of both truthful and untruthful information became available [23]. Along with social media and other amplifier algorithms (explained later), people became inundated with information that encouraged them to want to consume more similar reinforcing information, not other viewpoints. Fake news that is highly emotional attracts more social media clicks or likes generating greater profits and is reinforced by a targeted sophisticated computer program or algorithm, perpetuating echo chambers, inaccurate statements, and the extremes of thinking about medicine, vaccines, public health, politics, and sex. Enhanced by politicians, there became widespread distrust of media, science, and government by a large segment of the general voting population.

As COVID-19 evolved and strengthened in 2020, there subsequently came a flood of information as scientific disbelievers went to work, and anti-vaxxers evolved into political anti-maskers and mandate haters. This picked up steam at the end of 2020 coinciding with the presidential election and the release of COVID vaccine. Anti-quarantine advocates and anti-lockdown advocates proliferated. Finding better data became more and more irrelevant as extremes dug in. Mask mandates generated no-mask anti-laws. Too many people didn't know what to believe as economics was energized at the expense of science trying to do its thankless job. There was little compromise between believers and deniers, and getting closer to any scientific "truth" became harder and harder.

Add to this a plethora of information online and in multitudes of media sources, leading to too many sources of information that were not digestible quickly or easily. On top of that, science, especially in public health, is always evolving and changing, which creates confusion, especially for laypeople. For example, should we be wearing a mask outdoors or indoors; is it aerosol or droplet or contact transmission; and do a PCR test or a rapid antigen test?

Hearing or viewing limited published information coupled with a short attention span created a credibility gap between the general public, scientists, politicians, and economists. This led to the politicization and misunderstanding of public health's sometimes rapidly changing recommendations—some of which had to be recalled or modified. Stirred on by extremists, politicians, and computer algorithms, with poor counterarguments by scientists, the gap widened as there was no middle ground of compromise, only a misguided zero-sum game.

This crisis was exacerbated by echo chambers, which is an environment where a person only encounters information or opinions that reflect and reinforce their own, and filter bubbles, which is algorithm bias that skews or limits the information an individual user sees on social media, especially in cable news outlets, and in real life. In other words, there was gravitation to the extremes of inherent thinking. Those on each side of the political arena limited themselves to the same algorithm-directed one-dimensional news sources, leading to extremes of COVID-19 thinking. This generated more revenue from social media "hits" or "likes" for advertisers. Successful fake news also included spreading dissemination widely without any equivalent counterpunch, capturing inherent mistrust in people or their institutions, and intentionally repeating the false message over and over again.

There are other factors at play: the "newness" of widespread social media, inherent uncertainty, and all too easily lying to make a false point. When caught with a lie propagate a new one. With too many sources of information, much evades fact-checking (which is after the fact), and credibility is lost. Many, like Elon Musk and his changing comments on X formerly called Twitter, aren't concerned about truthfulness but rather with a political agenda. The misunderstood fact that science is always evolving creates confusion and a credibility gap for the layman. Non-peer-reviewed preprints abound. A widely circulated 2024 paper, for example, blamed excess mortality on Covid vaccine not the disease. This turned out to be a completely false premise actually plagiarized from previous work [25]. Marketing computer algorithms drive people to higher revenue-generating extremes. National and subnational enemies supply bots to confuse us even more. Fake news repeats itself over and over again. Inadvertent mistakes aren't forgiven as there were mistakes by institutions such as CDC and FDA adding to the credibility gap.

4 Major Platforms and Sources of Misinformation

To eventually counter misinformation, first we need to identify and understand where fake news is generated and where it thrives:

- Unregulated social media more than conventional media
- Disenfranchised minorities
- Conspiracy theorists
- Violent extremists (terrorists, white supremacists, racists, antisemites, islamophobes and other hate groups)
- Those that misconceive or misunderstand risk or how science works
- Those seeking financial, political, ego-driven, or religious gain

- Haters of America (or other governments)
- Those that feel pharmaceutical companies, healthcare workers, medical equipment manufacturers, and hospitals are corrupt
- Those who feel that scientists and others in leadership are arrogant and shouldn't be trusted

Fake news specific to COVID-19 includes ineffective harmful drugs touted as cures. This includes ingesting bleach and using hydroxychloroquine, which was promoted by former President Donald Trump, with no evidence of efficacy and great evidence of harm. Hydroxychloroquine, for example, is licensed for use in some autoimmune diseases, but has been found ineffective for COVID-19 in all clinical trials. Ivermectin [10] is another example. It is licensed for human use only for parasitic infections, but touted as a cure by some conservative extremists. Yearly ivermectin prescriptions hit 88,000 in August of 2021 compared to 3600 yearly prescriptions pre-pandemic. This ineffective treatment was used as an excuse and a substitute for not vaccinating by popular podcast host Joe Rogan and others. He touted ivermectin despite being refuted by many studies including the *New England Journal of Medicine* in a double-blind case-control study (the gold standard). These drugs also can have toxic side effects. Despite all this, there is still support for these drugs, and there have even been lawsuits filed against healthcare workers who refused to prescribe them.

Other issues specifically problematic include claims made about the ineffectiveness of COVID-19 vaccines [3], masks, testing, quarantine, mandates, and distancing. While some of these objections to public health policy have some credibility, the majority of scientific thought is on one side of the issue. To be fair, the highly respected Cochran Review looked at the effectiveness of masking for COVID-19 and found masks to be less effective than claimed—not ineffective if used properly.

Then there is media, right more than left, social more than mainstream, using an outdated part (Section 230) of a 1996 federal law which was meant to protect young Internet companies from liability. Thus, the "platform vs. content" controversy remains unchanged with the platform winning out. There are far-right extremists, militias (i.e., Faith Keepers, Proud Boys), some evangelicals, audio industry, podcasts like iHeart (600 podcasts with a vast online archive), Apple library, conspiracy theorists, 1st Amendment extremist individual rights advocates, and outright lying politicians. Although social media was originally understood as a friendly democratizing public discourse, a surprisingly small number of individuals can dominate a malignant conversation.

Perhaps more important is the Supreme Court ruling of May 2023, where it ruled for Google and Twitter over liability for terrorism-related content posted by users [17]. The court said families of terrorism victims had not shown the companies aided and abetted the attacks on their loved ones. The court's narrowly focused rulings passed on requests to limit a law that protects social media platforms from lawsuits over content posted by their users, even if the platform's algorithms promote videos that laud terrorist groups (Section 230). This was a win for social media; a loss for society. This was repeated in early 2024 when social media companies testified before Congress about what little they were doing regarding their detrimental effect on children. The Supreme Court took this up again in 2024 so we have not seen the end of litigation in this arena.

There were also specific people or organizations identified as sponsors of COVID-19 misinformation such as the "Big 12" and other prominent anti-vaxxers. According to a 2021 analysis, only 12 accounts, known as the "disinformation dozen," were responsible for 65% of anti-vaccine information on Twitter, Facebook, and Instagram" [14, 27].

These were identified as the culprits:

1. Joseph Mercola, osteopathic physician and major proponent of "natural health"; he sold ineffective, unproven medications from his office and refused interviews [2].
2. Robert F. Kennedy, Jr., son of Robert F. Kennedy, an environmental lawyer and founder of Children's Health Defense; He "regularly outperformed more reliable news outlets that met NewsGuard's criteria for trustworthiness, repeatedly publishing false information about COVID-19." He has a pre-COVID-19 history of promoting anti-vaxing. As of 2024, he was running for president.
3. Ty and Charlene Bollinger, self-acclaimed filmmakers and founders of the misleading docuseries *The Truth about Cancer.*
4. Sherri Tenpenny, osteopathic physician who believes vaccines cause autism, a theory universally disproven and propagated by discredited British physician Andrew Wakefield.
5. Rizza Islam, antisemitic anti-LGBTQ+ social media influencer, targets anti-vaccine messaging to the Black community.
6. Rashid Buttar, osteopathic physician known for using unproven and harmful chelation therapy for numerous conditions, including autism and cancer. Chelation is a treatment that uses medication to remove metals such as lead, mercury, iron, and arsenic from the body. It has limited therapeutic value in medicine and no value in COVID-19 treatment. He died in 2023 days after claiming he was poisoned with a substance in vaccine. Pre-pandemic he had been reprimanded by the North Carolina State Medical Board.
7. Erin Elizabeth runs an alternative health site called "Health Nut News." She has shared antisemitic posts.
8. Sayer Ji, founder of GreenMedInfo, a popular alternative medicine portal touting infective and sometime harmful COVID-19 treatments.
9. Kelly Brogan, author of books on alternative medicine, such as *Mind of Your Own.*
10. Christiane Northrup, OB/GYN physician and author who embraced pseudoscientific alternative medicine theories.
11. Ben Tapper, chiropractor who founded The Wellness Pointe.
12. Kevin Jenkins, social epidemiologist focused on inequities in health care.

As propagators of fake news become more sophisticated, the high-volume offenders are changing. Others instrumental in building COVID-19 conspiracy theories include Dr. Robert Malone, an mRNA "expert" who claims he invented mRNA technology (Katalin Kaliko appropriately received the Nobel Prize for this) and now has come to realize it is harmful and ineffective. He has received press interest and often gave interviews spreading fake news about the risks of COVID-19 vaccines.

Dr David Martin is an American financial analyst, former professor, self-help entrepreneur, and conspiracy theorist, known for speaking to a supposedly official European Union Summit. He played a starring role in the 2020 conspiracy theory-based documentary "Plandemic: Indoctrination" and has appeared as a guest analyst on networks including CNBC and Bloomberg. He is perhaps best known for his YouTube channel which has a cult-like following on Chinese social media. He suggests that COVID-19, along with SARS, are bioweapons engineered by the USA. He claims, without proof, that Pfizer nefariously manufactured and released a coronavirus before both the 2003 SARS pandemic and the COVID-19 pandemic in order to ensure that the world would accept a universal vaccine template.

Dr. Scott W. Atlas became a frequent adviser to Florida Governor DeSantis and President Trump with no infectious disease or epidemiology credentials. As a Stanford neuroradiologist, he argued that people who are not at risk of severe consequences should not face COVID-19 restrictions.

QAnon advocate and Congresswoman Margorie Taylor Greene has propagated a range of conspiracy theories, including "deep state"/Pizzagate/QAnon baseless election fraud and COVID-19 conspiracy theories. With misleading sympathetic sounding lines like "Save the Children," QAnon has a significant number of followers. In fact, 22% of Americans following the movement believe that a political "storm" is coming, 18% think violence might be necessary to save the country, and 16% hold that the government, media, and financial worlds are controlled by Satan-worshiping pedophiles.

Dr Joseph Ladapo, the Surgeon General of Florida, has a long history of false information dissemination in the Wall Street Journal's OpEd pages. More recently, he stated that men in the age group 18–39 years have an "increased risk of cardiac-related death" from the hybrid COVID-19 vaccine—a baseless claim. The reference he cited had no named authors and was published on Florida's own COVID-19 website. He has repeatedly advocated the use of ineffective and dangerous COVID-19 therapeutics over vaccine and claims COVID-19 vaccine contains harmful DNA fragments. During Florida's measles outbreak in 2023 he told parents to send their kids to school and not to vaccinate - both completely counter to public health policy [16].

Dr Ladapo was also an advocate of the "Plandemic," one of the most prominent conspiracy theories about COVID-19. This was an Internet book and video that had ten million hits before quickly being taken off all mainstream Internet sites. The video is filled with misleading and outright false information. However, people wearing white medical coats on the steps of the Supreme Court Building helped persuade vulnerable audiences of their messaging. After Dr. Ladapo issued misleading claims about the risks of COVID-19 shots for young men, the heads of the CDC and FDA sent a scathing four-page rebuttal. Such misinformation "puts people at risk of death or serious illness," they said.

Peter A. McCullough is a cardiologist with a history of spreading false or misleading claims about COVID-19 and vaccines. He promotes an unproven "spike protein detoxification" protocol on Instagram. It promotes unproven curcumin, bromelain, and nattokinase. As Chief Scientific Officer of The Wellness Company, he stands to gain financially from these products.

Another major platform for fake news is the group of popular conservative media sites and podcasts. Posts about vaccine mandates in particular have drawn significantly more engagement than conventional truthful scientific sources. These sites found success in pushing content that tapped into ongoing culture wars around the US COVID-19 policies and had more interactions—likes comments and shares—than the New York Times, Washington Post, Wall Street Journal, LA Times, and National Public Radio combined during the second half of 2021. Pandemic encompassed a series of films. Plandemic 3: The Great Awakening," for example, was a film that claimed AZT killed a majority of the AIDS population, a theory totally disclaimed.

The largest media platform that these organizations and individuals utilize to spread their message is YouTube, owned by Google. YouTube has taken down several video channels associated with high-profile anti-vaccine activists like Joseph Mercola and Robert F. Kennedy Jr., as part of a set of policies aimed at cutting down on anti-vaccine content. YouTube claims that it will ban any videos that claim that commonly used vaccines approved by health authorities are ineffective or dangerous. The company previously blocked videos that made those claims about coronavirus vaccines but not ones for other vaccines, such as measles or chickenpox. Conservative broadcasters touting anti-vaccination often refuse to answer when asked if they were vaccinated [12].

All told, seventy seven of the biggest spreaders of election fake news rose steeply in popularity in the 6 months before the January 6, 2021, attack on the US Capitol, gaining a stunning 25 million followers on Twitter and Facebook, according to a Washington Post data analysis. Many have since used their powerful megaphones to drive other culture war narratives, including critical race theory, transgender rights, and medical misinformation.

Sometimes, a misleading tweet comes out of nowhere, is about to go viral, and is quickly withdrawn. If the person has millions of followers, it makes it even worse. For example, Elan Musk, who has had more than 100 million Twitter followers, tweeted "There is a tiny possibility there might be more to this story (about Hillary Clinton and the attack on Nancy Pelosi's husband) than meets the eye." He included a link to an article from the Santa Monica Observer, a fringe website with a history of publishing false stories making extreme right-wing political lurid claims about the Capital attack without any evidence. The tweet was quickly withdrawn.

Posts about vaccine mandates in particular drew significantly more engagement than conventional ones, "indicating that the site found success in pushing content that tapped into ongoing culture wars around US COVID-19 policies" [12,15]. The messages regarding why not to vaccinate include:

1. Pharma is tricking us; vaccines don't work; they only want to make more money.
2. New mRNA vaccines have not been thoroughly tested like earlier vaccines.
3. Not enough data: It's irrelevant that hundreds of millions have been vaccinated with very, very minimal complications and that the vaccine works.
4. Government wants to control you. We have been lied to and can't trust the government.
5. Microchips, implanted by Bill Gates, are in the vaccine to monitor us.
6. COVID-19 causes only a mild illness, complications are rare, so why bother?
7. We are told that another booster and then another is needed; why couldn't they get it right in the first place?
8. "I've been infected. I'm immune."
9. The vaccine doesn't adequately prevent transmission.
10. My media source, doctor, politician, friend, and religious leader tell me what to do.
11. It's a free country, don't tell me what to do.
12. Individual concerns outweigh societal public health concerns.

Some are full truths; some are half-truths, and some have no truth whatsoever. There is the subtle misinterpretation of scientific studies to make a point. Take Simson's Paradox. If you are older, you are more likely to die from COVID-19. You are also more likely to have been vaccinated. Following along this path, the Epoch Times, a conservative publication incorrectly stated "Did you know that for the first time since the pandemic started, the majority of COVID-19 deaths are now coming from the vaccinated population?" This is because there are far more people vaccinated in this age group. Thus, it is false to imply that vaccination is more likely to lead to death [31].

There had been a debate in the public eye between Tucker Carlson and Dr. Anthony Fauci. Carlson, a former host on Fox News and MAGA promoter, argued without any evidence that big pharma and government only want control. They want to make money off people through their vaccines. He also claimed that the COVID-19 vaccines were not thoroughly tested like other vaccines. Carlson, who will not say whether or not he is vaccinated, claims the government wants to control you, the vaccine doesn't prevent infection, and COVID-19 is a mild illness anyway. Carlson, with no scientific background, has been critical of science as it has evolved, calling into question why scientists can't get it right in the first place. When another booster is recommended, this reinforces Carlson's viewpoint that the government doesn't know what it's doing and arbitrarily wants to control you. Dr Fauci ran the National Institute of Allergy and Infectious Diseases (NIAID) for decades and is thought to be the voice of reason in a divided world. Fauci initially had to defend the Trump administration. He had to make statements based on limited constantly changing information not intending to intentionally mislead. His nuanced comments generated not only criticism from the political right against him, but also death threats to his family. The incorrect implication was that he was intentionally misleading the public. His motive was never fully explored.

5 Consequences of Fake News

There are direct and indirect consequences of misleading or being misled about COVID-19. Direct consequences include drug reactions, increased hospitalizations, loss of faith in sound public health measures for other reasons, disenchanted scientists, a disheartened COVID-19 medical workforce, and excess deaths. Indirect consequences include extending anti-vaxing messaging to childhood vaccines, a greater number of more severe infections, disbelief of people more educated, more and more people leaving the healthcare field, and a greater lack of compromise continuing to divide us politically.

When CDC came out with the recommendation in October 2022 that COVID-19 vaccine should be incorporated into childhood vaccination schedules, this immediately incorrectly spread on conservative airwaves and political races that CDC was mandating, not recommending, this action. In truth, each state can make up its own mind, and very few states if any would issue a mandate. The effects of the anti-vaccination movement can be seen statistically. For example, global childhood vaccinations for 11 key diseases fell from 71% in 2019 to 68% in 2021. This is the worst backslide in a generation according to UNICEF and the WHO.

One of the biggest changes to news dissemination and consumption in the twenty-first century has been social media becoming a predominant source of information and fake news [1]. On social media with billions of users and followers on Twitter, TikTok, and Facebook, there is very little accountability for outright lying. There are certain fake news standards on social media platforms, but there is no strong regulation on preventing the spread of inaccurate information. With the exception of news outlets like the NY Times, Washington Post, Wall Street Journal, or CNN, the expensive investigative reporting of the past has been augmented by cheaper panels of talking heads on cable news. Anti-vaccine content is a driver for high engagement on social media. Sites that spread misleading or false medical information dramatically outperform reliable and often boring sources of health information on social media. Specific flagrant untruths include birtherism and the Big Lie. Increasing surveillance for taking down fake news is a cost center for social media outlets. In other words, it doesn't generate any income. Encouraging it sells advertising.

Misleading actions have consequences for scientists [11]. Hundreds protested government efforts to restrict educational access to Western science theories, including Darwin's theory of evolution, in June 2023 in India. Scientists in Mexico participated in a research strike in May 2023 to protest a national law they claimed would threaten the conditions for basic research. During the same month in Norway, three scientists were arrested for protesting the nation's slow-moving climate policy. In the past, scientists have also mobilized for social activism and political protest. This included the atomic bomb, pesticides, wars in SE Asia, genetic engineering, and the federal response to the AIDS epidemic. Add to this the fact that doctors who misled patients about vaccines, masks, and unproven, and potentially dangerous COVID-19 treatments during the pandemic and its aftermath have faced few repercussions from state medical boards, according to a Washington Post analysis of disciplinary records.

Fox News lost a $787.5M lawsuit falsely defaming a voting machine company, Dominion. Shockingly, legal discovery emails revealed that Fox News anchors clearly didn't believe what they were preaching and thus the mega-settlement.

One could call the spread of extreme false health information "disinformation terrorism," where the line is more and more blurred between opinion, fact, and harm. Conspiracy theories are repeated over and over again until they are ingrained and believed, as in the "Big Lie". regarding the 2020 presidential election The social media algorithms successfully target and move people to the extremes of the political spectrum and reinforce echo chambers by presenting more and more similar and engaging information. Add to these other sources of misinformation, aside from social media, like militias, evangelicals, the audio industry, and extreme individual rights advocates, then platform vs. content comes more into play.

6 Where We Currently Stand

The political-scientific divide specifically regarding COVID-19 was enhanced by former President Trump being dismissive of appropriate pandemic messaging. This caused a sizable portion of the USA to distrust politicians, scientists, pharma, and the media as a part of the arrogant elite class that some felt could not be trusted. After the first year of the pandemic, more liberal states and counties had more severe mitigation measures with lockdowns, mask and testing mandates, and penalties. Moderate states had less stringent rules such as indoor masking and transportation masking. Mild mitigation included the individual's right to refuse vaccination with few provisions requiring masking indoors. Many GOP voters and officials tended to downplay COVID-19 risk and the benefits of mask-wearing and vaccination, opting for milder mitigation. Some states even issued anti-mandate laws. In other words, the more left-wing the stricter the policy. There were 50 different US states acting like 50 different countries not always working together, thus making fake news more easily accessible. Interestingly, the population with more right-wing leanings tended to do more poorly when infected with the COVID-19 virus [32].

As mentioned, fake news and polarity are amplified by social media's computer algorithms developed to increase user engagement, which often leaves users craving for more emotionally amplified messaging that coincides with their previous biases. Profits from advertisers are maximized with more hits and likes. The business algorithm permeates into 70% of US social media platforms, such as Facebook and Instagram. Internationally, Facebook's algorithms have been maliciously used in India, China, and Burma. The "like" and "share" buttons allow users to provide feedback (and greater profit) from the algorithm, which will then show more and more extreme content. This breeds fake news, creates a lack of privacy, and promotes the extremes of thinking. *Users became a monetizing engine with the user as the data.*

Facebook (owned by Meta) has been a particularly bad actor [4, 9]. It has been unwilling to turn over data on how much fake news is spreading on its site. There has been public fallout over Cambridge Analytica's political analysis helping Republicans and foreign interference in the 2016 election, fake news in the 2020 election, misinformation about COVID-19 and the vaccines, and even an uptick in teen depression, suicide and loneliness from Facebook as shown in recent Congressional hearings. Frances Haugen, a Facebook whistleblower, shared details of the pernicious algorithm, spurring reinforcement of blame on Facebook. This exemplified profit over ethics, hiding behind a section of an old law initially used to protect Silicon Valley startups many decades ago. As mentioned, Section 230 states that "No provider or user of an interactive computer service shall be treated as the publisher or speaker of any information provided by another information content provider" (47 U.S.C. § 230). This provides social media companies with sweeping protections that let them choose what content they restrict, and how. In other words, platform over content, profit over ethics. This means social media platforms cannot be sued for leaving certain content up [26]. Examples include:

- Vicious hateful content against a racial group or the LGBTQ+ community
- White supremacy
- Self-degrading content leading to suicide
- Pornography, especially child
- Incitement to violence/sedition
- Antisemitism and islamophobia
- Content contradicting sound public health doctrine

Internationally, Facebook has also not responded well to policing fake news. Internal company documents revealed that Facebook meticulously studied its approach abroad—and was well aware that weaker moderation, especially in non-English-speaking countries, leaves the platform vulnerable to abuse by bad actors and authoritarian regimes. In India, Facebook's largest market, the company, had not developed algorithms that could detect hate speech in Hindi or Bengali. Myanmar's Facebook's Internet exclusivity bred overt racism and brutality against the Rohingya racial group. Even worse than Facebook, Twitter, as of the purchase by Elon Musk, stopped enforcing almost the entire COVID-19 misleading information universe creating a Wild West of information flow.

As previously mentioned, this can make things even worse, as when a federal judge in 2023 restricted some Biden administration officials from even communicating with social media companies. US District Judge Terry Doughty ordered several top officials and federal agencies—including the CDC and FBI—not to communicate with social media companies about taking down "content containing protected free speech" that's posted on its platforms. The judge, an appointee of former President Donald Trump, handed down the win to GOP states in a lawsuit accusing the government of going too far in efforts to combat COVID-19 disinformation.

7 Curbing Misinformation

What specifically can be done to combat misinformation without infringing on our 1st Amendment rights? [18] For example, what may seem as the innocuous publication of the book *The War on Ivermectin* was actually very harmful. "The medicine that saved millions and could have ended the COVID-19 pandemic" by Dr Pierre Kory was published by Simon and Shuster and gives false information by an otherwise respected publication about a scientifically unproven useless drug.

We need to expose these policy decisions and algorithms to the public. There needs to be education of the public starting at a very young age to regain lost scientific trust and to better understand the ethics of business v. science in a capitalist society. This is analogous to environmental, social, and governance (ESG) investing sometimes generating lower profits. Appropriate legislation is needed that's designed by those younger than those currently in power, in other words those who better understand the problem are more likely to solve the problem. Fact-checking should be emphasized under an artificial intelligence (AI) umbrella that combines disparate databases and catches misinformation early before it goes viral. The "liar's dividend" has to become more undervalued so fewer individuals or organizations use it.

There should be exposure of outdated policy decisions. Initially, Facebook and Instagram removed only three dozen (out of millions) of pages, groups, and accounts related to the "disinformation dozen," a very small start. In response, the platform pushed back on the White House's contention that the "12" described above were primarily responsible for the spread of vaccine misinformation. After the announcement, the White House continued to strongly criticize the company, implying Facebook needed to be more honest and transparent. Insiders said Facebook CEO Mark Zuckerberg chose growth and market share over public safety. Some of Zuckerberg's previously unreported decisions demonstrated his relentless determination to ensure Facebook's growing dominance, even at the expense of his publicly stated values, according to interviews with more than a dozen former employees. As mentioned, that ethos has come under fire in whistleblower complaints filed with the SEC by Frances Haugen. While it's unclear whether the SEC will eventually pursue action against the CEO personally, the allegations by the former Facebook product manager represent arguably the most profound challenge to Zuckerberg's ironclad leadership of the most powerful social media company in the world.

NYU's Center for Cybersecurity's Ad Observer looked at Facebook and demonstrated that extreme, unreliable news sources get more engagement or user interaction at the expense of accurate posts and reporting, which is sometimes boring and needs to be made more interesting. Interestingly, Facebook shut down their accounts when researchers had just begun studies intended to determine whether the platform was contributing to vaccine hesitancy and sowing distrust in elections. Current researchers are still trying to figure out the role the platform may have played leading up to the Capitol assault on January 6.

Exposing the algorithm's harm helps. SomeOfUs, an accountability advocacy group, tested TikTok's algorithm by creating an account and searching for and watching 20 widely viewed videos that sewed doubt about the election system [33]. Within an hour, the algorithm switched from neutral content to pushing polarizing fake news content, far-right extremism, QAnon conspiracy theories, and false COVID-19 narratives. When confronted, TikTok said they planned to do better.

New technology has given us better and better ways to spread all information. Deep fakes, for example, can easily convince us in a video that something that is completely wrong looks completely right [6]. Sophisticated AI algorithms can now be almost impossible to identify. This has gotten so problematic that a letter signed by hundreds of scientists in March 2023 called for a 6-month halt on AI. Many doubt this will have any effect. As we are getting more and more dependent on electronics, scams are getting more and more prevalent and sophisticated. Some have used the term "inoculating" against misinformation, while others have analyzed how obscure phenomona can become viral [5, 24].

There are also ways to counter this legislatively. There is detection software that can be installed and watermarks can be put on, but many gaps remain and will probably worsen in the short run. AI has the potential to both benefit society and weaken its grip on technology.

Appropriate legislation can help or hinder. Childhood (not including COVID-19) vaccinations in California are a requirement for public school entry. Section 230 needs to be rescinded. AB 2098 would make California the first state that could take legal action against healthcare professionals for conveying false information about COVID-19 and its treatments. While well-intentioned, this legislation could have a chilling effect on medical practice, with widespread repercussions that could paradoxically worsen patient care and make it more difficult for doctors to practice. It is a very fine line between helpful and harmful.

Educating the public, physicians, and other healthcare providers is essential. British researchers and a team from Google found that teaching people how to spot fake news made people more skeptical [24]. Cognitive bias needs more scientific study [8].

Fake news all too often boils down to mistrust. There is widespread mistrust of science, public health, conventional investigative news media, clerics, and people in leadership. Once lost, trust is not easily regained. Although this may be discouraging, science and scientists are still held in high esteem across many global publics, according to the Pew Research Center [11]. The majority of people have at least some trust in scientists, with India, Australia, and Spain leading in these metrics. Internationally, the political right distrusts scientists more than the left. Most people around the world also value government investment in scientific research.

Fact-checkers help—to a point. They can quickly research falsehoods, and a network of many fact-checkers augmented by AI technology can not only make stronger comments but can also potentially halt a falsehood from going viral or stop it in its early tracks. There are Twitter followers that number in the tens of millions for some celebrities. We could harness the more ethical of these celebrities to provide better education of good entertaining scientific messaging, repeated in an entertaining manner similar to how fake news gets traction.

How does AI actually enter into the dissemination of medical misinformation? [28] While current debate centers on AI taking over in expected and unexpected ways, good and bad, more important is the rapidly accelerating ease of using AI by writing computer code in English or some other language rather than by a coder using many lines of complex computer code. This will enable the malicious perpetrator to utilize misinformation in easier ways. There will be counterprograms, but this will be a war.

Unchecked lying can pay dividends as does repetition of that lie. We need to lessen the "liar's dividend" for those hoping to discredit reality and dodge accountability. If our future is predominantly made of fake news, global progress would be impeded in many ways. A nondemocratic autocratic government could emerge. Legitimate science would likely be frowned upon—at least for a political time. Specific to COVID-19, if the virus continues to spread in a harmful way, the unvaccinated would still continue to refuse vaccination, masking, and distancing. More unvaccinated than vaccinated would get hospitalized and die. Disproportionate cycles would continue, killing more of the unvaccinated than vaccinated, with ultimate survival of the fittest. Ironically, this may help the decline of COVID-19 fake news. Countering the repetition of a lie is important. Bigger than the US "big lie" was Nazi Germany's lying, where Hitler's propaganda minister Joseph Goebbels called attention to the fact that a repeated big lie told often enough will bring along more and more believers. History repeats itself.

References

1. Khullar D (2022) Social media and medical misinformation confronting new variants of an old problem. JAMA 328(14):1393–1394. https://doi.org/10.1001/jama.2022.17191
2. De Vynck G (2021) YouTube is banning prominent anti-vaccine activists and blocking all anti-vaccine content. 9/29/21. https://www.washingtonpost.com/technology/2021/09/29/youtube-ban-joseph-mercola/
3. Nania R, Crouch M (2022) Separating fact from fiction when it comes to getting vaccinated against COVID-19, AARP, Updated January 10, 2022. https://www.aarp.org/health/drugs-supplements/info-2020/covid-vaccine-myths.html?cmp=KNC-DSO-HEALTH-Health_COVID-NonBrand-Exact-36893-GOOG-COVID-DrugsSupplement-VaccineMyths-COVID-Exact-NonBrand&gclid=Cj0KCQiAm5ycBhCXARIsAPldzoUQtrGsg44XIMeQ4BQT4z5AMYIW9UMDLgquMuasx5DXZZFnhBy-Zj8aAsUIEALw_wcB&gclsrc=aw.ds
4. Zakrzewski C, De Vynck G, Masih N, Mahtani S (2021, Oct 24) How Facebook neglected the rest of the world, fueling hate speech and violence in India
5. Grant N, Hsu T (2022, Aug 24) Google finds 'Inoculating' people against misinformation helps blunt its power
6. Hsu T (2022) Deep fakes flourish on TikTok, Raising Alarms, 11/4/22 NY Times
7. Kerber R, Kavanagh J Individual differences in resistance to *Truth Decay*. https://www.rand.org/pubs/research_reports/RRA112-17.html
8. Kerber R, Kavanagh J Exploring the role of reasoning and cognitive biases. https://www.rand.org/pubs/research_reports/RRA112-17.html
9. Frier S (2021) The ugly truth about Facebook. NY Times 7/9/21. https://www.nytimes.com/2021/07/09/books/review/the-ugly-truth-sheera-frenkel-and-cecilia-kang.html
10. Cory P (2022) The War on Ivermectin. https://www.medpagetoday.com/special-reports/exclusives/102183?xid=nl_mpt_investigative2022-12-14&eun=g2128221d0r&utm_source=Sailthru&utm_medium=email&utm_campaign=InvestigativeMD_121422&utm_term=NL_Gen_Int_InvestigateMD_Active

11. Funk C, Tyson A, Kennedy B, Johnson C (2020) Pew Research Center. Science and Scientists Held in High Esteem Across Global Publics. 9/29/20. https://www.pewresearch.org/science/2020/09/29/science-and-scientists-held-in-high-esteem-across-global-publics/
12. Dyal NP (2020) 10 myths about Coronavirus vaccines 3/9/20 https://www.infectiousdiseaseadvisor.com/home/topics/prevention/10-myths-about-covid-19/
13. Henderson J Bill aimed at disciplining Docs for COVID Misinformation Approved by California Legislature. MedPage Today https://www.medpagetoday.com/special-reports/features/100471 8/31/22
14. Dwoskin E, Merrill J (2022) Trump's Big Lie fueled a new generation of social media influencers. Washington Post 9/20/22
15. Updates to our work on COVID-19 vaccine misinformation. Twitter Safety 3/1/21 https://blog.twitter.com/en_us/topics/company/2021/updates-to-our-work-on-covid-19-vaccine-misinformation
16. Sarkissian A (2023) Florida Surgeon General altered key findings in study on Covid-19 vaccine safety. Politico. 4/24/23. https://www.politico.com/news/2023/04/24/florida-surgeon-general-covid-vaccine-00093510
17. Supreme Court rules for Google, Twitter on terror-related content https://www.washingtonpost.com/politics/2023/05/18/gonzalez-v-google-twitter-section-230-supreme-court/?utm_source=alert&utm_medium=email&utm_campaign=wp_news_alert_revere&location=alert
18. Van Beusekom M (2023) Experts offer tips to combat rampant online COVID, other health misinformation. 7/18/23. https://www.cidrap.umn.edu/covid-19/experts-offer-tips-combat-rampant-online-covid-other-health-misinformation
19. DiResta R, Baron R, Scales D Addressing viral medical rumors and false or misleading information. Ann Intern Med. https://doi.org/10.7326/M23-1218
20. K Weintraub. Ask your doctor about misinformation: Most say it 'does actual harm'. USA Today. 3/29/23
21. G. Brumfiel. The U.S. Surgeon General Is Calling COVID-19 Misinformation An 'Urgent' Threat. NPR Radio. 7/16/21
22. Understanding the Fake News Universe. A Guide to Fake News terminology. Media Matters for America. 12/15/18
23. A Toffler. Future Shock 1970. Bantam Books
24. S.van der Linden. Foolproof: How Misinformation Infects Minds and How to build Immunity.
25. L.Simonsen, R Pedersen. COMMENTARY: Misleading BMJ Public Health paper on COVID-19 excess mortality needs to be retracted, CIDRAP. 6/17/24
26. Section 230. Wikipedia. Accessed 7/6/24
27. E.Salem. Majority of Covid misinformation came from 12 people, report finds. The Guardian. 7/17/21P.Bump. Tucker Carlson airs his most dishonest and dangerous pandemic segment yet. Washington Post. 1/26/22
28. B.Cartwright et.al. Deploying Artificial Intelligence to Combat Covid-19 Misinformation on Social MediaTechnological and Ethical Considerations. Proceedings of the 56th Hawaii International Conference on System Sciences. 2023
29. We are fighting an indemic. Gapeli.org.. https://www.gapeli.org/community/viewchop/id/1087
30. Slovakia's Election Deepfakes Show AI IS a Danger to Democracy. Wired. 10/3/23. https://www.wired.com/story/slovakias-election-deepfakes-show-ai-is-a-danger-to-democracy/
31. J von Kugelgen et.al. Simpson's Paradox in COVID-19 Case Fatality Rates: A Mediation Analysis of Age Related Causal Effects. Published online 2021 Apr 14. doi:10.1109/TAI.2021.3073088
32. J wallis et.al. Excess Death Rates for Republican and Democratic Registered Voters in Florida and Ohio During the COVID-19 Pandemic. JAMA Intern Med. 2023;183(9):916-923. doi:10.1001/jamainternmed.2023.1154
33. J.Brewster et.al. Beware the 'New Google:' TikTok's Search Engine Pumps Toxic Misinformation To Its Young Users. 9/14/22. NewsGuard. https://www.newsguardtech.com/misinformation-monitor/september-2022/

C. Everett Koop Healthcare System for Biosecurity and Defense

Haley R. Warzecha, Alison Podsednik, and Joseph M. Rosen

1 Introduction

1.1 C. Everett Koop (Surgeon General from 1982 to 1989)

Dr. C. Everett Koop was a visionary who pushed to eliminate health inequities, promote disease prevention, and ensure all Americans have access to equitable, high-quality health care. He was outspoken about the state of the American health-care system, and he believed health care should be a moral right for all Americans. In 2006, he argued that "our capability to prevent and treat disease seems to exceed our willingness to apply our interventions… obesity and its associated problems, including diabetes, cardiovascular disease, and orthopedic injury, are largely preventable" [1]. He also warned the United States was not prepared for a "potential influenza pandemic, bioterrorism, or the next global epidemic" [1].

In 1993, Dr. Koop and Dr. Joseph Rosen introduced cybercare: a groundbreaking telemedicine-based healthcare system combining telemedicine, telepresence, telesurgery, virtual reality simulators, and healthcare knowledge integrated into data cubes. The pair demonstrated how patients could receive high-level care in their communities by bringing specialists from distant hospitals to their homes via cyberspace [2]. They later co-authored several additional papers involving the cybercare concept model [3–5]. The COVID-19 pandemic swung the healthcare pendulum back to the patient's home, emphasizing the importance of in-home hospital care in

H. R. Warzecha
Dartmouth College, Hanover, NH, USA

A. Podsednik
Department of Medical School, University of Texas Rio Grande Valley, Edinburg, TX, USA

J. M. Rosen (✉)
Department of Surgery, Dartmouth Hitchcock Medical Center, Lebanon, NH, USA

J. M. Rosen, R. R. Colwell (eds.), *The COVID-19 Pandemic*,
https://doi.org/10.1007/978-3-031-62772-9_8

the twenty-first century. Telemedicine and tele-technology have provided a lifeline during this unprecedented time.

1.2 Historical Perspective in Pandemic Times

From 1956 to 1958, President Eisenhower launched monumental undertakings that forever changed American life. In the backdrop of the aftermath of WWII, he signed legislation to support the Interstate Highway System, establishing the Advanced Research Projects Agency (ARPA) which subsequently became DARPA (Defense Advanced Research Projects Agency), as well as the National Aeronautics and Space Administration (NASA) [6, 7]. These agencies and their robust funding indicate the need for colossal government involvement for the advancement of national security. Eisenhower's involvement during the post-WWII world and provision of government relief efforts set a precedent and expectation of government involvement in resolving the current pandemic. DARPA began funding nucleic acid vaccines in 2011 under the ADEPT (Autonomous Diagnostics to Enable Prevention and Therapeutics) and Pandemic Prevention Platform (P3) programs. By 2020, they had changed their focus to COVID-19, contributing to some of the first mRNA vaccines [8]. Subsequently, the US Department of Health and Human Services (HHS) began Operation Warp Speed which was a government collaboration to deliver 300 million vaccine doses by collaborating with industry to close 2021 [9]. The project was first set into action in an attempt to restructure vaccine development and distribution, which has traditionally been a gap for the United States [10].

2 Current American Healthcare System and Its Shortcomings

2.1 US Healthcare Delivery

The healthcare system in the United States is one of the most complex and expensive systems in the world. In 2006, C. Everett Koop argued that health care should be a moral right for all people and expressed concern about the expansion of the health inequity gap within the American population [1]. Unfortunately, the United States remains the only industrialized country that does not provide universal healthcare coverage for all its citizens [11]. Instead, American citizens are represented by a mix of public and private insurance providers, which include government-based programs (such as Medicare and Medicaid) or employer-based programs, each plan with different benefits [12]. Patients with health insurance must pay a minimum amount of money for their own care in the form of a deductible before insurance begins to cover costs [11]. This requires citizens to pay for at least a fraction of their health care out of pocket. However, even with these healthcare delivery systems in place, there are still more than 37 million Americans that lack health insurance entirely and 41 million Americans that are underinsured, lacking adequate access to care even if they do have insurance [13].

The United States' healthcare expenditure per capita is higher than all other countries [12]. However, spending is not translated into more coverage nor higher-quality services for Americans. An analysis of the Organization for Economic Cooperation and Development (OECD) data demonstrated that healthcare utilization in the United States is not greater than other countries, but that the price of healthcare goods and services is the culprit of spending differences [14]. The major areas of greater spending were revealed to be hospital services, prescription drugs, and administration. In 2016, US healthcare spending surpassed $10,000 per capita annually, more than double that of Australia, France, Canada, New Zealand, and the United Kingdom [15]. Then, in 2020, spending surpassed $12,000 per person, which is 19.7% of the Gross Domestic Product (GDP) [16]. Healthcare expenditures have continued to advance during the COVID-19 pandemic as federal funding increased in 2020 to aid vaccine development and drug stockpiling [17].

Percent (%) of GDP, adjusted for differences in cost of living
*Legend shows 2018 data**

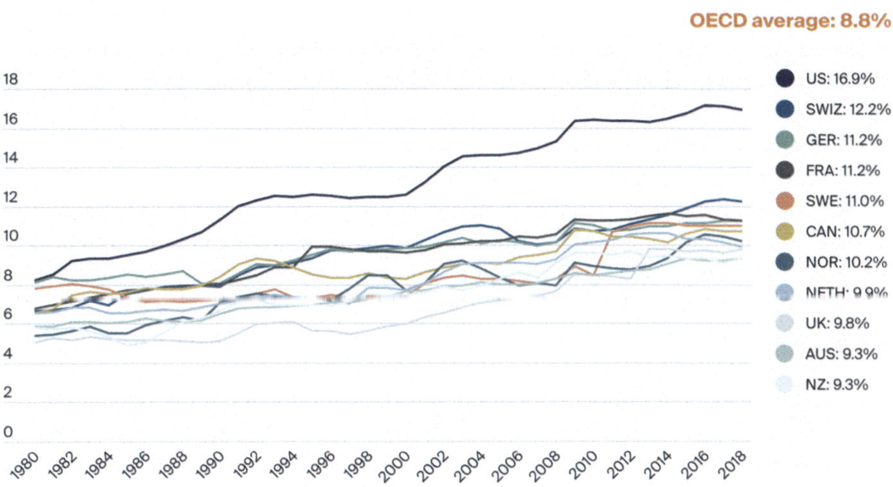

Source: Tikkanen and Abrams 2020 [15]

Despite massive healthcare spending compared to other OECD countries, the US fares much worse on several health metrics. For instance, the United States has the highest chronic disease burden of all OECD countries [15]. Diseases included in the count were heart disease, cancer, cerebrovascular disease (stroke), chronic lower respiratory diseases, Alzheimer's disease, and diabetes mellitus. Additionally, the nation's obesity rate is double the average of other OECD countries and the United States has the highest suicide rates among the 11 nations [15]. Based on a culmination of these factors, the United States has the lowest life expectancy of the OECD nations.

2.2 Long-Standing Inequity in Health Care

Chronic diseases have the potential to impact all people; however, long-standing structural and social inequities exist in America, leaving the most disadvantaged people with the highest burden of disease. According to the World Health Organization, social determinants of health include where people are born, grow up, live, and work [18]. Additionally, social determinants of health can be greatly influenced by structural determinants of health, even from past policies. Health outcomes tend to trend worse in populations with financial barriers. Poorer people are less likely to have access to a primary care physician, thereby decreasing likelihood to partake in preventative measures [19].

In terms of socioeconomic status (SES), disparities in preventable disease are evidenced in acute coronary syndrome outcomes. Low socioeconomic status (SES) individuals have a twofold increase in premature death resulting from myocardial infarction and coronary heart disease compared to individuals of average SES [20]. Increased mortality in low SES individuals may be related to a study finding demonstrating that uninsured individuals are 13% less likely to obtain routine physical exams and 12% more likely to possess undiagnosed hypertension than their insured counterparts [21].

Disparities in health outcomes can also be found along racial lines, particularly in American maternal mortality rates. Non-Hispanic black women have a 2.5–3.1 increased mortality relative to white or Hispanic women, which can partly be attributed to differences in insurance coverage and preventative care access that is known to decrease avoidable pregnancy complications [22].

Geographical location is another major contributing factor toward inequality relating to health outcomes. Many rural counties in America are classified as health professional shortage areas (HPSAs), with rural areas possessing a 19% decreased physician-to-resident ratio compared to urban areas [23]. Decreased access to health care leads to several implications, such as less preventative medicine check-ins, less access to urgent care, and less access to specialized and high-quality care.

Both the financial barriers to obtaining health insurance and decreased access to health care for those in rural areas lead to an increase in mortality for those populations. Unfortunately, most of the resultant negative outcomes are avoidable, but our healthcare system is not designed in a manner that prioritizes preventative care. In times of crisis, like a pandemic, these disparities and gaps in health care are highlighted. As so, populations with underlying health conditions have suffered more from COVID-19 and experienced worse health outcomes. Long-standing inequity in health care potentiates chronic disease through generations.

2.3 COVID-19 and Racial Disparities

The COVID-19 pandemic has evidenced the importance of addressing chronic diseases and focusing additional efforts on disease prevention rather than disease treatment. COVID-19 has ranked the third leading cause of death in the United States for two straight years, in 2020 and 2021 [24]. The disease is unlikely to be eradicated, as the virus mutates and produces new variants that evade natural and vaccine immunity. Instead, it will transition from pandemic to endemic [25]. As COVID-19 infections continue to rise, it is important to consider the secondary effects of each infection. COVID-19 has an extensive list of long-term health impacts coined "long COVID," or post-acute sequelae of COVID-19 (PASC), categorized as symptoms lasting greater than three months following infection and not explained by any other disease.

Long COVID leaves the public health system burdened with increased chronic disease secondary to COVID-19. Common PASC symptoms include chronic fatigue, chest pain, joint pain, weakness, and neurocognitive issues [26]. The symptoms affect a broad set of organ systems including the immune, cardiovascular, gastrointestinal, musculoskeletal, and nervous systems [27]. Several cardiovascular effects with relatively high morbidity include myocardial inflammation, myocardial infarction, and arrhythmias [28]. The long COVID symptoms will put an even greater burden on our healthcare system as more people continue to get infected, and the quality of life of these patients is significantly altered.

Preexisting conditions and certain chronic diseases put people at a higher risk of developing severe COVID-19 symptoms and long COVID symptoms, underscoring the need to promote the prevention of chronic diseases. The CDC issued a list of major risk factors for severe disease, including obesity, diabetes mellitus, and hypertension. Statistics published by the American College of Cardiology in 2020 state that for those with preexisting cardiovascular disease, fatality rates were 10.5%, followed by diabetes comorbidity having 7.3% mortality, and those with COPD having 6.3% mortality. In comparison, the fatality rate for people without preexisting conditions was less than 1% [25].

COVID-19 has disproportionately impacted certain racial underserved groups. Certain racial groups are disproportionately impacted by chronic disease, and the same systemic and social factors that predispose populations to chronic diseases also put them at risk for contracting COVID-19 and having worse health outcomes. A systematic review of the literature concluded that Black and Hispanic populations experience higher COVID-19 rates and higher mortality rates than white populations. Black and Hispanic populations also have an increased hospitalization rate [29]. Data from the American Public Media Research Lab indicate that Black populations are 3.2 times more likely to die from COVID-19 compared to White populations [30]. Similar to disparities seen in the H1N1 virus outbreak, certain groups have differences in disease exposure, susceptibility to infection, and access to health care [31]. Black and Hispanic populations tend to be most at risk. Additionally, low-income people can struggle with access to health care, dental care, and prescriptions. It is precisely these populations that also struggle with obtaining and sustaining health insurance coverage.

2.4 Prevention and Public Health Fund

Following the precedent set by Eisenhower in establishing strong governmental involvement in national security post-WWII, the US public continues to have high expectations for the government to protect them from various threats. In the case of COVID-19, the virus threatens US national security through significant morbidity and mortality, economic disruption, potential to weaken the military, and social and political unrest [32]. Public health agencies in the United States are expected to educate and protect the public from health threats, yet they have been strongly criticized in the wake of the COVID-19 pandemic.

On the national scale, agencies such as the CDC are losing public trust. According to a study by the Robert Wood Johnson Foundation and Harvard T.H. Chan School of Public Health, positive ratings of the public health system decreased from 43% in 2009 to 34% in 2021 [33]. This lack of trust and satisfaction has infiltrated state and local agencies as well. Only 41% of Americans trust state health departments, and 44% of Americans have faith in local health departments [33]. Dissatisfaction with public health departments is rooted in a lack of funding for public health financing compared to total health spending in the United States [34].

In 2008, Section 4002 of the Affordable Care Act (ACA) established the Prevention and Public Health Fund (PPHF). PPHF was the first mandatory funding stream dedicated to improving public health by establishing a consistent and growing investment in prevention, wellness, and public health initiatives. PPHF is required by law to be used "to provide for expanded and sustained national investment in prevention and public health programs to improve health and help restrain the rate of growth in private and public health care costs [35]."

The ACA authorization had originally granted a total of $18.75 billion to be dedicated toward the Prevention and Public Health Fund between FY2010 to FY2022, starting with only $500 million in 2010 and reaching $2 billion annually every year after. Instead, the PPHF has been subject to several amendments that have redirected the money to be used for non-public health priorities [36]. In 2012, Congress passed the Middle-Class Tax Relief Job Creation Act which decreased the fund by $6.25 billion over nine years to help pay for the extension of unemployment benefits and to delay cuts to physician Medicare reimbursements [37]. Similarly, in 2016, $3.5 billion was reduced from the fund over seven years (FY 2018–2024) to pay for the 21st Century Cures Act. Following directed cuts, two laws in 2018 cut over $2 billion from the fund with no specific redirection [37].

The figure below shows the original appropriated amounts for the Prevention and Public Health Fund and the reductions made through budget cuts to fund alternative items.

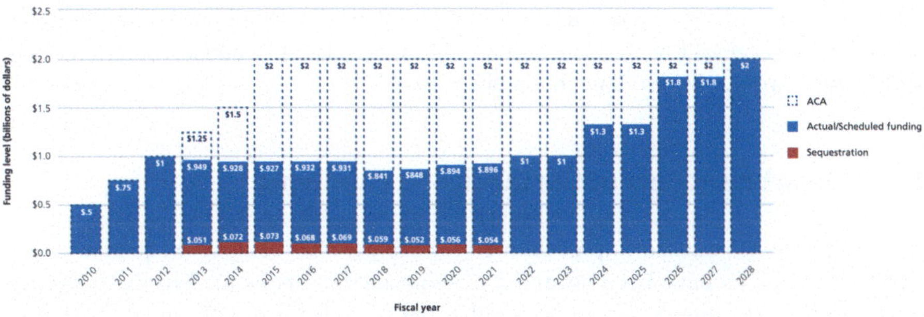

Public Health and Prevention Fund FY 2010–2028. (Source: APHA 2021 [37])

Given these budget cuts to PHPF and the general lack of spending toward public health, the American Public Health Association reports that only 2.5% of US health spending supports public health [37]. Advocates for public health investment argue that the United States is inefficient in its healthcare spending, evidenced by the fact that 90% of the $3.8 trillion in annual healthcare costs are allotted to treating chronic conditions, which are preventable. From an economic point of view, investing in public health to prevent chronic conditions offers high returns. From an infectious disease point of view, promoting public health may also be important in disease prevention, reduced morbidity, and reduced mortality.

The consequences of a weak public health infrastructure are apparent in the COVID-19 pandemic. Structurally, local public health agencies have lost almost 20% of their staffing since 2008 [38]. Since 2003, funds for the CDC's emergency preparedness program, which is the major source of federal funding for local public health agencies, have decreased by 30% [39]. Massive losses in funding directly translate to lower capabilities of these local public health agencies. For example, public health agencies have struggled to keep up with the demands for testing capabilities, contact tracing, health communications, occupational safety and health, and outbreak investigations [40]. Instead of having a comprehensive national public health structure, the US public health system is highly fragmented and underfunded, lacking personnel and coordination to strongly tackle the challenges of the COVID-19 pandemic.

The weakness of our public health system is evidenced by the burden of COVID-19 on the United States compared to other countries. By June 2020, the United States claimed four times the total confirmed cases of any other country [41]. It is important to note that experts attribute this inequality not solely to a lack of public health infrastructure but also to a long-term underinvestment in the health of the American population [41]. Although the third coronavirus stimulus bill provided $4.3 billion to public health organizations, the funds cannot make up for the chronic chasms in coordination and funding [41]. A secondary effect of the lack of public health funding and staffing is the neglected intended roles of our public health agencies as they divert all resources toward COVID-19. Significant reductions in public health services toward environmental health, substance abuse, maternal child health, and chronic disease have been observed [42].

2.5 Exorbitant Cost of the COVID-19 Pandemic

The pandemic has caused unprecedented economic and social consequences. Our healthcare systems have been strained, pushing medical personnel to the breaking point through the mental and emotional trauma of dealing with an unparalleled number of deaths. Multiple studies during the first year of the pandemic have confirmed that mental illnesses, including anxiety and depression, were significantly higher in healthcare workers as compared to the public [43]. As of August 2022, over six million people have died worldwide directly from COVID-19, while 14.9 million excess deaths are associated with COVID-19 in 2020 and 2021 [44, 45].

National stay-at-home orders have caused significant mental health concerns, educational disruptions for children, and economic losses [46]. Even when accounting for temporary output losses and fiscal support, the pandemic's economic toll on the Unites States is staggering amounting to a $7 trillion cost (2020–2021), which is roughly one-third of the US GDP [47]. When the country issued large-scale stay-at-home orders in Q2 of 2020, real GDP dropped nearly 9% or $1.8 trillion. As stay-at-home orders eased, output slowly recovered; nonetheless, output fell 3.4% by the close of 2021 due to dwindling demand and trade and supply chain disruptions [48]. For comparison, this was a larger decline in output and income than experienced during the economic collapse of the Great Recession of 2009.

Before the pandemic in January 2020, the International Monetary Fund (IMF) projected a 2% growth rate for the US economy [48]. The resulting loss in output in 2020 sums to more than $1 trillion. Although the US economy has rebounded powerfully due to the monumental fiscal stimulus—about $5.8 trillion in 2020–2021—and large-scale monetary expansion, the estimated deficit in 2021 remains around $300 billion [48].

Costs not incorporated into GDP numbers include the burden on healthcare systems and personnel, foregone medical operations, early deaths, population mental well-being, civic upheaval, and greater inequality due to many low-income people being unable to work from home. According to one analysis, the pandemic's health costs (premature death, long-term health impairments, and mental health costs) will be approximately $8.5 trillion for the United States [49]. In a similar vein, the Congressional Budget Office estimates there will be a $7.6 trillion loss in output over the next decade [50]. This brings the total cumulative economic and health costs to over $16 trillion, which is 90% of the annual Gross Domestic Product in the United States [49]. Fighting the pandemic with all available resources has a clear cost-benefit. Greater investment in public health infrastructure, vaccine development, telemedicine, and healthcare personnel would be costly, even in the hundreds of billions of dollars. However, investment in these systems upfront would still be only a fraction of the pandemic's trillions of dollars cost.

3 Advanced Technology and the Future of Health Care

3.1 The Cybercare Model

Dr. C. Everett Koop's vision was to create a dual-use healthcare system designed to both safeguard America from public health threats like pandemics and to prevent routine acute and chronic diseases [3]. Long before the COVID-19 pandemic, Koop advocated for a healthcare system that could protect against an existential biological threat while simultaneously laying the groundwork for a higher-quality healthcare system. This system would address health disparities and shift medicine's focus to prevention rather than treatment after disease progression. To do this, we can look at several aspects of Koop's vision of a healthcare system, termed cybercare. Cybercare is defined as "health care in cyberspace," where care is provided virtually and more locally, rather than patients traveling to a central hospital [51]. The goal of cybercare is to utilize technology with the aim of most healthcare provision occurring in the home setting, while hospitals remain distributors of specialized services [51].

One of the major advantages of cybercare is that it can improve access to health care by shifting focus from large hospitals that are often far distances for patients, to local healthcare centers and the home itself [51]. If care is provided in smaller, distributed clinics and homes rather than a centralized hospital, there are less barriers to treating more patients. The second advantage of decentralizing care involves system resilience. If care is centralized in one hospital, a singular natural disaster or terrorist event may disrupt the hospital, including all its patients and healthcare delivery [51]. In fact, this occurred in Toronto, Canada, in 2003 with the severe acute respiratory syndrome (SARS) epidemic. The ICU, emergency department, and outpatient clinics were all closed [52]. With a more diffuse spread of healthcare services that are not physically attached to one location, the system is becoming more flexible and adaptable.

The technology necessary for the implementation of the cybercare model includes genomics, telemedicine, robotics, simulation, artificial intelligence, electronic medical records, and smartphones [51]. In other words, the fundamental cybercare model is already present in the United States. The following sections will discuss how a range of technologies may advance cybercare into the twenty-first century and create a more equitable and resilient healthcare system for Americans.

1. *The National Emergency Tele-Critical Care Network*

The National Emergency Tele-Critical Care Network project (NETCCN) was created to connect those in need of critical care medical advice to expert specialists via a nationwide network. People can connect to providers through cell phones, tablets, and computers, shifting knowledge to be accessible and virtual. The organization was created in collaboration with the Assistant Secretary for Preparedness and Response and the Society for Critical Care Medicine [53]. The program allows patients to receive eConsults and remote monitoring at home. This can help people avoid traveling far distances, as well as aid in stopping the spread of disease. Additionally, NETCCN helps close the knowledge gap in tiered staffing models for healthcare providers, where noncritical care trained nurses or physicians may be caring for patients that are critically ill [53].

The idea of a specialized network performing cybercare has several advantages in a pandemic. First, the network can relieve task saturation for local care teams during surges. Remote NETCCN providers have previously helped on-site providers complete documentation and orders during high capacity, making patient flow more efficient [54]. Importantly, the experts in NETCCN provide necessary anxiety relief for caregivers that do not possess critical care training. For example, a patient requiring a transfer from a small, rural hospital to a higher-level center may not be able to be transferred immediately during surges. NETCCN experts are able to remotely guide the local providers through critical care procedures, such as maintaining continuous sedation, operating ventilator troubleshooting, treating multiple organ failure, and doing a thoracostomy [54]. The expert providers also aid in making diagnostic decisions and even order tests to help local noncritical care trained providers [54].

For hospitals that do have local critical care experts available, burnout may occur from providing constant supervision to the untrained staff. NETCCN can provide virtual coverage for the local team of supervisors, whether that is at night, during the day, or even on a weekend [54]. The power of the virtual network allows for the relief of on-site providers which helps recharge the staff. While critical care providers are on staff, NETCCN experts provide a second set of eyes to catch any lapses in decision-making in highly stressed environments, improving the quality of care [54]. If the in-person critical care providers are managing another emergency within the hospital, NETCCN can be called to handle additional emergent situations, filling yet another critical care gap.

In a pandemic, maintaining adequate staffing is a cardinal challenge due to healthcare workers falling ill. However, with NETCCN, local healthcare workers can join the remote NETCCN team and continue to advise their local team if needed when unable to provide services in person. The NETCCN onboarding process is less than 48 hours for remote experts [54]. The simplicity of the process is tailored toward clinicians that need to quarantine or isolate. This allows healthcare providers with critical care skills to continue to contribute to the healthcare field even when physically unable to assist. The NETCCN network represents a form of cybercare envisioned by Dr. Koop and has great value in providing a blueprint for other forms of virtual care.

NETCCN can be licensed to local hospitals and healthcare systems who wish to use the software internally [54]. For example, the state of Vermont collaborated with the Telemedicine and Advanced Technology Research Center (TATRC) in 2022 to make the NETCCN available to all state hospitals due to the lack of ICU and critical care experts in many of their rural hospitals. During surges, larger hospitals often reach their capacity and cannot accept more sick patients. A representative of the Vermont Association of Hospitals and Health Systems, Devon Green, stated that the virtual system helps "smaller hospitals to optimize care for these severely ill patients until transfer is possible [55]." The National Emergency Tele-Critical Care Network has supported 40 hospitals in 13 states and territories as of January 2022 [55].

As seen in Vermont, healthcare systems are starting to glimpse the value of cybercare, which will inevitably have several other applications. Healthcare providers that have retired or that are immunocompromised would be able to offer remote services if they wish. In the beginning of COVID-19, many elderly doctors came out of retirement to help the dire healthcare provider shortage, risking their own lives [56]. However, with a virtual network, they can join as remote experts and still make an impact. In the future, NETCCN proposes full integration with major Electronic Medical Records vendors, along with the capacity for remote control of IV pumps and mechanical ventilators. NETCCN demonstrates the value of cybercare for critical care patients; however, the concept of a virtual network of care can be applied in larger scales and to other fields. So far, psychiatry, cardiology, and radiology have the highest telehealth use and growth [57]. Virtual care networks can help alleviate health disparities along with provider burnout, increasing quality and access to care.

2. *Robotic Health Care*

Robotic healthcare provision has been created and theorized for the past few decades within biotechnology. Several surgical robotic systems have been designed with increasing technological complexity during the last 25 years [58]. Robots can provide a mechanical barrier between healthcare workers and patients, which is advantageous in terms of preventing the transmission of infectious disease. For example, robots can reduce pathogen spread during surgery, minimizing risk to both healthcare workers and the patient. They can also take vital signs and administer drugs and anesthetics [58]. During the pandemic, elective surgeries were paused amid concerns over the spread of COVID-19, which put a great strain on both patients who desired surgery along hospitals which lost core revenue [59]. With the use of robots in surgery, elective surgeries could continue, and the healthcare system would be able to work at full capacity.

Robotic surgery was designed to permit remote control over surgery without patient contact. In 1983, the Arthrobot was the first surgical robot that listened to voice commands in order to position a patient during orthopedic surgery [60]. Subsequently, the Unimation Puma 200 was utilized to reliably orient a needle during a brain biopsy in 1985 [61]. One of the biggest breakthroughs in robotic surgery occurred in 1994 when the FDA approved the use of an Automated Endoscopic System for Optimal Positioning (AESOP) [62]. AESOP is a robotic arm that controls a laparoscope via foot pedals and voice control. This development allowed for a stable view of the operating field, rather than relying on the steadiness of an assistant's hands which could be fatigued or inexperienced [62]. Following AESOP, the same company created ZEUS as the first complete robotic surgical system. ZEUS was designed to perform completely remote cardiac telesurgery. This invention could aid in overseas battlefields or even extraterrestrial settings [63]. The robotic system has seven degrees of freedom and imitates a surgeon's hand, designed with tremor elimination and motion scaling.

Marescaux et al. famously performed the first long-distance telesurgical procedure in 2001 called the Lindbergh operation, which was a laparoscopic cholecystectomy between New York and France using ZEUS [63]. Following ZEUS, the Da Vinci robotic system, which was released in 2000, became the favored robotic system. Since the beginning of the twenty-first century, new surgical robotic instruments have been continuously released for not only laparoscopy but for neurosurgery, endoluminal procedures, and orthopedic surgery [64]. Remote telerobotic surgery will be even faster and more powerful with the introduction of 5G, the next generation of the global wireless network. 5G latency is set to be minimized to merely 1–2 milliseconds, up to 100 times faster than previous 4G networks [65].

A study conducted in 2020 by Tian et al. has tested the efficacy of telerobotic surgery in 5G for a series of twelve spinal surgery cases [66]. Six hospitals throughout China participated in the study with the master control room in Beijing. The surgeons successfully completed a remote operation resulting in all patients gaining substantial relief from symptoms. It was concluded that remote 5G robotic spinal surgery was both accurate and reliable [66]. 5G technology also enables faster telementoring for surgeons that need extra guidance during a procedure [65]. Globally, thousands of surgical operations are not performed every year due to a shortage of available qualified surgeons [67]. The advent of 5G with robotic pairing can increase surgical access, which is advantageous in remote settings like the battlefield, natural disaster zones, and even extraterrestrial frontiers.

Robotic technology is not only used for surgeons but can function to carry out the role of nurses. In 2005, the military developed a robotic scrub nurse coined "Penelope." Penelope is voice-activated and can recognize surgical objects. When prompted by the surgeon, Penelope can hand the surgeon the correct instrument while simultaneously retrieving a discarded surgical instrument [68]. This would allow the human scrub nurse to engage in other tasks in the operating room, or alternatively, the robot could stand in when a scrub nurse is not available such as in remote settings. Since Penelope, robotic scrub nurses have been programmed with gesture recognition, and neural networks are being utilized to create more intelligent robotic scrub nurses in Japan [69, 70]. A robotic scrub nurse would be beneficial for telesurgery and specialized remote operations. Similarly, in the hospital setting, some of a nurse's functions can also be streamlined into robotic systems to reduce the healthcare worker's exposure to pathogens. For example, vital signs are often taken many times throughout the day by nurses. If robots were able to alleviate nurses from such a task, nurses would statistically reduce their contact with microbes.

Robots also present an economic advantage in a variety of healthcare settings. A study in rural Canada found that using a robot to record vitals in a clinic saved both time and money, with a favorable 2.3 cost-benefit ratio [71]. Notably, a field hospital in Wuhan, China, constructed a "smart" hospital staffed with robots that had the ability to deliver food and medications to patients [72]. Using robots for tasks involving patient contact also decreases the need for PPE and can free up staff labor to serve other purposes, which is more efficient and allows hospitals to work at a larger maximum capacity [58]. The use of robotics and automation are becoming less expensive as they are increased in use [73]. An economic analysis of ICU nurse task distribution theorized how automation could save both time and money. Tasks such as medication preparation, setting up diagnostic and treatment devices, preparing the patient's room, and blood test handling can all be enhanced by robotic technologies such as couriers and machines. The most time-intensive task for an ICU nurse is monitoring a patient. Automation of the task, such as using smart alarms, sensors, and artificial intelligence to analyze data, would reduce nurse workload for monitoring by an estimated 50% [73]. Robotics and automation for all tasks analyzed by the study have the potential to save up to 24% of total treatment cost.

Robots can further minimize COVID-19 risk by serving as disinfecting machines. International airports have employed continuous, noncontact UV disinfection devices (UVDs) to maintain cleanliness [58]. Hospitals have started to use UVDs, with high compliance and efficacy particularly for the extraordinarily contagious disease C. diff (Clostridioides difficile) [74]. Nursing homes have also begun to utilize UVDs and, in doing so, have found that they are superior to manual cleaning alone [75]. The use of ultraviolet disinfection has helped decrease infection rates and related hospitalizations in nursing homes, which indirectly reduces the burden on hospitals too [75].

3.2 Vaccines

Vaccinations are undoubtedly one of the strongest tools in preventative medicine. Since the eighteenth century, vaccines have been used to prevent severe disease in humans by providing induced immunity [76]. The earliest vaccines used the idea of live attenuation to expose patients to less virulent versions of microbes that the human immune system could fight and remember without harm. Louis Pasteur is widely known for his use of attenuation to create the cholera, anthrax, and rabies vaccines in the nineteenth century [77]. In the twentieth century, vaccines were being developed in the laboratory via cell cultures, such as the oral polio vaccine. Another technique used to develop vaccines is viral or bacterial inactivation, which today is utilized in vaccinations for influenza and encephalitis viruses [76].

The twenty-first century has featured vaccines that are produced using technology derived from molecular biology and genetic engineering. Engineering of pathogens in the laboratory has been noted to be more favorable toward mass production and, in certain respects, safer [76]. Recombinant vaccines are generated by inserting DNA from a pathogenic microbe of choice into either bacterial or yeast cells, which thereafter express recombinant genetics, taking advantage of the natural assembly line of cell organelles [78]. Recombinant vaccines can be either recombinant protein vaccines or DNA vaccines.

The various vaccine strategies have specific advantages and disadvantages. For example, recombinant protein vaccines present weak immunogenicity and are unable to evoke a strong bodily response when given alone and often require adjuvants to enhance the initially weak immune response. An example includes the Hepatitis B and HPV recombinant protein vaccines, which were successful due to the adjuvant aluminum salt [79]. Recombinant proteins have a lower production cost and pose less risk than traditional methods of attenuation or inactivation. In another strategy, DNA vaccines are advantageous for pathogens that are hypervariable. The engineered DNA can encode many specific antigens and offer wide immunity. However, DNA vaccines lack immunogenicity [80]. The vaccine that is optimal to protect against a particular disease is based on the mechanisms of infection of the pathogen along with the required immune response for adequate protection [79].

Perhaps most famous among new vaccine technology is the mRNA vaccine for COVID-19. Research concerning mRNA vaccines spans the past three decades [81]. The technology is based on the premise that the genetic code (DNA) is transcribed into RNA, which is then translated into proteins, such as the antigens found on foreign pathogens that trigger an immune response. The intermediate structure in the process of protein generation is messenger RNA or mRNA. Scientists have been researching mRNA as a novel technology to be utilized for antigen delivery based on its safety and its natural characteristics of efficacy, stability, and production ability [82]. With the ability to deliver proteins into cells, mRNA holds the potential to advance protein therapy, gene editing, and vaccination [83]. The first example of mRNA as a therapeutic occurred in 1989 when a company called Vical Incorporated packaged mRNA within a liposomal nanoparticle, transfecting mRNA into cells, which were translated into proteins [84]. mRNA is unique in that a single mRNA strand can code multiple antigens. This allows precise tailoring of mRNA to diseases and highly specific antigens on pathogens [81]. However, mRNA must be able to enter the cell membrane and avoid degradation. Other advancements in the past few years that have allowed the COVID-19 mRNA vaccines to be successful include the advancement of delivery technologies. The formulation of lipid nanoparticles (LNPs) for mRNA delivery has allowed for seamless intracellular delivery [83].

Furthermore, mRNA technology for vaccines received a major boost with the emergency authorization use of two mRNA vaccines for COVID-19, the Moderna and Pfizer-BioNTech vaccines [81]. These built off the past few decades of major technological advancement. Currently, mRNA technology is being researched and developed for several other vaccines against major diseases including but not limited to influenza, Zika, HIV, Ebola virus, and rabies virus [85].

Another major reason that the COVID-19 mRNA vaccines were so successful and developed at an impressively fast pace was that they used previous research about coronaviruses as a springboard, notably research regarding severe acute respiratory syndrome (SARS-CoV-1) and Middle East Respiratory System (MERS). SARS and MERS were reported in 2003 and 2012, respectively. Coronaviruses all contain one RNA genome that encodes four viral proteins: spike, envelope, membrane, and nucleocapsid. The spike (S) protein mediates viral entry by binding to a cellular receptor called angiotensin-converting enzyme 2 (ACE2) for SARS-CoV-1. This interaction prompts the S protein to alter its structure and ultimately fuse the viral and host cell membranes. Current antibodies against MERS-CoV and SARS-CoV-1 neutralize the S protein, so researchers had already spent significant time and effort to sequence the S protein by the time COVID-19, or SARS-CoV-2, was identified as the pathogen responsible for the pandemic [81].

1. *Future Vaccine Development*

The rapid speed at which the COVID-19 vaccines were produced as a benefit of the previously sequenced S protein underscores the need for vaccine preparedness. Future vaccine development can greatly rely on this lesson. An excellent example of this preparedness is through CEPI, the Coalition for Epidemic Preparedness Innovations. CEPI was founded following the Ebola outbreak of 2014–2015. The organization launched in 2017 with the objective of advancing vaccine candidates against "first priority pathogens" which include Lassa virus and MERS-CoV. CEPI established wide-ranging multiparty support from governments, NGOs, and research institutions. Creating a portfolio of the next most likely emerging infectious diseases (EIDs) requires immense upfront costs. At least $1 billion is required to bring four vaccines through phase 2a trials [86]. CEPI raised over $750 million by 2019, with contributions from governments and philanthropic organizations. Over 16 vaccine development partners signed agreements with CEPI by 2019 [87]. Following the COVID-19 pandemic, CEPI launched a 5-year $3.5 billion "pandemic-busting plan" that will continue to further invest in research and development for the next most concerning pathogens [88]. The plan includes developing vaccines against variants of concern for SARS-CoV-2 and advances vaccine development for other already known threats such as chikungunya, Lassa virus, and Nipah virus. The broad goal is to create a library of prototype vaccines for critical viral families that will be able to be employed in a time of outbreak. This preparedness would help CEPI reach its ambitious goal of rolling out a vaccine in just 100 days in the case of the next Disease X [86].

2. *Anti-vaccine Resistance*

Vaccinations have been highly effective at preventing morbidity resultant from infectious diseases. Approximately two to three million lives are saved per year globally by vaccines. The burden of disease has been significantly alleviated from children, who have traditionally been one of the most affected groups. Childhood diseases such as smallpox and polio were on a path to become eradicated as a result of global vaccine campaigns and widespread implementation [89]. However, vaccination rates for children have begun to decrease for the first time in decades [90]. The global vaccination rate against tuberculosis, diphtheria-tetanus-pertussis, polio, hepatitis B, and measles has all dropped significant percentage points in 2020. This decrease is due to a reduction in routine vaccination appointments along with anti-vaccination sentiment in certain areas of the world. Overall global vaccination rates in most countries have remained stable or reversed over the last decade [91].

In the United States, a reduction in vaccinations for measles and polio has resulted in outbreaks throughout the country. The size and number of measles outbreaks have increased in the United States, with unvaccinated individuals accounting for the vast majority of cases [92]. In July of 2022, polio, a disease once portended to be eradicated through widespread vaccination, was reported for the first time in nearly a decade in the United States in an unvaccinated individual [93]. Implications of vaccination refusal or delay are costly and include the lives of children. Before measles vaccination programs, over 6000 deaths were attributed to measles yearly. Rubella had caused over 2000 neonatal deaths and 11,000 fetal miscarriages. Based on data, over 100 million cases of preventable illnesses including measles and rubella have been prevented in the United States due to vaccination efforts. With decreasing trends in vaccination rates, these diseases are reemerging and threaten to contribute to premature mortalities once again [94].

Vaccine hesitancy consequences are observed in the disparities between vaccinated and non-vaccinated patients for COVID-19. Data from an Israeli study in 2021 show that two doses of the mRNA vaccines by Pfizer or Moderna have effectiveness of 94% after the second dose [95]. A case-control study of hospitalized adults in the United States between March and August of 2021 found that those that were fully vaccinated had a significantly lower likelihood of morbidity and mortality than those that were not vaccinated. Unvaccinated individuals comprised 84.2% of hospitalized patients and 93% of severe COVID-19 cases [96]. For all patients hospitalized with COVID-19, patients that were fully vaccinated were less likely to progress to critical disease. Additionally, outcomes including death and invasive mechanical ventilation were more frequent in unvaccinated individuals.

3.3 Therapeutics

Modern therapeutics to combat disease have a similar upward technological trajectory as vaccines. A greater understanding of biochemistry and genetics has spurred an explosion of innovation. An example of rapidly evolving and improving therapeutics includes antiviral drugs. Antivirals usually control the disease by taking advantage of the virus' replication process. Viral mutations, which enable a virus to become resistant to antiviral drugs, have forced researchers to identify new routes for antivirals to attack. Traditional antivirals target the enzymes that enable viral replication. Three main COVID-19 antivirals (molnupiravir, fluvoxamine, and Paxlovid) are oral medications that halt viral replication [97]. However, one weakness intrinsic to viruses that are relatively unexplored in the therapeutic realm is that viruses require a host to replicate. Researchers are now investigating "host-targeted" drugs that target the host's metabolic pathways, depleting the virus from the external energy sources that it uses to replicate. This may increase the threshold to viral resistance and may offer more broad-spectrum protection against multiple viruses [98].

Another emerging technology involves monoclonal antibodies. As a recently famous COVID-19 treatment, monoclonal antibodies are exogenous replicas of the antibodies found in the human body that can be used therapeutically to boost the immune system [99]. Monoclonals are engineered in a laboratory and can be derived from transgenic mice that have received parts of human DNA to ensure immune compatibility [99]. Advancement in genetic engineering has facilitated much of the progress in antibody development and utilization in treatment. In terms of COVID-19 treatment, the use of monoclonal antibodies demonstrated a 70% reduction in hospitalization or death in March of 2021 [100]. Monoclonals are used not only for COVID-19 treatment, but as therapy for other diseases, such as autoimmune diseases and cancer. Further research on antivirals and monoclonal antibodies will provide more tools in our arsenal to combat diseases of the future.

In certain emergency or remote operations, having pharmaceutical drugs immediately on hand is immensely beneficial and even lifesaving. Traditionally, pharmaceutical manufacturing employs batch processing at large locations. This entails large-scale production costs and long production times and is at risk of supply chain disruptions [101]. The US Defense Advanced Research Projects Agency (DARPA) recognizes these limitations and has financed several continuous-flow chemical devices that can synthesize pharmaceuticals in the field starting with only a few select raw components [102]. Through their Battlefield Medicine program, DARPA funded a study by MIT researchers to build a "pharmacy on demand" in 2016. The same research team previously created a larger prototype of a continuous manufacturing system in 2013 funded by Novartis [103]. The new system is a smaller and more portable pharmaceutical production device. Currently, the device can only produce solutions or suspensions, including the four liquid drugs Benadryl, lidocaine, Valium, and Prozac. The device can manufacture 1000 doses of a drug in 24 hours [104].

Continuous manufacturing techniques for pharmaceuticals have been endorsed by the FDA since April 2016 [105]. The technology has room for growth across more pharmaceutical manufacturers and products. Pharmacy on-demand systems can be customized to produce various medications [104]. One advantage of these small devices is avoidance of large-scale production costs incurred by a drug needed by only a few patients [106]. Additionally, in low-resource regions without large pharmaceutical storage facilities, drugs could be more easily made without the need for storage. The researchers are actively working on synthesizing new combinations of drugs and drugs in tablet form [106]. DARPA is primarily interested in the use of modular pharmaceutical devices in the remote battlefield setting. The compact nature of the device would also be advantageous for rapid production without major large-scale costs for new pharmaceuticals for emerging infectious diseases. Hard-hit areas would have access to drug development on a smaller and faster scale, which would decrease time and cost. The advanced development of modular and portable pharmaceutical production devices will help our healthcare system be more resilient through rapid and focused production of pharmaceuticals.

3.4 Diagnostic Capabilities

The ability to quickly diagnose an infectious disease, particularly pathogens that are on the WHO watch list, can be immensely valuable for future disease outbreaks and control strategy. The CDC defines a rapid diagnostic test (RDT) as a test that can be incorporated into a point-of-care testing protocol (i.e., outside the laboratory) that has a short performance time and is simple to use [107]. RDTs are often immunochromatographic tests consisting of a sample pad, absorption pad, conjugate pad, test, and control lines. Depending on the individual test, a color change on the test lines indicates either a positive or negative result [108]. In 2003, the World Health Organization published criteria for ideal tests to guide management of infections under the acronym ASSURED which stands for Affordable, Sensitive, Specific, User-friendly, Rapid, and Robust [109]. Rapid tests should be simple, fast, and accurate.

RDTs can be split into three categories: molecular, antigen detecting, and antibody testing. The polymerase chain reaction (PCR) tests for COVID-19 are examples of molecular diagnostics, whereas the 15-minute rapid tests are examples of antigen tests. The point-of-care antigen and molecular tests help reduce household and community transmission through simple and early detection of cases [110]. RDTs are also used for other infectious diseases in healthcare settings. A study conducted on emergency department RDTs found the following diseases currently available for testing: upper respiratory tract infection, lower respiratory tract infection, gastrointestinal infection, meningitis, encephalitis, fever in returning travelers, and sexually transmitted infections [111]. Implementation of the RDTs depends on test performance and the way the test is incorporated into patient flow. RDTs are also constrained in that they are not exhaustive and have performance limitations [111]. However, RDTs do have the power to rapidly detect infectious diseases, greatly aiding time-sensitive clinical decisions and enhancing outbreak control.

RDTs can be harnessed on a broader scale to prepare for future pandemics. Of the major pathogens announced by the World Health Organization, most diseases do not have tests at the rapid diagnostic level [108]. Researchers in Denmark analyzed the rapid test status of Crimean-Congo hemorrhagic fever, Ebola and Marburg viruses, MERS, SARS, Nipah disease, Rift Valley fever, Zika, Lassa fever virus, and COVID-19. They found that Crimean-Congo hemorrhagic fever has one commercially available diagnostic test, Ebola virus has six RDTs, Lassa fever virus has two, and Zika virus has over ten. There are no other rapid diagnostic tests that are commercially available and certified for the other pathogens. The researchers argue that the quick rate of rapid test development and production for COVID-19 could be replicated for other pathogens of epidemic potential. However, the development of COVID-19 rapid tests required mass funding from the government and private organizations [108]. The US federal government provided funding to develop hundreds of new rapid diagnostic tests among different companies, showcasing the ability to accelerate production if needed [108]. To be prepared for the next epidemic, it would be advantageous to bring other pathogens up to a rapid diagnostic level. This is in alignment with CEPI's goal of preparing a library of prototype vaccines for the most dangerous pathogens. A faster production pipeline for a RDT for the next pathogen will allow for quicker quarantine and decreased containment timelines, potentially preventing the next epidemic from becoming a pandemic.

3.5 Extended Reality and Medical Simulation

The advancement of computing and display technologies has brought about extended reality (XR) that is increasingly more realistic and accessible. Extended reality refers to technology that enhances human senses, such as virtual reality (VR), augmented reality (AR), and mixed reality (MR). Virtual reality immerses users in a fully artificial digital environment, augmented reality overlays virtual objects in the real-world while maintaining user visibility, and mixed reality overlays and anchors virtual objects to the actual environment, allowing the real word and digital world to interact [112].

Extended reality has enormous potential in medical education and medical practice. Already, XR has been utilized to a certain extent by students and practitioners. For example, virtual reality has become a popular tool during medical education. VR has been utilized in surgical training, helping neurosurgery residents learn important brain structures and vasculature in an interactive 3D experience [113]. Devices can be used to visualize human anatomy during preclinical years too, and students report VR and AR is more engaging than textbooks or cadavers [112]. For example, some medical schools utilize Microsoft's Hololens technology, called HoloAnatomy, to study anatomy in augmented reality [112]. The simulation technology can be used to develop technical clinical competencies (hard skills) but also develop a caregiver's soft skills like empathy and communication [114]. For example, students also may learn how to handle stressful situations and gain practice in a low-stakes environment through the use of VR, such as CPR and mass casualty response [112]. Virtual reality has great ability to transform medical education into a more engaging and immersive process, supplementing clinical learning and visualization.

Virtual reality as an educational tool may also address a major global healthcare shortage in the next decade. There is a shortage of 18 million healthcare workers in the world at present, and one major contributing factor is a lack of effective medical education that utilizes hospital-based learning. The hospital model is limiting in that it can be too expensive and impractical for low-resource settings [115]. The US Department of Defense (DoD) has recognized the use of extended reality in low-resource settings. The Telemedicine and Advanced Technology Research Center (TATRC), an agency of the DoD, established the Medical Modeling, Simulation, Informatics and Visualization (MMSIV) Division in order to study how high-fidelity simulation technology can aid in casualty care in the military [116]. The division created a portfolio of simulation-based training made for personnel ranging from combat medics to specialty surgeons. Using this technology, the US Army conducted a study using augmented reality telestration devices (AR glasses in this case) to guide non-surgeons through damage control procedures. Six non-surgeon providers were able to free fascial compartments and gain control of the femoral artery in anatomically precise surgical models during a simulated damage control procedure [117]. This demonstrates the feasibility of remote diagnosis and treatment using AR telestration in challenging settings with little training. AR can be used for low-resource settings outside of the military as well, such as in low-resource countries and even space travel.

3.6 Artificial Intelligence

1. *Biobanking*

Biobanking is a field that uses mass data from biological repositories to conduct large-scale research and clinical care [118]. Virtual biobanks have become readily accessible, with computerized databases and software developed for biobank packages related to inventory, expenses, consent, and data storage [118]. Advancements in the fields of genomics, proteomics, and personalized medicine have also contributed to the processing of biobank data. Over the past two decades, large-scale population biobanks have become common around the world. The UK Biobank opened in 2006 and the Danish National Biobank opened in 2012. In the United States, one of the largest and earliest biobanks was the AIDS Specimen Bank at the University of California, San Francisco, established in 1983. This was an early example of a disease-specific application of a biobank [119]. Another major biobank that propelled the science forward was the establishment of the US National Cancer Human Biobank (caHUB) in 2009. CaHUB was designed to propose evidence-based standard operating procedures and quality standards for biospecimens intended for genome tissue expression [120]. The findings from caHUB helped streamline processing across all biobanks ensuring that data collected were high quality and ready for analysis.

Since then, biobanks have been established in several countries for various research purposes. Biobanking has been instrumental in analyzing epidemiology and infectious disease trends for COVID-19. The European Union has utilized data from the pan-EU Biobanking and Biomolecular Resources Research Infrastructure in their research [121]. Countries in the Asia Pacific region have also established COVID-19 biobanks. For example, Taiwan quickly established the National Biobank Consortium of Taiwan, collecting blood samples from COVID-19 patients starting in February of 2020. DNA and RNA genome sequencing along with analysis for antibody titers can be conducted on the samples. The analysis is paired with available clinical data (such as symptoms, disease duration, and treatments) to identify significant medical relationships and large-scale patterns [122]. Many studies have shown differences in clinical response to COVID-19 based on genetic determinants. For example, studies from Italy, which was a major epicenter of the pandemic, found several genetic candidates responsible for differential susceptibility to COVID-19 in the Italian population [123]. The power of biobanking can help clinicians uncover genetic predispositions and protect population health.

Analysis of large-scale biobanks is greatly assisted by artificial intelligence (AI). AI models can diagnose diseases and predict risk for disease. A systematic literature review of AI in biobank development found that the pair had a large impact on medical systems. The use of personalized medicine, also known as precision medicine, was prompted through this fine-tuned health analytics service [124]. AI has been able to help detect Alzheimer's disease, cardiovascular diseases, and other hard-hitting chronic diseases including cancer, obesity, and diabetes. In terms of infectious disease, machine learning AI models have been used to predict COVID-19 risk. One study using the UK and Danish biobank developed a machine learning model that could predict the risk of death using common risk factors with fair success [125]. Another study by Dabbah et al. applied machine learning to the UK Biobank in modeling COVID-19 mortality. The researchers identified novel predictors of mortality including prior acute kidney failure, urinary tract infections, and pneumonias [126]. Another large-scale machine learning study analyzed the risk factors leading to severe COVID-19 and death. Wong et al. classified the major contributing factors for severity as age, number of drugs received, renal function, waist-to-hip ratio, and Townsend deprivation index. In predicting mortality, the top three factors were found to be age, testosterone levels, and number of drugs received [127]. These precise findings help clinicians identify risk factors for developing severe disease and can be useful in making clinical decisions. Biobanking combined with the analyzing capability of artificial intelligence provides a powerful tool for combatting disease and pandemics.

2. *Contact Tracing*

Another tool essential to stopping the spread of disease is contact tracing. Contact tracing collects data on people's location and movement in relation to their health status. Manual contact tracing requires human labor, whereas digital contact tracing is an app-based form utilizing smartphones and other technologies. A comprehensive literature review found that digital contact tracing has a higher capacity for the number of contacts identified than manual tracing and reduces response time to track exposed individuals [128]. The review found that four out of eight studies concluded that digital contact tracing alone can significantly reduce infections more than manual tracing, and two reported a small improvement [128]. The use of a combination of digital and manual contact tracing would reduce epidemic size by 80% compared to manual tracing alone reducing epidemic size by 60% [129].

Over 36 countries have employed applied machine learning and artificial intelligence-based contact tracing apps [130]. For example, South Korea's contact tracing system, the COVID-19 Smart Management System, uses data from security cameras, credit card transactions, and GPS locations from cell phones and cars to track small-scale movements of people. Similarly, China, Hong Kong, and Taiwan used AI in contact tracing and enforcement. There is strong public support for collectivism in those countries, enabling such wide-scale surveillance and enforcement [131]. However, the use of such invasive surveillance is a point of concern for privacy rights in other countries such as the Unite States. There are currently legal precedents and protections in place, and researchers are improving cryptography to safeguard privacy. Regardless of the level of data being collected, the use of AI for digital contact tracing has proven efficient. Employing contact tracing to stop the spread of disease, combined with other guidelines such as wearing masks, testing, and social distancing, has proven to reduce infections and is a critical surveillance tool that will help in future outbreaks. Artificial intelligence serves as a powerful supplementary tool to analyze these mass data samples and provide important epidemiological knowledge to a community.

4 Discussion

The model outlined in this chapter is a concept model of what cybercare fundamentals offer to future healthcare delivery. The current delivery system is rife with complexity and high cost. Health inequities plague our citizens, especially in relation to COVID-19 outcomes. Greater investment into public health and the emerging technologies laid out in this chapter can help increase healthcare access in remote settings, cut down the time to receive care, and increase quality. Advancements in vaccines, therapeutics, and diagnostics will improve public health through prevention and precise detection and treatment. The use of extended reality can deliver care to remote settings, with the National Emergency Tele-Critical Care Network providing a blueprint for incorporating cybercare into other fields. Robotic health care paired with pharmacies on demand can make procedures and drugs available on demand in low-access settings. Lastly, the utilization of data through biobanking in conjunction with artificial intelligence will help clinicians find patterns and treatments for our most pressing conditions and infectious diseases.

Emerging technologies are starting to make this vision a reality in the United States, but a truly integrated model requires coordinated effort, leadership, and funding to increase the presence of these efforts. Initiatives such as CEPI's vaccine preparedness goal still require more research, development, and funding. Similarly, rapid diagnostic tests should be brought to the level of pandemic readiness for pathogens of concern for the world to be prepared for the next pandemic. We recognize the need for detailed implementation and financial plans to serve as a roadmap for transitioning to this model. Widespread coordination on all fronts of healthcare delivery must occur for the ambitious implementation of many of these ideas. Bipartisan cooperation and action are required to make this a reality. As Dr. C. Everett Koop pointed out in his 2006 paper in the *American Journal of Public Health*, "Politics should serve health, not the other way around... our growing health care crisis can be resolved only with the concerted efforts of both major political parties" [1]. It is more pressing than ever to move our healthcare system forward. COVID-19 validated worries and predictions made nearly a decade ago about the inadequate state of US planning and readiness for a pandemic or biowarfare attack, specifically around timely pathogen identification, detection, screening, surge capacity, quarantine, and isolation procedures to contain the viral spread.

The White House's *American Pandemic Preparedness: Transforming Our Capabilities* document states that another major pandemic worse than COVID-19 is anticipated in the next decade, emphasizing the significance of investing in pandemic preparedness today [132]. COVID-19 revealed that our country's political divisions can greatly hinder response to a pandemic. However, both sides of the aisle agree that the US response was inadequate and faulty in a variety of ways. CDC director Dr. Rachel Walensky sent messages within the CDC admitting missteps during the pandemic, noting that the agency must develop better procedures in order to respond more quickly and efficiently in the next public health emergency [133]. In the future, the United States and the world must be better equipped to handle epidemics and pandemics. We must heed the warnings made by experts and the lessons previous epidemics or the current pandemic has taught us. We have the tools to improve health systems, to address inequities in prevention and care, and to mitigate the next public health pathogen or threat. Dr. C. Everett Koop's Healthcare System for Biosecurity and Defense can become a reality if we recognize its value and invest in novel technologies to make the healthcare system more accessible and resilient.

5 Summary and Conclusion

Predictions made by Dr. Koop and colleagues occurred two decades before COVID-19 [2, 4, 5]. Although the United States cannot undo the tragic events from the last two years, we can learn from this dark period and invest in and embrace a dual-use healthcare system to meet the unmet needs the healthcare systems are currently experiencing. Cybercare in the virtual sphere can decentralize care and add surge capacity, benefiting both patients and healthcare providers. We already have technologies that can convert our healthcare system into our proposed model, including those identified and presented in this chapter. These technologies can help address the social and racial health disparities, the need for increased focus on prevention rather than treatment alone, and the country's lack of preparedness to deal with pandemics. Unprecedented times call for unprecedented actions, and the United States must unite. Bipartisan action is required to transform health care to meet ongoing demands during this pandemic, prepare for future pandemics (natural or man-made), address unmet needs of all populations, and shift medicine's focus from disease treatment to disease prevention.

References

1. Koop CE (2006) Health and health care for the 21st century: for all the people. Am J Public Health 96(12):2090–2092
2. Rosen JM, Simpson MK (2001) Cybercare: combining healthcare and cyberspace in the 21st century. In: 2001 conference proceedings of the 23rd annual international conference of the IEEE engineering in medicine and biology society. Istanbul, Turkey
3. Koop CE et al (2008) Future delivery of health care: cybercare. IEEE Eng Med Biol Mag 27(6):29–38
4. Rosen J et al (2002) The future of command and control for disaster response. Utilizing information and virtual reality technology, the Cybercare system can link resources throughout the country for distributed yet coordinated command and control. IEEE Eng Med Biol Mag 21(5):56–68
5. Rosen JM, Koop CE, Grigg EB (2002) Cybercare: a system for confronting bioterrorism. Bridge Spring 32.1(Spring):34–40
6. Lacy L (2018) Dwight D. Eisenhower and the birth of the Interstate Highway System. In: Army sustainment
7. Editors, H.c (2022) Cold war history. 2009 June 22, [cited 2022 July 23]; Available from: https://www.history.com/topics/cold-war/cold-war-history
8. Sempowski GD et al (2020) Pandemic preparedness: developing vaccines and therapeutic antibodies for COVID-19. Cell 181(7):1458–1463

9. O'Callaghan KP, Blatz AM, Offit PA (2020) Developing a SARS-CoV-2 vaccine at warp speed. JAMA 324(5):437–438
10. Slaoui M, Hepburn M (2020) Developing safe and effective Covid vaccines - operation warp speed's strategy and approach. N Engl J Med 383(18):1701–1703
11. Crowley R et al (2020) Envisioning a better U.S. Health Care System for all: coverage and cost of care. Ann Intern Med 172(2 Suppl):S7–S32
12. Galvani AP et al (2020) Improving the prognosis of health care in the USA. Lancet 395(10223):524–533
13. Galea S, Maani N (2020) The cost of preventable disease in the USA. Lancet Public Health 5(10):e513–e514
14. Papanicolas I, Woskie LR, Jha AK (2018) Health care spending in the United States and other high-income countries. JAMA 319(10):1024–1039
15. Tikkanen RAM (2020) U.S. Health Care from a global perspective, 2019: higher spending, worse outcomes? In: January 2020. The Commonwealth Fund
16. Services, C.f.M.M (2021) National Health Expenditure Data: Historical. [cited 2022 July 23]; Available from: https://www.cms.gov/Research-Statistics-Data-and-Systems/Statistics-Trends-and-Reports/NationalHealthExpendData/NationalHealthAccountsHistorical
17. Kurani NOJ, Wager E, Fox L, Amin K (2022) How has U.S. spending on healthcare changed over time?. Health Spending 2022 [cited 2022; Available from: https://www.healthsystemtracker.org/chart-collection/u-s-spending-healthcare-changed-time/#Total%20national%20health%20expenditures,%20US%20$%20Billions,%201970-2020
18. World Health Organization, W.H.O. Social Determinants of Health. Health Topics [cited 2022 October 11]; Available from: https://www.who.int/health-topics/social-determinants-of-health#tab=tab_1
19. Crear-Perry J et al (2021) Social and structural determinants of health inequities in maternal health. J Womens Health (Larchmt) 30(2):230–235
20. Hamad R et al (2020) Association of Low Socioeconomic Status with premature coronary heart disease in US adults. JAMA Cardiol 5(8):899–908
21. Fowler-Brown A et al (2007) Risk of cardiovascular events and death--does insurance matter? J Gen Intern Med 22(4):502–507
22. Hoyert D (2022) Maternal mortality rates in the United States, 2020. National Center for Health Statistics Health E-Stats
23. Petterson SM et al (2013) Unequal distribution of the U.S. primary care workforce. Am Fam Physician 87(11):Online
24. Shiels MS et al (2022) Leading causes of death in the US during the COVID-19 pandemic, March 2020 to October 2021. JAMA Intern Med 182(8):883–886
25. Mullen B (2020) COVID-19 clinical guidance for the cardiovascular care team. American College of Cardiology
26. Raveendran AV, Jayadevan R, Sashidharan S (2021) Long COVID: an overview. Diabetes Metab Syndr 15(3):869–875
27. Silva Andrade B et al (2021) Long-COVID and post-COVID health complications: an up-to-date review on clinical conditions and their possible molecular mechanisms. Viruses 13(4):700
28. Raman B et al (2022) Long COVID: post-acute sequelae of COVID-19 with a cardiovascular focus. Eur Heart J 43(11):1157–1172
29. Mackey K et al (2021) Racial and ethnic disparities in COVID-19-related infections, hospitalizations, and deaths : a systematic review. Ann Intern Med 174(3):362–373
30. Gawthrop E (2022) The color of Coronavirus: COVID-19 deaths by race and ethnicity in the US. [cited 2022; Available from: https://www.apmresearchlab.org/covid/deaths-by-race#reporting
31. Quinn SC et al (2011) Racial disparities in exposure, susceptibility, and access to health care in the US H1N1 influenza pandemic. Am J Public Health 101(2):285–293

32. Oshewolo S, Nwozor A (2020) COVID-19: projecting the National Security Dimensions of pandemics. Strateg Anal 44(3):269–275
33. The Public's Perspective on the United States Public Health System (2021) Harvard T.H. Chan School of Public Health
34. (OFR), O.o.F.R (2021) Prevention and public health fund. [cited 2021 December 30]; Available from: https://www.cdc.gov/funding/documents/the-prevention-and-public-health-fund.pdf
35. Lister SA (2017) The ACA prevention and public health fund: in brief. Congressional Research Service
36. Fraser MR (2019) A brief history of the prevention and public health fund: implications for public health advocates. Am J Public Health 109(4):572–577
37. Prevention and Public Health Fund Fact Sheet 2021; Available from: https://www.apha.org/-/media/files/pdf/factsheets/200129_pphf_factsheet.ashx
38. (NACCHO), N.A.o.C.C.H.O (2020) NACCHO's 2019 profile study: changes in Local Health Department Workforce and Finance Capacity Since 2008
39. Faberman RKB (2020) The impact of chronic underfunding on America's Public Health System: trends, risks, and recommendations, [cited 2022 October 11]; Available from: https://www.tfah.org/report-details/publichealthfunding2020/
40. Dirlikov E et al (2020) CDC deployments to state, tribal, local, and territorial Health Departments for COVID-19 emergency public health response - United States, January 21-July 25, 2020. MMWR Morb Mortal Wkly Rep 69(39):1398–1403
41. Maani N, Galea S (2020) COVID-19 and underinvestment in the public health infrastructure of the United States. Milbank Q 98(2):250–259
42. Kintziger KW et al (2021) The impact of the COVID-19 response on the provision of other public health services in the U.S.: a cross sectional study. PLoS One 16(10):e0255844
43. Smallwood N, Willis K (2021) Mental health among healthcare workers during the COVID-19 pandemic. Respirology 26(11):1016–1017
44. World Health Organization, W.H.O (2022) WHO Coronavirus (COVID-19) dashboard
45. World Health Organization, W.H.O (2022) In: Bradshaw D, McCormack K (eds) 14.9 million excess deaths associated with the COVID-19 pandemic in 2020 and 2021
46. Tull MT et al (2020) Psychological outcomes associated with stay-at-home orders and the perceived impact of COVID-19 on daily life. Psychiatry Res 289:113098
47. Cherif RHF This is why we should test everyone for COVID-19. COVID Action Platform 2020 [cited 2021 Dec 30]
48. International Monetary Fund, I.M.F (2021) World economic outlook: recovery during a pandemic—health concerns, supply disruptions, price pressures, Washington, DC
49. Cutler DM, Summers LH (2020) The COVID-19 pandemic and the $16 trillion virus. JAMA 324(15):1495–1496
50. Congressional Budget Office, C.B.O (2020, Jan) The budget and economic outlook: 2020 to 2030
51. Rosen JM et al (2016) Cybercare 2.0: meeting the challenge of the global burden of disease in 2030. Health Technol (Berl) 6:35–51
52. Varia M et al (2003) Investigation of a nosocomial outbreak of severe acute respiratory syndrome (SARS) in Toronto, Canada. Can Med Assoc J 169(4):285
53. TATRC (2022) National Emergency Tele-critical Care Network; Available from: https://www.tatrc.org/netccn/
54. TATRC (2022) NETCCN overview. Available from: https://www.tatrc.org/netccn/overview.html
55. DeBernardis L (2022) Vermont taps NETCCN for tele-critical care in fight against Omicron. TATRC
56. Farrell C (2020) Covid-19 draws retired doctors back to work [cited 2022]
57. Rangachari P, Mushiana SS, Herbert K (2021) A narrative review of factors historically influencing telehealth use across six medical specialties in the United States. Int J Environ Res Public Health 18(9):4995

58. Zemmar A, Lozano AM, Nelson BJ (2020) The rise of robots in surgical environments during COVID-19. Nat Mach Intell 2(10):566–572
59. Singh S, Coleman-Lochner L Hospitals burn through cash while Congress weighs next stimulus business. Prognosis 2020 [cited 2022]
60. Smith JA et al (2016) 30 years of neurosurgical robots: review and trends for manipulators and associated navigational systems. Ann Biomed Eng 44(4):836–846
61. Kwoh YS et al (1988) A robot with improved absolute positioning accuracy for CT guided stereotactic brain surgery. IEEE Trans Biomed Eng 35(2):153–160
62. Marescaux J, Rubino F (2003) The ZEUS robotic system: experimental and clinical applications. Surg Clin N Am 83(6):1305–1315
63. Marescaux J et al (2001) Transatlantic robot-assisted telesurgery. Nature 413(6854):379–380
64. Troccaz J, Dagnino G, Yang GZ (2019) Frontiers of medical robotics: from concept to systems to clinical translation. Annu Rev Biomed Eng 21:193–218
65. Lacy AM et al (2019) 5G-assisted telementored surgery. Br J Surg 106(12):1576–1579
66. Tian W et al (2020) Telerobotic spinal surgery based on 5G network: the first 12 cases. Neurospine 17(1):114–120
67. Meara JG et al (2015) Global surgery 2030: evidence and solutions for achieving health, welfare, and economic development. Lancet 386(9993):569–624
68. Satava RM (2005) Telesurgery, robotics, and the future of telemedicine. Eur Surg 37/5:304–307
69. Jacob MG, Li YT, Wachs JP (2011) A gesture driven robotic scrub nurse. In: 2011 IEEE international conference on systems, man, and cybernetics
70. Nakano A, Nagamune K (2022) A development of robotic scrub nurse system - detection for surgical instruments using faster region-based convolutional neural network. J Adv Comput Intell Intell Inform 26(1):74–82
71. Broadbent E et al (2015) The cost-effectiveness of a robot measuring vital signs in a rural medical practice. In: 2015 24th IEEE international symposium on robot and human interactive communication (RO-MAN)
72. Wittbold K et al (2020) How hospitals are using AI to battle Covid-19. Health and Behavioral Science
73. Kosa G et al (2022) Robots and intelligent medical devices in the intensive care unit: vision, state of the art and economic analysis. TechRxiv
74. Fleming M et al (2018) Deployment of a touchless ultraviolet light robot for terminal room disinfection: the importance of audit and feedback. Am J Infect Control 46(2):241–243
75. Kovach CR et al (2017) Evaluation of an ultraviolet room disinfection protocol to decrease nursing home microbial burden, infection and hospitalization rates. BMC Infect Dis 17(1):186
76. Plotkin S (2014) History of vaccination. Proc Natl Acad Sci USA 111(34):12283–12287
77. Cavaillon JM, Legout S (2022) Louis Pasteur: between myth and reality. Biomol Ther 12(4):596
78. Soler E, Houdebine LM (2007) Preparation of recombinant vaccines. Biotechnol Annu Rev 13:65–94
79. Nascimento IP, Leite LC (2012) Recombinant vaccines and the development of new vaccine strategies. Braz J Med Biol Res 45(12):1102–1111
80. Lee J et al (2018) Engineering DNA vaccines against infectious diseases. Acta Biomater 80:31–47
81. Verbeke R et al (2019) Three decades of messenger RNA vaccine development. Nano Today 28:100766
82. Pardi N et al (2018) mRNA vaccines - a new era in vaccinology. Nat Rev Drug Discov 17(4):261–279
83. Kowalski PS et al (2019) Delivering the messenger: advances in technologies for therapeutic mRNA delivery. Mol Ther 27(4):710–728
84. Malone RW, Felgner PL, Verma IM (1989) Cationic liposome-mediated RNA transfection. Proc Natl Acad Sci 86(16):6077–6081

85. Chaudhary N, Weissman D, Whitehead KA (2021) mRNA vaccines for infectious diseases: principles, delivery and clinical translation. Nat Rev Drug Discov 20(11):817–838
86. (CEPI), C.f.E.P.I. Preliminary business plan 2017–2021. [cited 2022 October 11]
87. Gouglas D et al (2019) CEPI: driving progress toward epidemic preparedness and response. Epidemiol Rev 41(1):28–33
88. Samarasekera U (2021) CEPI prepares for future pandemics and epidemics. Lancet Infect Dis 21(5):608
89. Delany I, Rappuoli R, De Gregorio E (2014) Vaccines for the 21st century. EMBO Mol Med 6(6):708–720
90. Maltezou HC et al (2022) Decreasing routine vaccination rates in children in the COVID-19 era. Vaccine 40(18):2525–2527
91. GBD 2020, Release 1, Vaccine Coverage Collaborators (2021) Measuring routine childhood vaccination coverage in 204 countries and territories, 1980-2019: a systematic analysis for the Global Burden of Disease Study 2020, Release 1. Lancet 398(10299):503–521
92. Phadke VK, Bednarczyk RA, Omer SB (2020) Vaccine refusal and measles outbreaks in the US. JAMA 324(13):1344–1345
93. Russo GB et al (2022) Re-emergence of poliovirus in the United States: considerations and implications. Ann Neurol 92(5):725–728
94. Olusanya OA et al (2021) Addressing parental vaccine hesitancy and other barriers to childhood/adolescent vaccination uptake during the Coronavirus (COVID-19) pandemic. Front Immunol 12:663074
95. Haas EJ et al (2021) Impact and effectiveness of mRNA BNT162b2 vaccine against SARS-CoV-2 infections and COVID-19 cases, hospitalisations, and deaths following a nationwide vaccination campaign in Israel: an observational study using national surveillance data. Lancet 397(10287):1819–1829
96. Tenforde MW et al (2021) Association between mRNA vaccination and COVID-19 hospitalization and disease severity. JAMA 326(20):2043–2054
97. Wen W et al (2022) Efficacy and safety of three new oral antiviral treatment (molnupiravir, fluvoxamine and Paxlovid) for COVID-19a meta-analysis. Ann Med 54(1):516–523
98. Mahajan S et al (2021) Antiviral strategies targeting host factors and mechanisms obliging +ssRNA viral pathogens. Bioorg Med Chem 46:116356
99. Stryjewska A et al (2013) Biotechnology and genetic engineering in the new drug development. Part II. Monoclonal antibodies, modern vaccines and gene therapy. Pharmacol Rep 65(5):1086–1101
100. Corti D et al (2021) Tackling COVID-19 with neutralizing monoclonal antibodies. Cell 184(12):3086–3108
101. Am Ende DJ, Ende MTA (2019) Chemical engineering in the pharmaceutical industry. In: Chemical engineering in the pharmaceutical industry, pp 1–17
102. Markarian J (2021) Advanced manufacturing technologies shift outside the box. Pharm Technol 45.4:16–19
103. Mascia S et al (2013) End-to-end continuous manufacturing of pharmaceuticals: integrated synthesis, purification, and final dosage formation. Angew Chem Int Ed 52(47):12359–12363
104. Adamo A et al (2016) On-demand continuous-flow production of pharmaceuticals in a compact, reconfigurable system. Science 352(6281):61–67
105. Brennan Z FDA allows first switch from batch to continuous manufacturing for HIV drug. Regulatory Focus 2016 [cited 2022]
106. Trafton A Pharmacy on demand 2016 [cited 2021 December 30]; Available from: https://news.mit.edu/2016/portable-pharmacy-on-demand-0331
107. Rabold E, Waggoner J (2019) Rapid diagnostic tests for infectious diseases. In: CDC yellow book. Centers for Disease Control and Prevention Traveler's Health
108. Castillo-Leon J et al (2021) Commercially available rapid diagnostic tests for the detection of high priority pathogens: status and challenges. Analyst 146(12):3750–3776
109. Mabey D et al (2004) Diagnostics for the developing world. Nat Rev Microbiol 2(3):231–240

110. Dinnes J et al (2020) Rapid, point-of-care antigen and molecular-based tests for diagnosis of SARS-CoV-2 infection. Cochrane Database Syst Rev 8(8):Cd013705
111. Bouzid D et al (2021) Rapid diagnostic tests for infectious diseases in the emergency department. Clin Microbiol Infect 27(2):182–191
112. Andrews C et al (2019) Extended reality in medical practice. Curr Treat Options Cardiovasc Med 21(4):18
113. Bernardo A (2017) Virtual reality and simulation in neurosurgical training. World Neurosurg 106:1015–1029
114. Haowen J et al (2021) Virtual reality in medical students' education: a scoping review protocol. BMJ Open 11(5):e046986
115. Barteit S et al (2021) Augmented, mixed, and virtual reality-based head-mounted devices for medical education: systematic review. JMIR Ser Games 9(3):e29080
116. TATRC Medical modeling, simulation, informatics and visualization 2022 [cited 2022]; Available from: https://www.tatrc.org/www/divisions/mmsiv/
117. Harris TE et al (2020) Augmented reality forward damage control procedures for nonsurgeons: a feasibility demonstration. Mil Med 185(Suppl 1):521–525
118. De Souza YG, Greenspan JS (2013) Biobanking past, present and future: responsibilities and benefits. AIDS 27(3):303–312
119. Greenspan JS et al (1991) The UCSF AIDS Specimen Bank. Lab Med 22(11):790–792
120. National Cancer Institute, N.I.H. Division of Cancer Treatment & Diagnosis Standard Operating Procedures 2015 [cited 2022 October 11]; Available from: https://biospecimens.cancer.gov/resources/SOPs/
121. Holub P et al (2020) BBMRI-ERIC's contributions to research and knowledge exchange on COVID-19. Eur J Hum Genet 28(6):728–731
122. Huang SF et al (2020) Rapid establishment of a COVID-19 biobank in NHRI by National Biobank Consortium of Taiwan. Biom J 43(4):314–317
123. Anastassopoulou C et al (2020) Human genetic factors associated with susceptibility to SARS-CoV-2 infection and COVID-19 disease severity. Hum Genomics 14(1):40
124. Battineni G et al (2022) A survey on the role of artificial intelligence in biobanking studies: a systematic review. Diagnostics 12(5):1179
125. Jimenez-Solem E et al (2021) Developing and validating COVID-19 adverse outcome risk prediction models from a bi-national European cohort of 5594 patients. Sci Rep 11(1):3246
126. Dabbah MA et al (2021) Machine learning approach to dynamic risk modeling of mortality in COVID-19: a UK Biobank study. Sci Rep 11(1):16936
127. Wong KC-Y et al (2021) Uncovering clinical risk factors and predicting severe COVID-19 cases using UK Biobank data: machine learning approach. JMIR Public Health Surveill 7(9):e29544
128. Grekousis G, Liu Y (2021) Digital contact tracing, community uptake, and proximity awareness technology to fight COVID-19: a systematic review. Sustain Cities Soc 71:102995
129. Barrat A et al (2020) Effect of manual and digital contact tracing on COVID-19 outbreaks: a study on empirical contact data. medRxiv:2020.07.24.20159947
130. Lalmuanawma S, Hussain J, Chhakchhuak L (2020) Applications of machine learning and artificial intelligence for Covid-19 (SARS-CoV-2) pandemic: a review. Chaos, Solitons Fractals 139:110059
131. Lin L, Hou Z (2020) Combat COVID-19 with artificial intelligence and big data. J Travel Med 27(5):taaa080
132. Lander E, Sullivan J (2021) American pandemic preparedness: transforming our capabilities. The White House
133. Edwards E (2022 Aug 17) CDC director calls for drastic changes to the agency following pandemic missteps. [cited 2022]

Coronavirus Variant Emergence and Epidemiology

Joseph M. Rosen and Haley R. Warzecha

1 What Are Viral Lineage, Variant, and Strain?

A "lineage" refers to viruses that are closely related to each other sharing a common ancestor [10]. These signify direct relationships between ancestors and descendants over time. The subfamily *Coronavirinae* has four genera: *Alphacoronavirus*, *Betacoronavirus*, *Gammacoronavirus*, and *Deltacoronavirus*. The *Betacoronavirus* genus includes four lineages named A, B, C, and D. Severe acute respiratory syndrome coronavirus (SARS-CoV) is a member of lineage B and includes both SARS-CoV-1 (emerged in 2003) and SARS-CoV-2 (emerged in 2019 and causes COVID-19) [55]. SARS-CoV and Middle East Respiratory Syndrome coronavirus (MERS-CoV) are in different lineages within the genus *Betacoronavirus* and share a common ancestor before the lineages split.

J. M. Rosen (✉)
Department of Surgery, Dartmouth Hitchcock Medical Center, Lebanon, NH, USA

H. R. Warzecha
Dartmouth College, Hanover, NH, USA

J. M. Rosen, R. R. Colwell (eds.), *The COVID-19 Pandemic*,
https://doi.org/10.1007/978-3-031-62772-9_9

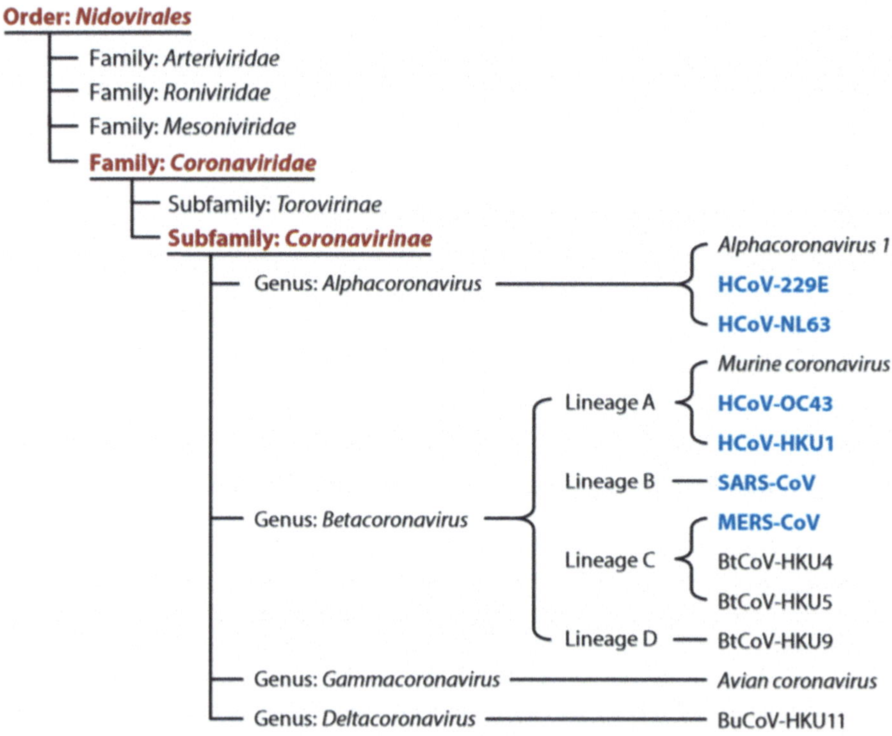

Source: Yan et al. [55]

A "variant" is a virus with a different genomic sequence from the original virus in the form of a genetic mutation. If a variant starts presenting different physical properties compared to the original virus, it can then be classified as a new "strain" [10]. For example, SARS-CoV-1 and SARS-CoV-2 represent different strains of coronaviruses because they have different properties including symptom presentation, transmission, and receptor binding. A comparative review of the two strains found that the spike protein (S), an important structural protein, is slightly different between the two strains, contributing to the increased transmissibility of SARS-CoV-2 [37]. Within the viruses that are classified as SARS-CoV-2, there may also be variants that have developed properties meriting classifying them as new strains, which will be covered in this chapter. In public and scientific discussions, variants and strains have been used interchangeably.

2 What Mechanism Influences the Infectivity of SARS-CoV-2?

The surface of the SARS-CoV-2 is coated with spike proteins (S), which attach to angiotensin-converting enzyme-2 (ACE2) receptors on the host cell to fuse and allow for viral entry, starting the infection process. The strength of the binding and fusion influences the infectivity of SARS-CoV-2. The spike protein is a transmembrane protein that is cleaved into subunit 1 (S1) and subunit 2 (S2). S1 contains the N-terminal domain (NTD), a receptor-binding domain (RBD), and several other subdomains [23]. During viral entry, the RBD on S1 recognizes the host cell's ACE2 receptor and binds to the host cell. S2 is the part of the protein that initiates membrane fusion [40]. Studies have shown that the affinity and stability of the RBD with the ACE2 receptor on SARS-CoV-2 are stronger than the affinity and strength of other coronaviruses. Cong et al. propose an anchor-locker binding mechanism through computational analysis [13].

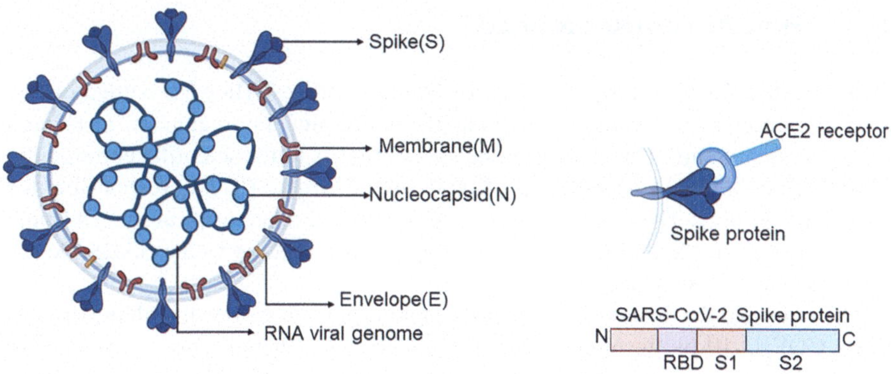

Source: Min and Sun [32]

3 How Do New SARS-CoV-2 Variants Become More Dangerous than Previous Variants?

Over time, genetic mutations will arise as a natural result of repeated viral replication, resulting in new variants. Mutations in the spike protein that help the virus evade the human immune defense systems are especially beneficial. The mutated virus with advantageous properties is more likely to replicate than other variants [1]. The new strain starts to spread faster than others and then can pose a serious threat to public health.

The strength of a variant is determined by host cell entry (S protein binding to ACE2 receptor) and antibody-mediated neutralization. For example, in terms of entering cells, the Alpha and Delta variants have a mutation in the spike protein that enhances furin cleavage. The enzyme furin activates the spike protein through a series of cuts at the S1/S2 cleavage site [36]. In the two variants, the mutations increase furin cleavage and are hypothesized to make the virus more effective at entering cells through greater activation of the S protein [56].

Another example of variant strength comes in the form of immune evasion. In the case of variant A.30, the S protein contains five deletions and ten amino acid substitutions, most changes within the S1 subunit. These mutations help the S protein evade recognition from targeted antibodies that recognize the original amino acids [3, para. 2]. One of the mutations, E484K, specifically reduces antibody-mediated neutralization. The mutation language "E484K" indicates that the amino acid glutamic acid (E) was changed to lysine (K) at position 484 [3, para. 2]. Antibodies normally recognize E instead of K and therefore are less effective against the new variant. This decreases the efficacy of the immune system's antibodies, which may come from vaccine immunity or previous natural immunity. The new virus is able to replicate more successfully since it can escape the body's defense systems.

4 How Are Viruses Named?

Generally, the WHO follows a pattern for naming viruses. They are named on the basis of their genetic structure to facilitate the development of diagnostic tests, vaccines, and treatments. This work is done by virologists in the scientific community, with the International Committee on Taxonomy of Viruses (ICTV) responsible for naming the viruses. The ICTV named the virus "severe acute respiratory syndrome coronavirus 2" or "SARS-CoV-2" based on its close relation to the SARS-CoV-1 outbreak in 2003 [49, para. 3].

The virus and the disease that the virus causes are differentiated, with the disease called COVID-19 by the WHO. Director-General Dr. Ghebreyesus explained that the "CO" stands for "corona," the "VI" stands for "virus," the "D" stands for "disease," and the "19" stands for 2019, the year the virus was identified [49]. By referring to the disease separately from the virus itself, it is easier to communicate epidemiological principles of prevention, transmission, and treatment in a clinical manner. However, SARS-CoV-2 and COVID-19 are often used interchangeably, referring to either the virus or disease based on context.

5 How Are the Major SARS-CoV-2 Strains and Variants Named?

One way of discussing the viral strains is by clades according to GISAID, the Global Initiative on Sharing Avian Influenza Data [21]. Clades are a group of organisms or viruses that share a common ancestor. A clade includes a last common ancestor with all the lineages that branch off from it, showing all descendants [11]. The original strain from Wuhan, China, appeared in late 2019 and was scientifically named the L clade, used interchangeably as the L strain. At the beginning of 2020, the S strain branched off from the original L strain. The L strain further diverged into the V and G strains. The G strain next evolved into substrains of GR, GH, and GV. Recently, the GR strain branched off into GRY. Several other mutations were grouped together as the O strain [12, para. 3, 21].

By the end of 2020, the most predominant strain in North America was the GH strain, and in South America, it was the GR strain. The G, GH, and GR strains were identified in Asia in March 2020, a month later than their emergence in Europe [17, para. 8, 31]. The GISAID clade definitions utilize genome distances of the phylogenetic clusters, in combination with marker variants, to categorize the strains [21]. Using the letter strains was more common earlier in the pandemic.

Another way of scientifically naming the strains was provided by Rambaut et al. in July 2020. The researchers proposed a dynamic nomenclature framework called Pango lineages, which stands for Phylogenetic Assignment of Named Global Outbreak LINeages. The major lineage labels begin with a letter, with the two major lineages named A and B [38]. The A lineage includes the original Wuhan virus. Following the first letter is a number such as A.1. The lineage designation is made using phylogenetic evidence that the new variant branched off from an ancestral lineage and has been established in a new geographic population. This process is iterative, allowing for a lineage to give rise to sublineages up to three sublevels. After the fourth sublevel, the first letter is replaced by a new letter combination [38]. For example, after A, the next label would be AA, ending in AZ. After B's third sublevel, the next level is BA going to BZ.

In May 2021, the World Health Organization announced a new naming system for the variants for the general public to use. The previously discussed naming systems helped scientists systematically classify all variants that exist. The new Greek naming system was designed to convey which variants were important for the public to know. They decided each variant would be given a name from the Greek alphabet. This would simplify naming for the general population and remove stigma arising from the origin countries. This decision aimed to encourage countries to come forward with new outbreaks without fear of discrimination [51, 52].

The new naming system was important in communicating the spread of one of the first major variants, the Alpha variant (B.1.1.7), which was first detected in the United Kingdom and designated a variant of concern in December 2020. Once the WHO began assigning variant letters of the Greek alphabet in May 2021, Alpha was easily understood to signify an emerging, dangerous COVID-19 variant that the public should be aware of [51].

6 How Are the Variants Classified?

The U.S. Department of Health and Human Services established the SARS-CoV-2 Interagency Group (SIG) during the onset of the pandemic. The SIG and the WHO both use the following classifications for virus variants: variants of interest, variants of concern, and variants being monitored [46].

7 What Is a Variant of Interest (VOI)?

A SARS-CoV-2 variant of interest (VOI) is classified as a variant that has genetic changes that affect transmissibility, severity of disease, immune evasion, or therapeutic escape [50, para. 9]. VOIs can cause significant community transmission, with "other apparent epidemiological impacts to suggest an emerging risk to global public health" [50, para. 9]. A variant of interest also must cause community transmission in multiple countries. The assessment of VOIs is conducted by the WHO in consultation with the Virus Evolution Working Group. Below is a table of variants of interest. Since then, the WHO has announced that there are no currently circulating VOIs and all are previously circulating.

8 Variants of Interest

Name of variant	Lineage	Origin	Major mutations	Date recognized as VOI
Epsilon[a]	B.1.429, B.1.427	USA	S13I, W152C, and L452R on S protein	March 2021
Eta[a]	B.1.525	Multiple countries	E484K and F888L on S protein	March 2021
Kappa[a]	B.1.617.1	India	E154K, L452R, E484Q, and P681R on S protein	April 2021
Lambda[a]	C.37	Peru	G75V, T76I, L452Q, F490S, D614G, and T859N on S protein	June 2021
Iota[a]	B.1.526	USA	E484K on S protein	March 2021
Theta[a]	P.3	Philippines	E484K, N501Y, and P681H on S protein	March 2021
Zeta[a]	P.2	Brazil	L18F, T20N, P26S, F157L, E484K, D614G, S929I, and V1176F on S protein	March 2021
Mu[a]	B.1.621	Colombia	N501Y, D614G, P681H, and D950N on S protein	August 2021

Aleem et al. [1], World Health Organization [51]
[a]Indicates previously circulating

9 What Is a Variant of Concern (VOC)?

A SARS-CoV-2 variant of concern is the next level of threat above VOI. It is defined as a variant that meets the definition of a VOI but also has increased transmissibility or "detrimental change" in epidemiology. It possesses increased virulence or decreases the effectiveness of public health measures, such as diagnostics, vaccines, and therapeutics [50, para. 11]. These pose a higher level of concern to public health. All VOCs in the table except for Omicron variants have been classified as previously circulating.

10 Variants of Concern

Name of variant	Lineage	Origin	Major mutations	Date recognized as VOC
Alpha[a]	B.1.1.7	United Kingdom	N501Y on S protein	December 2020
Beta[a]	B.1.351	South Africa	K417N, E484K, and N501Y on S protein	December 2020
Gamma[a]	P.1	Brazil	N501Y, E484K, and K417T on S protein	January 2021
Delta[a]	B.1.617.2	India	P681R and L452R on S protein	May 2021
Delta Plus[a]	AY.4.2	UK	Y145H and A222V on S protein	October 2021
Omicron	B.1.1.529 AKA BA.1	South Africa	32 mutations on S protein	November 2021
Omicron BA.2	BA.2	South Africa	31 mutations on S protein	November 2021
Omicron BA.4 and BA.5	BA.4 and BA.5	South Africa	L452R and F486V on S protein [16]	January and February 2022, respectively [16]

Galloway et al. [19], World Health Organization [51]
[a]Indicates previously circulating

11 Previously Circulating VOCs

11.1 Alpha Variant

Scientists have found no evidence of increased virulence (severity) of the Alpha variant. Alpha did spread rapidly with remarkably high transmissibility, and the variant had been detected in around 120 countries by May 2021 [19].

11.2 Beta Variant

Prevalence of the Beta variant was higher among young people with no underlying health conditions and was more frequently resulting in serious illness in comparison with other variants [19]. The Beta variant was rapidly replaced by the Delta variant in South Africa in 2021, with much higher transmission rates [42].

11.3 Gamma Variant

The Gamma variant's spike protein mutations showed 2.2 times higher transmissibility than the original lineage with the same ability to infect both adults and elderly people [15].

11.4 Delta Variant

For the Delta variant, the K417N mutation at the spike protein indicated escape of antibody binding but may have also decreased binding to the ACE2 receptor. Delta was shown to spread almost twice as fast as the Alpha variant and mutated further to form the Delta plus variant [51].

11.5 Delta Plus Variant

Delta Plus variant was predicted to be slightly more transmissible than Delta. However, this was a marginal difference that experts said was more of a "headache" than a significant game-changer [26, 29, para. 8, 45] .

12 Currently Circulating VOCs

12.1 Omicron Variant

Thirty-two mutations in the spike protein give Omicron enhanced immune escape compared to the original strain. The variant started to become dominant around the globe by early 2022, spread faster than other strains, and caused a higher reinfection risk than other variants [44, 50, paras. 1–3].

A Note on Omicron Subvariants

The Pango and Greek systems are altered with the emergence of new Omicron subvariants. The original B.1.1.529 virus exceeded three sublevels, so the first part, "B," was replaced with a new letter, which was "BA." Rather than calling the variant B.1.1.529.2, the Pango system has designated the subvariant BA.2. The other two subvariants are BA.1 and BA.3. The subvariants are easy for people to understand as Omicron subvariants, so a combination of the Pango and Greek systems is being utilized.

12.2 Omicron BA.2 ("Stealth")

BA.2 emerged shortly after the spread of BA.1 in late 2021. Scientists found that the spike protein is more pathogenic than BA.1 and has a higher replication efficacy than the previous Omicron variant. Notably, BA.2 was better able to avoid immunity, particularly from vaccines, contributing to its rapid global spread [54]. BA.2 was highly resurgent not only because of higher transmissibility, but people were gathering unmasked indoors in late 2021, and there was waning immunity from previous vaccinations or prior infections compounding decreased vaccine efficacy [4].

12.3 Omicron Variants BA.4 and BA.5

Omicron subvariant BA.2 was the current dominant form of COVID-19 in the spring of 2022. However, the share of BA.4 and BA.5 has been on the rise globally in new COVID-19 cases and the percentage is growing [6]. As of August 2022, BA.5 accounts for the majority of the world's COVID-19 cases, and BA.4 represents over 10% of cases [30]. The variants are similar to BA.2 except for a deletion. In comparison with each other, they have identical spike proteins, only differing outside of S. BA.4 and BA.5 are hypothesized to be related through a recombination event [43].

13 What Is a Variant Being Monitored (VBM)?

Variant being monitored is a class of variants at a level below VOI and VOC that has reduced national or regional significance. These variants pose much less of a risk to public health but are still being surveilled for indications of increased transmissibility or severity. These variants include those that are no longer detected or have been designated as previously circulating [10].

14 Variants Being Monitored (VBM)

Name of variant	Lineage	Date recognized		
		VOC	VOI	VBM
Alpha	B.1.1.7 and Q lineages	December 29, 2020		September 21, 2021
Beta	B.1.351 and descendent lineages	December 29, 2020		September 21, 2021
Gamma	P.1 and descendent lineages	December 29, 2020		September 21, 2021
Delta	B.1.617.2 and AY lineages	June 15, 2021		April 14, 2022
Epsilon	B.1.427 B.1.429	March 19, 2021	February 26, 2021 June 29, 2021	September 21, 2021
Eta	B.1.525		February 26, 2021	September 21, 2021
Iota	B.1.526		February 26, 2021	September 21, 2021
Kappa	B.1.617.1		May 7, 2021	September 21, 2021
N/A	B.1.617.3		May 7, 2021	September 21, 2021
Zeta	P.2		February 26, 2021	September 21, 2021
Mu	B.1.621, B.1.621.1			September 21, 2021

CDC [10]

15 What Is a Variant of High Consequence (VOHC)?

A VOHC has attributes of VOCs and additionally shows evidence that medical countermeasures against the variant would have significantly reduced effectiveness compared to previously circulating variants [10]. Examples of this level of warning would include diagnostic test failure, significantly reduced vaccine or therapeutic effectiveness, and increased disease severity. No variants have met the CDC's criteria for VOHC yet. The WHO does not use variants of high consequence classification like the SARS-CoV-2 Interagency Group, yet it does use the terms VBM, VOI, and VOC [10].

16 How Do We Stay Ahead of Variants?

We must maintain and improve surveillance and sequencing efforts of SARS-CoV-2 in order to stay up-to-date on viral strains and new mutations. These genomic sequences should be continuously submitted to large databases such as GISAID and the open-source project Nextstrain, which collects pathogen genome data [35]. Additionally, as per the WHO International Health Regulations, all countries are legally bound to report events of international public health importance. If a significant outbreak occurs, it should be immediately reported to facilitate rapid response [48]. Lastly, scientific studies on COVID-19 epidemiology, diagnostics, immunity, medical countermeasures, and vaccinations must continue to stay ahead of variants. Public health and social measures to contain the virus and reduce its spread will help reduce the availability of viral hosts, reducing the amount of viral replication and therefore viral mutation. The odds of the virus mutating into a dangerous variant are lower when the virus is contained and spreading slower [51].

17 What Could Happen Next?

Scenario 1: Elimination
This scenario is the least likely. Take the case study of measles. The measles virus is present in a few reported cases but does not have a high capacity to evolve to evade immunity. This example of elimination requires no community transmission of measles for at least a year, but cases do reappear. The success of measles elimination was due to a widespread vaccination program in the USA and the Americas as a whole [7].

Scenario 2: Endemic
This scenario could follow the "flu" virus path where every year, a new variant of the virus is different from past strains, and vaccines may not effectively contain the virus. COVID-19 could become seasonal. The CDC Director Dr. Rachel Wallensy has warned that the virus could be similar to other respiratory diseases, appearing in winter months [39].

Scenario 3: Worst Case
SARS-CoV-2 evolves into a deadlier virus by recombining with other variants or evolving into a new animal host(s). This situation would be the worst outcome and is something we should prepare for. At the level of community transmission in the world, the emergence of a more serious variant is a possibility.

18 Conclusion

Billions of people around the world are still unvaccinated, providing plenty of opportunities for the virus to evolve into new variants that can evade vaccines and natural immunity. Even with vaccines and antivirals, we cannot presume that life has returned to normal. Most likely, we have moved to a new phase of the pandemic. A new variant of concern with increased transmissibility and immune escape properties would be a reversal of waves of infection and hospitalization. Omicron offers perhaps a perfect example. We must remain vigilant and stay ahead of new variants through surveillance, sequencing, and research.

Bibliography

1. Aleem A et al (2022) Emerging variants of SARS-COV-2 and novel therapeutics against coronavirus (COVID-19). In: StatPearls [Internet]. StatPearls Publishing, Treasure Island. https://pubmed.ncbi.nlm.nih.gov/34033342/
2. Altarawneh HN et al (2022) Protection of SARS-COV-2 natural infection against reinfection with the Omicron BA.4 or BA.5 subvariants. MedRxiv, Cold Spring Harbor Laboratory Press, New York. https://www.medrxiv.org/content/10.1101/2022.07.11.22277448v1
3. Arora P et al (2021) The spike protein of SARS-COV-2 variant A.30 is heavily mutated and evades vaccine-induced antibodies with high efficiency. Nature News, Nature Publishing Group. https://www.nature.com/articles/s41423-021-00779-5
4. Berg S (2022) What doctors wish patients knew about the BA.2 Omicron subvariant. American Medical Association. https://www.ama-assn.org/delivering-care/public-health/what-doctors-wish-patients-knew-about-ba2-omicron-subvariant#:~:text=It's%20more%20contagious%2C%20but%20not%20more%20severe,-%E2%80%9CThe%20concern%20is&text=That's%20because%20the%20BA.,of%20the%20number%20of%20cases
5. Broggi A et al (2020) Type III interferons disrupt the lung epithelial barrier upon viral recognition. Science 369:706–712. https://www.ncbi.nlm.nih.gov/pmc/articles/PMC7292499/
6. Callaway E (2022) What Omicron's BA.4 and BA.5 variants mean for the pandemic. Nature News, Nature Publishing Group. https://www.nature.com/articles/d41586-022-01730-y
7. CDC (2020) History of measles. Centers for Disease Control and Prevention. https://www.cdc.gov/measles/about/history.html#:~:text=Measles%20was%20declared%20eliminated%20(absence,control%20in%20the%20Americas%20region
8. CDC (2022) Reinfection. Centers for Disease Control and Prevention. https://www.cdc.gov/coronavirus/2019-ncov/hcp/clinical-care/clinical-considerations-reinfection.html
9. CDC (2022) Reinfections and COVID-19. Centers for Disease Control and Prevention. https://www.cdc.gov/coronavirus/2019-ncov/your-health/reinfection.html
10. CDC (2022) SARS-COV-2 variant classifications and definitions. Centers for Disease Control and Prevention. https://www.cdc.gov/coronavirus/2019-ncov/variants/variant-classifications.html

11. Cellinese N, Baum DA, Mishler BD (2012) Species and phylogenetic nomenclature. Syst Biol 61(5):885–891. https://academic.oup.com/sysbio/article/61/5/885/1735952
12. Chowdhury J et al (2020) How the novel coronavirus evolved. Reuters, Thomson Reuters. https://graphics.reuters.com/HEALTH-CORONAVIRUS/EVOLUTION/yxmpjqkdzvr/
13. Cong Y et al (2021) Anchor-locker binding mechanism of the coronavirus spike protein to human ACE2: insights from computational analysis. J Chem Inf Model 61(7):3529–3542
14. Cornwall W (2020) Just 50% of Americans plan to get a COVID-19 vaccine. Here's how to win over the rest. Science. https://www.science.org/content/article/just-50-americans-plan-get-covid-19-vaccine-here-s-how-win-over-rest
15. Coutinho RM et al (2021) Model-based estimation of transmissibility and reinfection of SARS-CoV-2 P. 1 variant. Commun Med 1(1):1–8
16. ECDC (2022) Epidemiological update: SARS-COV-2 Omicron sub-lineages BA.4 and BA.5. European Centre for Disease Prevention and Control. https://www.ecdc.europa.eu/en/news-events/epidemiological-update-sars-cov-2-omicron-sub-lineages-ba4-and-ba5#:~:text=BA.-,4%20and%20BA.,binding%20domain%20compared%20to%20BA
17. EurekAlert! (2020) The six strains of SARS-COV-2. EurekAlert!. https://www.eurekalert.org/news-releases/612634
18. Evans C (2021) NIH scientists identify mechanism that may influence infectivity of SARS-COV-2 variants. National Institutes of Health, U.S. Department of Health and Human Services. https://www.nih.gov/news-events/news-releases/nih-scientists-identify-mechanism-may-influence-infectivity-sars-cov-2-variants#:~:text=The%20outer%20surface%20of%20SARS,starting%20with%20the%20furin%20enzyme
19. Galloway SE et al (2021) Emergence of SARS-COV-2 B.1.1.7 lineage – United States, December 29, 2020–January 12, 2021. Centers for Disease Control and Prevention. https://www.cdc.gov/mmwr/volumes/70/wr/mm7003e2.htm
20. Ghosh P et al (2020) COVID-19 in India: statewise analysis and prediction. JMIR Public Health Surveill 6(3):e20341. https://doi.org/10.2196/20341
21. GISAID (2021) Clade and lineage nomenclature aids in genomic epidemiology of active HCoV-19 viruses. GISAID. https://gisaid.org/resources/statements-clarifications/clade-and-lineage-nomenclature-aids-in-genomic-epidemiology-of-active-hcov-19-viruses/
22. India Today (2020) Is India entering stage 3 of covid-19 outbreak? India Today. https://www.indiatoday.in/programme/to-the-point/video/is-india-entering-stage-3-of-covid-19-outbreak-1661544-2020-03-30
23. Jackson CB et al (2021) Mechanisms of SARS-COV-2 entry into cells. Nature News, Nature Publishing Group. https://www.nature.com/articles/s41580-021-00418-x
24. Javaid A (2022) SARS-COV-2 variants list: how many variants of covid-19 are there in the world? Jagranjosh.com. https://www.jagranjosh.com/general-knowledge/variants-of-sars-cov-2-in-the-world-covid-19-variants-1623846555-1
25. Kannan SR et al (2022) Omicron SARS-COV-2 variant: unique features and their impact on pre-existing antibodies. J Autoimmun 126:102779. https://www.sciencedirect.com/science/article/pii/S0896841121001876?via=ihub%252523sec1
26. Katella K (2022) Omicron, delta, alpha, and more: what to know about the coronavirus variants. Yale Medicine. https://www.yalemedicine.org/news/covid-19-variants-of-concern-omicron
27. KFF (2022) KFF COVID-19 Vaccine Monitor dashboard. KFF. https://www.kff.org/coronavirus-covid-19/dashboard/kff-covid-19-vaccine-monitor-dashboard/
28. Lanese N (2021) Coronavirus variants to be named using greek letters, who says. LiveScience, Purch. https://www.livescience.com/who-new-coronavirus-variant-names.html
29. Lucey DR (2021) Ay.4.2 Delta subvariant spreading in UK (>15,000 cases) and found in USA: get vaccinated. IDSA Home. https://www.idsociety.org/science-speaks-blog/2021/ay.4.2-delta-subvariant-spreading-in-uk%2D%2D15000-cases-and-found-in-usa-get-vaccinated/#/+/0/publishedDate_na_dt/desc/
30. Mahase E (2022) Covid-19: what we know about the BA.4 and BA.5 omicron variants. BMJ 378:o1969

31. Mercatelli D, Giorgi FM (2020) Geographic and genomic distribution of SARS-COV-2 mutations. Frontiers. https://www.frontiersin.org/articles/10.3389/fmicb.2020.01800/full
32. Min L, Sun Q (2021) Antibodies and vaccines target RBD of SARS-COV-2. Frontiers. https://www.frontiersin.org/articles/10.3389/fmolb.2021.671633/full
33. Mitropoulos A (2022) Americans keep getting reinfected with COVID-19 as new variants emerge, data shows. ABC News, ABC News Network. https://abcnews.go.com/Health/americans-reinfected-covid-19-variants-emerge-data-shows/story?id=85114588
34. National Herald (2020) Stage 3 begins, we are running out of time: COVID-19 task force. National Herald. https://www.nationalheraldindia.com/flick-past/stage-3-begins-we-are-running-out-of-time-covid-19-task-force
35. Nextstrain (2022) Genomic epidemiology of SARS-CoV-2 with subsampling focused globally over the past 6 months. Nextstrain. https://nextstrain.org/ncov/gisaid/global/6m
36. Peacock TP et al (2021) The furin cleavage site in the SARS-CoV-2 spike protein is required for transmission in ferrets. Nat Microbiol 6(7):899–909
37. Rabaan AA et al (2020) SARS-CoV-2, SARS-CoV, and MERS-COV: a comparative overview. Infez Med 28(2):174–184
38. Rambaut A et al (2020) A dynamic nomenclature proposal for SARS-CoV-2 lineages to assist genomic epidemiology. Nat Microbiol 5(11):1403–1407
39. Ries J (2022) Covid will become a seasonal disease, experts say. Healthline, Healthline Media. https://www.healthline.com/health-news/covid-19-will-likely-become-a-seasonal-disease-cdc-director-says
40. Sui Y et al (2021) SARS-CoV-2 spike protein suppresses ACE2 and type I interferon expression in primary cells from macaque lung bronchoalveolar lavage. Front Immunol 12:658428
41. Talesnik D (2022) Fauci discusses covid control, recovery. National Institutes of Health, U.S. Department of Health and Human Services. https://nihrecord.nih.gov/2022/07/22/fauci-discusses-covid-control-recovery
42. Tegally H et al (2021) Rapid replacement of the Beta variant by the Delta variant in South Africa. MedRxiv. https://www.medrxiv.org/content/10.1101/2021.09.23.21264018v1
43. Tegally H et al (2022) Emergence of SARS-CoV-2 Omicron lineages BA.4 and BA.5 in South Africa. Nat Med 28(9):1785–1790. https://www.nature.com/articles/s41591-022-01911-2
44. Tian D et al (2022) The emergence and epidemic characteristics of the highly mutated SARS-CoV-2 Omicron variant. J Med Virol 94(6):2376–2383
45. UK Health Security Agency (2021) SARS-CoV-2 variants of concern and variants under investigation in England. Technical briefing 26. UK Health Security Agency. https://assets.publishing.service.gov.uk/government/uploads/system/uploads/attachment_data/file/1028113/Technical_Briefing_26.pdf
46. Walensky RP, Walke HT, Fauci AS (2021) SARS-CoV-2 variants of concern in the United States—challenges and opportunities. JAMA 325(11):1037–1038
47. Wendler R (2022) Covid-19's BA.4 and BA.5 Omicron subvariants: 12 questions, answered. MD Anderson Cancer Center. https://www.mdanderson.org/cancerwise/covid-19-s-ba-4-and-ba-5-omicron-subvariants%2D%2D12-questions%2D%2Danswered.h00-159541323.html#:~:text=5%20Omicron%20subvariants%3F,19%20cases%20in%20the%20U.S
48. World Health Organization (2008) International health regulations (2005). World Health Organization
49. World Health Organization (2019) Naming the coronavirus disease (COVID-19) and the virus that causes it. World Health Organization. https://www.who.int/emergencies/diseases/novel-coronavirus-2019/technical-guidance/naming-the-coronavirus-disease-(covid-2019)-and-the-virus-that-causes-it
50. World Health Organization (2021) Classification of Omicron (B.1.1.529): SARS-COV-2 variant of concern. World Health Organization. https://www.who.int/news/item/26-11-2021-classification-of-omicron-(b.1.1.529)-sars-cov-2-variant-of-concern
51. World Health Organization (2021) Tracking SARS-COV-2 variants. World Health Organization. https://www.who.int/activities/tracking-SARS-CoV-2-variants

52. World Health Organization (2021) Who announces simple, easy-to-say labels for SARS-COV-2 variants of interest and concern. World Health Organization. https://www.emro.who.int/media/news/who-announces-simple-easy-to-say-labels-for-sars-cov-2-variants-of-interest-and-concern.html
53. World Health Organization (2022) Previously circulating VOIs. World Health Organization. https://www.who.int/activities/tracking-SARS-CoV-2-variants/previously-circulating-vois
54. Yamasoba D et al (2022) Virological characteristics of the SARS-CoV-2 Omicron BA. 2 spike. Cell 185:2103–2115.e19
55. Yan Y et al (2020) The first 75 days of novel coronavirus (SARS-CoV-2) outbreak: recent advances, prevention, and treatment. Int J Environ Res Public Health 17(7):2323
56. Zhang L et al (2021) Furin cleavage of the SARS-CoV-2 spike is modulated by O-glycosylation. Proc Natl Acad Sci U S A 118(47):e2109905118

Vietnam's Response to COVID-19: January 2020 to July 2022

Kamila Zakowicz and Linh Vuong

1 Introduction

A large number of governments, particularly those of the Western world, have been heavily criticized for untimely or ill-fitting containment policies addressing the COVID-19 pandemic. Yet, in the face of it all, Vietnam and other Asian countries appear to have had the firmest grasp on how to initially contain the twenty-first century's worst public health disaster to date. Until the highly virulent Delta variant arrived in Vietnam in April 2021, Vietnam had kept its cases and death rate exceedingly low since the start of the pandemic. Even in the midst of the virulent "fourth wave," Vietnam reported a relatively low rate of 44.27 deaths per 100,000 people when compared to the United States, which reported 308.76 deaths per 100,000 [74]. Strategies of pandemic containment in Vietnam changed dramatically over time, from a zero-new-case approach based on strict non-pharmaceutical interventions (NPIs), i.e., lockdowns, tracing, and mass isolation, to immunization, employing vaccine diplomacy and public support.

The healthcare system in Vietnam is multilateral. Direct healthcare delivery comprises four levels: Level I includes large central hospitals, such as Bach Mai in Hanoi, and Levels II-IV include the smaller provincial, district, and commune healthcare facilities and hospitals [50]. These levels are led and connected by the Ministry of Health (MOH), which has a centralized online platform and network, not unlike the Center for Disease Control (CDC) in the United States. The MOH works closely with other actors, namely, the government and prime minister, as well as the various vice prime ministers, military, health companies, and schools. The Vietnamese government also coordinates with international forces, notably the World Health Organization (WHO) and other countries, to procure funding and navigate diplomacy in the health sector. These players are complemented by

K. Zakowicz (✉) · L. Vuong
Dartmouth College, Hanover, NH, USA

J. M. Rosen, R. R. Colwell (eds.), *The COVID-19 Pandemic*,
https://doi.org/10.1007/978-3-031-62772-9_10

volunteer and non-profit organizations, as well as the public. Evolution of COVID-19 in Vietnam proved to be complex but mappable and included all of these actors and more.

To map its progression, the pandemic in Vietnam was divided into four waves. Before the first wave, the history of pandemics in Vietnam provided the baseline support for Vietnam's "readiness" with respect to the COVID-19 pandemic. The first and second waves covered January 2020 to August 2020 and are focused on the early containment strategies that were largely successful for a zero-new-case approach in Vietnam. The third and fourth waves cover January 2021 to July 2022 and comprise a distinct shift in strategy to immunization, employing vaccination, as cases soared due to the Delta variant and other factors. Finally, this report concludes with major lessons learned from Vietnam's past and present pandemic responses.

2 Credibility of Vietnam Coronavirus Statistics

Vietnam's status as a one-party Communist state inevitably elicits a reaction in Western, non-Communist societies to deny credibility of its COVID-19 statistics. It is true that there is a lack of transparency with respect to official details of the medical process for Vietnamese COVID-19 patients and infected case numbers, as the MOH has not released data to those not directly involved in the medical sector. Thus, credibility of the number of infected cases remains uncertain. Lack of public access to data during the pandemic neither supports nor denies the successful results the Vietnamese government achieved in containing outbreaks.

Today, confirmation from independent international sources argues for Vietnam's credibility. Dr. John MacArthur, Director of the Centers for Disease Control and Prevention in Thailand, stated his faith in the Vietnam case numbers in a 2020 telephonic briefing with Dr. Barbara Marston, CDC COVID-19 International Task Force Lead (US Department of State) [59]:

> [Our] team that's up in Hanoi working very, very closely with MOH counterparts on many, many aspects of this outbreak response, providing technical assistance in the areas of surveillance, data analysis, laboratory testing, the actual going into the field and doing investigations, contact tracing with the Vietnamese counterparts, and really, sort of trying to support them and their approach. [...] And so those relationships are strong, and it kind of allows us to get a sense of whether the numbers are real, the numbers are not real, only because of deficiencies in testing and some such, or whether there's some reason to hide those numbers. And from the communications I've had with my Vietnam team, is they, at this point in time, don't have any indication that those numbers are false.

Although the number of cases, deaths, and vaccinations may be somewhat biased from the Vietnamese MOH, international news outlets, i.e., Reuters, generally acknowledge effectiveness of Vietnam's approach, regardless of questions concerning numerical transparency.

3 Historical Pandemic Response: Severe Acute Respiratory Syndrome (SARS), 2003

The COVID-19 global outbreak was not the first coronavirus to sweep through Vietnam in the twenty-first century. In February 2003, SARS coronavirus initially appeared in Guangdong Province, southern China, around mid-November 2002. Over a 3–4-month period, it spread to 26 countries and resulted in roughly 8098 infections and 774 reported deaths [71]. Among affected countries outside of China, Vietnam was one of the first four to report alarmingly atypical cases of pneumonia, following hospitalization of a male traveler in Hanoi, who had recently traveled to Hong Kong [48]. These cases were later confirmed to be SARS, and all were treated in the Hanoi Vietnam-French Hospital. Of 63 SARS cases in Hanoi, 36 were healthcare workers in direct contact with the first patient, making 57 percent of infected medical personnel during the pandemic, the highest of all affected countries [17]. Six recorded deaths in Vietnam comprised doctors and nurses at the French Hospital [57].

Two important factors contributed to Vietnam's success in becoming SARS-free after just 2 months: timely response by the hospital responsible for SARS patients and governmental quick and transparent efforts to enforce policies that ensured effective containment of the disease.

The commitment and resourcefulness that went into controlling the SARS infections were especially impressive because, at the time, medical resources in Vietnam were relatively scarce. Only a few days after an initial report to the WHO by Dr. Carlo Urbani, WHO representative who traveled to Hanoi to do a situational assessment, the French Hospital issued a hospital-wide quarantine order on March 5, 2003. When interviewed about the situation that led to the hospital lockdown, nurse Xuân, one of the first responders, shared the following about her experience:

> I vaguely guessed that the outbreak was becoming more and more dangerous. That night [March 5, 2003], I bought a lot of groceries and instructed my husband to cook and take care of the kids. As expected, the hospital sent out the order to close down the hospital and [everyone] must stay there. [28] (translated by Linh Vuong)

All working hospital staff and residing patients were required to remain at the hospital for a period of more than 30 days. This swift move to isolate SARS cases largely prevented spread of the disease within the community and minimized possible damage from the outbreak.

According to Dr. Võ Văn Bản, Vice Director of the French Hospital during the SARS outbreak, most of the hospital medical personnel were unaware of the impending SARS pandemic when they came into contact with patient zero on February 26, 2003. As a result, they were not able to take needed drastic preventive measures, such as isolating the patient and wearing personal protective equipment. As a consequence, an unfortunately disproportionate number of hospital staff became infected and died. This devastating unintended consequence prompted a response from the hospital to protect staff and control the infection. First, after identifying the threat to be SARS, the hospital allowed high-risk personnel, i.e., those who were pregnant or had young dependents at home, to go on leave. The hospital provided staff with additional PPE, specifically N95 masks [36].

The hospital addressed the urgent need to provide sufficient essential life-saving equipment by cooperating with Bach Mai Hospital and its French associates to acquire ventilators and other essential equipment [71]. After the first few frantic days spent coordinating additional medical supplies and working intensively to minimize further spread of the disease, the Vietnamese MOH directed the French Hospital to move patients to the National Hospital of Tropical Diseases, where medical personnel and resources were better equipped for the emergency [28]. Fast and efficient cooperation by the network of hospitals enabled Vietnam to mitigate the consequences of the medical institutions having been unprepared for such an unprecedented and deadly medical emergency.

By March 15, CDC issued warnings and guidelines to global health departments about the SARS pandemic [1]. Immediately thereafter, the Vietnamese government strengthened its border control and enforced extensive medical screening at ports of entry, including the airports, across the country, deploying existing resources, e.g., medical personnel from the National Institute of Hygiene and Epidemiology and the National Hospital for Tropical Diseases. Despite limited funds, Vietnam committed to funding efforts to fight the highly transmittable disease. The Finance Ministry allocated $2 million for medical equipment and activities related to SARS prevention and targeted approximately $1 million for Vietnam border provinces tasked with preventing the SARS agent from leaving or entering Vietnam [45].

The MOH directed formation of a National Steering Committee for the Prevention of SARS, with assistance of governmental sectors with authority for national security and government communications. The Committee comprised divisions responsible for community tracing, as well as regional disease-monitoring task forces. According to the General Department of Preventive Medicine in Vietnam, task forces included the following sub-committees:

1. Disease monitoring
2. Disease treatment
3. Public health education and media
4. Logistics

Among the countries most affected by SARS, Vietnam had the least resources [45]. The country acknowledged it lacked necessary medical and monetary resources, making it not feasible to fight the SARS pandemic alone. To overcome its vulnerability, Vietnam actively sought international aid. As soon as the first severe SARS case had appeared, the French Hospital was assisted financially by France and received ca. $100,000 to sterilize and disinfect hospital equipment. In addition, Vietnam asked Japan to dispatch medical experts. Japan concurred and also provided two ventilators and medical supplies. In addition to requesting assistance from other nations, Vietnam cooperated extensively with international public health organizations and was supplied with PPE, e.g., masks and gowns, by WHO and CDC. Direct consultation was provided by a team of medical professionals sent by Doctors Without Borders [45].

On April 28, 2003, only 2 months after its first SARS case, Vietnam became the first country in the world to contain the outbreak [72]. Vietnam's holistic approach to battling a public health emergency made it possible to minimize spread of a deadly virus. Vietnam's success in 2003 would become a blueprint for how the country could deal with the COVID-19 pandemic in 2020.

4 January 2020–August 2020: Zero-New-Case Approach of the First and Second Waves

Official COVID-19 statistics for Vietnam August 2020 showed 1029 confirmed cases and 27 deaths. The Vietnamese government eased lockdown restrictions on April 23, 2020, and entered a "new normal"—a step few affected countries were able to take at that point [25]. It is useful to assess how Vietnam achieved this outcome at such an early date, considering its population of 98 million, and shares a ca. 1400-km (870-mile) border with China [52].

After the first case of COVID-19 was reported on January 23, 2020, Vietnam had gone almost 100 days without a locally transmitted case until an outbreak in Da Nang Province in July 2020. That outbreak caused the number of positive cases reported by the Vietnamese MOH to rise from 324 in May 2020 to 1009 in August 2020 and was caused in large part by domestic Vietnamese tourists at a popular resort in Da Nang Province [7]. Yet CDC reported ~12,000 new cases per day in the United States, the lowest point in July 2020, and nearly 115,000 cases per day by August 2022 [2, 3]. How did Vietnam effect containment of the pandemic so quickly? The Journal of Travel Medicine reported Vietnam's response was similar in other countries, but Vietnam's strategies were particularly effective because implementation was much earlier [9].

Here, we examine several of the early systemic strategies employed by the Vietnamese government to prevent initial spread of COVID-19, focusing on NPIs that aided the government commitment to "zero-new-case." These included early and enforced quarantine, government tracing, distribution of supplies and testing, utilization of technological tools, and leadership organization.

4.1 Budget-Friendly Approach and Comparison to South Korea

Much of Vietnam's success from December 2019 to August 2020 can be attributed to proactive and effective lockdown measures. Vietnam took a preventative approach because, with $2,715.3 GDP per capita as was reported in 2019, Vietnam could not afford comprehensive testing and generous spending. To illustrate economic constraints on the Vietnam pandemic response, both Vietnam and South Korea had effective early COVID-19 mitigation outcomes, but adopted different approaches. With a significantly higher GDP per capita ($31,597 in 2020), South Korea could afford to employ comprehensive testing and generous spending in its pandemic response [76].

South Korea reported just under 18,000 cases on August 25, 2020, and ca. 1.8 million tests were performed [24]. Early in January 2020, however, officials of the South Korean health ministry convened a meeting of representatives from more than 20 medical companies on production of COVID-19 testing kits. One week after the January 27th meeting, a company was approved to produce the kits, with others following soon after [51]. Asia Times reported that the South Korea national health insurance system had spent 310 million US dollars on COVID-19 treatment at that point. This paled in comparison with the trillions of dollars the United States had spent on its COVID-19 response at the time, but Dr. Kim Sun-min, President of the Health Insurance Review and Assessment Service (HIRA), nonetheless called COVID-19 a "relatively cheap disease" because it does not require MRIs, surgery, or other expensive equipment for treatment, other than ventilators to treat extreme cases [46]. For comparison, Vietnam had spent approximately 3 million US dollars on the SARS outbreak in 2003 and could not afford the same so-called "low-cost, high-tech" strategy. Vietnamese test kits were launched in March 2020, only after several grants had been awarded to the Vietnamese government by WHO. Thus, daily testing was introduced more slowly in Vietnam than in South Korea (Fig. 1).

The Vietnamese government opted to launch a prevention plan in line with the resources available. Travel restrictions, quarantine centers, school closures, and lockdowns were executed early to prevent, not mitigate, spread of the virus.

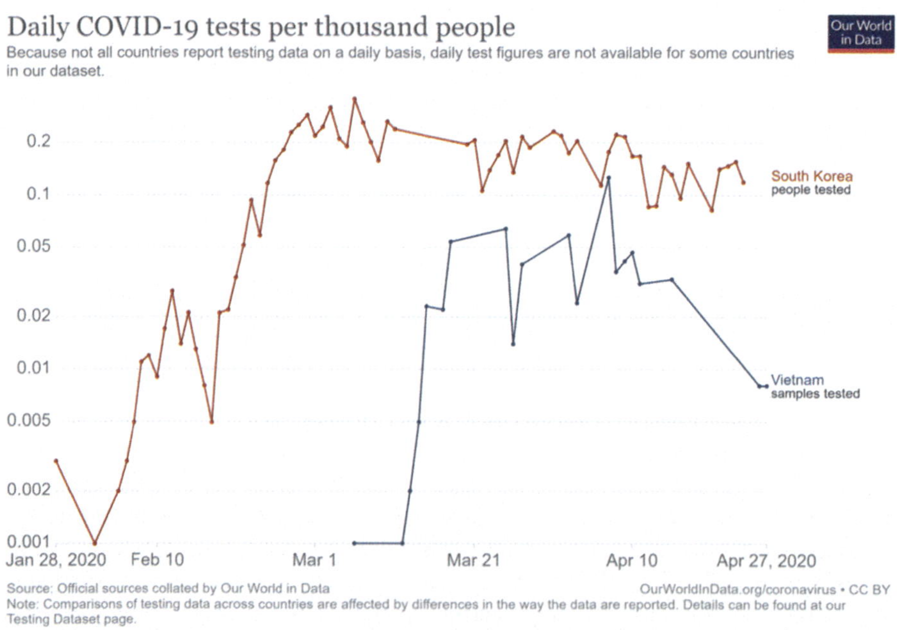

Fig. 1 Graph depicting daily COVID-19 tests in South Korea (people tested) and Vietnam (samples tested) from January 28 to April 27, 2020. Daily testing rose significantly in South Korea starting in February, while in Vietnam daily testing did not start until March. Note that the discrepancy between using people tested in Korea and samples tested in Vietnam is due to limited data published by each government. (© CC BY Our World in Data)

4.2 Travel

Immediately after its first case on January 23, 2020, Vietnam canceled flights to and from Wuhan, China, and established health screenings at airports [54]. A 14-day government-sanctioned quarantine was instituted for travelers on February 15, and by March 22, all foreign travelers were banned from entering Vietnam [27, 67]. Health declaration forms were mandatory upon entry into the country, and foreigners who arrived before March 1 were considered for automatic extension of stay to August 31. The country went into limited national lockdown on April 1, shutting down borders, public transportation, and gatherings of more than 10 people [6, 14, 49, 60]. These measures were a preventative strategy and likely contributed to Vietnam's minimal number of positive cases and deaths. After March, domestic restrictions loosened and airlines resumed domestic travel until the reinstatement of restrictions after the outbreak in Da Nang. The foreign travel ban, however, remained in effect in August, and citizens returning from abroad had to abide by the mandatory government quarantine.

4.3 Isolation Centers

The MOH listed several sites as candidates for a centralized isolation facility. These included army and police barracks, school dormitories, factories and enterprises, new and unused apartment buildings, hotels and resorts, schools, and communal health facilities [25].

Returning citizens and travelers were restricted to centralized isolation facilities for 14 days, where they were given government-subsidized food and accommodation. The MOH listed requirements for isolation facilities, including essential living conditions (electricity, water, and bathrooms), ventilation, security and safety, fire prevention, and a location away from residential areas yet convenient for transportation and waste removal. Where possible, rooms were equipped with television or internet. They were allowed to receive packages from friends and loved ones, but no alcohol. Conditions were reported to be minimal but comfortable, with volunteer translators for foreigners and fences between the bunk beds [31]. The People's Committees of each district or province were responsible for implementing the facilities, which were staffed and guarded by health workers, military, and law enforcement [30, 39].

4.4 School Closings

Hanoi was the first province to cancel in-person schooling in late January, after the first COVID-19 case in Vietnam on January 23rd 2020. Other provinces and cities followed suit, and by February 16th, all 63 districts opted to extend school closures. After either a three-month school break or three-months online learning, schoolchildren began to return to school with physical distancing measures in place starting early May and gradual reopening in phases [44, 56, 63, 70].

4.5 Government Tracing

The government resourced a comprehensive tracing program of positive cases. The MOH outlined a system and protocol for carrying out tracing, most of which relied on collaboration between individual citizens and local health department officials. The system classified cases on a scale from F0 to F5. F5 cases were notified of possible exposure and asked to self-monitor, F1–F4 cases were isolated and quarantined in the home, and contacts of F0 cases were traced for 14–28 days. F0 cases were infected patients, F1 cases were those who had direct contact with F0 cases, F2 cases had contact with F1 cases, etc [27] (Fig. 2).

Different levels of action were necessary for each level of tracing. The Vietnamese MOH reported the following protocols of response from the MOH for each level. F0 individuals needed to be hospitalized, treated, and isolated to prevent further infection. F1 individuals had to wear a mask immediately, notify the local county health department, isolate at a hospital, and notify F2 individuals. F2-F3 individuals had to wear a mask immediately, notify the local health department, follow isolation instructions by department staff, and notify the next level of individuals. F4 individuals had to enact home isolation and notify the local health department [66, 9].

Vietnam had one of the most comprehensive tracing policies in the world. While some parts of Europe and other countries were stepping up to trace as thoroughly as Vietnam, the United States employed limited tracing measures (Fig. 3).

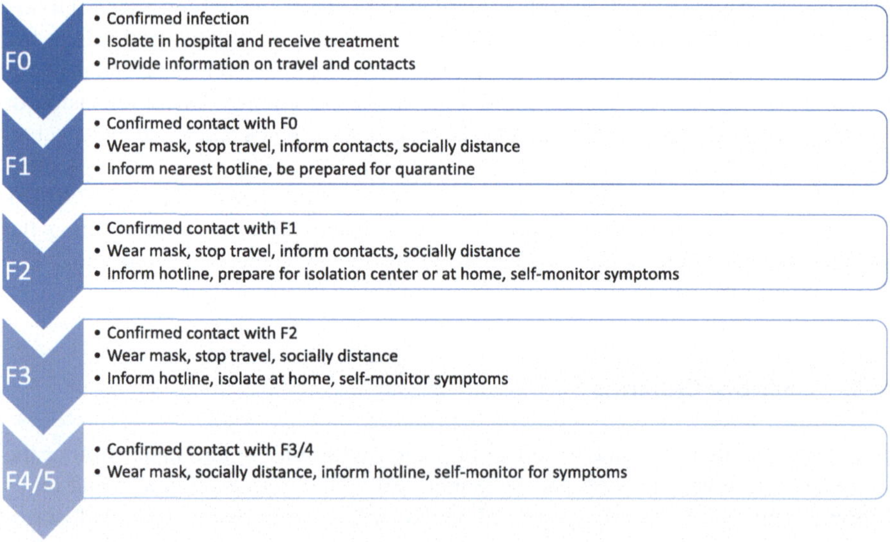

Fig. 2 Chart depicting Vietnamese tracing levels F0–F5

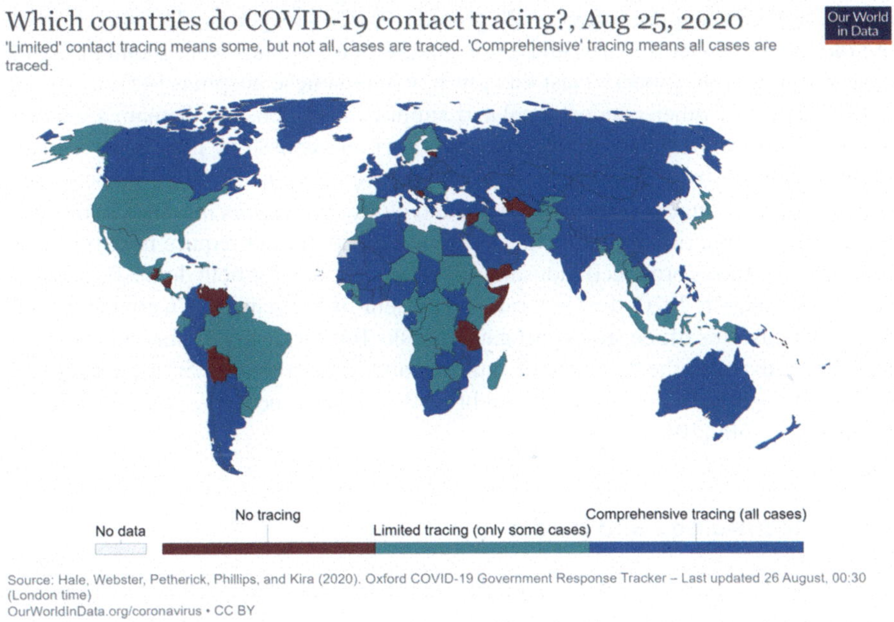

Fig. 3 Map depicting tracing levels of different countries in August 2020. (© CC BY Our World in Data)

4.6 Supplies, PPE, and Testing

Reuters and The Japan Times reported that collaboration between the Vietnamese government, medical supply company Viet A Corps, and research facilities at the state-run Military Medical University (MMU) was key to Vietnam's comprehensive testing and supply strategy. By late February, MMU and Viet A Corps had designed a mass-producible test kit for COVID-19. The government issued a license, and by the end of March, 250,000 kits had been issued in Vietnam. At the same time, the MOH worked to increase the number of laboratories testing for COVID-19 in Vietnam [16]. By the end of April, the World Bank reported that Vietnam conducted 967 tests for every positive case found [22]. The same week, Vietnam received a seal of approval by WHO to export tests. Two months earlier, WHO had sent laboratory supplies to Vietnam to start developing the kit. Vietnam developed two internationally used tests that avoided false negatives: COVID-19 antibody and polymerase chain reaction (PCR) tests to detect the presence of the virus [68].

The kits had been researched and developed by multiple organizations and political bodies in Vietnam working together. The Vietnam Academy of Science and Technology (VAST), consulting business IMM, and Vietnamese University of Technology (UoT) were consulted by the government and worked together to develop publicly funded test kits. Vietnam used a collaborative top-down approach, and the IMM kit was developed by the military and mass produced by Viet A Corps. Vietnam's kits have been tested and found to be on par with those distributed by the

WHO and CDC, and producers could provide 10,000 kits per day [18, 40]. It must be noted that the Viet A Corps and a few government officials were found guilty of price gouging with respect to test kits given to Vietnamese hospitals [42].

Production of other pandemic-related supplies also boosted Vietnam's exports and its economy. With the global economic downturn caused by the virus, Vietnam found a solution by producing personal protective equipment (PPE), namely face masks. The government converted textile and clothing factories into face mask producers, production of which was quick, efficient, and did not require import of raw materials. Factories produced thousands of masks per day for both domestic use and exports. By March 2020, the General Department of Vietnamese Customs allowed provincial customs offices to export gauze masks for international humanitarian aid [65]. By mid-April, the Ministry of Industry and Trade reported 50 producers had combined capacity to manufacture eight million masks per day, i.e., 200 million masks per month [10].

4.7 Technology and Media

In addition to information from the CDC and WHO, Vietnamese citizens accessed information and announcements from the government through the MOH website. The site was updated daily, with announcements of new cases and links to news about COVID-19. The homepage featured a live database crucial for citizens to stay up to date on infections and status of the different regions of the country. There were additional tabs for articles on the latest COVID research, recommendations from the MOH on a variety of topics, and updates on industry support. Questions could be submitted, health support locations found, instructions obtained, and quizzes taken on health protocols. The MOH also created digital posters on subjects such as hygiene, social distancing measures, tracing protocols, ways to manage stress, testing protocols, and advice on maintaining good health during a pandemic. The information is centralized, organized, and easy to access. The live webpage contained a plethora of information connecting citizens to the government directly through the MOH.

The MOH and Ministry of Information and Communications launched the NCOVI application, a platform for users to find recommendations on COVID-19 protocols and send voluntary health declarations. The application asked users to provide contact information, family details, and a health and disease survey response. By March, the app was in the #1 spot on the Vietnam Apple Store. By August 13, 2020, there had been 41,293 downloads of the Google Play Store app. Many left comments finding the application useful, informative, and accessible [15]. On April 21, technology firm Bkav and the Ministry of Information and Communications launched an additional application, Bluezone (Fig. 4). Bluezone operates using Bluetooth to link smartphones within two meters and notifies users if they have had contact with an infected individual in the last 14 days. Ca. 42,849 users downloaded the app from Google Play as of August 13, 2020 [29]. The benefit

Fig. 4 BlueZone application logo. (Image by Bluezone Global, distributed under a GNU General Public License)

of the app was its efficiency since the government did not have to collect information or expend valuable resources for tracing.

4.8 Leadership and Culture

Strong, centralized leadership was vital to Vietnam's swift and effective response to COVID-19. Key players in Vietnam's coordinated pandemic response were the vice prime minister, the military, and the MOH. These leaders covered a lot of ground in a very short time, which was needed when COVID-19 surfaced in Vietnam.

Vice Prime Minister Vu Duc Dam had been delegated to lead the MOH after the previous MOH leader stepped down in 2019. His tenure continued through the first wave of the pandemic [64], working with scientists, leaders in the MOH, and other parts of the government to position MOH in spearheading the campaign for tracing, quarantine, and health resources. The MOH was delegated to former MOH Vice Minister Nguyen Thanh Long on July 7, 2020 [55]. Mr. Long served as Minister of Health until June 2022, at which time he was dismissed because of involvement in a COVID-19 test kit price gouging scandal [42]. The MOH then appointed Deputy Minister Do Xuan Tuyen as Minister of Health.

Military resources were utilized and supplemented with police resources, and staffing was provided by MOH preventative medicine officials. Local and federal troops maintained isolation centers, the military research center MMU assisted in developing the testing kit, and military student volunteers were mobilized to carry out contact tracing protocols and collect samples in Da Nang [69].

The MOH in Vietnam had essentially been preparing for a pandemic long before COVID-19. In 1961, the Direction of Healthcare Activities (DOHA) was established to facilitate central communication and guidance between higher tier administrative health care and lower tier hospital care. A report published in the Journal of Environmental Health and Preventive Medicine describes the DOHA's two missions [50].

Dr. Lily Hue, a doctor at the central Bac Mai hospital, described to the authors how the initiative established "a strong preventative medicine sector from [a] central level...down to [a] provincial, district, and communal level" (2020). Effective communication between central hospitals such as the Bac Mai and smaller hospitals allowed the Vietnamese healthcare network to act quickly and effectively with the MOH to respond to COVID-19.

Dr. Nguyen Thi Kim Tien, Minister of Health for 8 years before stepping down in 2019, raised capacity for disease prevention by investing in the healthcare sector. She implemented innovative technology in the hospitals, raised healthcare worker salaries, and provided patients with amenities such as expanded hospital rooms and furnished waiting rooms [38].

5 January 2021–July 2022: Vaccination and Immunization Approach of the Third and Fourth Waves

The relatively low transmission rates in Vietnam rose steeply on January 28, 2021, after a 28-year-old man in Ho Chi Minh City tested positive for the UK Alpha variant after attending a wedding party in another province. By the end of the same day, 88 additional positive tests were recorded. This outbreak, along with those of workers at Ho Chi Minh City airport who tested positive for an African variant, resulted in a third wave of transmission and 910 new cases until March 25, 2021 [23]. Although the outbreak was relatively mild and death rates were low, health officials and government leaders turned to vaccination as the next step in the Vietnam COVID-19 containment strategy. The vaccination strategy, however well-planned, was disrupted by the largest and most deadly wave of COVID-19 transmission in Vietnam, the fourth wave.

During the first three waves of the pandemic, Vietnam was able to use low-cost solutions and effective policies to contain COVID-19. The fourth wave, however, demonstrated "limited hard power" was needed to contain the new Delta variant [8]. On April 27, 2021, the Delta variant, known to be highly contagious, was confirmed in Vietnam on the MOH website. It was found to be 60% more transmissible than the UK Alpha variant, already 60% more transmissible than the original strain [20, 34]. This variant was deadly for Vietnam, where total COVID-19-related deaths increased from 35 in April to 15,018 by September 2021 and 43,092 by July 2022 [20, 23]. As a result, the case fatality rate (CFR) in Vietnam rose above the global average of 2.06% to 2.50% [23]. These statistics suggest NPI alone was not sufficient to contain the Delta variant (Fig. 5).

Due to the severity of the Delta variant, Vietnam had to vaccinate high priority groups as quickly as possible. Initiation of the vaccine rollout was slow, with less than 1% of the population vaccinated by May 2021, and by August 30, only ~20% of the population had been vaccinated [34]. This may have been due to several factors, including reduced staff due to transmission and strict measures implemented in the fourth wave, insufficient funds to rapidly develop domestic vaccines, and a global vaccine shortage stemming from an insufficient supply of vaccine

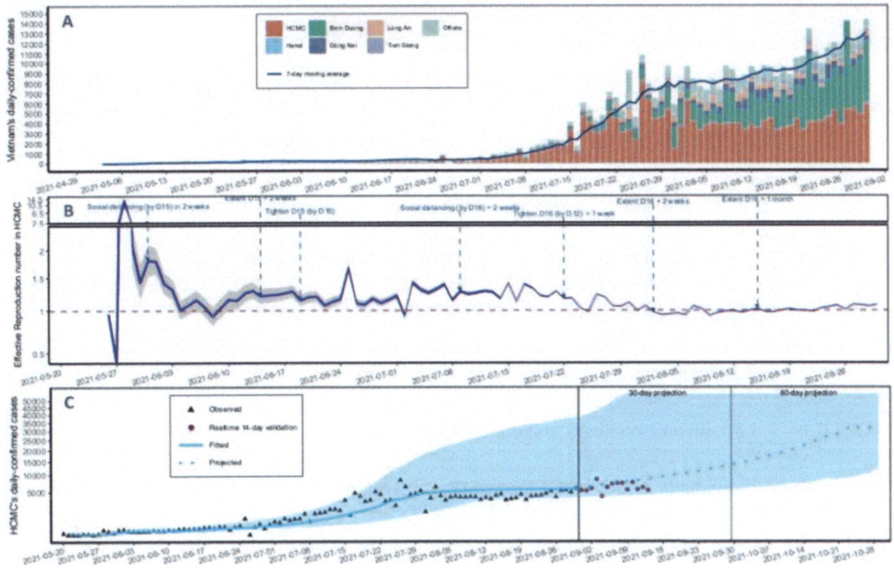

Fig. 5 The fourth wave of COVID-19 in Vietnam. First is the epidemic curve by province (**a**). Next is the time-varying reproduction number (*R*) of the virus and the outcome of various public health measures (D15, D16, and D12) in Ho Chi Minh City (HCMC). The red dotted horizontal line is set to 1, where $R < 1$ indicates the epidemic is controlled (**b**). Next shown are 30-day and 60-day projections of daily new confirmed cases in HCMC (**c**). (Figure taken from [34] © copyright permission granted by Dr. Nguyen The Phuong)

components and high-income countries willing to pay premiums to pre-order more doses than they needed [13]. However, the campaign was quickly ramped up, and by December 2021, 70% of the population was vaccinated [75]. The vaccination rate ranged from 80% to 90% by 2022 [19, 43].

While the government was preparing for vaccination by approving vaccines for use and strategically procuring doses, public health officials drafted an implementation plan. On January 29, 2021, a day after the outbreak, Vietnam announced approval of the AstraZeneca vaccine and the company's pledge of 30 million doses for Vietnam [79]. Together with 39.8 million doses donated by the global vaccine access initiative COVAX, Vietnam began implementing a vaccination training program for healthcare workers to enact a tiered vaccination approach starting in July 2021 [26]. The tiered vaccination program proved useful in many countries by prioritizing the most at-risk populations to receive vaccinations first. Israel, for example, had one of the lowest case fatality rates (CFR) of 2021 at 0.6%. This achievement was supported by an expertly tiered vaccination program that reduced the burden on hospitals, decreased vaccination delays, and increased logistical efficiency through fewer groups and shorter age intervals [23]. Tiered vaccination plans and training were thus important first steps to Vietnam's vaccination strategy (Fig. 6).

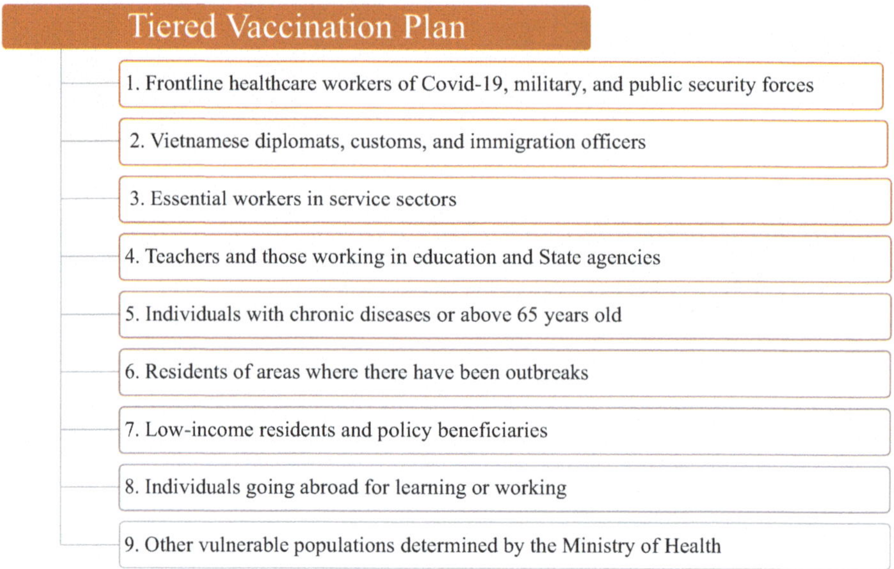

Fig. 6 Tiered vaccination plan created by the Vietnamese government and MOH

Vietnam created a tier framework personalized to outbreak demographics. Healthcare workers were the first priority, followed by diplomats and immigration personnel, essential workers, teachers and educators, people with chronic illnesses, people living in areas with outbreaks, low-income workers, etc. [5]. Some prioritized groups, e.g., low-income workers, were clearly linked to outbreak demographics. The third and fourth waves centered around crowded, poorly ventilated industrial areas and factories such as the Hosiden Bac Giang Company, where nearly 1000 out of 4300 workers were infected [20]. The fourth wave, however, also coincided with a global vaccine shortage.

As a low-middle-income country, Vietnam was not prepared to buy or internally manufacture the doses it needed as the fourth wave of the pandemic hit the country in April 2021. In fact, Vietnam was approved for a phase 3 clinical trial of its domestic vaccine Nanocovax when the Delta wave hit and it could no longer afford to wait [53]. Vietnam relied on donations through vaccine diplomacy [4]. Vaccine diplomacy has been defined as "almost any aspect of global health diplomacy that relies on the use or delivery of vaccines and encompasses the important work of the GAVI Alliance, as well as elements of the WHO, the Gates Foundation, and other important international organizations" with an emphasis on "humanitarian intervention" [77]. The COVID-19 Vaccine Global Access (COVAX) coalition was created in 2021 through a partnership between the WHO, Gavi, and the Coalition for Epidemic Preparedness Innovations (CEPI) to ensure equitable access to COVID-19 vaccination by providing low-cost vaccines to all countries. Through COVAX, Vietnam initially received 4.1 million vaccine doses by May 2021 to help combat the Delta surge free of charge and continued diplomatic efforts to receive over 50 million doses through COVAX since [61, 62]. Vietnam also procured millions of donations from many countries directly.

Vietnam's Western allies, particularly the United States and Europe, were significant patrons of Vietnam's vaccine supply. The United States was not only the biggest contributor to COVAX, but USAID reported donating 40 million doses in total to Vietnam, along with $23.46 million in COVID-19 assistance. Europe was the second biggest contributor to COVAX and contributed supplies and vaccine doses to Vietnam. Germany donated ten million doses to Vietnam by the turn of 2022, solidifying the Western presence in Vietnam during COVID-19 [12].

Vietnam has been particularly strategic in vaccine diplomacy and generally surrounding COVID-19 policy with China. Vietnam welcomed vaccine donations from China despite geopolitical conflicts concerning disputed territory in the South China Sea [8]. After approval of the UK-based AstraZeneca vaccine and the Russian Sputnik vaccine in March 2021 and donations from the United States, EU, and Russia, Vietnam approved the Chinese Sinopharm vaccine in early June 2021 and received 500,000 donated doses from China the same month [32]. There was speculation that approval was made reluctantly and that the global shortage of vaccines at the time and severity of the Delta outbreak in Vietnam prompted Vietnamese officials to consider Chinese vaccines, while the Vietnamese people remained skeptical of China's vaccine products [33, 37, 58]. However, Vietnamese officials had showcased diplomacy with China before the vaccine shortage [8]. Vietnam's approach to China and COVID-19, along with its relations with the United States and EU, undoubtedly served to benefit Vietnam in expanding its vaccination campaign.

One crucial aspect of Vietnam's successful vaccination campaign was acceptance of vaccination by the general public. Vaccine acceptance has generally been high in Vietnam since 2020, with acceptance rates ranging as high as 98% and as many as 89.2% of a study's participants in 2022 agreeing that mass vaccination will help end the pandemic [35, 78]. A high vaccine acceptance does not necessarily correlate with thorough knowledge of the vaccination, however. A recent population study found that only 46.7% of 1708 study participants had adequate or thorough knowledge about COVID-19 vaccination [11]. This type of high acceptance may thus be attributed to more of a cultural phenomenon.

Evidence of cultural forces influencing vaccine acceptance over economical or educational factors can be seen in the United States. One study found that early vaccine skepticism among "essential workers" in the United States, 68% of which were healthcare workers, was centered not around vaccine safety and efficacy but more a freedom of health choices [47]. Another study featured in BMC Public Health found that conservatism and political party were the two factors most positively correlated with vaccine skepticism, not socioeconomic factors [21]. It is clear that in the United States, cultural and political ideas have contributed to vaccine hesitancy in the wake of COVID-19.

Vietnam, for its part, has shown public acceptance of the vaccine and efficient healthcare delivery can have lasting effects, according to UNICEF representative Rana Flowers:

> There is no doubt in my mind, that COVAX, that the many donors of vaccine to Viet Nam chose to give to the people of Viet Nam because they were taking the virus seriously, and the delivery of the vaccines was efficient and safe [61].

Taking the virus seriously has taken many forms in Vietnam, from public health to the individual. NPIs such as active tracing, isolation, and lockdowns were swift and effective in containing the virus in the early days of the pandemic—analysis has shown that the Rt of the virus decreased below 1 and as low as 0.02 in 2020 (95% CI: 0–0.12) [41]. Media agencies have been active during the pandemic, with social platforms sending ~20 billion COVID-related messages to 98 million people [20]. Provincial rapid response teams collaborated with military, medical students, and public health officials to implement NPIs, testing, and vaccination in a multilateral approach to generate local human willpower and resources. Public acceptance of the measures was not only important for their effectiveness but also for how the world and public health agencies viewed Vietnam.

Pandemic fatigue, however, may have also been a driver for the reduced containment of the most recent fourth wave [34]. Vietnam implemented some of the strictest lockdowns and restrictions worldwide. So, it is not surprising that the patience of the Vietnamese people had worn thin and adherence to strict containment was reduced. The WHO recognized pandemic fatigue as a legitimate global public health concern, especially regarding COVID-19, and advised countries to combat the phenomenon by engaging the public and loosening stricter measures to allow people to live their life of choice once again [73]. It is possible that pandemic fatigue and severity of the Delta variant were catalysts for Vietnam's pivot to its current immunization COVID-19 response strategy.

Vietnam shifted from "zero-new-case-approach" to immunization. Bloomberg reported Vietnam Prime Minister Pham Minh Chinh directed the MOH to treat COVID-19 as an endemic disease in March 2022. Not only did this contrast with Vietnam's previous lockdown strategies, but also the directive was given in the midst of a seven-day average of 95 deaths per day (albeit decreased from a peak seven-day average of 300 deaths per day in August 2021) [5]. This may seem surprising given Vietnam's previous containment strategy, but factors such as availability of vaccinations and pandemic fatigue may help explain this change. Along with an almost 90% reported adult vaccination rate, the Vietnamese MOH encouraged its citizens to get third and fourth doses to increase their immunity and forged ahead with an ambitious child vaccination campaign. The MOH website targeted those who do not agree to receive the next vaccinations because they think they are protected by immunity and emphasized that immunity starts to decline after vaccination and new strains are still forming. In addition to its vaccination program, the MOH implemented policies such as pooling 10–16 samples together to optimize cost and increase testing capacity of hospitals and other facilities [20]. These practices show that Vietnam shifted to an immunization-heavy approach to COVID-19 in 2022.

Countries around the world, including the United States, and the global scientific community have much to learn from Vietnam and its response to COVID-19. First, Vietnam had previously built a pandemic preparedness framework due to previous outbreaks of related infectious diseases, i.e., SARS in 2003. Vietnam employed a preventative approach to COVID-19 that required effective coordination between government, science, technology, military resources, and the MOH. Vietnam serves as a model in building an early pandemic response plan that serves the needs of each country and best utilizes its resources. Adaptability also played a role in Vietnam's later pandemic strategies in drafting a strategic vaccination plan and encouraging continued immunizations during pandemic fatigue. The key takeaways from the Vietnamese response are a review and improvement of previous pandemic response tactics, early and enforced quarantine, utilization of testing and technology, an organized leadership team, effective diplomacy, and adapting to approaches as needed.

Of course, the citizens of Vietnam played a major role in the response strategies. In the end, as Dr. Thanh Quang of the Vietnamese National Children's Hospital stated, "everyone, from a 5 [year] old to a 90 [year] old, [it didn't] matter what social-economical background, understood what COVID-19 [is] and how dangerous it is." Without the compliance and understanding of its citizens, none of what Vietnam did would have been possible.

Cross References

Pandemics Prior to COVID-19, Coronavirus Variant Emergence and Epidemiology, COVID-19 Vaccine Development, The US Federal COVID-19 Response, Global US response from USAID

References

1. CDC (2013) CDC SARS response timeline | About | CDC. Centers for Disease Control and Prevention. https://www.cdc.gov/about/history/sars/timeline.htm. Accessed 6 Aug 2022
2. CDC (2020a) COVID data tracker. Centers for Disease Control and Prevention. https://covid.cdc.gov/covid-data-tracker. Accessed 2 Aug 2022
3. CDC (2020b) Coronavirus Disease 2019 (COVID-19) in the U.S. Centers for Disease Control and Prevention. https://www.cdc.gov/coronavirus/2019-ncov/cases-updates/cases-in-us.html. Accessed 6 July 2020
4. CDC (2021) CDC Funding | COVID-19 | Budget. https://www.cdc.gov/budget/fact-sheets/COVID-19/funding/index.html. Accessed 8 Aug 2022
5. Chau MN (2022) Vietnam mulls shift to Covid endemic path as virus surges. Bloomberg.com
6. Dezan Shira & Associates (2020a) COVID-19 in Vietnam: travel updates and restrictions. Vietnam Briefing News. https://www.vietnam-briefing.com/news/COVID-19-vietnam-travel-updates-restrictions.html/. Accessed 20 May 2020
7. Dezan Shira & Associates (2020b) Vietnam business operations and the coronavirus: updates. Vietnam Briefing News. https://www.vietnam-briefing.com/news/vietnam-business-operations-and-the-coronavirus-updates.html/. Accessed 13 Aug 2020
8. Dinh Tinh L, Thu Ngan VT (2022) The COVID-19 pandemic and the emergence of Vietnam as a middle power. J Curr Southeast Asian Aff 41(2):303–325. https://doi.org/10.1177/18681034211057569
9. Dinh L, Dinh P, Nguyen PDM, Nguyen DHN, Hoang T (2020) Vietnam's response to COVID-19: prompt and proactive actions. J Travel Med 27(3). https://doi.org/10.1093/jtm/taaa047
10. Dougn D (2020) Coronavirus offers opportunity for face mask production business. Vietnam Insider. https://vietnaminsider.vn/coronavirus-offers-opportunity-for-face-mask-production-business/. Accessed 20 May 2020

11. Duong MC, Duong BT, Nguyen HT, Nguyen Thi Quynh T, Nguyen DP (2022) Knowledge about COVID-19 vaccine and vaccination in Vietnam: a population survey. J Am Pharm Assoc (2003) 62(4):1197–1205.e4. https://doi.org/10.1016/j.japh.2022.01.014
12. European Commission (2022) Top Trading Partners. Retrieved from: https://trade.ec.europa.eu/doclib/docs/2006/september/tradoc_122530.pdf
13. Feinmann J (2021) COVID-19: global vaccine production is a mess and shortages are down to more than just hoarding. BMJ 375:n2375. https://doi.org/10.1136/bmj.n2375
14. Gardaworld (2020) Vietnam: government issues new coronavirus-related travel restrictions February 15/update 8. GardaWorld. https://www.garda.com/crisis24/news-alerts/314431/vietnam-government-issues-new-coronavirus-related-travel-restrictions-february-15-update-8. Accessed 18 May 2020
15. Google Play NCOVI – Ứng dụng trên Google Play. https://play.google.com/store/apps/details?id=com.vnptit.innovation.ncovi&hl=vi. Accessed 27 May 2020
16. Hoang VM, Hoang HH, Khuong QL, La NQ, Tran TTH (2020) Describing the pattern of the COVID-19 epidemic in Vietnam. Glob Health Action 13(1):1776526. https://doi.org/10.1080/16549716.2020.1776526
17. Hörmansdorfer S, Campe H, Sing A (2008) SARS – pandemie und emerging disease. J Verbr Lebensm 3(4):417–420. https://doi.org/10.1007/s00003-008-0374-0
18. Klingler-Vidra R, Linh Tran B, Uusikyla I (2020) Testing capacity: state capacity and COVID-19 testing. Global Policy J. https://www.globalpolicyjournal.com/blog/09/04/2020/testing-capacity-state-capacity-and-COVID-19-testing. Accessed 20 May 2020
19. Le L (2022) Vietnam vaccinates 90% of its population against COVID-19. https://www.gavi.org/vaccineswork/vietnam-vaccinates-90-its-population-against-COVID-19. Accessed 3 Aug 2022
20. Le AM, Dao NG, Latkin CA, Vu LG, Dam VAT, Vu GT, Ho CSH, Ho RCM (2022) Confronting rapid reemergence of COVID-19 in metropolitans of Vietnam: updated Vietnam's policies and response measures. Front Public Health 10:790370. https://doi.org/10.3389/fpubh.2022.790370
21. Levin J, Bradshaw M (2022) Determinants of COVID-19 skepticism and SARS-CoV-2 vaccine hesitancy: findings from a national population survey of U.S. adults. BMC Public Health 22(1):1047. https://doi.org/10.1186/s12889-022-13477-2
22. Minh Le S (2020) Containing the coronavirus (COVID-19): lessons from Vietnam. https://blogs.worldbank.org/health/containing-coronavirus-COVID-19-lessons-vietnam. Accessed 20 May 2020
23. Minh LHN, Khoi Quan N, Le TN, Khanh PNQ, Huy NT (2021) COVID-19 timeline of Vietnam: important milestones through four waves of the pandemic and lesson learned. Front Public Health 9:709067. https://doi.org/10.3389/fpubh.2021.709067
24. Ministry of Health and Welfare South Korea Coronavirus disease 19 (COVID-19). Coronavirus disease 19(COVID-19). http://ncov.mohw.go.kr/en/. Accessed 21 Aug 2020
25. MOH (2020) The Ministry of Health issues guidelines on medical isolation at concentrated COVID-19 disease isolation facilities – details – Ministry of Health – COVID-19 acute respiratory disease epidemic page. Ministry of Health Vietnam. https://ncov.moh.gov.vn/web/guest/-/bo-y-te-ban-hanh-huong-dan-cach-ly-y-te-tai-co-so-cach-ly-tap-trung-phong-chong-dich-COVID-19. Accessed 19 May 2020
26. MOH (2022) Vaccine – Tiêm chủng – Bộ Y tế – Cổng thông tin của Bộ Y tế về đại dịch COVID-19. Ministry of Health Vietnam. https://covid19.gov.vn/vaccine-tiem-chung.htm. Accessed 5 Aug 2022
27. Nam NH, Do Le B-T, Huy NT (2021) Community-based measures to against the COVID-19: an experience from Vietnam with consecutive 3 months of no new infection in the community during the first wave of pandemic. Front Public Health 9:583655. https://doi.org/10.3389/fpubh.2021.583655
28. National Hospital for Tropical Diseases (2013) Hồi ức 45 ngày kinh hoàng chống dịch SARS. National Hospital for Tropical Diseases. http://benhnhietdoi.vn/tin-tuc/chi-tiet/hoi-uc-45-ngay-kinh-hoang-chong-dich-sars/220. Accessed 25 Aug 2020

29. Nguyen D (2020a) Vietnam launches COVID-19 contact tracing app. Vietnam Insider. https://vietnaminsider.vn/vietnam-launches-COVID-19-contact-tracing-app/. Accessed 27 May 2020
30. Nguyen K (2020b) Quarantined in Vietnam: scenes from inside a center for returning citizens. NPR.org. https://www.npr.org/sections/pictureshow/2020/04/06/823963731/quarantined-in-vietnam-scenes-from-inside-a-center-for-returning-citizens. Accessed 19 May 2020
31. Nguyen S (2020c) Coronavirus: life inside Vietnam's army-run quarantine camps. South China Morning Post. https://www.scmp.com/week-asia/health-environment/article/3076734/coronavirus-life-inside-vietnams-army-run-quarantine. Accessed 19 May 2020
32. Nguyen P (2021a) Vietnam receives 500,000 Sinopharm COVID-19 vaccine doses donation from China. Reuters. https://www.reuters.com/world/asia-pacific/vietnam-receives-500000-sinopharm-COVID-19-vaccine-doses-donation-china-2021-06-20/. Accessed 30 July 2022
33. Nguyen S (2021b) Vietnam approves Sinopharm's vaccine, but will people take it? South China Morning Post. https://www.scmp.com/week-asia/health-environment/article/3136137/coronavirus-vietnam-approves-sinopharms-vaccine-will. Accessed 30 July 2022
34. Nguyen T-P, Wong ZS, Wang L, Thanh TT, Van Nguyen H, Gilmour S (2021) Rapid impact assessments of COVID-19 control measures against the Delta variant and short-term projections of new confirmed cases in Vietnam. J Glob Health 11:03118. https://doi.org/10.7189/jogh.11.03118
35. Nhu HV, Tuyet-Hanh TT, Quang LN, Linh TNQ, Tien TQ (2022) COVID-19 vaccine acceptance in Vietnam: an online cross-sectional study. Asia Pac J Public Health 34(1):131–133. https://doi.org/10.1177/10105395211053732
36. Nishiura H, Kuratsuji T, Quy T, Phi NC, Van Ban V, Ha LED, Long HT, Yanai H, Keicho N, Kirikae T, Sasazuki T, Anderson RM (2005) Rapid awareness and transmission of severe acute respiratory syndrome in Hanoi French Hospital, Vietnam. Am J Trop Med Hyg 73(1):17–25
37. Onishi T (2021) Vietnam takes Sinopharm vaccines to ease Ho Chi Minh City's woes. Nikkei Asia. https://asia.nikkei.com/Spotlight/Asia-Insight/Vietnam-takes-Sinopharm-vaccines-to-ease-Ho-Chi-Minh-City-s-woes. Accessed 30 July 2022
38. Online TT (2019) Bộ trưởng Bộ Y tế Nguyễn Thị Kim Tiến: "Tôi cảm ơn những lời chỉ trích" – Tuổi Trẻ Online. Tuoi Tre Online. https://tuoitre.vn/bo-truong-bo-y-te-nguyen-thi-kim-tien-toi-cam-on-nhung-loi-chi-trich-20191121083316875.htm. Accessed 13 Aug 2020
39. Pearson J, Nguyen P (2020) Vietnam quarantines tens of thousands in camps amid vigorous attack on coronavirus. Reuters
40. Pham L (2020b) How Vietnam becomes exporter of coronavirus test kits? hanoitimes.vn. http://hanoitimes.vn/how-vietnam-becomes-exporter-of-coronavirus-test-kits-311953.html. Accessed 20 May 2020
41. Quach H-L, Nguyen KC, Hoang N-A, Pham TQ, Tran DN, Le MTQ, Do HT, Vien CC, Phan LT, Ngu ND, Tran TA, Phung DC, Tran QD, Dang TQ, Dang D-A, Vogt F (2021) Association of public health interventions and COVID-19 incidence in Vietnam, January to December 2020. Int J Infect Dis 110:S28–S43. https://doi.org/10.1016/j.ijid.2021.07.044
42. Reuters (2022) Vietnam's health minister, Hanoi city chief arrested amid anti-graft drive. Reuters
43. Ritchie H, Mathieu E, Rodés-Guirao L, Appel C, Giattino C, Ortiz-Ospina E, Hasell J, Macdonald B, Beltekian D, Roser M (2020) Coronavirus pandemic (COVID-19). Our World in Data
44. Saigoneer (2020) Vietnam continues nationwide school shutdown due to COVID-19. Saigoneer. https://saigoneer.com/saigon-health/18327-vietnam-continues-nationwide-school-shutdown-due-to-COVID-19. Accessed 18 May 2020
45. Salaam T, Redacted (2003) Severe acute respiratory syndrome (SARS): the international response. Every CRS Report. https://www.everycrsreport.com/reports/RL32072.html. Accessed 25 Aug 2020
46. Salmon A (2020) Inside Korea's low-cost, high-tech COVID-19 strategy. Asia Times. https://asiatimes.com/2020/06/the-secrets-behind-south-koreas-COVID-19-success/. Accessed 25 Aug 2020

47. Savoia E, Su M, Piltch-Loeb R, Masterson E, Testa MA (2021) COVID-19 vaccine early skepticism, misinformation and informational needs among essential workers in the USA. Int J Environ Res Public Health 18(24):13244. https://doi.org/10.3390/ijerph182413244
48. Schwirian KP (2006) The political ecology of plague in the global network of cities: the Sars epidemic of 2002–2003. Community Ecol 10:241–268. https://doi.org/10.1016/S1479-3520(06)10010-0
49. Sullivan M (2020) In Vietnam, there have been fewer than 300 COVID-19 cases and no deaths. Here's why. NPR.org. https://www.npr.org/sections/coronavirus-live-updates/2020/04/16/835748673/in-vietnam-there-have-been-fewer-than-300-COVID-19-cases-and-no-deaths-heres-why. Accessed 20 May 2020
50. Takashima K, Wada K, Tra TT, Smith DR (2017) A review of Vietnam's healthcare reform through the Direction of Healthcare Activities (DOHA). Environ Health Prev Med 22(1):74. https://doi.org/10.1186/s12199-017-0682-z
51. Terhune C, Levine D, Jin H, Lee JL (2020) Special report: how Korea trounced U.S. in race to test people for coronavirus. Reuters
52. Thao NH (2009) A line runs through it: Vietnam and China complete boundary marking process. Vietnam Law and Legal Forum. https://vietnamlawmagazine.vn/a-line-runs-through-it-vietnam-and-china-complete-boundary-marking-process-3227.html. Accessed 25 Aug 2020
53. Thu HL (2021) Delta variant outbreak challenges Vietnam's COVID-19 response strategy. Brookings. https://www.brookings.edu/blog/order-from-chaos/2021/08/11/delta-variant-outbreak-challenges-vietnams-COVID-19-response-strategy/. Accessed 4 Aug 2022
54. Toui Tre News (2020) Vietnam aviation authority ceases all flights to and from coronavirus-stricken Wuhan. Tuoi Tre News. http://tuoitrenews.vn/news/business/20200124/vietnam-aviation-authority-ceases-all-flights-to-and-from-coronavirusstricken-wuhan/52707.html. Accessed 18 May 2020
55. Toui Tre Online (2020) Ông Nguyễn Thanh Long làm quyền Bộ trưởng Bộ Y tế. TUOI TRE ONLINE. https://tuoitre.vn/news-20200706161424781.htm. Accessed 25 Aug 2020
56. Tuan V, Tung M (2020) Vietnam schools set to reopen in June after four-month break – VnExpress International. In: VnExpress International – Latest news, business, travel and analysis from Vietnam. https://e.vnexpress.net/news/news/vietnam-schools-set-to-reopen-in-june-after-four-month-break-4083091.html. Accessed 18 May 2020
57. Tuan PA, Horby P, Dinh PN, Mai LTQ, Zambon M, Shah J, Huy VQ, Bloom S, Gopal R, Comer J, Plant A (2007) SARS transmission in Vietnam outside of the health-care setting. Epidemiol Infect 135(3):392–401. https://doi.org/10.1017/S0950268806006996
58. Tung NC (2022) US-China COVID-19 vaccine diplomacy competition in Vietnam: where vaccines go, influence may follow. Aust J Int Aff 76:653
59. U.S. Department of State (2020) Telephonic briefing with Dr. Barbara Marston, CDC COVID-19 International Task Force Lead; and Dr. John MacArthur, CDC Thailand Country Director. United States Department of State. https://2017-2021.state.gov/telephonic-briefing-with-dr-barbara-marston-cdc-COVID-19-international-task-force-lead-and-dr-john-macarthur-cdc-thailand-country-director/. Accessed 7 Aug 2022
60. U.S. Embassy & Consulate COVID-19 Information. U.S. Embassy & Consulate in Vietnam. https://vn.usembassy.gov/u-s-citizen-services/COVID-19-information/. Accessed 18 May 2020
61. UNICEF (2022) Viet Nam receives an additional 6.27 million COVID-19 vaccine doses donated by Germany, Luxembourg, Portugal and the United Kingdom through the COVAX Facility. https://www.unicef.org/vietnam/press-releases/viet-nam-receives-additional-627-million-COVID-19-vaccine-doses-donated-germany. Accessed 4 Aug 2022
62. UNICEF, WHO (2021) First shipment of COVID-19 vaccine from COVAX Facility arrives in Viet Nam. https://www.who.int/vietnam/news/detail/01-04-2021-first-shipment-of-COVID-19-vaccine-from-covax-facility-arrives-in-viet-nam. Accessed 4 Aug 2022
63. Vietnam Insider (2020) Hanoi extends school closure for another week over coronavirus concerns. Vietnam Insider. https://vietnaminsider.vn/hanoi-extends-school-closure-for-another-week-over-coronavirus-concerns/. Accessed 18 May 2020

64. Vietnam Investment Review (2019) DPM Vu Duc Dam appointed as Secretary of MoH Party Affairs Committee. Vietnam Investment Review – VIR. https://www.vir.com.vn/dpm-vu-duc-dam-appointed-as-secretary-of-moh-party-affairs-committee-71144.html. Accessed 25 Aug 2020
65. Vietnam Times (2020) Prevention of Covid 19: Vietnamese Customs continues giving directives about exporting face masks. Vietnam Times. https://vietnamtimes.org.vn/prevention-of-COVID-19-vietnamese-customs-continues-giving-directives-about-exporting-face-masks-18450.html. Accessed 19 May 2020
66. VietnamPlus (2020) Phân loại cách ly người nhiễm, nghi nhiễm COVID-19 | Sức khỏe | Vietnam+ (VietnamPlus). VietnamPlus. https://www.vietnamplus.vn/infographics-phan-loai-cach-ly-nguoi-nhiem-nghi-nhiem-covid19/627447.vnp. Accessed 19 May 2020
67. Vo H (2022) COVID-19 information. U.S. Embassy & Consulate in Vietnam. https://vn.usembassy.gov/u-s-citizen-services/COVID-19-information/. Accessed 2 Aug 2022
68. Voice of America (2020) Vietnam poised to export COVID-19 test kits. Voice of America – English. https://www.voanews.com/COVID-19-pandemic/vietnam-poised-export-COVID-19-test-kits. Accessed 20 May 2020
69. Vu K (2020) Vietnam says origin of Danang outbreak hard to track as virus cases rise. Reuters
70. Vu K, Nguyen M (2020) Vietnam reopens schools after easing coronavirus curbs. Reuters
71. Vu HT, Leitmeyer KC, Le DH, Miller MJ, Nguyen QH, Uyeki TM, Reynolds MG, Aagesen J, Nicholson KG, Vu QH, Bach HA, Plant AJ (2004) Clinical description of a completed outbreak of SARS in Vietnam, February–May, 2003. Emerg Infect Dis 10(2):334–338. https://doi.org/10.3201/eid1002.030761
72. WHO (2003) Update 95 – SARS: chronology of a serial killer. World Health Organization. https://www.who.int/emergencies/disease-outbreak-news/item/2003_07_04-en. Accessed 2 Aug 2022
73. WHO (2020) Pandemic fatigue: reinvigorating the public to prevent COVID-19: policy considerations for Member States in the WHO European Region. World Health Organization. Regional Office for Europe
74. WHO (2022a) Viet Nam: WHO Coronavirus Disease (COVID-19) Dashboard with vaccination data. https://covid19.who.int. Accessed 5 Aug 2022
75. WHO. Scaling up COVID-19 vaccination rates in Viet Nam through vaccine diplomacy, efficient vaccine rollout and enhancing effective service delivery. https://www.who.int/about/accountability/results/who-results-report-2020-mtr/country-story/2021/vietnam. Accessed 3 Aug 2022b
76. World Bank (1961) GDP per capita (current US$) – Singapore, Taiwan, China, Hong Kong SAR, China, Korea, Rep. | Data. World Bank. https://data.worldbank.org/indicator/ny.gdp.pcap.cd?locations=sg-tw-hk-kr&name_desc=false. Accessed 19 May 2020
77. Hotez PJ (2014) Vaccine Diplomacy: Historical Perspectives and Future Directions PLoS. Neglected Tropical Diseases 8(6):e2808. https://doi.org/10.1371/journal.pntd.0002808
78. Wouters OJ, Shadlen KC, Salcher-Konrad M, Pollard AJ, Larson HJ, Teerawattananon Y, Jit M (2021) Challenges in ensuring global access to COVID-19 vaccines: production affordability allocation and deployment. The Lancet 397(10278):1023–1034
79. Nguyen, P. Vietnam says to buy AstraZeneca vaccine, in talks with other producers, *Reuters*, URL.

Supplemental Reading

Duong DM, Le VT, Ha B (2020) Controlling the COVID-19 pandemic in Vietnam: lessons from a limited resource country. Asia Pac J Public Health 32(4):161–162. https://doi.org/10.1177/1010539520927290

Escandón K, Rasmussen AL, Bogoch II, Murray EJ, Escandón K, Popescu SV, Kindrachuk J (2021) COVID-19 false dichotomies and a comprehensive review of the evidence regarding public health, COVID-19 symptomatology, SARS-CoV-2 transmission, mask wearing, and reinfection. BMC Infect Dis 21(1):710. https://doi.org/10.1186/s12879-021-06357-4

Tam LT, Ho HX, Nguyen DP, Elias A, Le A (2021) Receptivity of governmental communication and its effectiveness during COVID-19 pandemic emergency in Vietnam: a qualitative study. Glob J Flex Syst Manag 22(Suppl 1):45–64. https://doi.org/10.1007/s40171-021-00269-7

Van Nguyen Q, Cao DA, Nghiem SH (2021) Spread of COVID-19 and policy responses in Vietnam: an overview. Int J Infect Dis 103:157–161. https://doi.org/10.1016/j.ijid.2020.11.154

Afghanistan and COVID-19

Meena Said and Amanda Charlotte Foshag

1 Background

The COVID-19 pandemic had a significant effect on global health and the world economy. However, preexisting healthcare and the strength of the economy of each nation were important factors in determining how populations fared. This chapter illustrates how the COVID-19 pandemic affected Afghanistan, a country that has one of the world's weakest healthcare infrastructures and economies. Given a lack of adequate data, the experience of a nonprofit organization, *Wellness Worldwide for Sustainable Communities*, is the lens through which the COVID-19 pandemic's effect on Afghanistan is viewed.

Afghanistan's population is approximately 38.6 million. It is the sixth poorest country in the world according to recent rankings [1, 2]. The GDP per capita is 516.7 USD [2]. Corruption, war, and drought resulted in a poverty rate of ca. 50% during the first three waves of the COVID-19 pandemic and prior to the withdrawal of American forces in August 2021. The Taliban takeover was followed by sanctions and the freezing of Afghan assets by the international community, which caused a significant decline in the economy, with preliminary data showing a contraction of the Afghan economy by 21% in 2021 [3]. According to the UNDP, 97% of Afghans would live in poverty by mid-2022 [4]. Approximately 18.9 million Afghans are acutely food insecure. All 34 provinces are facing emergency or crisis levels of food insecurity, and 3.9 million children suffer from malnutrition [5].

Health metrics mirror the grim economic data, and health data starkly indicate that healthcare for Afghans is among the worst in the world. Three statistics that are sharply indicative of the overall state of the healthcare system are maternal

M. Said (✉)
Saint Johns Cancer Institute, Santa Monica, CA, USA

A. C. Foshag
Georgetown University School of Medicine, Washington, DC, USA

J. M. Rosen, R. R. Colwell (eds.), *The COVID-19 Pandemic*,
https://doi.org/10.1007/978-3-031-62772-9_11

mortality, child and infant mortality, and access to health facilities (defined as living within 1 h of walking from the nearest health facility) [6]. The maternal mortality ratio was 680 deaths per 100,000 live births in 2017 [7], and the infant mortality rate in 2020 was 45 deaths per 1000 live births [8]. Less than 10% of the population lives within 1 h of walking distance from a health facility [9]. It is within this severely limited healthcare infrastructure that the COVID-19 pandemic first hit in March 2020.

There were three waves of COVID-19 cases from March 2020 through September 2021 in Afghanistan. The first wave occurred in March 2020 and continued through August 2020. The second wave began in November 2020 and continued through the end of January 2021. The third wave began in May 2021 and continued through August 2021 [10]. The first, second, and third waves had the following peak daily new case numbers: 758.57, 221.29, and 2095.57, respectively [10]. Daily death peaks for the first, second, and third waves were as follows: 26.43, 18.29, and 97.57, respectively [10]. As of November 2022, the cumulative number of deaths was 7829 [10]. The actual death toll and case rates from COVID-19 are likely higher than the numbers reported here because of limited testing and incomplete records of death rates.

In the absence of reliable data, the real-world experience of organizations on the ground provides valuable insight into how the COVID-19 pandemic affected Afghanistan. In this chapter, we evaluate the effects of the COVID-19 pandemic on activities of the nonprofit organization, *Wellness Worldwide for Sustainable Communities*. The main initial aim of this organization was to afford care for patients with cancer or a complex surgical disease and to provide primary care. However, the COVID-19 pandemic affected the operations of the organization in many ways. It is through the changes occurring during the pandemic that we can understand its true ramifications for one of the most vulnerable populations in the world. In particular, the experience of *Wellness Worldwide for Sustainable Communities* in Afghanistan during COVID-19 illustrates how the pandemic affected a conflict zone region with a vulnerable healthcare system. Three programs evaluated in the case of *Wellness Worldwide* are comprised of the following: (1) provision of cancer and complex surgical care; (2) food aid in the setting of a drastic economic decline due to the COVID-19 pandemic; and (3) crisis in primary care clinics during the COVID-19 pandemic.

2 Provision of Cancer and Complex Surgical Care

Wellness Worldwide was initially created in response to the need for cancer and complex surgical care in Afghanistan. There is not a single cancer center or radiation facility in the entire country. Those patients who are able to afford the expense go to neighboring countries for care. However, given acute poverty levels, this is an option for only a small minority of patients. There is limited surgical care and chemotherapy available in various hospitals and centers in Kabul. Moreover, these services are also expensive and often below standard care.

Pediatric cancer patients are perhaps the most vulnerable population within this already very underserved population. Both global and local data are indicative of the disparity between high- and middle-income countries with respect to cancer care. In high-income countries (HIC), 80% of children will survive a cancer diagnosis [11]. In low- and middle-income countries (LMIC), there is a survival rate of 20% for children with cancer [11]. The World Health Organization has an initiative to increase the survival rate for children in LMIC to 60% by 2030 [12]. Despite this, the global incidence of pediatric cancer is increasing. In 2009, there were 175,058 new pediatric cancers, and in 2015, there were 416,500 new pediatric cancers [13]. In Afghanistan, the annual new pediatric cancer cases in 2020 were 1730 [14]. With respect to the overall population, there were 22,817 new cancer cases per year [14]. The mortality rate is 81.2% [14]. Statistics in Afghanistan are incomplete and inconsistent. However, the Institute for Health Metrics and Evaluation (IHME) at the University of Washington suggests an approximate incidence of 68,000 in 2017 and a prevalence of between 87,681 and 181,500 people with cancer [15].

The severe deficit in cancer care is also illustrated by data regarding radiation treatment facilities. For the Afghan population to have care consistent with standards in the United States, the nation would require 200 radiation facilities [16]. However, there are no radiation treatment facilities in the country, and patients often present with advanced disease. Furthermore, the lack of palliative radiation for pain control creates an agonizing experience for impoverished patients in Afghanistan [17, 18]. In response to these factors, *Wellness Worldwide for Sustainable Communities* created a novel scheme, International and Regional Tertiary Care Partnerships for Afghan Children, to address the cancer needs of Afghan children (Fig. 1). The patient care plan involved the coordination of three teams located in three respective countries with each team having a distinct role. The first team, designated Team A, is comprised of *Wellness Worldwide for Sustainable Communities* team members in Afghanistan. Team A meets with the patient and their family, collects available medical data, communicates data with Team B members, and arranges the transportation of the patient either internally or to facilities abroad. The second team, Team B, is composed of surgeons and medical oncologists in the United States. Team B has the following responsibilities: initial evaluation of available studies; formulating clinical recommendations; management based on regional resources; communication with doctors in Afghanistan, Pakistan, or India; and cost assessment. The third team, Team C, is located at the point of treatment (usually in India or Pakistan). Team C is comprised of volunteers who ensure patients have appropriate lodging and transportation as well as being evaluated and treated at the proper facilities. It also serves as an intermediary between the United States and Indian or Pakistani doctors (Fig. 1). Depending on disease complexity and severity, the total cost (including travel and lodging) to treat one child with cancer is estimated to be between 1000 and 5000 USD.

The COVID-19 pandemic interfered with *Wellness Worldwide for Sustainable Communities*' ability to deliver care to these very sick patients. Regional hospitals in Pakistan, used as a key resource to provide care for Afghan cancer patients, have adequate radiation facilities, surgical services, and chemotherapy services. However,

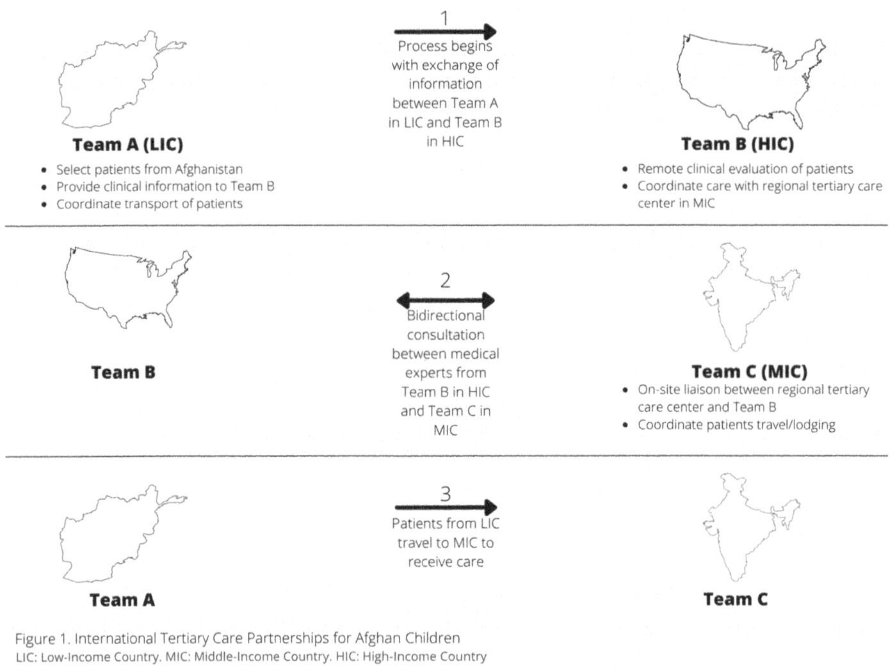

Fig. 1 International tertiary care partnerships for Afghan children. LIC Low-income country, MIC Middle-income country, HIC High-income country

in March 2020, Torkham, the major land border crossing between Afghanistan and Pakistan, was closed, with no traffic allowed in either direction. A 15-year-old male patient with severe and advanced sarcoma of his face presented to the organization for help. The patient had a prior evaluation determining that he had a sarcoma. However, the patient's family could not afford medical care, and his disease progressed to the extent that the tumor involved the entire right side of his face, leaving the patient in severe pain. *Wellness Worldwide for Sustainable Communities* was able to transport the patient from his home village in Ghazni Province to Kabul. In Kabul, he was able to get intravenous pain medication. However, palliative radiation therapy was required to achieve adequate control of the patient's pain. Despite multiple efforts to obtain permission from the authorities for the patient to cross into Pakistan, he was not given permission. There were also no passenger planes flying at the time that would have allowed him to be transported to India or Pakistan. The patient wanted to be with his family and elected to return to his home. Unfortunately, he never had proper pain control and died a few months later, after suffering a prolonged period of extreme agony. This tragedy illustrates how COVID-19 interrupted critical healthcare delivery to Afghans due to restrictions on regional cross-border travel brought on by the pandemic.

3 COVID-19 Pandemic, Economic Decline, and Food Aid

The COVID-19 pandemic hit Afghanistan at a vulnerable time. In the spring of 2020, Afghanistan was in the midst of political transition and increasing conflict. COVID-19 was first identified in Herat Province and Nimroz Province (bordering Iran) and was brought into the country by Afghan returnees. By May 2020, the virus had spread to at least 33 of Afghanistan's 34 provinces.

The pandemic also disrupted economic activity and regional trade. Afghanistan's poverty level increased to 72% during 2020 [19]. In the first half of 2020, prices for basic food items increased by 20% [20]. Border closures negatively impacted exports. Household incomes declined, and demand for goods decreased. Contraction of the economy had the most severe effect on households depending on activities affected by lockdowns. Approximately 15 million Afghans live in households that obtain more than half of their income from small-scale retail businesses and daily labor in construction, agriculture, or personal services [21]. In this context, there was a drastic increase in hunger. Local tribal elders and community leaders reached out to *Wellness Worldwide for Sustainable Communities* for food aid when the lockdown occurred in March 2020. With the closure of the border between Afghanistan and Pakistan, trade came to a halt. Farmers' goods could not be transported across the border and were spoiled as a result. Many child laborers who were reliant on brick-making for their livelihood, producing bricks transported to Pakistan, lost their source of income. Thus, lockdowns severely affected the local economy and tribal elders pleaded for help to feed their people. There was no assistance available from United Nations-affiliated organizations or the Afghan government at the time.

In response, *Wellness Worldwide* implemented food distribution programs systematically. Since the need was immense and resources were limited, all of those in need could not be helped. The organization's food distribution occurred in the provinces of Nangarhar and Kunar. Food was available in local markets, but the price was high and income had decreased, prohibiting families from obtaining food. Wellness Worldwide identified the most vulnerable aid recipients by consulting local tribal and religious leaders to determine those most in need. For example, household heads who were widows were prioritized for aid. Food was distributed in packages estimated to feed a family of between 7 and 8 for 1 month (average family size in Afghanistan is 7.8), and the food packages contained flour, oil, and beans, modeled after the World Food Program (WFP) food packages [22]. Sugar and tea were added to the food packages when resources permitted. Approximately 2370 people received food aid, at a cost of 0.37 USD per day. The WFP cost is ca. 0.43 USD per person per day [23].

4 COVID-19 Pandemic and Primary Care

Primary care available to the Afghan population is limited with 0.3 physicians per 1000 in Afghanistan [24]. Prior to the Taliban takeover, there were 4.9 physicians per 1000 [24]. WHO estimates at least 2.5 medical staff (physicians, nurses, and

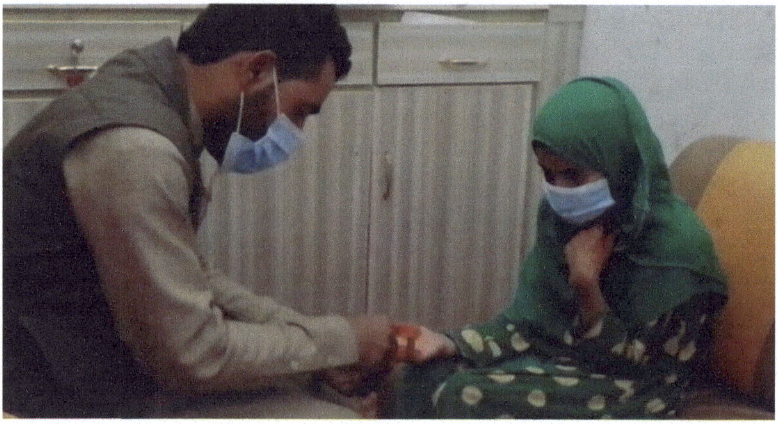

Fig. 2 Afghan child receives medical care at Bibi Rukhshana Women and Children's Clinic

midwives) per 1000 people are necessary to provide adequate coverage, with primary care intervention [25]. To address the lack of primary care, *Wellness Worldwide for Sustainable Communities* opened a free clinic, the Bibi Rukhshana Women and Children's Clinic in 2021. The BiBi Rukhshana clinic provides primary care and serves internally displaced refugees from Kabul Camp in Rodat District, Nangarhar Province. According to local tribal leader estimates, prior to the establishment of the Bibi Rukhshana clinic, approximately 35,000 people lived in the camp without primary care and only a single pharmacist was available. Events of the third and deadliest COVID-19 wave in Afghanistan were disastrous for an already extremely limited healthcare system (Fig. 2).

During the third wave of the COVID-19 pandemic in June 2021, the situation in Afghanistan had become dire. The Ministry of Public Health reported that all 34 provinces were confirmed to have COVID-19 cases. A total of 71,012 patients recovered, and 4962 patients died. The testing rate for COVID-19 was 61,112 out of an entire population of approximately 40 million. The country did not have facilities to test for the Delta variant, but given the regional spread, it was presumed to be the dominant variant. The positive test rate was 42%. Because of limited public health resources and the absence of a national death register, the number of confirmed cases and deaths from COVID-19 is likely to be a severe underestimation. At the time, there were 30 laboratories, each with testing capacity of 8500 per day. Schools, government offices, and mass gatherings were banned and under lockdown to decrease the spread of disease. The country received 1.6 million vaccination doses, available from health facilities and mobile teams. A total of 898,000 were vaccinated via the Ministry of Public Health program [26].

Wellness Worldwide for Sustainable Communities' primary care clinic was staffed by one family medicine physician, as well as a medical assistant, and provided basic laboratory services. All services and medications were free of charge to patients (medications were sometimes also offered at a rate of 60–70% discount depending on available funds and patient volume). During the third wave of the

COVID-19 pandemic, approximately 20–46 patients were evaluated per day, most presenting with symptoms consistent with COVID-19 illness. However, the clinic lacked COVID-19 testing and a limited oxygen supply. Countrywide, there were 30 oxygen plants [27], and hospitals and clinics throughout Afghanistan were overwhelmed with patients. Patients requiring oxygen and higher levels of care were referred to the main COVID-19 hospital in Jalalabad, Nangarhar, where there was also only a limited supply of oxygen. Since hospitals were in short supply of equipment and materials in general (not only oxygen), many patients were cared for at home. The families of patients obtained oxygen either from oxygen plants or from the open market. A 45-liter cylinder of oxygen serves one patient 8–10 days, at a low flow rate. The oxygen supply depletes much more quickly when the oxygen requirement is higher. Oxygen-producing plants filled oxygen tanks at a cost of 5 USD per cylinder, which would cost $50 in the market [28], and profiteers charged as much as 110 USD for those in desperate need in Nangarhar Province. With an average per capita income of 517 USD, this was unaffordable for most Afghans [29]. Locals described an organized "oxygen mafia" colluding to keep prices high. This was reported in India also, and Indian Congress officials called for state government intervention [30]. In response to the oxygen crisis, *Wellness Worldwide for Sustainable Communities* contacted the eastern Afghanistan division of the United Nations Office for Humanitarian Affairs. Unfortunately, they were unable to provide assistance. Oxygen concentrators, which could generate oxygen from air for medical use, were available abroad. *Wellness Worldwide* coordinated a complex procurement effort and obtained 20 oxygen concentrators from Vietnam, but the oxygen concentrators arrived after the number of severe cases declined significantly, at the end of the third COVID-19 wave.

5 Conclusion

The experience of one nonprofit organization in Afghanistan, *Wellness Worldwide for Sustainable Communities*, illustrates how the COVID-19 pandemic affected a war-torn and impoverished nation. The organization, focused on healthcare, engaged in food aid because of severe food insecurity in Afghanistan during the COVID-19 pandemic. Economic decline associated with the pandemic exacerbated an already weak economy and contributed to widespread hunger. Insufficient healthcare for those afflicted with COVID-19 in Afghanistan was worsened, notably when the number of cases increased and the need for oxygen exceeded capacity. There were scores of preventable deaths due to the lack of oxygen for critical COVID-19 patients. In the case of cancer patients, children with cancer were among the most vulnerable populations in the country, and the lack of cancer care in Afghanistan rendered cross-border travel a necessity for obtaining treatment. Unfortunately, border closures during the pandemic prevented timely care for these patients, causing deaths. The catastrophic consequences of COVID-19 in Afghanistan illustrate that an underlying lack of healthcare infrastructure and a weak economy render a pandemic even more tragic. Since the third wave of COVID-19 in the summer of 2021, a new political reality has emerged with the Taliban in power. Resulting sanctions and the freezing of Afghan state funds, coupled with the withdrawal of foreign aid (70% of the public sector depends on international aid), are indicative of an even worse humanitarian catastrophe [31]. What lies ahead for Afghanistan is unclear as events destabilize an already impoverished polity emerging from over four decades of war.

References

1. Poorest Countries in the World (2022) World population review. https://worldpopulationreview.com/country-rankings/poorest-countries-in-the-world
2. Afghanistan GDP (2022) The World Bank. https://data.worldbank.org/country/AF
3. Afghanistan Economy (2022) The World Bank. https://www.worldbank.org/en/country/afghanistan/overview#:~:text=GDP%20(current%20US%24),1960%201980%20 2000&text=The%20preliminary%20official%20GDP%20statistics,financial%20sector%20 remains%20in%20crisis
4. Millions of Afghan Lives and Livelihoods in Danger Without Support- Says United Nations Development Program Chief (2022) https://www.undp.org/press-releases/millions-afghan-lives-and-livelihoods-danger-without-support-says-un-development-programme-chief
5. OCHA Services (2022) Reliefweb. "WFP Afghanistan: Situation Report 19 July 2022." 23 July. https://reliefweb.int/report/afghanistan/wfp-afghanistan-situation-report-19-july-2022
6. Newbrander W, Ickx P, Feroz F, Stanekzai H (2014) Afghanistan's basic package of health services: its development and effects on rebuilding the health system. Glob Public Health 9:S6
7. Maternal Mortality, Afghanistan (2022) World Data Atlas. https://knoema.com/atlas/Afghanistan/Maternal-mortality-ratio
8. Infant Mortality, Afghanistan (2022) statista.com/statistics/1072357/child-mortality-rate-afghanistan18002020/#:~:text=Afghanistan's%20infant%20mortality%20rate%20 (among,child%20deaths%20occur%20during%20infancy
9. Tharwani GH, Kumar P, Shaeen SK et al (2022) Maternal mortality in Afghanistan: challenges, efforts, and recommendations. Clin Epidemiol Glob Health 15:101038
10. Afghanistan COVID Data (2022) https://ourworldindata.org/coronavirus/country/afghanistan
11. Bhakta N, Force LM, Allemani C, Atun R, Bray F, Coleman MP et al (2019) Childhood cancer burden: a review of global estimates. Lancet Oncol 20:e42
12. WHO Childhood Cancer Cure Rate Initiative (2022) https://www.who.int/activities/improving-childhood-cancer-cure-rate
13. Force LM, Abdollahpour I, Advani S, Aguis D, Ahmadian E, Alahdab F et al (2019) The global burden of childhood and adolescent cancer in 2017: an analysis of the global burden of disease study 2017. Lancet Oncol 20:1211
14. Afghanistan WHO Cancer Country Profiles (2022) https://cdn.who.int/media/docs/default-source/country-profiles/cancer/afg_2020.pdf?sfvrsn=194f8d07_5&download=true
15. Global Burden of Disease, Cancer (2022) https://ourworldindata.org/search?q=cancer
16. Meadenwald D, Vordermark D, Dietzel C (2018) Number of radiation machines per population and cancer mortality: an ecological study. Clin Epidiemiol 10:1249–1273
17. Sasaki H, Bouesseau M, Marston J, Mori R (2017) A scoping review of palliative care for children in low- and middle-income countries. BMC Palliat Care 16:60
18. Caruso Brown AE, Howard SC, Baker JN, Ribeiro RC, Lam CG (2014) Reported availability and gaps of pediatric palliative care in low- and middle-income countries: a systematic review of published data. J Palliat Med 17:1369–1383
19. Afghanistan Poverty Rate (2022) https://reliefweb.int/report/afghanistan/afghanistan-COVID-19-situation-report-05-may-2020

20. World Bank, Food Prices (2022) https://documents1.worldbank.org/curated/en/132851594655294015/pdf/Afghanistan-Development-Update-Surviving-the-Storm.pdf
21. Afghanistan development update World Bank (2022) https://www.worldbank.org/en/country/afghanistan/overview
22. World Food Program, Food basket (2022) https://www.wfp.org/wfp-food-basket
23. World Food Program, how much would it cost to end hunger (2022) https://www.wfpusa.org/articles/how-much-would-it-cost-to-end-world-hunger/
24. Physicians per population, Afghanistan (2022) https://data.worldbank.org/indicator/SH.MED.PHYS.ZS?locations=AF
25. Physicians needed per population (2022) https://databank.worldbank.org/metadataglossary/health-nutrition-and-population-statistics/series/SH.MED.PHYS.ZS#:~:text=The%20WHO%20estimates%20that%20at,%2C%20World%20Health%20Report%202006
26. WHO (2021) Afghanistan Vaccinations, Strategic situation report: COVID-19 no 100. WHO
27. Oxygen Shortage Afghanistan (2022) https://apnews.com/article/afghanistan-coronavirus-pandemic-health-oxygen-shortage-a5f1b9cf47479cf38cae6b765d7c9250
28. Essar MY, Hasan MM, Islam Z et al (2021) COVID-19 and multiple crises in Afghanistan: an urgent battle. Conflict Health 15(1):70
29. GDP Afghanistan (2022) https://data.worldbank.org/indicator/NY.GDP.PCAP.CD?locations=AF
30. Oxygen Mafia, India (2022) https://timesofindia.indiatimes.com/city/hyderabad/act-against-oxygen-mafia-cong/articleshow/82783674.cms
31. WHO (2022) Afghanistan's health care system is on the brink of collapse. https://www.cnbc.com/2021/09/24/afghanistans-health-care-system-is-on-the-brink-of-collapse-who.html

Crisis Management with Chinese Characteristics: From SARS to COVID-19

Yu Luo Rioux

1 Introduction: from SARS to COVID-19

The symptomatic onset of the first patient can be traced back to 1 December 2019 [5, 23, 47, 58, 60] while the origin of COVID-19 may still be debatable. Li Wenliang, the ophthalmologist in Wuhan who later died of COVID-19, warned his classmates of coronavirus via the most popular Chinese social media WeChat on 30 December 2019. On the following day, Chinese authorities alerted the World Health Organization (WHO) of pneumonia cases in Wuhan, Hubei, with the treatment for COVID-19 cases, then termed as Wuhan pneumonia, an infectious disease caused by the SARS-CoV-2 virus, namely, the 2019 novel coronavirus. The disease was renamed as 2019-nCoV and later known as COVID-19 [78].

The connection between the COVID-19 pandemic and the SARS (severe acute respiratory syndrome) epidemic is clear. COVID-19 hit the world like a "return of the coronavirus" 17 years later. The SARS epidemic, or *Feidian* (atypical pneumonia) in Chinese, assailed China between November 2002 and June 2003 [6]. According to the WHO, as of 7 August 2003, total cases of SARS reached 8422 over 32 nations and states, with 919 fatalities, at the rate of 11%. Among them, 5327 cases (349 fatalities) occurred in mainland China, 1755 (300 fatalities) in Hong Kong, and 665 (180 fatalities) in Taiwan ([7]; SARS Expert Committee (2003). Report. https://www.sars-expertcom.gov.hk/english/reports/reports/reports_fullrpt.html; [72]: 186).

In terms of the geographic distribution in China (Fig. 1), SARS cases were proportionally higher in Guangdong Province, where the initial outburst was believed to have originated (in Hong Kong nearby), relative to the rest of China, with the

Y. L. Rioux (✉)
St. Joseph Hospital, Covenant Health, Nashua, NH, USA

University of Massachusetts, Lowell, MA, USA
e-mail: yluo@colorado.edu

J. M. Rosen, R. R. Colwell (eds.), *The COVID-19 Pandemic*,
https://doi.org/10.1007/978-3-031-62772-9_12

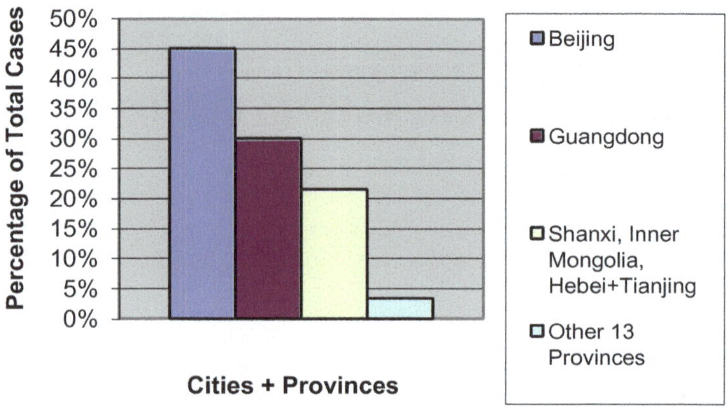

Fig. 1 Geographic distribution of SARS cases in China (as of 10 May 2023). (Source: Xinhua News Agency)

highest incidence occurring in the capital city Beijing. The high rate in Beijing was attributed to the government-led establishment of a specialized SARS hospital there to focus on the treatment and the prevention of further spread. The pattern is confirmed in a detailed discussion of the spatiotemporal diffusion of SARS in mainland China [15].

In comparison, the current COVID-19 pandemic has affected the entire world on a much larger scale with significantly greater fatalities than the SARS epidemic, which has lasted for 3 years and may still be ongoing. As of July 2023, there have been a total of 767,726,861 confirmed COVID-19 cases globally, with 6,948,764 total deaths, among which there have been 103,436,829 confirmed cases with 1,127,152 deaths in the United States and 99,292,081 confirmed cases with 121,490 deaths in China [27, 75]. The dynamic of confirmed COVID-19 cases in the world is made explicit in the chart shown in Fig. 2, presenting daily confirmed new cases with a "7-day moving average" [27]. The original chart is interactive, illustrating the outbreak evolution for the ten most affected nation-states currently (as of 10 March 2023). Specific numbers corresponding to each country appear with the cursor moving along the time axle, and the curve of each country appears with a click on its name at the bottom of the chart. Although Fig. 2 is merely a snapshot of the overall scenario, it still captures the global situation of COVID-19 from the beginning.

For the purpose of examining pandemic responses, the trajectory of COVID-19 outbursts is roughly categorized into four stages globally: first, December 2019 to March 2020—the initial stage; second, March to October 2020—global transmission with spikes and the Delta variant; third, October 2020 to November 2021—from the Delta to the Omicron variant; fourth, November 2021 to May 2023—the Omicron variant to conclusion.

In general, the United States has the most confirmed cases of COVID-19 and by far the most at the peak wave around January 2022 [27]. COVID-19 statistics provides details in terms of total reported cases, total deaths, and total recovered by country or region (Table 1):

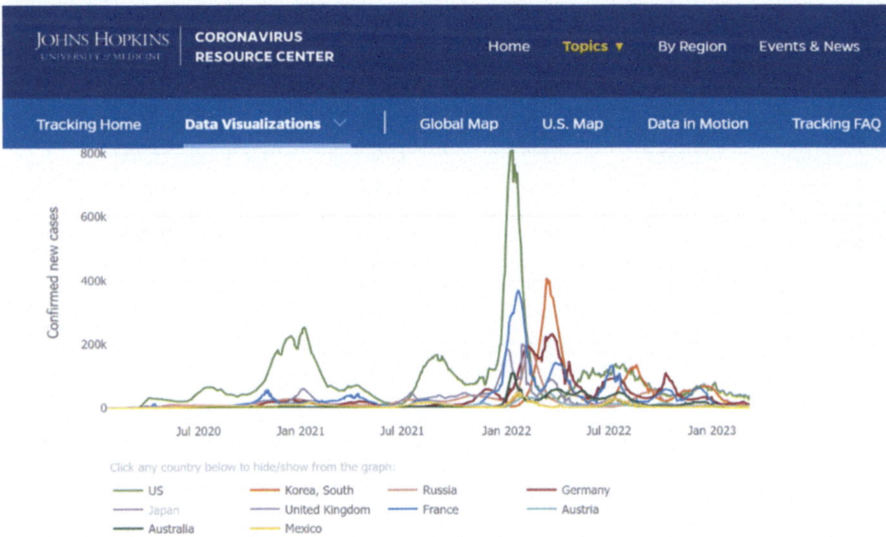

Fig. 2 Confirmed new COVID-19 cases worldwide (up to 10 March 2023). (Source: JHU Coronavirus Resource Center)

Table 1 Reported COVID-19 cases by country or territory

County	Total Cases	New Cases	Total Deaths	Total Recovered	New Recovered	Active Cases	Population
World	691,456,414	+397	6,899,448	663,999,074	+23,302	20,557,892	
USA	107,375,693		1,168,621	105,551,254		655,818	334,805,269
India	44,994,765		531,914	44,461,455		1,396	1,406,631,776
France	40,138,560		167,612	39,964,889	+1,217	6,029	65,584,518
Germany	38,428,685		174,352	38,240,600		13,733	83,883,596
Brazil	37,693,506		704,320	36,249,161		740,025	215,353,593
Japan	33,803,572		74,694	N/A	N/A	N/A	125,584,838
S. Korea	32,415,857		35,111	32,117,344	+20,849	263,402	51,329,899
Italy	25,897,801		190,868	25,593,568		113,365	60,262,770
UK	24,636,637		227,524	24,408,117	+372	996	68,497,907
Russia	22,967,718		399,715	22,414,985		153,018	145,805,947
China	503,302		5,272	379,053		118,977	1,448,471,400
Vietnam	11,621,320	+32	43,206	10,640,238	+7	937,876	98,953,541

Source: JHU; WHO; Worldometer (up to 15 July [77])

Based on the numbers of China and the United States in Table 1, the ratio of total deaths over total cases is 1.05% and 1.09%, respectively; similarly, the ratio of total recovered over total cases is 75.31% and 98.30%, respectively (much lower for China), while the ratio of total cases over the population is 0.03% and 32.07%, respectively (much lower for China) (Table 2).

It is hard to say that the level of medical treatment for COVID-19 is close between the two countries because the death ratio over the reported cases is close as the

Table 2 Statistical comparison of COVID cases between China and the United States

County	Total Cases	Total Deaths	Total Recovered	Active Cases	Population
USA	107,375,693	1,168,621	105,551,254	655,818	334,805,269
China	503,302	5,272	379,053	118,977	1,448,471,400
USA ratios		1.09%	98.30%		32.07%
China ratios		1.05%	75.31%		0.03%

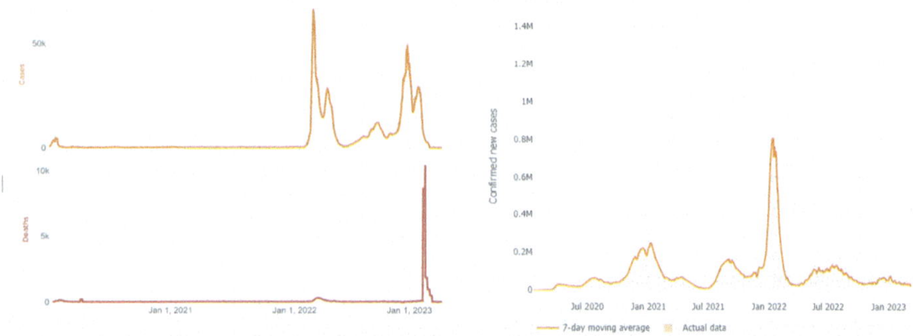

Fig. 3 Comparison of COVID-19 cases between China (left) and the United States (right). (Source: JHU Coronavirus Resource Center [27])

recovered rate for China is much lower. Apparently, the ratio for COVID cases over the population is much lower for China. What is the correlation between COVID statistics and pandemic response approaches or public health crisis management?

A closer look at the confirmed cases with the timeline between the United States and China may be more explicit in terms of the different pandemic situation. A comparison of the confirmed cases in Fig. 3 indicates that there have been several waves of the confirmed cases in the United States, with the spike occurring around January of 2022; conversely, there have been only two waves of the confirmed cases in China in early 2022 and early 2023, respectively, with deaths concentrated and peaked around January of 2023. This pattern indicates how China as a nation successfully contained the spread of COVID-19 until the emergence of the less deadly Omicron variant at the later stage.

While each nation or state may vary in terms of timeline, number of confirmed cases, and number of deaths for COVID-19 outbreaks, China and the United States have taken different approaches in response to the pandemic. For example, one has the advantage of a central government, while the other relies on the power of an independent state; one focuses more on a strict quarantine, while the other promotes vaccination.

Scrutiny over the correlation between pandemic responses and spikes of confirmed cases as well as that of fatalities may reveal the efficacy of measures taken by each nation or state, lending lessons for the future. More specifically, in what ways have the Chinese responded differently to the COVID-19 pandemic? How was such a large-scale public health crisis managed in China within varying levels of governance—national, provincial or municipal, and local? How effective was the crisis management with Chinese characteristic in waging this "smokeless war without gunpowder?" What lessons for the present pandemic has China learned from fighting against the related SARS epidemic? This chapter endeavors to address some of these issues. Here, governance is used in the sense expanded by Rhodes, including the process or act of governing by governmental institutions as well as self-organized actors ([28]: 317). It is well known how governance in China is highly centralized, featuring a multilevel bureaucracy. Lin et al. [36] analyzes the mechanism of governance in China as follows:

> the Chinese bureaucracy is an elaborate network that extends to all levels of society down to the neighborhood and working unit, and exhibits a high degree of discipline by international standards (Parish and Whyte 1978, 1984). Within each level, an impressive organizational apparatus is able to transmit state policies down to the lower-level government agencies step-by-step through several layers of government bureaucracy [46].

Given its unique form of governance, China has responded differently to the public health crisis of COVID-19, particularly different from the approaches of the United States. This is because the Chinese governance features central leadership and top-down management, with what Bardhan [3] specified as an "authoritarianism-democracy distinction." Simply put, the Chinese government is under the leadership of the Chinese Communist Party (CCP), with the CCP Politburo Standing Committee as the top leadership, and the National People's Congress (NPC) as the national legislature, guaranteed by the PRC Constitution. The central government oversees the State Council, the Central Military Commission, the national Supervisory Commission, Supreme People's Court and Supreme People's Procuratorate, and the provincial and local government as well [59].

The Chinese authorities have summarized their successful experience and efforts in fighting against the COVID-19 pandemic comprised of ten facets [12], including the following:

- Strong command system
- Lockdown
- Testing, isolation and treatment
- Pairing assistance
- Makeshift hospitals
- Combined use of traditional Chinese medicine and western medicine
- Community-based epidemic prevention and control
- Information release
- Dynamic monitoring
- Universal participation

Peng et al. [44] summarize the Chinese experience of COVID-19 treatment and control strategies at the early stage. Other scholars applaud China's digital technology, all-staff vaccination, all-staff Nucleic Acid Amplification Testing, lockdowns, zero-COVID policy, psychological counseling, and its accountability system as successful strategies against the COVID-19 pandemic [11].

There is ample literature reviewing China's crisis management for the COVID-19 pandemic, and there are also summaries and analysis of how China responded to the SARS epidemic. Studies of the SARS epidemic in China written in English are considerable, ranging from essays, books, and edited volumes. Topics include law and legislation [37], crisis management ([1, 40, 45, 48, 56]: 5–54; [76]), coping mechanisms ([25]: 160–74; [34, 38, 52]), health system [61], and freedom of press [62]. The edited volumes by Kleinman and Watson [30], Koh et al. [32], and McLean [40] explore a wide variety of issues pertaining to the SARS epidemic.

Learning from SARS: Preparing for the Next Disease Outbreak [31] is an important work compiled of workshops on SARS with the support of the National Academies of Sciences, Engineering, and Medicine, summarizing implications of the SARS epidemic for the future, with Huang's piece ([22]: 116–136) offering a political perspective. Though Fidler [18] and Schwirian [49] investigate the implication of SARS on a global scale, Weber and Lu [70] scrutinize the interactions between SARS and youth as online civic participants.

On the tenth anniversary of the SARS epidemic, the *Journal of Thoracic Disease, Supplement* published a special issue on the "anti-SARS symposium," warning the public of the ongoing battle against infectious diseases ([20], 5: S94–S102) while analyzing chances missed and lessons learned for future references ([80], 5: S90–3).

Most of these pieces, however, focus either on the SARS epidemic or on the COVID-19 pandemic. This chapter, therefore, intends to connect the dots, bridging the gap between the crisis management with Chinese characteristic at the later stage of the SARS epidemic and the early stage of the COVID-19 pandemic. With secondary research materials on COVID-19 and personal experiences in the United States, this chapter will analyze the above-mentioned measures in light of primary resources gathered during the author's yearlong field research on the SARS epidemic between 2002 and 2003.

Simply put, crisis management for the SARS epidemic involved the detection and reporting of suspicious cases, isolation or quarantine, and treatment of confirmed cases (Fig. 4).

Connection to and comparison of the SARS epidemic to the COVID-19 pandemic in China, therefore, shed light upon the crisis management employed during the current COVID-19 pandemic.

Accordingly, this chapter focuses on the sociopolitical aspect of epidemic and/or pandemic control in the midst of a public health crisis rather than on the medical science spectrum.

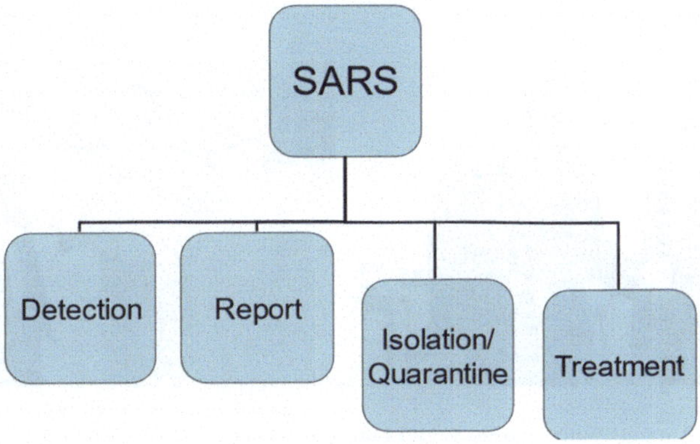

Fig. 4 Simplistic model of crisis management for SARS

2 Crisis Management at the National Level

With centralized top-down governance, China as a nation responded to the COVID-19 pandemic in ways similar to the approach taken during the SARS epidemic at its later stage, albeit COVID-19 presented a monumental health crisis worldwide greater in scale and complexity.

Thanks to experiences from the SARS epidemic 17 years earlier, the Chinese government responded efficiently and decisively to the emerging 2019-nCov cases, which is necessary for COVID-19 responses [47]. On 1 January 2020, 1 month after the first novel coronavirus infection occurred, the WHO was alerted and the Huanan Seafood Wholesale Market with most of the initial COVID-19 cases was closed [4]. *Protocol for Prevention and Control of COVID-19 (Edition 6)* was issued by the National Health Commission of the People's Republic of China [9] in February, providing timely guidance for virus containment in response to the public health crisis. Based on the newly issued *Law on the Prevention and Control of Infectious Diseases of the People's Republic of China* (mandated by the National People's Congress, passed in 1989 and updated in 2004 and later in 2013) and *Regulations on Preparedness for and Response to Emergent Public Health Hazards*, various state departments and governments at different levels came up with specific measures implementing the law and the regulations. Wang Qishan [64–66], Mayor of Beijing (2008–2013), consequently announced *Regulations on the Implementation of Prevention and Control Measures for Infectious SARS in Beijing* on 28 May 2023.

China was commended for quick action in the initial stage of COVID-19 [43]. While China responded to COVID-19 in less than a month, rather than 4 months later as it did with the SARS epidemic, the government carried over many of the strategies, including detection, quarantine, and treatment, from the previous experience, which was not in "stark contrast" to the approach to SARS. The similarities and improvements are discussed in the following sections.

Transparency and cooperation with international communities were valuable lessons learned from China's experiences during the SARS epidemic. Top leaders

Fig. 5 Xi Jinping meets with visiting WHO Director-General Tedros Adhanom Ghebreyesus, 29 January 2020 (left). Zhong Nanshan meets in Guangzhou with Walter Ian Lipkin who joins the COVID-19 battle, 30 January 2020 (right). (Sources: Xinhuanet and China Economic Net https://www.fmprc.gov.cn/mfa_eng/zxxx_662805/t1737014.shtml, https://news.ifeng.com/c/7v7mUOlBWQS. Accessed 19 September 2020)

were in touch with the WHO, while medical professionals worked together with experts from the international community (Fig. 5).

It is noteworthy that Dr. Zhong Nanshan, a Chinese pulmonologist, has been a national icon and medical authority for infectious disease control since the fight against the SARS epidemic. He frequently appeared on TV and in the news media, providing insights into SARS and later COVID-19 as a reliable source for the public [16, 17]. His contributions were instrumental in managing the SARS outbreak and consequently earned him national respect and an international reputation as a recipient of the Order of the Republic, the highest order of honor in China. He once served as president of the Chinese Medical Association (2005~2009) and editor-in-chief for *the Journal of Thoracic Disease* [26]. His work on the SARS epidemic [85] offers valuable advice for meeting the challenges of future emerging infectious diseases. Zhong's meeting with Dr. Walter Ian Lipkin at the beginning of the COVID-19 outbreak was significant, symbolizing joint efforts at the international level against the pandemic.

Government officials are visible in the public eye at the center of the crisis. Top leaders are instrumental in boosting the confidence of a nation battling the pandemic, creating a public image of being at the forefront of the "battlefield." Representing the State Council and the CCP central leadership, Premier (2003–2013) Wen Jiabao and Vice Premier Wu Yi made visits to construction sites, shopping centers, college campuses, and residential homes (Fig. 6).

By the same token, Li Keqiang, Governor (1998–2003) of Henan Province, made a research and investigation trip to Xuchang, together with a group of provincial officials, regarding the prevention and treatment of SARS and emphasizing central leadership determination to be "United as one, defeating SARS" [55, 64–66]. Large and imposing posters on the streets were designed to reinforce the spirit (Fig. 7).

The giant posters seen across the urban landscape were erected by government agencies and aimed at consoling and motivating the populace to be proactive by

Fig. 6 Premier Wen Jiabao and Vice Premier Wu Yi encouraged combat engineers (top left) and military medical providers (top right). They visited Peking University (bottom left) and the home of a retiree (bottom right). (Source: Xinhua News Agency http://img1.jiansuo.net/cms/upload/picture/1/1755.jpg. Accessed 9 April 2010)

acknowledging those who were fighting the disease. Messages in the posters, whose images made their way to the Internet, conveyed persistent optimism:

> **United** as one, defeating "SARS"
> SARS virus will collapse in the end; only the dignity of human being stands
> Utmost Salute to white-clothed angels who are fighting in the first front against "SARS"
> A doctor says, this "SARS" incidence is unprecedented. We've experienced the test of life and death. We have no choice except for gaining victory. Our only choice is to declare war against SARS, winning the battle of guarding our Capital.—top
> **Bless you**: Leaders and comrades on the forefront against SARS.—bottom

With leadership of China's central government, a series of protocols, regulations, and guidelines were published in response to the SARS epidemic. *Regulations on Contagious SARS Prevention and Management*, for instance, was issued by the Ministry of Health, PRC on 14 May 2003 (cf. [64–66]).

Bulletin boards have been an effective means of communication and information dissemination in China to draw the public attention to news, events, seasonal topics, or popular science. Figure 8 shows two administrative office workers updating the posting inside one of the weatherproof freestanding bulletin boards in front of the administrative building of the Jinggangshan Municipal Party Committee. The title line reads: "Strengthen the prevention of SARS," with information on SARS and "Health in Summer."

Fig. 7 Giant posters of the SARS epidemic on the street. (Source: People's Daily Online Accessed 5 October 2010)

The advantages of a centralized and top-down power structure in public health crisis management are best exemplified in the construction of Beijing Xiaotangshan Hospital (Fig. 9)—a health facility specializing in SARS treatment. The Beijing Xiaotangshan Hospital, comprising 508 rooms, 1000 beds, and 1200 medical staff, was constructed in 7 days and nights with support from all over China, treating SARS patients as of 30 April 2003 [21, 39]. With the direct leadership of the party central working closely with the State Council, resources and manpower of the entire country were mobilized. Medicine and medical supplies were gathered extensively from all provinces, while doctors and nurses were temporarily recruited from different regions (Fig. 10).

Fig. 8 Bulletin boards in Jinggangshan, Jiangxi. (Photograph by the Author, April 2003)

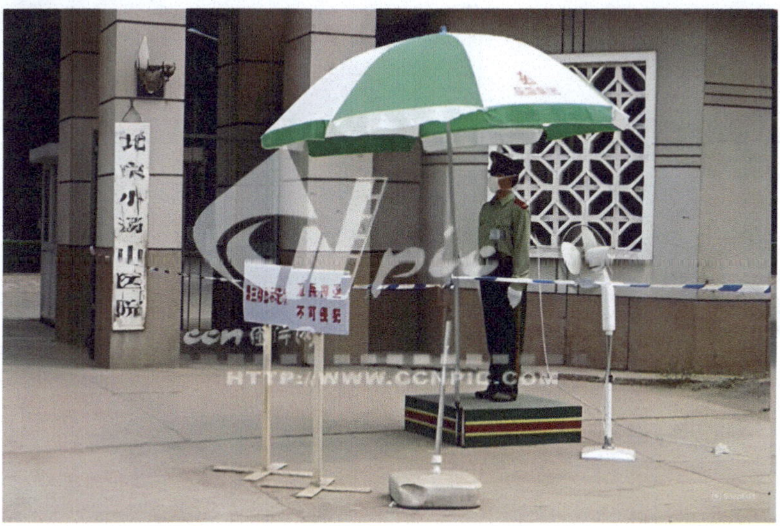

Fig. 9 Beijing Xiaotangshan Hospital. (Source: CCNPIC http://img.hexun.com/2008-09-12/108745188.jpg. Accessed 10 April 2010)

The Xiaotangshan Hospital re-opened in February 2020 as a designated hospital for screening international arrivals during the COVID-19 pandemic [39].

While isolation and testing for confirmation are important in controlling spread of the coronavirus, treatment is crucial to lower fatality and contain the epidemic.

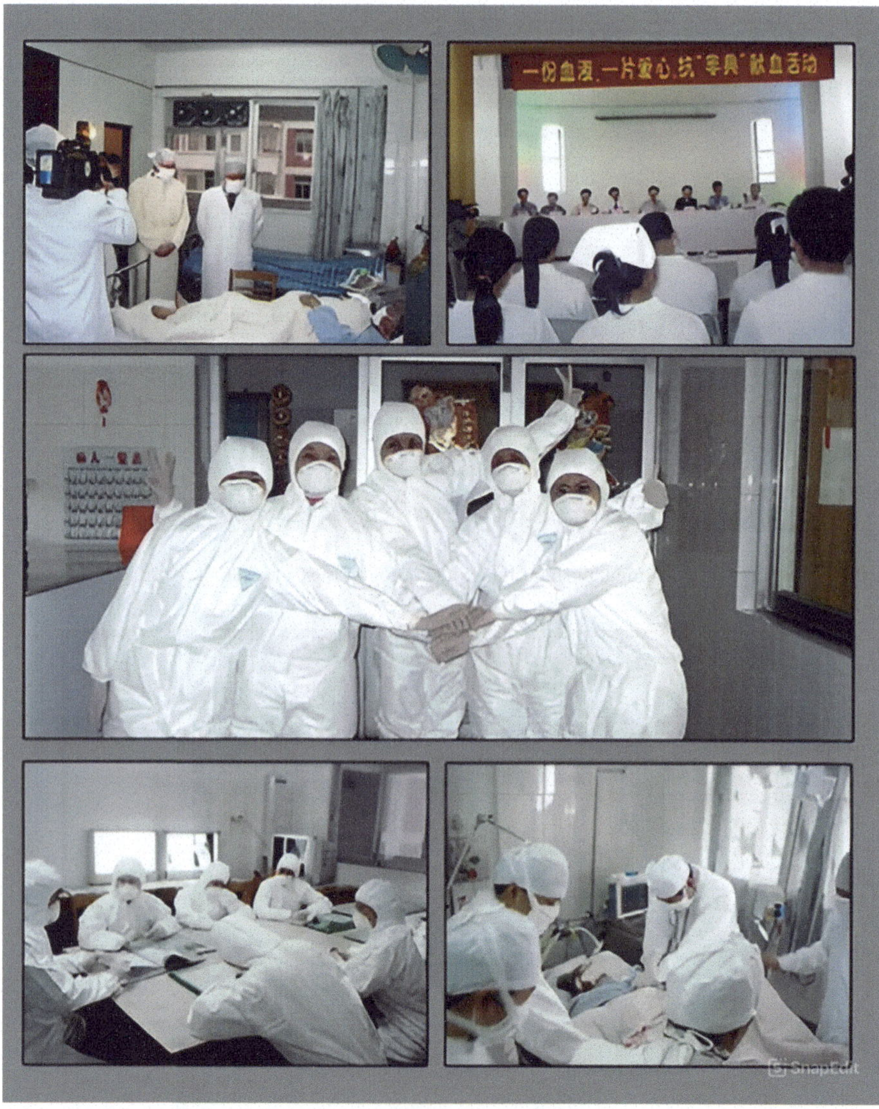

Fig. 10 Recruitment and motivation of medical professionals. (Source: Dayoo http://images.dayoo.com/gov/59476/59486/img/attachement/jpg/site1/20080912/001372af7ade0a3446f305.jpg. Accessed 10 April 2010)

In summary, top Chinese leadership played a central role in the battle against SARS and the COVID-19 pandemic, with support from the international community [73], issuing urgent protocols and regulations, providing administrative guidance, and coordinating healthcare resource delivery nationwide.

3 Crisis Management at the Provincial or Municipal Level

With direct leadership of the central government, the Chinese provincial or municipal government unifies the upper and lower levels of governance to facilitate implementation of central government policies. There are 22 provinces and four municipalities (Beijing, Chongqing, Shanghai, and Tianjin) in addition to five Autonomous Regions (Inner Mongolia, Guangxi, Tibet, Ningxia, and Xinjiang) and two Special Administrative Regions (SAR) (Hong Kong and Macau). The provincial government is usually located in the provincial capital city, which also applies to the five Autonomous Regions. The Hubei provincial government and Wuhan city government, for instance, are both located in the city of Wuhan. The provincial government supervises all city governments within Hubei, implementing the newly issued urgent public healthcare regulations during the SARS and COVID-19 outbreaks.

Detection and isolation seem to be an effective administrative measure for controlling the spread of viruses. During the SARS epidemic between 2002 and 2003, in contrast, isolation or quarantine was at a much smaller scale, usually for the unit of a family, a few hospital wards, or a suite in an apartment building.

During the COVID-19 pandemic, however, lockdown was administered in several cities in China. With a strong command system, Wuhan was locked down at the onset of the pandemic, 23 January 2020 [14, 84]. Thus were fourteen other cities in Hubei Province [41]. Travel bans or restrictions prevented outbreaks initially. For the first time in Chinese history, Wuhan was locked down for 76 days, until 8 April 2020 [69].

Following the lockdown of Wuhan and fourteen other cities in Hubei, authorities administered strict lockdown measures, from 23 December 2021 to 24 January 2022, in Xi'an [68, 81]. Shanghai was locked down for 2 months in 2022 due to a new wave of COVID-19 confirmed cases [24, 83]. Other major cities locked down due to COVID-19 include Chengdu and Beijing, later in 2022.

While lockdown was not necessary for SARS in 2003, roadblocks and screening stations were set up to identify potential patient "suspects" in a timely fashion. The expedient measure of disease control was enacted by the government with synchronized efforts. An outdoor checkpoint, for instance, was placed in Jiezidun, a small town bordering Hubei and Jiangxi Provinces (Fig. 11, left). All passengers, without exception, were required to go through screening, day or night. Specially appointed doctors or healthcare professionals from a local Epidemic Prevention Station took temperatures and related vital signs of each passenger on buses (Fig. 11, right). Travelers with fever or elevated temperatures were retained in isolation at a

Fig. 11 Checkpoint and screening along provincial borders. (Photograph by the Author, June 2003)

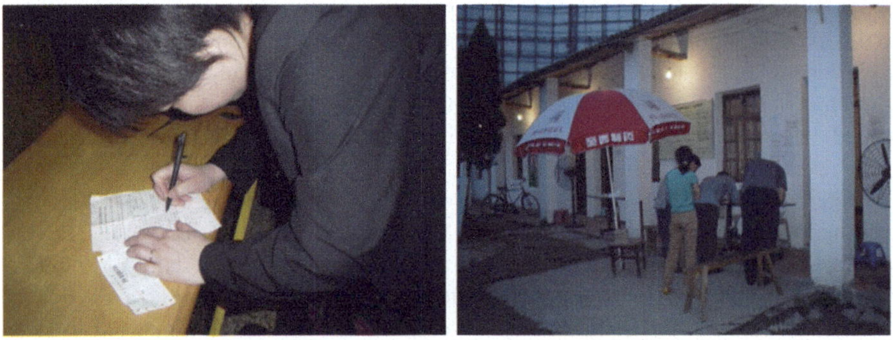

Fig. 12 Registration in the Nanjing Bus Station (Jiangsu) upon departure (left). Registration in Nanchang (Jiangxi) at a temporary stop in the suburbs prior to entering a bus station (right) in the city. (Photograph by the Author, June 2003)

designated facility operated by the government and observed for up to 14 days and treated, if necessary, before being discharged.

Temperature screening during the SARS epidemic was deemed an effective strategy for cross-infection prevention. It was also done diligently upon entering a station, before boarding, and after arrival. Extra screening was required for interprovincial travelers, as shown in the above photographs.

Apart from screening at checkpoints between cities or provinces, tracing was another important strategy used to prevent cross-infection during the peak of the SARS epidemic. A two-way system of tracing was administered, both at departure and on arrival. That is, a passenger had to register at the original station and do so again upon arrival at the destination (Fig. 12), leaving identity, contact information, and travel information from the ticket; such personal data proved to be helpful in tracking travelers who had been in close contact with infected cases.

The margin of error can be disruptive for this type of tracing, that is, mandatory registration. An unintentional error may be excused, but some were intentional.

Based on the author's observation, people were usually cooperative for temperature checks while traveling. Passenger registration, however, was not, for some individuals. A middle school teacher in Hubei, a former classmate, for example, shared that he deliberately gave inaccurate contact information—either spelling of his name or his phone number—at passenger information collection points. His rationale was that he was uncomfortable disclosing personal information. "I'm not traveling far, and Hubei is not a SARS infected area. Collecting passenger information is just one of many bureaucratic practices; it's useless," he reasoned. According to him, he was not alone in providing inaccurate contact information. This example implies temperature measurements and passenger registration are best used as a means of governance for preliminary SARS screening, complemented with other methods for effective disease control and prevention. Health code, an improvement in information collection for tracing in the COVID-19 era of smart phones, is discussed in the following section.

Registration, as well as temperature screening, was mandatory if one wished to travel. Temporary booths or tents were established for this purpose as illustrated in the plaza outside Nanchang Railway Station (Fig. 13, right), a common scene at a public transportation space during the SARS epidemic.

Provincial and municipal governments, however, were committed to informing travelers of the significance of tracing by registration. Upon arrival, for example, temperatures were taken again at the SARS Prevention and Treatment Headquarters on the outskirts of Nanchang (Fig. 12, right). A full-sized bulletin board was prominent, announcing temporary regulations issued by Nanchang SARS Prevention and Treatment Headquarters (Fig. 13, left):

1. Those who had contact with SARS patients or suspects have to be retained for 14 days, one person per room.
2. All who had contact with SARS suspects have to be retained until the day after the release of the suspect(s).
3. All who come (return) to Nanchang from areas with relatively more SARS patients have to be retained for 7 days.

Fig. 13 Bulletin board for temporary regulation announcement at the outskirts of Nanchang (left). Plaza in front of Nanchang Railway Station, Jiangxi (right). (Photograph by the Author, June 2003)

4. Apart from the above-mentioned persons, all who come (return) to Nanchang have to be retained for 3 days.
5. Retainers from executive, enterprising, and institutional work-units in Nanchang area will be in the care of their respective work-units. All others will be in the care of local county (district), village (town, farm, street), and village (residential) committee(s).

Evidently, these SARS Prevention and Treatment Headquarters were temporary crisis management institutions authorized by the (central, provincial, or municipal) government, forming the frontline for battling SARS. The above-mentioned announcement indicated how institutional governance was backed by work-units, villages, and residential committees at the local level. A network or disease screening and control were thus closely knitted, uniting the most critical elements in the war on SARS. Details of how local communities complemented provincial and municipal government efforts in detection and prevention are provided in the following chapter.

Large-scale lockdown may have effectively contained the spread of COVID-19, but treatment of confirmed cases represents the other side of the coin for crisis management. In Wuhan, two field hospitals, *Leishenshan* and *Huoshenshan*, modeled after the Beijing *Xiaotangshan* Hospital, were swiftly established in early 2020 (Fig. 14). This time, the two hospitals were built from scratch, with up to 7000 people working around the clock in three shifts, also in record time—10 days. *Huoshenshan*, literally meaning the Vulcan Mountain, and *Leishenshan*, the Thunder Mountain—both related to mythological gods of fire and thunder—imply

Fig. 14 Hospital construction site on 23 January 2020 (top left) and Huoshenshan Hospital between 3 February and 15 April 2020 (top right). Hospital construction site on 25 January 2020 (bottom left) and Leishenshan Hospital between 8 February and 15 April 2020 (bottom right). (Source: https://en.wikipedia.org/wiki/Huoshenshan_Hospital, https://en.wikipedia.org/wiki/Leishenshan_Hospital. Accessed 20 September 2021)

the power of fire overcoming lung disease from infection with the coronavirus. There were 1000 beds over 60,000 m^2 on 2 floors, 30 intensive care units, medical equipment rooms, and quarantine wards, treating severe patients at Huoshenshan Hospital and ca. 1500 beds for Leishenshan. Again, administrative and financial power at the national level was remarkable. The National Development and Reform Commission allocated ¥300 million *Yuan* on 27 January 2020, and materials worth ¥60.28 million *Yuan* were donated by the State Grid Corporation of China on the same day.

In addition to the two field hospitals, the municipal government of Wuhan also began establishing makeshift hospitals, some of which were named *Fangcang* (meaning mobile cabin) Shelter Hospitals [8, 13, 51, 79]. The concept of *Fangcang* hospital originated from and applied in battlefields was used in the fight against COVID-19. Some of these hospitals include ones converted from conference centers, sports facilities such as gymnasiums, and exhibition centers [42, 53, 54, 67]. Three of the *Fangcang* hospitals were set up as early as 3 February 2020 providing a total of 3400 beds, initially for patients suffering from mild symptoms and then admitting all SARS patients 2 days later. A total of 16 such makeshift hospitals were built in Wuhan with support from all over China as of 10 March 2020 and treated a total of over 12,000 COVID-19 patients [35].

Without facilitation or coordination of provincial or municipal governments, implementation of the central government policies would have been impossible. It would have been very difficult to rapidly establish large hospitals with the capacity to isolate and treat all coronavirus patients, consequently lowering infection and fatality rates.

4 Crisis Management at the Local Level

Local governments, institutions, businesses, and residential committees were responsible for motivating individuals, making sure everyone complied with the urgent public healthcare regulations without exception.

To put into practice the protocols, guidelines and regulations for SARS prevention and treatment issued by central and provincial authorities, local governments, and institutes organized study sessions, discussions, and meetings. Fig. 15 shows a notice issued on 30 April 2003 as one example of many meetings conveying and implementing decisions concerning SARS prevention and treatment made by upper-tier agencies under the administration of central and provincial governments.

Horizontal red banners have been characteristic of the Chinese popular means of propaganda to publicize and promote government policies. One at the entrance gate of Jiangxi Revolutionary Martyrs' Memorial Hall (JRMMH) in Nanchang, the provincial capital of Jiangxi, reads: "United as one in fighting SARS" (Fig. 16, left) and one above a hotpot restaurant states (Fig. 16, right): "Informants of SARS cases are heroes." Both banners echo the government's endeavors in fighting the epidemic and the national policy to report and trace infected cases.

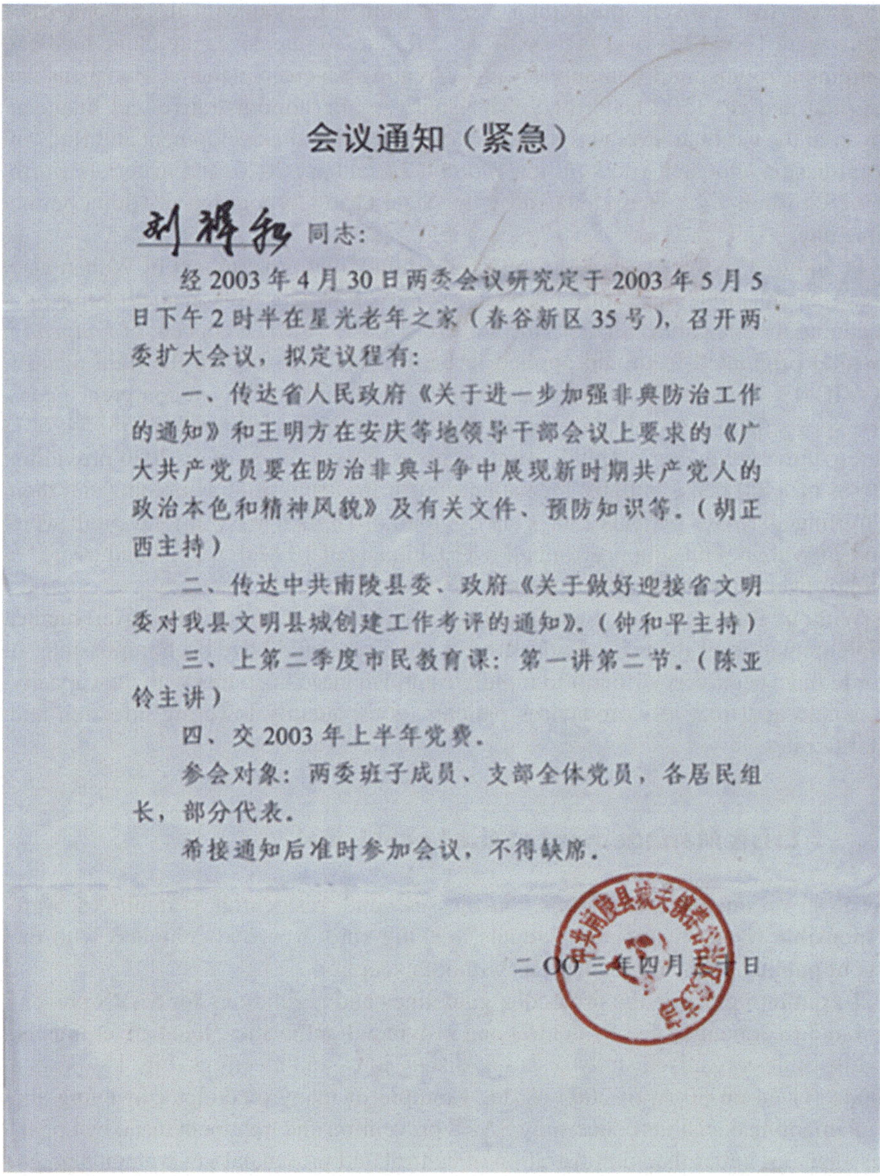

会议通知（紧急）

刘祥和 同志:

经2003年4月30日两委会议研究定于2003年5月5日下午2时半在星光老年之家（春谷新区35号），召开两委扩大会议，拟定议程有:

一、传达省人民政府《关于进一步加强非典防治工作的通知》和王明方在安庆等地领导干部会议上要求的《广大共产党员要在防治非典斗争中展现新时期共产党人的政治本色和精神风貌》及有关文件、预防知识等。（胡正西主持）

二、传达中共南陵县委、政府《关于做好迎接省文明委对我县文明县城创建工作考评的通知》。（钟和平主持）

三、上第二季度市民教育课：第一讲第二节。（陈亚铃主讲）

四、交2003年上半年党费。

参会对象：两委班子成员、支部全体党员，各居民组长，部分代表。

希接通知后准时参加会议，不得缺席。

二OO三年四月三十日

Fig. 15 Notice for (urgent) meeting issued by a local community CCP party branch. (Source: http://www.ahnllxhzs.com/wz/5052/%E6%8A%97%E5%87%BB%E9%9D%9E%E5%85%B8%E7%B4%A7%E6%80%A5%E9%80%9A%E7%9F%A5.jpg. Accessed 3 September 2009)

Fig. 16 JRMMH and a hotpot restaurant in Nanchang, Jiangxi. (Photograph by the Author, May 2003)

Fig. 17 Jiangxi University of Finance and Economics in Nanchang, Jiangxi. (Photograph by the Author, May 2003)

In general, local communities and organizations, including schools, work-units, residential committees, hospitals, and pharmacies were cooperative with government agencies during both the COVID-19 and SARS outbreaks. The Jiangxi University of Finance and Economics, for instance, posted *Entry and Exit Instructions* at its entrance gate on 6 May 2003, reiterating policies and regulations for SARS prevention and treatment issued by national as well as local governments (Fig. 17).

A university professor in Nanchang shared his view of China's ability to efficiently mobilize the entire nation in its fight against SARS, claiming that success was attributable to China's highly centralized form of government. The centralized system, in his opinion, was key to effective and successful control of disease diffusion. The individual was motivated to take action, forming a grassroots network of monitoring to prevent disease. This systematic approach represented the strength of what Mao Zedong termed *renmin zhanzheng de haiyang* (the sea of war by the people). Vietnam, also with a highly centralized system of governance, was much

Fig. 18 A guarded gate in Nanchang, Jiangxi. (Photograph by the Author, June 2003)

more effective than Canada, for example, in getting the pandemic under control, albeit having much inferior health care systems and less resources ([19]: 28).

The assumption of the professor from Nanchang is correct in that a network for disease prevention can be effective. Within such a grassroots network, volunteers are motivated at the personal level. In 2003, the author witnessed many instances of local communities or organizations fighting SARS in Jiangxi. Figure 18 shows one such gated community in Nanchang. At the entrance of a residential complex, a woman volunteer was stationed to guard the complex against possible SARS virus carriers. Part of the woman's responsibility, in her role as entrance guard, was to register visitors and to watch for suspicious persons attempting to gain entry into the community.

As claimed by the red-cloth horizontal banner above the gate between the two buildings: "All returnees to Nanchang must report to SARS prevention institutions, or they will be held accountable and face serious consequences." The sign on the table, which includes the contact phone number for both the SARS Prevention District Office and local Street Office, requests: "For the health of others and the family, visitors please voluntarily register." As a resident, the guard on duty is usually familiar with all residents who are either neighbors or longtime colleagues from the same work-unit. A checkpoint at the gate screened SARS "suspects," keeping a record of contact information of the visitors.

The scene in the picture above is one example of the many gates at a typical work-unit or walled community where "insiders" were organized for self-protection to check the spread of the disease from infected "outsiders." Such measures of grassroots governance effectively protected guarded communities from being affected by non-residents. In the meantime, visitor registration made it much easier and more efficient to trace infectors as soon as they were detected.

The work-unit played an important role in facilitating the individual as an integral part of the fight against SARS in urban areas, an influence made possible because members of a typical work-unit tend to live in the same apartment complex. Up until the end of the century, most clusters of apartment buildings had been built by a work-unit that was, in turn, responsible for the social welfare of its employees. A relative and a classmate of the author residing in Beijing recalled that their work-units handed them herbal medicine and incense sticks, respectively, against the SARS virus, in addition to providing information on preventative action. Measures such as these were the norm for numerous work-units during the SARS outbreak.

Vigilance at a local level for a gated community—a work-unit—is embedded in the efforts to discipline members with respect to wearing masks. While wearing masks was recommended during the SARS epidemic, mass masking was mandatory during the COVID-19 pandemic in China. Those going outside of their homes to public spaces, whether inside or outside, had to wear a mask. Even non-healthcare facilities including supermarkets, for example, would not allow customers without masks to enter. When a man (Fig. 19) was caught not wearing a mask outside, he was escorted home by a team of security guards stationed at a checkpoint near the gate of a residential community.

Due to the severity and increasing prevalence of COVID-19, isolation as a response measure was implemented in far more stringent ways for quarantine or self-quarantine, particularly during lockdowns, than was the case during the SARS outbreak. Internet writers published a myriad of slogans in the format of couplets based on government policies or regulations. Volunteers broadcast slogans with loudspeakers around residential communities or villages. Apart from horizontal or vertical banners, some of the slogans promoting isolation were painted directly on the wall of a house or public building (Fig. 20).

The bottom two lines of Chinese characters painted on the wall of this independent village house, *jintian daochu luan pao, mingnian fenshang zhang cao*, mean "Running around today everywhere, and grass will grow on your grave next year." WeChat, the most popular social media platform in China, popularized the image and message, discouraging residents from leaving their homes. In this era of smart phones, countless similar slogans created by Internet writers have been popularized by various social media in the form of texts, images, or video clips.

WeChat, on the other hand, provided a virtual venue for information exchange and emotional support at the community as well as individual level. People in isolation or quarantine rely on WeChat to conduct business, shop, exchange information, seek help, or socialize during the lockdown. WeChat groups formed based on residence, with people from all walks of life, including group leaders—sometimes volunteers or government officials were appointed to be in charge of volunteers representing local government agencies, while serving local communities. Members ordered food items, PPE (personal protective equipment), and other necessities (Fig. 21) as a group, for instance, which were delivered in bulk to group leaders or volunteers who, in turn, helped with the distribution.

The author's family members residing in a medium-sized city in Hubei received free vegetables, rice, and noodles several times that were delivered by volunteers

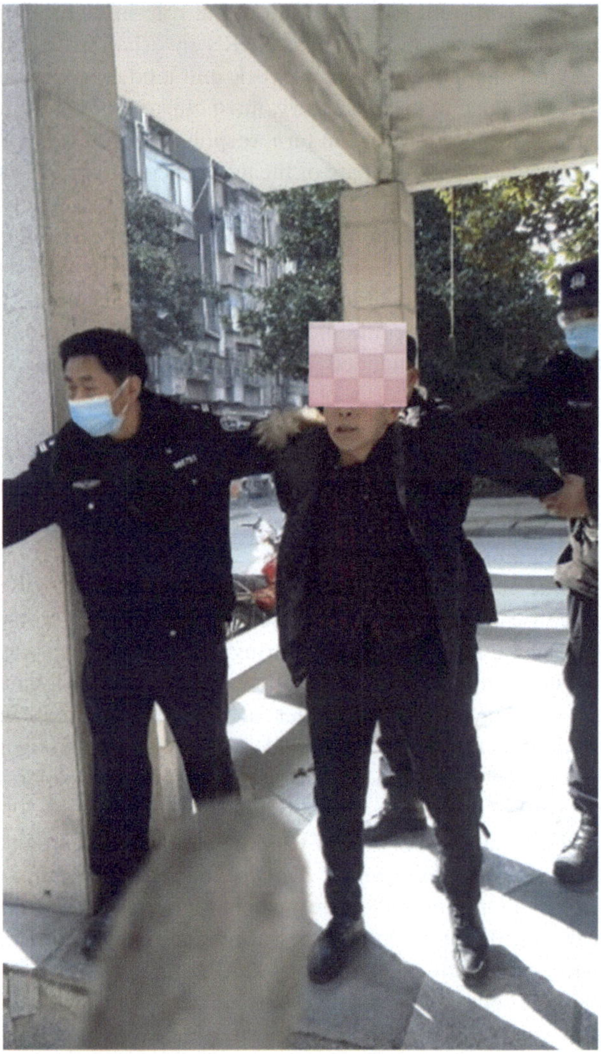

Fig. 19 Residential complex, Zhijiang, Hubei. (Source: WeChat group posting, October 2020)

during the first lockdown in 2020. The city government set aside special funds and ordered food items from local businesses, grocery stores, and vegetable farmers. Civil servants were responsible for distribution in one neighborhood or block at a time. During the COVID-19 lockdown period, government officials, civil servants, administrators, and clerks served as volunteers. Many were assigned either in charge of temperature checks from home to home, on duty at checkpoints, or distributed PPE and food items.

The zero-COVID policy followed by the Chinese government necessitated the promotion of mass vaccination and mass nucleic acid amplification testing (RT-PCR) or nucleic acid testing. The policy emerged in late 2021 in response to the highly

Fig. 20 Slogans painted on the wall of a village house. (Source: WeChat group posting, October 2020)

transmissible Delta and Omicron variants. Officials recommended mass vaccination should follow the zero-COVID policy [82]. Incentives were given particularly to citizens taking the second or third booster of the Sinovac COVID-19 Vaccine—CoronaVac [74]. Different from the vaccines—Pfizer-BioNTech and Moderna made in the United States—based on messenger RNA (mRNA) technology, the Sinovac COVID 19 vaccine made in China was based on the traditional inactivated pathogen technology [2, 10]. While the efficacy of Pfizer and Moderna reached up to 96% and 94%, respectively, efficacy of CoronaVac was only up to 51%. That is probably part of the reason why the Chinese government had to appeal to lockdowns between late 2022 and early 2023.

Nucleic acid testing provided another option for early detection and isolation for the Chinese government. Nucleic acid testing represented a step forward from temperature screening, since the latter detected symptomatic COVID-19 cases, that is, those with symptoms, but not effective for identifying asymptomatic COVID-19 infections. One of the specific actions taken to follow the zero-COVID strategy was, therefore, to administer mass nucleic acid testing. Nucleic acid testing centers were established in many neighborhoods or communities. Residents waited in long lines to take the test. For a period of time, people had to repeat the test every 2 days. Everyone, including returnees and travelers, were required to take the test. Visitors from outside China had to provide a negative nucleic acid test report within 48 h up to 29 April 2023. It was an exhausting process and expensive for the average person. Yet, it did not prevent the outbreak that occurred in January 2023.

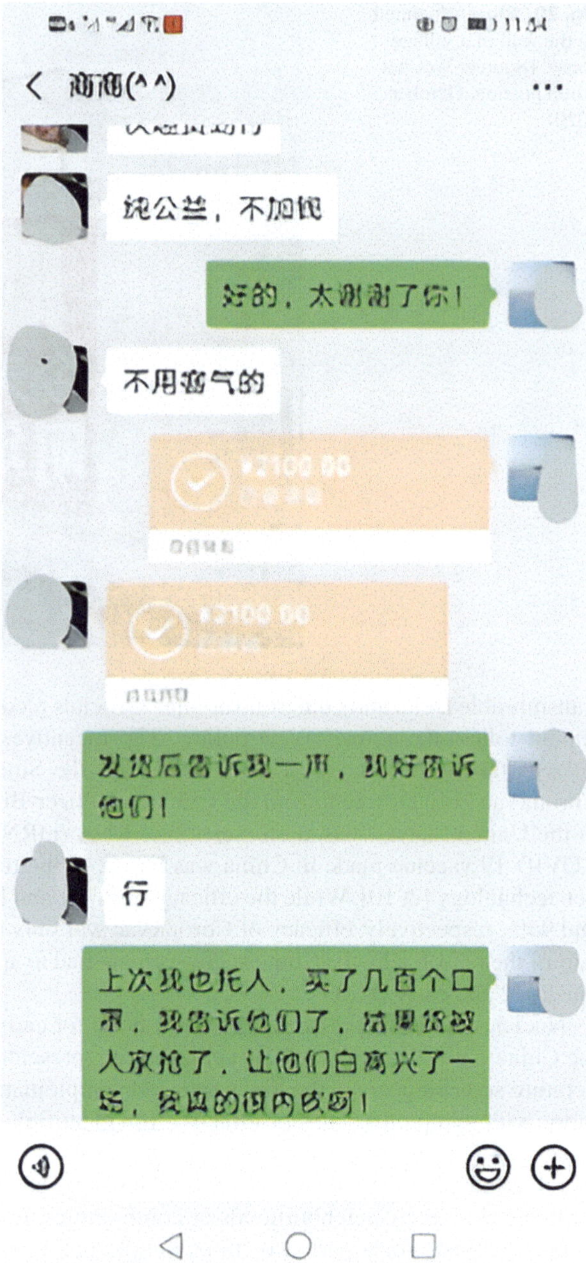

Fig. 21 WeChat transaction of masks. (Source: WeChat group posting, c. 2020)

Health code, also known as the green code, is a smartphone application for health information management. The phone application was applied initially in Shenzhen in February 2020 and adopted by over 200 cities in China by April that year. In terms of bolstering prevention and control efforts, the health code smart phone application represented a significant development for the tracing approach to the COVID pandemic. In general, a green QR code with no travel restrictions would be issued to the smartphone owner if the required coronavirus screening results of the person were standard. A yellow QR code with some travel restriction and surveillance would be issued if some of the coronavirus screening results of the person were not up to the standard. A red QR code would be issued for travel ban or quarantine if the coronavirus screening results of the person did not meet the requirement or were missing. Figure 22 presents a few health codes, all green, from Tianjin, Hubei, and Nanjing, each slightly different.

The health code is a much more timely, dynamic, convenient, and accurate tracker than the mandatory registration form manually filled out at each transportation facility.

By the same token, advancements of AI and the Internet brought the preliminary COVID-19 response to a new height compared to SARS control strategies. Virtual visits and online video appointments were possible with the sophisticated digital technology, eliminating direct contact while providing quality individualized medical service during the pandemic in China. WeDoctor® (Fig. 23), for instance, established lifetime health databases, medical record management, and health management [63]:

Taking advantage of the online hospital system, WeDoctor® treated patients with chronic diseases in Wuhan to prevent cross-infection during the pandemic, delivering medicine and providing psychological consultation for those who stayed home. There were ten million users of the online system per day at peak. There were 6000 online doctors selected from AAA hospitals through the request of the Chinese government to help overseas Chinese, particularly Chinese students, with one million users per day at peak. WeDoctor® also held health seminars to educate the public on COVID-19 [71].

While WeDoctor® offers a platform of professional medical care on the users' existing devices for convenient and timely care during a time of urgent need or public health crisis, intelligent hospitals are available with purchase of a mini robot. Roby-mini®, a home care robot system, is capable of providing services a hospital may be able to promise. Roby-mini® is applicable in a public space to help (potential) patients in a public space. To name a few functions, Roby-mini® is useful for the following:

- Front desk: face recognition and mask recognition at the entrance or exit for information confirmation.
- Temperature screening: real-time detection, collection, and recording of temperature in a public space with busy flow of passers-by.

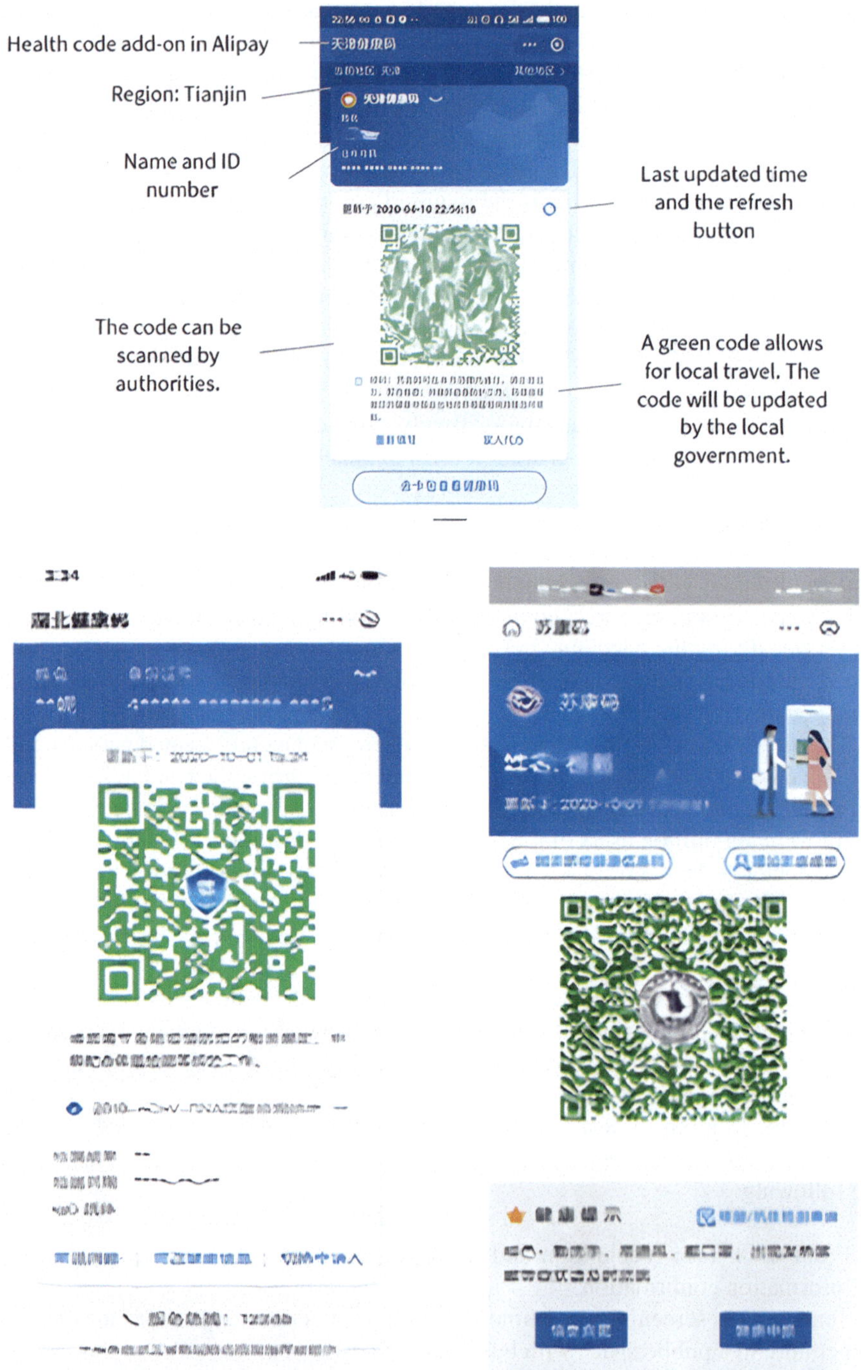

Fig. 22 Health code from Tianjin (top). (Source: [57] Health code from Hubei (bottom left) and Jiangsu (bottom right). Source: Screenshots provided by friends of the author in 2020)

Fig. 23 WeDoctor interface on digital devices. (Source: WeDoctor)

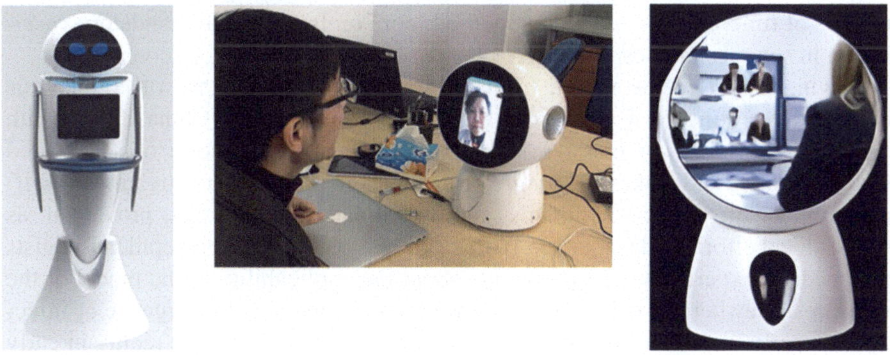

Fig. 24 Cybercare medicine: Roby-mini medical care robots. (Resource: Shanghai PT Info Co., Ltd)

- Intelligent warning: warning signals for people with and abnormal body temperature to notify relevant personnel.
- Remote assistance: saving manpower and avoiding cross-infection (Fig. 24).

On the one hand, the pandemic gave rise to cloud medicine, but on the other hand, digital technology holds a pivotal key to future healthcare and safety, particularly in the midst of a pandemic when in-person contact becomes problematic.

5 Concluding Remarks

Several years into the combat against the COVID-19 pandemic, almost all nations of the world have had confirmed cases and fatalities, varying in number and spikes. Given its unique form of governance, China responded differently to this public health crisis, particularly when compared to the United States. China as a nation successfully contained the current pandemic early on and lowered the death rate in the early stages, doing so in part as a result of lessons learned from the SARS epidemic.

Research and analysis indicate that the centralized, top-down model of governance for crisis management was effective in preventing and confining the spread of the SARS epidemic in its later stage in China. Ethnographic field research on SARS shows that governance, at all levels, from the national, provincial, or municipal to the local and the individual, was motivated to work in unison against the SARS crisis at its later stage. China's centralized power was effective in maneuvering manpower and resources to treat patients relying upon national support and eventually conquering the epidemic. Lessons learned from the SARS crisis have had a far-reaching impact on China's public health crisis management, shedding light on approaches used to contain H5N1 flu in China in 2005, H1N1 flu in 2009, H7N9 flu in 2013, and currently on controlling the COVID-19 pandemic.

In comparison to the United States, China administered strict lockdown in a few large cities including Wuhan, Xi'an, Guangzhou, and Shanghai for shorter or longer periods of time. The United States focused more on freedom but mandatory vaccination in schools and hospital where clusters of people are in close contact.

Having surprised the world, China contained the initial wave of coronavirus infection, resulting in a rapid plummet of confirmed cases. China did so with "extreme measures" of massive lockdown of cities; construction of specialized medical centers and makeshift *Fangcang* hospitals for concentrated treatment; (self) quarantine with public surveillance; and coordination of national, provincial, as well as local efforts and resources. With its unique political system, epidemiologists and medical experts had serious doubts about the applicability of the model or the measures elsewhere in the world. They were concerned about resurgence of infections as early as 2020 [33]. Their concern corresponded to China's reality in early 2023. Nonetheless, the health crisis management approach with Chinese characteristics achieved its goal of efficiently containing spread, and thus saved thousands of lives during the pandemic.

The COVID-19 virus has undergone a series of mutations, from the Delta variant to the Omicron, to the XBB, and other subvariants [29]. It is not certain what the future holds for the world and we are not out of the woods yet. We have to be prepared to live with COVID-19 with its future variants and subvariants, just as some scientists warned us [50].

Currently, China and the United States still have substantial numbers of active COVID-19 cases (Tables 1 and 2). China faces the challenges of "repeated pandemic waves, increased downward pressure on the economy and social exhaustion" [11].

We have paid expensive tolls against COVID-19 at the cost of millions of lives, financial loss, economic downturn, and emotional challenges. We fought the battles, and we are still standing! We have learned good and bad lessons, sometimes the hard way. Is one approach of response to public health crisis more efficient than the other? Is one model of governance more advantageous than the other? A definitive answer depends on the historical, political, socioeconomic, geographical, and cultural situation of the nation or the state. The most suitable model of governance is the best. How do we know? Connection and comparison help identify patterns and learn better. The global community needs to work together, to learn from each other, to help each other, and to come to terms with one's comparative advantage, rather than lamenting the globalization of infectious disease. There is no point of return.

The future lies in a united effort.

Acknowledgments Thanks to the National Science Foundation (award # BCS-0202018) for funding the author's year-long dissertation field research in China beginning in late 2002 and for a Developing Areas Research and Teaching (DART) pre-dissertation fellowship at the University of Colorado at Boulder for resumed field research during the summer of 2004.

The initial field trip enabled the author to obtain first-hand experience and research materials regarding the SARS crisis between 2002 and 2003, without which this chapter could not have been written. Assistance from institutes and individuals in China, including families and friends, was indispensable.

Thanks to the author's soulmate David Rioux for being the first reader and the best cheerleader and to the author's doctoral advisor, Dr. Tim Oakes, for his valuable guidance and assistance.

Special thanks to Dr. Joseph Rosen for inviting the author to guest lecture year after year, from SARS to COVID-19. Yearly lectures and discussions with him and his Dartmouth College students have kept the author's research relevant and up to date. Working with Dr. Rosen in potential ways to help with COVID-19 relief proves rewarding.

Cross References

Johns Hopkins University Coronavirus Resource Center. https://coronavirus.jhu.edu/data/new-cases

Parish WL, Whyte MK (1978) Village and family in contemporary China. University of Chicago Press, Chicago

Parish WL, Whyte MK (1984) Urban life in contemporary China. University of Chicago Press, Chicago

SARS Expert Committee (2003). Report. https://www.sars-expertcom.gov.hk/english/reports/reports/reports_fullrpt.html

WHO Coronavirus (COVID-19) Dashboard. https://covid19.who.int/

References

1. Ahmad A, Krumkamp R, Reintjes R (2009) Controlling SARS: a review on China's response compared with other SARS-affected countries. Trop Med Int Health Suppl 1(14):36–45
2. Asundi A, O'Leary C, Bhadelia N (2021) Global COVID-19 vaccine inequity: the scope, the impact, and the challenges. Cell Host Microbe 29(7):1036–1039. https://doi.org/10.1016/j.chom.2021.06.007
3. Bardhan P (2020) The Chinese governance system: its strengths and weaknesses in a comparative development perspective. China Econ Rev 61:101430. https://www.ncbi.nlm.nih.gov/pmc/articles/PMC7147895/
4. CDC (2020) COVID data tracker. Centers for Disease Control and Prevention. https://covid.cdc.gov/covid-data-tracker. Accessed 2 Aug 2022
5. CDC Control and Prevention (2023) CDC museum COVID-19 timeline. https://www.cdc.gov/museum/timeline/covid19.html. Accessed 3 June 2023
6. Centers for Disease Control and Prevention (CDC) (2013). CDC SARS response timeline. https://www.cdc.gov/about/history/sars/timeline.htm. Accessed 6 Aug 2022 and in June of 2023
7. Chan-Yeung M, Xu R-H (2003) SARS: epidemiology. Respirology 8(Suppl 1):S9–S14. https://www.ncbi.nlm.nih.gov/pmc/articles/PMC7169193/#b3

8. Chen S, Zhang Z, Yang J, Wang J, Zhai X, Bärnighausen T, Wang C (2020) Fangcang shelter hospitals: a novel concept for responding to public health emergencies. Lancet 395(10232):1305–1314. https://www.thelancet.com/journals/lancet/article/PIIS0140-6736(20)30744-3/fulltext. Accessed 5 Oct 2020
9. China CDC Weekly (2020) Protocol for prevention and control of COVID-19 (Edition 6). National Health Commission of the People's Republic of China. https://doi.org/10.46234/ccdcw2020.082
10. Corum J, Zimmer C (2021) How the Sinovac vaccine works. The New York Times, 5/7/2021. https://www.nytimes.com/interactive/2020/health/sinovac-covid-19-vaccine.html. Accessed 15 Sept 2021
11. Ding D, Zhang R (2022) China's COVID-19 control strategy and its impact on the global pandemic. Front Public Health 10:857003. https://doi.org/10.3389/fpubh.2022.857003. Accessed 15 June 2023
12. Editorial (2020) 10 key phrases in China's fight against COVID-19. China Daily, 4/23/2020. https://covid-19.chinadaily.com.cn/a/202004/23/WS5ea16007a3105d50a3d18668_1.html. Accessed 15 Sept 2020
13. Editorial, Xinhua News Agency (2020) Wuhan Rushes to Build "Fangcang Hospitals." 新华社。《武汉抢建"方舱医院"》。http://www.gov.cn/xinwen/2020-02/04/content_5474534.htm#1. Accessed 10 Sept 2020
14. Engle J (2020) Lesson of the day: 'To Tame Coronavirus, Mao-Style Social Control Blankets China' The New York Times, Februay 19. https://www.nytimes.com/2020/02/19/learning/lesson-of-the-day-to-tame-coronavirus-mao-style-social-control-blankets-china.html. Accessed 15 Mar 2020
15. Fang L-Q, Vlas D, Sake J, Feng D, Liang S, Xu Y-F, Zhou J-P, Richardus JH, Cao W-C (2009) Geographical spread of SARS in mainland China. Trop Med Int Health 14(Suppl 1):14–20. https://www.ncbi.nlm.nih.gov/pmc/articles/PMC7169839/
16. Feng E (2020) Meet Dr. Zhong Nanshan, the public face of the COVID-19 fight in China. NPR, 15 April. https://www.npr.org/2020/04/15/835308147/meet-dr-zhong-nanshan-the-public-face-of-the-covid-19-fight-in-china. Accessed 14 Sept 2021
17. Feng E, Cheng A (2020) They call him a hero: Dr. Zhong is the public face of China's War against coronavirus. NPR, 2 April. https://www.npr.org/sections/goatsandsoda/2020/04/02/825957192/dr-zhong-is-the-supreme-commander-in-china-s-war-against-coronavirus. Accessed 14 Sept 2021
18. Fidler DP (2004) SARS, governance, and the globalization of disease. Palgrave Macmillan, New York
19. Grigg E, Rosen J, Koop E (2006) The biological disaster challenge: why we are least prepared for the most devastating threat and what we need to do about it. J Emerg Manag 4(1):23–35
20. He MC, Li G, Zeng G-Q (2013) The war against infectious respiratory disease goes on: report on the 10th anniversary symposium of anti-SARS. J Thorac Dis 5(Supplement):S94–S102
21. Hou J (2019) Liu Changyong the witness: construction of Xiaotangshan hospital within seven days and nights. China Econ Wkly 18:161–163. 新中国经济70年·抗击非典|亲历者刘长永:七昼夜建成小汤山医院。《中国经济周刊》2019年9月30日第18期, 161–163 页。
22. Huang Y (2004) The SARS epidemic and its aftermath in China: a political perspective. In: Knobler S, Mahmoud A, Lemon S, Mack A, Sivitz L, Oberholtzer K (eds) Learning from SARS: preparing for the next disease outbreak—workshop summary. The National Academies Press, Washington, DC, pp 116–136. https://doi.org/10.17226/10915. Accessed 19 Sept 2009
23. Huaxia (2020) China publishes timeline on COVID-19 information sharing, int'l cooperation. Xinhuanet, 6 April. http://www.xinhuanet.com/english/2020-04/06/c_138951662.htm. Accessed 12 Sept 2021
24. Jackson P, Jalil AZ (2022) Shanghai lockdown: China eases COVID restrictions after two months. BBC News, 1 June. https://www.bbc.com/news/world-asia-china-61647687. Accessed 14 May 2023
25. Jeffries I (2006) China: a guide to economic and political developments. Routledge, London/New York

26. Jiang Y (2020) Zhong Nanshan himself. 姜永斌。《钟南山其人》。https://www.ccdi.gov.cn/yaowen/202001/t20200122_210211.html. Accessed 20 Sept 2021
27. Johns Hopkins University (JHU) (2023) Confirmed new cases of COVID-19 worldwide. https://coronavirus.jhu.edu/data/new-cases. Accessed 17 June 2023
28. Johnston RJ, Gregory D, Pratt G, Watts M (eds) (2000) The dictionary of human geography, 4th edn. Blackwell Publishers Ltd.
29. Katella K (2023) Omicron, delta, alpha, and more: what to know about the Coronavirus variants. Yale Medicine, (2/3/2023). https://www.yalemedicine.org/news/covid-19-variants-of-concern-omicron. Accessed 21 June 2023
30. Kleinman A, Watson JL eds (2006) SARS in China: prelude to pandemic? Stanford, Calif.: Stanford University Press
31. Knobler S, Mahmoud A, Lemon S, Mack A, Sivitz L, Oberholtzer K (eds) (2004) Learning from SARS: preparing for the next disease outbreak: workshop summary. The National Academies Press, Washington, DC. https://doi.org/10.17226/10915. Accessed 19 Sept 2009
32. Koh T, Plant A, Lee EH (2003) The new global threat: severe acute respiratory syndrome and its impacts. River Edge, NJ: World Scientific.
33. Kupferschmidt K, Cohen J (2020) Can China's COVID-19 strategy work elsewhere? Science 367(6482):1061–1062. https://www.science.org/doi/full/10.1126/science.367.6482.1061. Accessed 31 May 2023
34. Lee SH (2006) SARS in China and Hong Kong. New York: Novinka
35. Liao J, Li C (2020) A total of over 12,000 people treated in 16 Wuhan Fangcang Hospitals. Xinhua News Agency, 10 March. 廖君,黎昌政。《16家武汉方舱医院休舱累计收治1.2万余人》。新华社 2020-03-11。http://www.gov.cn/xinwen/2020-03/11/content_5489856.htm. Accessed 15 Mar 2020
36. Lin JY, Tao R, Liu M (2006) Decentralization and local governance in China's economic transition. In: Bardhan P, Mookherjee D (eds) Decentralization and local governance in developing countries: a comparative perspective. The MIT Press, Cambridge/London, pp 305–327
37. Liu C (2004) Chinese law on SARS. W.S. Hein, Buffalo
38. Loh C, Exchange C (eds) (2004) At the epicentre: Hong Kong and the SARS outbreak. Hong Kong University Press, Hong Kong
39. Luo Z, Zhang Y, Zhang Y, Zheng Y, MacIntyre CR, Liang Y, Wang Q, Ma Y (2021) Prevention of SARS-CoV-2 transmission from international arrivals: Xiaotangshan Designated Hospital, China. Bull World Health Organ 99(5):374–380. https://www.ncbi.nlm.nih.gov/pmc/articles/PMC8061664/. Accessed 20 Sept 2021
40. McLean A (ed) (2005) SARS: a case study in emerging infections. Oxford University Press, Oxford
41. Mu W, Yang W (2020) Wuhan Pneumonia virus continues to spread: Hubei orders 15 cities to be locked down. Associated Press and Reuters, 24 January. 文木, 杨威廉。《武汉肺炎病毒持续扩散 湖北下令封15个城市》。美联社,路透社等, 2020-1-24. https://www.dw.com/zh/%E6%AD%A6%E6%B1%89%E8%82%BA%E7%82%8E%E7%97%85%E6%AF%92%E6%8C%81%E7%BB%AD%E6%89%A9%E6%95%A3-%E6%B9%96%E5%8C%97%E4%B8%8B%E4%BB%A4%E5%B0%8115%E4%B8%AA%E5%9F%8E%E5%B8%82/a-52132769. Accessed 30 Jan 2020
42. News Desk (Xinhua) (2020) Wuhan to convert gymnasium, exhibition center into temporary hospitals. The Jakarta Post, 2/4/2020. https://www.thejakartapost.com/news/2020/02/04/wuhan-to-convert-gymnasium-exhibition-center-into-temporary-hospitals.html. Accessed 13 Sept 2020
43. Nkengasong J (2020) China's response to a novel coronavirus stands in stark contrast to the 2002 SARS outbreak response. Nat Med 26:310–311. https://www.nature.com/articles/s41591-020-0771-1. Accessed 22 Aug 2021
44. Peng F, Tu L, Yang Y, Peng H, Wang R, Hu Q, Cao F, Jiang T, Sun J, Xu G, Chang C (2020) Management and Treatment of COVID-19: The Chinese Experience. Can J Cardiol 36(6):915–930. https://doi.org/10.1016/j.cjca.2020.04.010

45. Puska SM (2005) SARS 2002-2003: a case study in crisis management. In: Scobell A, Wortzel LM (eds) Chinese national security: decisionmaking under stress. Strategic Studies Institute, U.S. Army War College, Carlisle Barracks, pp 85–134
46. Qi J (1995) The role of local state in China's transitional economy. China Q 144(Special Issue on China's Transitional Economy):1132–1149
47. Saidi AMO, Nur FA, Al-Mandhari AS, El Rabbat M, Hafeez A, Abubakar A (2020) Al Decisive leadership is a necessity in the COVID-19 response. Lancet 396(10247):295–298. https://www.thelancet.com/journals/lancet/article/PIIS0140-6736(20)31493-8/fulltext. Accessed 2 July 2023
48. Schwartz J (2012) Compensating for the 'authoritarian advantage' in crisis response: a comparative case study of SARS pandemic responses in China and Taiwan. J Chin Polit Sci 17(3):313–331
49. Schwirian KP (2006) Political ecology of plague in the global network of cities: the SARS epidemic of 2002-2003. In: McCright AM, Clark TN (eds) Community and ecology: dynamics of place, sustainability and politics. Elsevier JAI, Amsterdam/Oxford, pp 241–268
50. Servick K (2022) Is it time to live with COVID-19? Some scientists warn of 'endemic delusion.' Science, 15 February 2022. https://www.science.org/content/article/it-time-live-covid-19-some-scientists-warn-endemic-delusion. Accessed 2 Apr 2023
51. Shang L, Xu J, Cao B (2020) Fangcang shelter hospitals in COVID-19 pandemic: the practice and its significance. Clin Microbiol Infect 26:976–978. https://www.clinicalmicrobiologyandinfection.com/article/S1198-743X(20)30266-4/fulltext. Accessed 2 Oct 2020
52. Starling AE, Hong Kong Museum of Medical Sciences (2006) Plague, SARS and the story of medicine in Hong Kong. Hong Kong University Press, Hong Kong
53. Steinbuch Y (2020) Wuhan rushes to open makeshift hospitals for coronavirus victims. New York Post, 2/5/2020. https://nypost.com/2020/02/05/wuhan-rushes-to-open-makeshift-hospitals-for-coronavirus-victims/. Accessed 2 Sept 2020
54. Steinbuch Y (2020) Wuhan rushes to open makeshift hospitals for coronavirus victims. New York Post, 5 February. https://nypost.com/2020/02/05/wuhan-rushes-to-open-makeshift-hospitals-for-coronavirus-victims/. Accessed 10 Sept 2020
55. Sun J (2003) United as one to defeat SARS with comradery, prevention and treatment. The Guangming Daily, 4/27/2003. 孙杰。《团结一致 同舟共济 群防群治 战胜非典: 温家宝在考察北京非典型肺炎防治工作时强调》。https://www.gmw.cn/01gmrb/2003-04/27/02-D0F6F73F4E4F719748256D1500017EA1.htm. Accessed 7 Oct 2009
56. Swaine MD (2005) Chinese crisis management: framework for analysis, tentative observations, and questions for the future. In: Scobell A, Wortzel LM (eds) Chinese national security: decision making under stress. Strategic Studies Institute, U.S. Army War College, Carlisle Barracks, PA, pp 5–54
57. Tan S (2020) China's Novel Health Tracker: Green on Public Health, Red on Data Surveillance. CSIS Blog Post, 4 May 2020. https://www.csis.org/blogs/trustee-china-hand/chinas-novel-health-tracker-green-public-health-red-data-surveillance. Accessed 12 Sept 2021
58. Taylor DB (2020) A timeline of the coronavirus pandemic. The New York Times, 14 February. https://www.nytimes.com/article/coronavirus-timeline.html. Accessed 4 Sept 2021
59. The Chinese Central Government's Official Web Portal. https://www.gov.cn/english/. Accessed 1 June 2023
60. Tweeten L, Barone E, Wolfson E (2020) A timeline of how the Wuhan coronavirus has spread—and how the world has reacted. Time, 30 January. https://time.com/5774366/how-coronavirus-spread-china/. Accessed 27 Sept 2021
61. United States, Congressional-Executive Commission on China (USCECC) (2003) Dangerous secrets: SARS and China's healthcare system: roundtable before the Congressional-Executive Commission on China, One Hundred Eighth Congress, first session, May 12, 2003. U.S. G.P.O, Washington
62. United States, Congressional-Executive Commission on China (USCECC) (2003) Freedom of the press in China after SARS: reform and retrenchment: roundtable before the Congressional-

Executive Commission on China, One Hundred Eighth Congress, first session, September 22, 2003. U.S. G.P.O, Washington
63. Vo H (2019) China's healthcare revolution: WeDoctor. https://digital.hbs.edu/platform-digit/submission/chinas-healthcare-revolution-wedoctor/. Accessed 9 Sept 2021
64. Wang L (2003) Li Keqiang: United as one, defeating "SARS" with science. Henan Daily Online, 5/4/2003. 王琳。《李克强:团结一致依靠科学战胜"非典"》。《河南报业网》2003年5月4日。https://www.hncourt.gov.cn/public/detail.php?id=1262. Accessed 9 Oct 2009
65. Wang M (2003) General Secretary of the Provincial Party Committee Li Keqiang: Establish a national epidemic prevention network. People's Daily Online, 5/5/2003. 王明浩。《河南省委书记李克强:构筑全民防疫网》。《人民网》2003年5月5日。https://news.sina.com.cn/o/2003-05-05/0914104937s.shtml. Accessed 9 Oct 2009
66. Wang Q (2003) 王岐山 (2003)。Regulations on the implementation of prevention and control measures for infectious SARS in Beijing.《北京市实施传染性非典型肺炎预防控制措施若干规定》。https://www.beijing.gov.cn/zhengce/gfxwj/201905/t20190522_56542.html. Accessed 11 Oct 2009
67. Wang X (2020) Makeshift hospitals point the way. China Daily Global, 31 March. https://www.chinadaily.com.cn/a/202003/31/WS5e82a7e4a3101282172833c7.html. Accessed 10 Sept 2020
68. Wang V, Dong J (2021) China, holding to its 'zero Covid' strategy, keeps a city of 13 million locked down. The New York Times, 12/30/2021. https://www.nytimes.com/2021/12/30/world/asia/china-xian-lockdown-covid.html. Accessed 9 Jan 2022
69. Wang F, Li Zs (2020) Pneumonia epidemic: Wuhan "locked down" for 76 days, and life in the city was "terrifying." BBC News Chinese, 8 April. 王凡、李宗憲 (2020)。《肺炎疫情:武漢「封城」76天,城内生活「膽戰心驚」》。BBC News中文, 2020年4月8日。https://www.bbc.com/zhongwen/trad/chinese-news-52196359. Accessed 10 Apr 2020
70. Weber I, Lu J (2006) SARS, Youth and Online Civic Participation in China. In: Holden TJM, Scrase TJ (eds) Medi@sia: global media/tion in and out of context. Routledge, New York, pp 158–199
71. WeDoctor. WeDoctor (2020) https://www.crunchbase.com/organization/guahao#section-overview/. Accessed 23 Sept 2021
72. WHO (2006) SARS: how a global epidemic was stopped. WHO Regional Office for the Western Pacific Region, Manila
73. WHO (2020) Timeline of WHO's response to COVID-19. WHO, June 29. https://www.who.int/news-room/detail/29-06-2020-covidtimeline. Accessed 13 Sept 2021
74. WHO (2021) WHO validates Sinovac COVID-19 vaccine for emergency use and issues interim policy recommendations. WHO, 6/1/2022. https://www.who.int/news/item/01-06-2021-who-validates-sinovac-covid-19-vaccine-for-emergency-use-and-issues-interim-policy-recommendations. Accessed 18 Sept 2021
75. WHO (2023) WHO Coronavirus (COVID-19) dashboard. https://covid19.who.int/. Accessed 2 July 2023
76. Wong J, Zheng Y (eds) (2004) The SARS epidemic: challenges to China's crisis management. World Scientific, Singapore/Hong Kong
77. Worldometer (2023) COVID-19 Coronavirus pandemic. https://www.worldometers.info/coronavirus/. Accessed 15 July 2023
78. Xinhuanet (2020) China publishes timeline on COVID-19 information sharing, int'l cooperation. http://www.xinhuanet.com/english/2020-04/06/c_138951662.htm. Accessed 23 Sept 2020
79. Xinhuanet (2020) 20 "Fangcang" Hospitals Rush to Assist Wuhan. What are "Fangcang" Hospitals? Xinhuanet, 2/5/2020. 《20家"方舱医院"驰援武汉!"方舱医院"是什么?》. 新华网2020-2-2. http://www.xinhuanet.com/politics/2020-02/05/c_1125532816.htm. Accessed 6 Sept 2020
80. Xu R (2013) Chance missed, but still there! Memoirs at the 10th anniversary of 2003 SARS outbreak. J Thorac Dis Suppl 5:S90–S93

81. Yeung J (2021) This Chinese city found a Covid cluster. All 13 million residents are now under lockdown. CNN (Hong Kong), 12/23/2021. https://www.cnn.com/2021/12/23/china/xian-china-covid-lockdown-intl-hnk/index.html. Accessed 9 Jan 2022
82. Yuan S (2022) Zero COVID in China: what next? Lancet 399(10338):1856–1857. https://doi.org/10.1016/S0140-6736(22)00873-X
83. Zhang X, Zhang W, Chen S (2022) Shanghai's life-saving efforts against the current omicron wave of the COVID-19 pandemic. Lancet 399(10340):2011–2012. https://www.thelancet.com/journals/lancet/article/PIIS0140-6736(22)00838-8/fulltext. Accessed 17 June 2023
84. Zhong R, Mozur P (2020) To tame coronavirus, Mao-style social control blankets China The New York Times, February 15. https://www.nytimes.com/2020/02/15/business/china-coronavirus-lockdown.html. Accessed 15 Mar 2020
85. Zhong N, Zeng G (2006) What we have learnt from SARS epidemics in China. BMJ 333(7564):389–391. https://doi.org/10.1136/bmj.333.7564.389

Federal Pandemic Response

Ellen Embrey

1 Introduction

For more than a decade, the US health system (separate systems/sectors loosely interconnected and inadequately integrated for optimal health outcomes for the US population as a whole) has fallen far short of its potential to fully support individual and population health. US healthcare spending exceeds that of any other Organization for Economic Cooperation and Development (OECD) nation but paradoxically experiences lower life expectancies, higher suicide rates, higher chronic disease burdens, higher obesity rates, and higher hospitalization rates from preventable causes than any of its peers [1].

Gaps in healthcare and public health services, capacity, and infrastructure, particularly in rural and underserved areas of the USA, existed long before the first cases of COVID-19 in the USA were identified in January 2020. These gaps underpinned and significantly complicated the US experience and success in responding to the COVID-19 pandemic from December 2019 to its present endemic state [1]. At present, the nation's focus has shifted to mitigate more effectively the spread of endemic influenza, COVID-19, and respiratory syncytial virus (RSV), characterizing and researching definitive diagnosis and treatment guidelines for an increasingly growing segment of the US population with persistent, long-term health issues following COVID-19 infection, popularly referred to as "Long-COVID."

Throughout the last 3 years, many peer reviewed academic, scientific, and Federal agency reports, studies, and research initiatives have been funded, and results published, several of which were used early in the pandemic to inform Federal response guidelines to reduce COVID-19 incidence and prevalence and morbidity and mortality [4]. In most cases, findings and conclusions led to

E. Embrey (✉)
Stratitia, Inc., Washington, DC, USA
e-mail: eembrey@stratitia.com

J. M. Rosen, R. R. Colwell (eds.), *The COVID-19 Pandemic*,
https://doi.org/10.1007/978-3-031-62772-9_13

recommendations for improved response, with calls for further review, research, and study. With the exception of actions taken by the Federal government to sponsor and oversee the rapid development, determination of safety and efficacy, and authorization of emergency use, production, distribution, and prioritization for administration of COVID-19 vaccines, antiviral and other treatments to significantly reduce the morbidity and mortality rates of COVID-19 infection, most of what have been learned from COVID-19 pandemic response over the last 3 years, have been learned in far smaller scale US public health emergencies in the last 20 years.

2 Background

The US COVID-19 response experience reinforced and deepened the nation's understanding of longstanding critical vulnerabilities in our systems of healthcare and highlighted the consequences of failing to ensure health equity and access to care in communities throughout our nation in a sustainable way [1]. Many workable public health emergency frameworks, standards, and metrics have been proposed and piloted to provide data-driven awareness to states and Federal agencies on their areas of strengths and vulnerabilities so that when public health emergencies overwhelm state and local capacities, all stakeholders would know specifically where to focus on providing surge resources and funding. Unfortunately, the lack of sufficient and sustained levels of funding for state and Federal health programs has effectively hampered the nation to implement standards, metrics, and capability/capacity framework of community-based health services. The Department of Homeland Security's *National Response Framework* [3], published by the Federal Emergency Management Agency, was most recently updated September 28, 2019. It describes the importance of a whole community response plan with guidelines, common concepts of operation, and a standard, scalable national response framework, (supported by sector specific plans and annexes) to enable Federal, state, and local authorities to plan, work, and continuously update approaches that enhance the effectiveness of Federal state and local roles, responsibilities, and authorities, as emergency response scope and complexity increase in size and impact economic and essential government services.

This chapter outlines the documented actions of the US Federal government in shaping and guiding the US COVID-19 pandemic response as the virus spread and evolved, from identification/characterization in early January 2020 to its evolution to endemic status today. The Federal government undertook a series of actions to support the nation's states, localities, and territories response, informed by the World Health Organization, global pandemic incidence, prevalence, morbidity and mortality, experiences and scientific discoveries reported by states, localities, and territories of the USA and by other nations around the globe, including US and international scientific, academic, business, news networks/outlets (professional), and social media communities at large.

Broadly speaking, the US Congress and the Executive Branch of the Federal government provided initial, expanded, and ongoing support to Federal programs,

US states, localities, and territories, and international countries for COVID-19 response in two critical ways: (1) by providing funding to Federal government, states, localities, and territories to fill gaps in capability and effect a consistent and equitable response to the health and economic impacts of the COVID 19 pandemic; and (2) by issuing US Regulatory, Scientific, and Policy Guidelines for States, localities, territories, and individuals to mitigate the spread and improve the nation's capability to effectively respond to the emerging and changing COVID-19 virus and its impacts on all aspects of the nation's economy and quality of life.

Response 1: Federal Funding to US Government, States, Localities, and Territories to Respond to the Health, Economic, and National Security Impacts of the COVID-19 Pandemic

The US Congress, with varying degrees of coordination with the White House and 44 Federal agencies within the Executive Branch, appropriated funds at different intervals in the first 2 years of the pandemic to facilitate the national response to the COVID-19 pandemic. These funds were made available through enactment of the Coronavirus Aid, Relief, and Economic Security (CARES) Act (March 27, 2020) and other supplemental legislation. In March 2021, additional funds were appropriated through the American Rescue Plan Act. Forty-four Federal agencies received funding from Congress, and they in turn issued awards (primarily as financial assistance to state agencies and individuals. Federal COVID-19 outlays to states, localities, and territories, (see chart below) exceeded $2.3 Trillion, covering a wide range of needs and delivered as grants, direct payments, contracts, infrastructure, staffing, loans, and other services [5]. In total, of the $4.7 Trillion authorized by Congress for COVID-19 response, the Federal government reported [5] $4.3 Trillion in outlays (as of 23 September 2023) for responding to COVID-19 pandemic challenges.

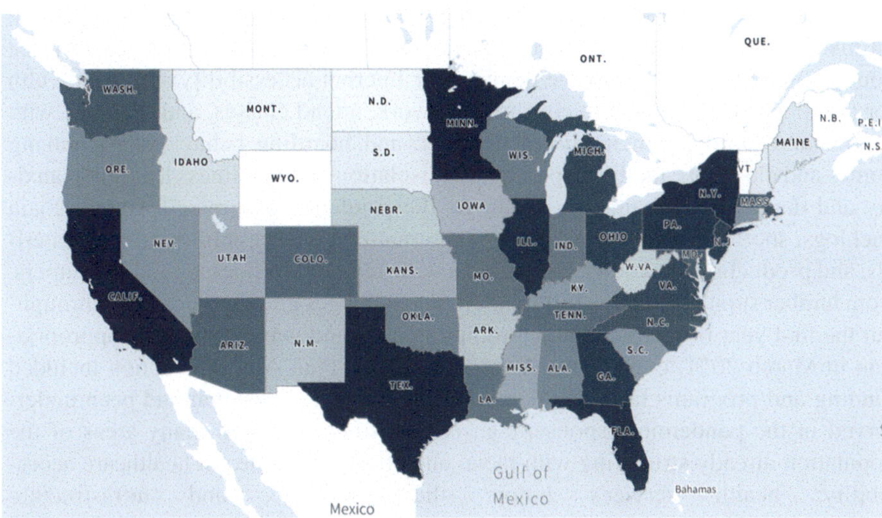

https://www.usaspending.gov/disaster/covid-19/?section=award_spending (hover over each state to get funding outlays awarded for COVID-19). Colors of states on map reflect amount awarded to each state on a range from bright (low) to dark(high), a low of $5.6B (white) to a high of $280.8B (black)

Number of recipients	Award obligations	Award outlays	Number of awards
14.73 million	**$2.50 trillion**	**$2.37 trillion**	**22.00 million**

Within the first 6 months of the pandemic, the US Federal government with supporting guidance from the office of the White House (Executive Branch) and the US Congress (Legislative Branch) developed and passed legislation that provided funding to Federal agencies and states, localities, and territories to (a) facilitate the expansion of local public health and health system services (e.g. contact tracing, crises care public health and healthcare delivery workforce and staffing models), (b) facilitate education, training, and crisis response guidelines to mitigate staff and PPE shortages at locations of high-risk vulnerability(nursing homes, long-term care, and skilled nursing facilities, and rural and underserved communities of the country), (c) expand local infrastructure and associated response services (e.g. reference labs, testing and validation resources, disease reporting systems/platforms), and (d) mitigate and stabilize economic challenges resulting from the implementation of pandemic mitigation strategies (especially between mid-January 2020 and late December 2020, when EUA vaccines to prevent infection were not available) such as temporary social distancing guidelines and mask wearing when being outside the household. These also included temporary restrictions limiting indoor gathering of people in all nonessential sites. Federal time–limited restrictions or complete shutdown recommendations impacted transportation hubs/subways, airlines/buses/ships (air/land/sea), institutions/businesses, such as office buildings, restaurants/bars, schools (preschool, elementary, intermediate, secondary and colleges), daycare, hotels, resorts, and indoor and outdoor sporting events. These closures led to widespread levels of unemployment, business revenue losses, and, ultimately, the shuttering of many businesses. These actions in turn led to price gouging and inflation, explosive demand for Internet accessibility, and bandwidth for families at home trying to concurrently work, attend classes, and socialize with friends and family, principally via Internet, and hoarding behaviors. At nursing homes and elder care facilities, particularly, isolation and loneliness leading to anxiety and depression became the norm. As the pandemic progressed, supply chain backlogs, shortages due to domestic and international inaccessibility to raw materials, and production slowdowns impacted a wide range of businesses and consumers, from lumber supplies to baby formula. These challenges grew and persisted throughout the first year of the pandemic, prompting a second wave of funding appropriation in March 2021, entitled the American Rescue Plan Act. This action included funding and programs targeting populations within the USA that had been underserved in the pandemic response, e.g., native Americans, and many areas of the population already struggling with racial and ethnic inequities in healthcare accessibility, health services before the pandemic, and infrastructure support—particularly for inadequate levels of food and health services in rural and underserved areas of the USA [1, 4].

As the pandemic evolved in three waves over a 2.5-year period, the economic impact varied widely across the states and relief funding to mitigate these impacts helped pave the way for adopting new or expanded models for delivering services (crises staffing, low-contact workflows, virtual/remote consultations/collaborations, technology innovations that enhanced understanding and utilization of data-driven analytics, projections, trends, data validation/management, and decision-making) giving rise to new perspectives and studies of the business case to make permanent some of the innovation and efficiencies applied in all sectors as needed after the pandemic ended.

Response 2: Issuing US Regulatory, Scientific, and Policy Guidelines for Consistent, Equitable, and Science-Driven Health Response

Federal government responsibility for leading and shaping a coordinated and effective national public health emergency responses via support to states and territories falls primarily to the Department of Health and Human Services (DHHS), consistent with the National Response Framework. Although the White House and its interagency task forces and the Departments of Homeland Security, Labor, Commerce, State, Veterans Affairs, Agriculture, and Defense, to name only a few, also played important roles in supporting the health services response, the HHS Centers for Disease Prevention and Control, National Institutes of Health, Federal Drug Administration, Public Health Service, Indian Health Service, Substance Abuse and Mental Health Services Administration, and Office of the Assistant Secretary for Preparedness and Response played leadership roles in preparing, planning, overseeing, and helping to set policies, regulatory pathways, guidelines, and priorities for allocating scarce resources to optimize the responses to the highest risks and vulnerabilities emerging from the US collective COVID-19 pandemic experience.

These HHS entities have long-established collaborative relationships with respect to disease prevention, detection, reporting and treatment, and mechanisms for delivering services, support, guidelines, and funding-associated public health service initiatives necessary to ensure effective community-based care, all within the scope, strategies, and plans of the National Response Framework and Emergency Support Function #8, Public Health, and Medical Services. Because each US state, locality, and territory has unique demographic, governance, and economic drivers impacting their public health and health services emergency response capabilities, the Federal government focuses on issuing best practice and scientifically validated models, frameworks, guidelines, planning tools, and information that serves as a dynamic national resource to request and receive assistance, exchange best practices and approaches, and leverage the latest scientifically validated information to promote the most effective approaches in responding to common challenges. Even if receptive to Federal assistance, guidelines, and resources, inconsistent and widely varying individual state, locality, and territory capacities to implement Federal public health and medical services guidelines and reporting systems are impacted by various factors, including state leadership acceptance of Federal guidelines, policies and risk mitigation approaches, shortages of testing laboratories, backlogs of FDA reviews of EUA requests, global and national shortages in PPE, shortages of qualified medical service and public health staff, and poorly integrated data systems between and among state medical and public health systems and other social services impacted by the pandemic.

Some Federal issuances are regulatory, as in the FDA's role in reviewing evidence necessary to authorize time-limited emergency use authorization, (and ultimately permanent FDA approval) for safe and effective new diagnostic laboratory and at home tests, vaccines, therapeutics, and other treatments that help to reduce COVID-19 morbidity and mortality. Most Federal–state relationships prior to January 2020 involved support on a much smaller scale than experienced in the COVID-19 pandemic. A recent example of effective collaboration between US Federal and state governments during a pandemic occurred during the H1N1 pandemic in 2009. The 2009 H1N1 pandemic was estimated to be associated with 151,700 to 575,400 deaths worldwide during the first year it circulated. Ten years later, the same core approaches were initiated and plans updated to facilitate coordination and collaboration around the stages of COVID-19 pandemic as it evolved.

The Federal government, primarily the departments of HHS and DHS (and their subordinate agencies/institutes/centers), began tracking the novel virus after its first report from Wuhan, China, and took iterative actions over time to prepare and respond to pandemic as it spread rapidly into the USA. These actions, highlighted in the timeline paragraphs below, involved taking actions within the international community, issuing alerts, setting Federal guidelines, and releasing scientific and technical information to the nation and its news outlets to enable the states, their leaders, and their citizens to anticipate and learn from the COVID-19 experiences of others, at both home and international level. The novel virus spread from less than 10 US cases confirmed in January 2020 to 60 cases confirmed in twelve states 1 month later, and by mid-April, 2020, 500,000 cases with 18,600 deaths in the USA had been attributed to COVID-19. By that date, the CDC reported numbers of US COVID-19 cases and deaths were higher than any other country in the world [4].

The David J. Sencer CDC Museum, in association with the Smithsonian Institution, published a digital exhibit of key US Federal actions taken between January 2020 and June 2022 in a "COVID-19 Timeline" [4], that detailed, by day, the specific actions taken by the Federal government, particularly the Department of Health and Human Services and its key agencies, with respect to public health matters, ensuring the USA complies with its obligations under the International Health Regulations and to ensure information from the states in response to public health emergencies of international concern is collected and shared with the WHO. While the Federal government represents the USA and reports required information to the WHO about the US national response and pandemic status, its key role is to provide law, policy, resources, guidance, and information to state and local governments. It is the state and local governments that are responsible for implementing response efforts on the ground and making decisions that are best for their communities.

Many of the challenges experienced in the US pandemic response were exacerbated by the different COVID-19 response approaches taken in each of the fifty-four states and territories of the USA, and the inability to gain national consensus when political and scientific options differed. The following paragraphs are highlights of Federal actions taken based on the David J. Sencer CDC Museum exhibit timelines from January 2020 to July 2022 [4].

Key Actions Taken by the US Federal Government [4]

Federal Actions Taken on January 2020:

In response to the rapid spread of a novel virus identified during November–December 2019 in Wuhan, China, the first US Federal response was initiated on **5 January 2020,** when CDC's National Center for Immunization and Respiratory Diseases was activated to investigate that a novel virus was causing difficulty to treat severe pneumonia cases. The genomic sequence of the virus was submitted by the Wuhan public health organization to the Chinese government CDC equivalent on that same day, but the sequence was not published for WHO and other countries' use until 5 days later: **January 10**. On **17 January**, the CDC initiated travel screenings for symptoms of 2019 novel coronavirus infection of all passengers arriving in San Francisco, CA, from Wuhan, China. On **20 January**, CDC reported the first case of laboratory-confirmed "2019 Novel Coronavirus" in the state of Washington, the same day it also activated its Emergency Operations Center (EOC) to actively track and respond to the emerging outbreak. **On January 21**, the Secretary of HHS formally declared the 2019 Novel Coronavirus (2019-nCoV) outbreak in the USA a public health emergency, triggering a series of actions that enabled states, localities, and territories to request and receive Federal funding and other support in response to/mitigating the emergency in their states. The CDC confirmed travel-related SARS-COV-2 cases in AZ and CA on **26 January**. On **28 January**, the CDC issued a Level 3 Travel Health update advising travelers to avoid all nonessential travel to China due to the 2019 Novel Coronavirus outbreak, while the US state department relocated US citizens from Wuhan, China, back to the USA due to the 2019 Novel Coronavirus outbreak (2019-nCoV). The CDC confirmed on **30 January** that the SARS-COV-2 virus had spread between two people in IL, with no history of recent travel.

Federal Actions Taken on February 2020 [4]

On **3 February**, the CDC submitted an emergency use authorization (EUA) to FDA to expedite approval for a CDC-developed SARS-COV-2 diagnostic test for use by the national laboratory response network. FDA approved an EUA for the diagnostic test kit on **4 February** and CDC began shipping its laboratory test kits ("CDC 2019-nCov Real Time RT-PCR") to select domestic and international laboratories. On **February 11**, the World Health Organization (WHO) announced "COVID-19" as the official name of the disease causing the 2019 Novel Coronavirus outbreak. On **February 13**, the CDC confirmed the 15th case of COVID-19 in the USA. The CDC's COVID-19 Response Incident Manager held a press briefing on **18 February**, advising the nation to expect that aggressive mitigation efforts might be needed to contain the spread of SARS-COV-2 virus in the USA, which might include school closings, workplace shutdowns, and cancelling large gatherings and public events, adding "the disruption to everyday life may be severe." On **28 February**, the CDC confirmed that 3 of 4 new cases in California, Oregon, and Washington were likely due to community spread of SARS-CoV-2 virus in the USA rather than travel-related infections. On **February 29**, the CDC announced revision to its *Criteria to Guide Evaluation and Testing of Patients Under Investigation (PUI) for COVID-19* to include any patients with a severe respiratory illness (even those with no travel history to affected areas or known exposure to other cases) in anticipation of widespread person-to-person transmission. Also on **February 29**, the FDA announced its policy decision to allow certain US laboratories to develop and begin to use validated COVID-19 diagnostics before FDA had completed its review of their EUA requests, facilitating rapid expansion of laboratory testing to address high demand for COVID-19 testing across the USA. Lastly, on **February 29**, the CDC, and the Washington Department of Public Health, reported the laboratory-confirmed first death in the USA of an individual with COVID-19.

Federal Actions Taken on March 2020 [4]:

On March 1, CDC modified an existing national respiratory virus surveillance network used to track hospitalizations associated with influenza and respiratory syncytial virus (RSV) across the USA to create a familiar network for reporting and tracking the numbers and rates of COVID-19 hospitalizations, called COVID-NET. On **3 March**, CDC reported sixty cases in 12 states, 2 travel related, 11 person-to-person spread, and 27 unknown. On **12 March**, FDA discontinued the requirement for CDC to perform confirmatory testing for positive COVID-19 diagnoses. On **13 March**, the US President declared a nationwide emergency and issued a travel ban restricting non-US citizens from traveling to the USA from 26 European countries, due to COVID-19. On **15 March**, states began to implement shutdowns to mitigate the spread of COVID-19. In NYC, schools were shut down and in OH, restaurants and bars were closed. **17 March**, federally funded Moderna Therapeutics began human trials of its vaccine to protect against COVID-19 at a research facility in Seattle, WA. Also on **17 March**, The Centers for Medicaid and Medicare (CMS) temporarily expanded telehealth benefits to enable beneficiaries to receive a wider range of health services from their doctors without having to travel to a healthcare facility. On **27 March**, the US President signed the Coronavirus Aid, Relief, and Economic Security (CARES) Act into law. The Act provided funding of $1200 per adult and more for families with children, as well as expanded unemployment benefits, forgivable small business loans, loans to major industries and corporations, and expanded funding to state and local governments in response to the economic impact caused by COVID-19. On **28 March**, the US President extended all existing CDC-recommended social distancing guidelines through the end of April 2020. Also on **28 March**, the CDC issued a domestic travel advisory for NY, NJ, and CT due to high community transmission rates of COVID-19 infection in those states, urging residents to refrain from all nonessential domestic travel for at least 14 days, effective immediately. On **March 31**, Dr. Anthony Fauci, and Dr. Deborah Brix, at a White House briefing, advised that between 100,000 and 240,000 deaths in the

USA were anticipated, even if social distancing and public health measures were perfectly enacted.

Federal Actions Taken on April 2020 [4]

At a White House briefing on **3 April**, the CDC announced new mask-wearing guidelines and recommended all people wear masks when outside their home. On **4 April**, CDC launched the new weekly CARS-COV-2 virus surveillance report, "COVIDView", summarizing weekly data on COVID hospitalizations, deaths, and testing. On **7 April,** the HHS announced $186 Million in additional funding to state and local jurisdictions for COVID-19 response. On **8 April**, the HHS announced its first contract for ventilator production under the Defense Production Act with General Motors. On **13 April**, most states in the USA reported widespread cases of COVID-19. On **16 April**, the White House released a plan recommending steps the states should take to reopen, calling for states or metropolitan areas to meet benchmarks, such as reducing COVID-19 cases or deaths before reopening or stopping mitigation strategies. On **20 April**, as the COVID-19 pandemic grew, dire shortages of personal protective equipment (PPE) (gowns, eye shields, masks, and even body bags) were reported. On **22 April**, the CDC recommended people to restrict their pets' interactions with other people or animals outside their household, after cats in two separate locations tested positive for COVID-19. On **April 23**, using funds from the CARES Act, the CDC announced $631 million to fund and expand the existing Epidemiology and Laboratory Capacity for Prevention and Control of Emerging Infectious Diseases (ELC), allowing state health departments to expand their capacity for testing, contact tracing, and containment of COVID-19. On **30 April**, the White House announced the launching of "Operation Warp Speed", a Federal government interagency initiative to produce a vaccine against the SARS-COV-2 virus as quickly as possible. The operation funded development of six promising vaccine candidates while they were still in clinical trial phase, including Pfizer-BioNTech and Moderna mRNA vaccines.

Federal Actions Taken on May 2020 [4]:

On **May 1**, the FDA issued an emergency use authorization for the use of the antiviral drug remdesivir to treat those hospitalized with severe COVID-19. On the same day, the CDC issued guidelines to determine a "PPE Burn rate Calculator" to help healthcare facilities plan and optimize PPE via Apple and Android app stores. It also launched **on May 1** the SARS-CoV2 Sequencing for Public Health Emergency Response, Epidemiology and Surveillance (SPHERES), a national network to provide real-time genomic sequencing data to public health response teams investigating COVID-19 cases, permitting a standard routine for tracking the virus as it evolved. On **May 9**, the Department of Labor reported the US unemployment rate at 14.7%, the highest since the Great Depression, with over 20.5 million people out of work, with leisure, hospitality, and healthcare sector workers taking the biggest hit. On **May 12**, the NIH Director Fauci testified before the US Senate that "experts believe more people have died from COVID-19 than have been officially reported" and warned against reopening too quickly. On **May 18**, the HHS awarded states, territories, and local jurisdictions $11Billion in new funding to support testing for COVID-19. On the same day, the CDC Epidemiology and Laboratory Capacity for

Prevention and Control of Emerging Infectious Diseases (ELC) provided $10.25 Billion to states, territories, and local jurisdictions, and the Indian Health Service (IHS) provided $750 Million for IHS and tribal and urban Indian Health Programs to expand testing capacity, as the Navajo Nation had reported the highest COVID-19 infection rate per capita in the USA.

Federal Actions Taken on June 2020 [4]

On **June 4**, the HHS released new laboratory guidance for data reporting COVID-19, including new data on race, ethnicity, age, and sex demographics. The CDC released that same day its consolidated guidelines for COVID-19 testing, with a focus on nursing homes, long-term care facilities, and high-density critical infrastructure workplaces, e.g., food production facilities. On **June 16,** the HHS announced COVID-19 vaccines would be free of charge for older adults and other groups experiencing disproportionate impact of the COVID-19 pandemic. On **June 25, the** CDC expanded the list of those people at increased risk for severe COVID-19, noting the risk increased with age and included individuals with chronic kidney disease, Chronic Obstructive Pulmonary Disease (COPD), obesity, serious heart conditions, sickle cell disease, type 2 diabetics, and people who are immunocompromised as a result of solid organ transplants. On **June 30,** the NIH Director Fauci warned a Senate committee that the number of new cases each day in the USA could rise from 40,000 to 100,000, likely overwhelming an already strained US healthcare system.

Federal Actions Taken on July 2020 [4]

On July 9, the White House issued guidance to states to redirect reporting COVID-19 hospitalization rates and equipment availability from the CDC to HHS, into a system outsourced to a contractor, raising concerns over politicization of public health data and privacy. On **July 22**, DoD and HHS funded Pfizer-BioNTech to deliver and distribute 100 million doses of COVID-19 vaccine in December 2020, upon confirmation that the vaccine was safe and effective. That same day, the CDC released antibody data showing ten times more SARS-CoV-2 infections occurred than initially reported, between March 2020 and May 2020, noting the rates varied by region, from 6 to 24 times more cases than initially reported. On **July 23**, the CDC released resources for school administrators, teachers, parents, guardians, and caregivers to help build appropriate public health strategies to slow the spread of COVID-19 in schools.

Federal Actions Taken on August 2020 [4]

On **August 11**, the White House announced an agreement to pay $1.5 Billion to Moderna to deliver 100 million doses of their COVID-19 vaccine. On **August 14**, the CDC released a report from a survey on mental health conducted among adults in the USA in June 2020. Ca. forty-one percent of responders indicated they were struggling with mental health, and 11% reported seriously considering suicide. Essential workers, unpaid caregivers, young adults, and racial and ethnic minority groups were found to be at higher risk for experiencing mental health struggles, with 31% of unpaid caregivers reporting considering suicide. On the same day, the CDC released data indicating most COVID-19-positive individuals were infectious to others for up to 10 days after symptoms first appeared, but those with severe illness or who were immunocompromised could be infectious for up to 20 days. On **August 15**, the FDA issued an EUA to Yale Public Health for its rapid diagnostic test for COVID-19 SalivaDirect, which enabled laboratories to increase the capacity and efficiency in testing for COVID-19. On **August 23**, the FDA issued an EUA for convalescent plasma to treat individuals hospitalized with severe COVID-19. On **August 26**, the FDA issued an EUA to Abbot for its BinaxNOW Covid-19 Test Kit, a rapid antigen test that can detect COVID-19 infection within 15 minutes, employing the same technology as an influenza test.

Federal Actions Taken on September 2020 [4]

On **September 11**, the CDC released data in its MMWR (Morbidity and Mortality Weekly Report) publication indicating an estimated 41% of US adults had delayed or avoided seeking medical care, including urgent or emergency care. Unpaid caregivers, individuals with underlying medical conditions, Black adults, non-White Hispanic adults, young adults, and individuals with disabilities were the most affected. On **September 14**, the CDC ended its requirement to screen travelers from China, Iran, and the Schengen Area of Europe for COVID-19 symptoms. The next day, **September 15**, the CDC released data in the MMWR listing individuals who had tested positive were more than twice as likely to have recently dined out and more than four times as likely to have recently visited a bar or cafe. The CDC also released an infographic guide to help schools mitigate COVID-19 transmission. On **September 16**, the HHS announced a plan to make COVID-19 vaccines free in the USA. On **September 23**, the HHS and CDC awarded states, territories, and local jurisdictions $200 Million, allocated from the CARES Act, to support COVID-19 vaccine distribution.

Federal Actions Taken on October 2020 [4]

On **October 5**, the CDC updated its "How COVID-19 is Spread" guidelines to acknowledge potential airborne spread of the virus—even when six feet of social distance was maintained, if the area was poorly ventilated or enclosed, with activities that required heavy breathing, e.g., singing or physical exercise.

Federal Actions Taken on November 2020 [4]

On **November 6**, the FDA issued an EUA for the Eli Lilly bamlanivimab, a monoclonal antibody treatment that mimics the immune system response to infection with SARS-COV-2 and appeared to protect patients at increased risk of severe infection from COVID-19. On **November 17**, the NIH Director Fauci announced the need to understand "long COVID" symptoms, such as persistent fatigue, shortness of breath, muscle aches, sporadic fevers, and loss of concentration, experienced in ca. one-third of patients for weeks or months after contracting COVID-19. On **November 20**, as cases of COVID-19 surged in the USA beyond 11,000,000, the CDC recommended Americans to stay home for Thanksgiving and avoid contact with all individuals not living in their household for the last 14 days. On **November 21**, the CDC revised its Travel Health Notice system for COVID-19, adding a fourth alert level: "VERY HIGH", recommending all travelers test 1–3 days before and 3–5 days after all international air travel, in addition to staying home for 7–14 days after travel. On **23 November**, the FDA granted an EUA for the COVID-19 antibody treatment manufactured by Regeneron.

Federal Actions Taken on December 2020 [4]

On **December 3**, CDC's ACIP recommended healthcare professionals and older people living in long-term care facilities be offered vaccine first in the initial phase of the program. The CDC also noted that individuals aged seventy years and older who live in multigenerational homes should be given priority as soon as more doses of the vaccine are available. On **December 11**, the FDA issued an EUA for the Pfizer-BioNTech COVID-19 vaccine for everyone aged sixteen years and older. The next week, on **December 18**, the FDA issued an EUA for the Moderna COVID-19 vaccine. On **December 22**, the CDC released a report in the MMWR outlining ACIP recommendations for COVID-19 vaccination prioritization allocation while the supply was limited (Phase 1a) namely healthcare personnel and residents of long-term care facilities. Phase 1b included essential workers and all who were 75 years and older. Phase 1c included people aged 65–74 years and all individuals aged 16–64 years old who had a medical condition that increased the risk of severe disease from COVID-19. Phase 2 included all persons ages 16 years and older not already covered in earlier phases. On **December 24**, the White House announced purchase of an additional 100 million doses of Pfizer-BioNTech COVID 19 vaccine. On **December 27**, the US President signed a second COVID-19 relief act into law, (CARES Act) which included $900 Billion funding for enhanced unemployment benefits, business loans, purchase and distribution of COVID-19 vaccines and testing kits, and direct cash payments of $600.

Federal Actions Taken on January 2021 [4]

On **January 6**, the CDC and HHS announced plans to deliver $22 billion in funding to states, localities, and territories to support expanded COVID-19 testing and vaccine distribution (from the CARES Act). On **January 12**, the CDC expands COVID-19-negative test requirement to include all air passengers entering the USA. **January 21**, the White House, with a new President, announces the National Strategy for the COVID-19 Response, identifies 7 goals to restore trust, vaccinate, test, and treat COVID-19 while protecting schools, businesses, and workers, in addition to advancing health equity and building the nation's preparedness for future pandemics, calling it a wartime undertaking. The White House also announced on January 21, continued US membership in the WHO and that the USA would join the COVID-19 Vaccine Global Access Facility (COVAX) to support vaccination in low to middle income countries. On **22 January**, the CDC released data in the MMWR on the emerging and more transmissible COVID-19 "Alpha" variant, recommending universal and increased compliance with mitigation strategies, e.g., social distancing, masking, and higher vaccination coverage. On **January 26**, the CDC required face masks to be worn by all travelers while on public transportation and inside transportation hubs, effective February 2, 2021.

Federal Actions Taken on February 2021 [4]

On **February 1**, the White House, supported by the DoD, HHS, and CDC, announced a $231.8 million award to the Australian company Ellume to supply rapid at-home tests in the USA, committing shipment of 100,000 test kits per month from February through July, 2021. On **February 27**, the FDA approved an EUA for the Johnson & Johnson's single dose COVID-19 vaccine for everyone aged eighteen and older.

Federal Actions Taken on March 2021 [4]

On **March 2**, the White House issued a directive to all states to make pre-K-12 teachers, school staff, and childcare workers eligible for COVID-19 vaccinations and to prioritize this group during the month of March within the Federal Retail Pharmacy Program. On **March 8**, the CDC issued guidance recommending individuals fully vaccinated against COVID-19 could safely gather with other fully vaccinated individuals indoors without masks and without social distancing. On **11 March**, the White House announced plans for all adult Americans to be eligible and able to receive a COVID-19 vaccine by May 1, 2021. The plan involved accessibility to vaccines at seven hundred community health centers in under-resourced communities, doubling the number of pharmacies providing vaccines and federally run mass vaccination centers, deploying more than 4000 active-duty troops to support the effort launching "Find a Vaccination Website" and a 1–800 Number. The White House also signed a $1.9 Trillion "American Rescue Plan" into law, which included stimulus funding for expanded unemployment benefits, rental assistance, and COVID-19 vaccinations, as well as extending the child tax credit for 1 year and providing direct cash payments of up to $1400 per individual. The White House also announced $1.75 Billion investment in expanding genomic sequencing to identify and track new variants, as well as another $50 Billion to expand the nation's testing capabilities. On **March 17**, the CDC announced guidance on social distancing in K-12 schools, namely that most elementary students could safely socially distance 3 feet instead of 6 feet inside a classroom with universal masking, but middle and high school students should maintain at least 6 feet in communities where transmission level was high. The CDC also announced $2.25 Billion in spending over 2 years to address COVID-19-related health disparities for individuals living in rural areas and for racial and ethnic minority groups. On **25 March**, the CDC announced $300 million in funding for states, localities, and territories, and tribes and tribal organizations for community health worker services to address disparities in access to COVID-19-related services, such as testing, contract tracing, immunizations and factors that increased the risk of severe COVID-19 illness, such as chronic disease, smoking, and pregnancy and 30 community public health needs exacerbated by COVID-19, such as health and mental healthcare access and food insecurity. The CDC also announced a plan to provide patients and staff of dialysis clinics easy access to COVID-19 vaccinations on site. On **31 March,** the CDC, with NIH, launched a community health testing program, "Say Yes! COVID-19 Test", in Pitt County, North Carolina, and Chattanooga/Hamilton County, TN, providing 160,000 residents with access to free, at home COVID-19 tests to use up to three times a week for 1 month to determine if frequent self-testing would reduce the community spread of COVID-19.

Federal Actions Taken on April 2021 [4]

On **April 2**, the CDC recommended individuals fully vaccinated against COVID-19 could safely travel with lower risk to themselves. On **April 3,** the CDC announced $3 Billion in additional funding for expanded COVID-19 vaccination programs. On **April 6**, the CDC announced an estimate that nearly 80% of pre-K through twelve teachers, school staff, and childcare working in the USA have received at least the first dose of COVID-19 vaccine. On **April 8**, the CDC Director released a statement on racism and health amid the COVID-19 pandemic that "Yet, the disparities seen over the past year were not a result of COVID-19. Instead, the pandemic illuminated inequities that have existed for all of America a known, but often unaddressed, epidemic impacting public health: racism." On **April 13**, the CDC and FDA jointly issued a statement pausing the use of the Johnson & Johnson COVID-19 vaccine for all individuals aged eighteen and older in the USA. Six cases of a rare and serious blood clot had occurred in individuals who had received the J7J vaccine and those cases were investigated. On **April 23**, the FDA recommended the continued use of the Johnson & Johnson COVID-19 vaccine for individuals older than 18, after a thorough safety review had been conducted. On **April 28,** the CDC found that the Pfizer-BioNTech and Moderna mRNA COVID-19 vaccines reduced risk of hospitalization with SARS-CoV-2 in individuals 65 and older by 94%.

Federal Actions Taken on May 2021 [4]

On **May 10**, the FDA expanded the EUA for the Pfizer-BioNTech COVID-19 vaccine to include all adolescents aged 12–15. On **May 14**, the CDC found the mRNA COVID-19 vaccines, Pfizer-BioNTech and Moderna, reduced risk of infection with SARS-CoV-2 virus by approximately 94%.

Federal Actions Taken on June 2021 [4]

On **June 7,** the CDC reported that the Pfizer-BioNTech and Moderna mRNA COVID-19 vaccines reduced risk of infection with SARS-CoV-2 virus by 91% and protected against severe illness and hospitalization if a breakthrough infection occurred. On **June 11**, the CDC published data in MMWR showing that while most COVID-19-associated hospitalizations occurred in older adults, severe disease requiring hospitalization could occur in all age groups, including adolescents aged 12–17 years. On **June 30, the** CDC National Breast and Cervical Cancer Early Detection Program reported the total number of cancer screenings had decreased by an average of 87% for breast cancer, and 84% for cervical cancer during early 2020 due to COVID-19 care, especially among racial and ethnic minority groups and lower income individuals for whom breast cancer screenings had declined 98%, notably among American Indian and Alaskan Native women.

Federal Actions Taken on July 2021 [4]

On **July 16**, the CDC issued an exception for unaccompanied non-citizen children into the USA under Title 42. On **July 27**, amid a Delta variant surge, the CDC released updated masking guidance, recommending everyone in areas with substantial or high transmission rates wear a mask indoors. On **July 30**, the CDC published data in MMWR showing an increase in breakthrough infections of COVID-19 in Barnstable County, Massachusetts, had occurred in July. Early data showed high viral loads in individuals infected with the Delta variant of COVID-19, prompting concern that (unlike other variants up to that time) vaccinated people infected with Delta could transmit the virus to others.

Federal Actions Taken on August 2021 [4]

On **August 6**, the CDC published data in MMWR indicating unvaccinated individuals were more than twice as likely to be reinfected with COVID-19 than those who were fully vaccinated, after initially contracting the virus, demonstrating COVID-19 vaccines offered stronger protection than natural immunity alone. On **August 13**, CDC's ACIP recommended an additional dose of COVID-19 vaccine after the two dose vaccine series for everyone who had moderately to severely compromised immune systems. On **August 11**, the CDC released a statement assuring the public that COVID-19 vaccination was safe for those who were pregnant and breastfeeding, noting there was no evidence that any vaccine, including COVID-19 vaccines, caused fertility problems in women or men. On **August 23**, the FDA fully approved the Pfizer-BioNTech COVID-19 vaccine for everyone aged eighteen and older. Full FDA approval reinforced that the Pfizer-BioNTech vaccine had met agency's exacting standards for safety, effectiveness, and consistent quality in manufacturing. On **August 18**, the CDC announced a new center, Center for Forecasting and Outbreak Analytics (CFA), aimed to improve the nation's ability to forecast and model emerging health threats, including pandemics like COVID-19, using data analytics. The HHS, CDC, and FDA issued a statement concluding that booster vaccination with Pfizer-BioNTech, Moderna, and Johnson & Johnson COVID-19 vaccines would be needed to protect against severe disease, hospitalization, and death.

Federal Actions Taken on September 2021 [4]

On **September 1**, the CDC issued a digital toolkit for individuals with intellectual and developmental disabilities and their caregivers to aid in navigating the COVID-19 pandemic, such as videos, stories, posters, and interactive activities related to being tested, vaccinated, masked, and social distanced. By that date, the CDC had issued forty videos and twenty-five web resources in American Sign Language about COVID-19. On **September 3**, the CDC announced an additional $300 million for community health worker services to support COVID-19 prevention and control. On **September 17**, the White House, with CDC, provided $2.1 Billion for state, local, and territorial public health departments for resources needed to prevent infections in healthcare settings, detect and contain infectious organisms, enhance laboratory capacity, and combat infectious disease threats, including COVID-19. On **September 24**, CDC's ACIP recommended Pfizer-BioNTech vaccine boosters for everyone aged 65 and older, residents of long-term care settings, individuals aged 50–64 years with underlying medical conditions, and those aged 18–49 years with underlying medical conditions and/or who lived or worked in high-risk settings to be given at least 6 months after their primary vaccination series. The CDC published three studies in MMWR on the effect of COVID-19 pandemic on education. Despite school closures in some areas, ca. 96% of K-12 schools remained open for in-person learning. Schools without universal indoor mask mandates were three times more likely to have COVID-19 outbreaks than K-12 schools that required universal masking.

Federal Actions Taken on October 2021 [4]

On **October 7**, a Federal report indicated more than 140,000 children in the USA lost their primary or secondary caregiver to the COVID-19 pandemic. One in every 168 American Indian and Alaska Native children, 1 in every 310 Black children,

1 in every 412 non-White Hispanic children, 1 in every 612 Asian children, and 1 in every 753 White children had experienced orphanhood or death of caregivers. The CDC added mental health conditions as a risk factor associated with severe illness from COVID-19. On **October 26**, CDC's ACIP recommended Moderna or Pfizer-BioNTech COVID-19 vaccine boosters for those who were 65 and older and all aged 18 years and older who were residents of long-term care settings and had underlying medical conditions, and those who lived or worked in high-risk settings, to be given at least 6 months after the primary vaccination series. ACIP also recommended booster vaccination for everyone who had received Johnson & Johnson COVID-19 vaccine more than 2 months previously (before August 25).

Federal Actions Taken on November 2021 [4]

On **November 3**, the CDC released a statement on the COVID-19 pandemic, noting more than four hundred different animals had been found to be infected with COVID-19 and, as a zoonotic virus, SARS-COV-2 could spread between people and animals. On **November 6**, all non-citizens traveling to the USA were required to be fully vaccinated and to provide proof of vaccination status to fly to the USA, as well as demonstrate a pre-departure negative COVID-19 test within 3 days of boarding their flights. On **November 10**, the CDC reported that more than twenty-two million infants missed their first dose of the measles vaccine in 2020, the largest global increase of unvaccinated children in two decades, attributed partially to disruption caused by the COVID-19 pandemic.

Federal Actions Taken on December 2021 [4]

On **December 3**, the FDA revised the EUA for the Eli Lilly monoclonal antibodies, bamlanivimab and etesevimab, allowing the drugs to be used together for treatment of mild to moderate COVID-19 in pediatric patients, including newborns, with a positive COVID-19 test and at elevated risk for progression to severe COVID-19. On **December 4**, the CDC introduced a new one-day testing policy requiring international travelers to have a negative pre-departure COVID-19 test 24 hours before boarding their flight. On **December 8**, the FDA issued an EUA for the AstraZeneca Evusheld, two monoclonal antibodies, for pre-exposure prevention of COVID-19 in adults and children ages 12 and older who weighed at least eighty-eight pounds and had moderate to severely compromised immune function or a history of a severe adverse reaction to COVID-19 vaccine. On **December 9**, the CDC and FDA expanded COVID-19 booster recommendations to include everyone aged 16 years and older. On **December 16**, the ACIP updated its recommendation, expressing clinical preference for individuals to receive, when possible, an mRNA COVID-19 vaccine, Pfizer-BioNTech or Moderna, rather than the Johnson & Johnson COVID-19 vaccine. On **December 17**, the CDC issued two reports in MMWR highlighting the use of test-to-stay practices to help keep students in school during the pandemic. On **December 20**, the CDC released data estimating the Omicron variant of COVID-19 was ca. 1.5 times more transmissible than the Delta variant. On **December 22**, the FDA authorized the Pfizer antiviral Paxlovid to treat COVID-19 under an EUA for adults and children aged 12 years and older who weighed at least 88 pounds, tested positive and was at elevated risk for progression to severe disease. It was the first treatment for COVID-19 that was taken orally and used at home. On **December 23**, the FDA authorized the Merck antiviral Molnupiravir for treatment of COVID-19, issued under an EUA, for all adults and children aged 18 years and older who tested positive and was at elevated risk for progression to severe disease. It was the second treatment for COVID-19 taken orally and used at home. Despite supply concerns, Paxlovid remained the preferred oral antiviral treatment for COVID-19. On **December 27**, the CDC shortened the recommended isolation period for individuals with COVID-19 to 5 days, followed by 5 days of wearing a mask around others if asymptomatic or if symptoms were resolving, i.e., no fever for 24 h. The CDC updated the quarantine period for individuals exposed to COVID-19 to wear a mask around others for 10 days and be tested on day 5 if boosted or vaccinated within the previous 6 months. If the exposed individual was unvaccinated, the CDC recommended a quarantine period of 5 days, followed by a strict mask use for an additional 5 days.

Federal Actions Taken on January 2022 [4]

On **January 3**, the USA reported nearly one million new COVID-19 infections—the highest daily total of any country in the world. The number of hospitalizations rose nearly 50% in just 1 week. The FDA amended the EUA for Pfizer-BioNTech COVID-19 vaccine to allow a single booster dose for all individuals aged 12–15 years, shortened the period between completion of the primary vaccination series of the Pfizer-BioNTech COVID-19 vaccine, and booster dose to at least 5 months and allowed a third primary series dose for immunocompromised children aged 5–11 years. On **January 4**, the ACIP recommended moderately or severely immunocompromised children aged 5–11 years receive an additional primary dose of Pfizer BioNTech vaccine 28 days after their second shot. On **January 7**, the FDA amended the EUA for Moderna COVID-19 vaccine to shorten the time between primary series of the vaccine and booster dose, from at least 5 months to 6 months for individuals aged 18 years and older. That same day, the CDC endorsed the FDA recommendation. On **January 11**, the White House purchased 600,000 doses of the GSK and Vir Biotechnology antibody Sotrovimab, after the drug had been found to be effective against both Delta and Omicron variants. On **January 14**, the CDC updated guidelines on masks to emphasize fit, comfort, and consistent wear, considering the daily average of new COVID-19 infections reported in the USA spiked from 119,215 to 805,062 in a single month. On **January 19**, the White House purchased one billion COVID-19 tests and created an online portal where individuals could order free at-home rapid antigen tests through the US Postal Service. On **January 20**, the CDC published data in MMWR indicating that during the Delta surge, both COVID-19 vaccination and surviving a prior infection provided protection against COVID-19 infection and hospitalization. On **January 24**, the FDA revised the EUA for the Eli Lilly monoclonal antibodies to limit their use after the treatments were shown to be ineffective against the Omicron variant. The FDA revised the EUA for the Regeneron monoclonal antibody, Regn-cov, to limit its use after the treatment was found to be ineffective against the Omicron variant. The Omicron variant accounted for 99% of all COVID-19 cases in the USA (January 2022). On **January 31**, the FDA fully approved Moderna COVID-19 vaccine for everyone aged 18 and older. The full FDA approval reinforced that the Moderna

COVID-19 vaccine met the agency high standards for safety, effectiveness, and consistent quality in manufacturing. By the end of January 2022, the US COVID-19 vaccines were estimated to have saved at least 250,000 lives and prevented more than a million hospitalizations.

Federal Actions Taken on February 2022 [4]

On **February 4**, the US death rate had climbed 30% in 2 weeks amid an Omicron surge, with more than 2600 deaths from COVID-19 each day and the number of recorded deaths in the USA caused by COVID-19 surpassing 900,000. On **February 11**, the CDC released data showing the COVID-19 boosters remained safe and were highly effective for everyone aged 5 years and older against severe disease caused by the Omicron and Delta variant surges. The CDC published data in the MMWR showing the Omicron variant incidence had risen from 1% in the USA to 99% of all infections in just 6 weeks. On **February 15**, the CDC published data in MMWR showing maternal COVID-19 vaccination during pregnancy with an mRNA vaccine, Pfizer-BioNTech or Moderna, reduced risk of COVID-19 hospitalization in infants (babies under 6 months) by approximately 60%. On **February 18**, the CDC published two reports in MMWR on pediatric emergency department visits during the pandemic, recommending increased awareness of health concerns among children and adolescents due to delayed medical care and heightened emotional distress, especially for adolescent girls aged 12–17 years.

Federal Actions Taken on March 2022 [4]

On **March 3**, the CDC updated the "COVID-19 Community Level" status, showing that more than 90% of the US population was in a location with either low or medium COVID-19 community transmission level. The USA provided more than 480 million COVID-19 vaccine doses as donations to 110 countries; On **March 4**, the CDC reported that, while access to COVID-19 vaccines had expanded, the gap in COVID-19 vaccination rates between urban and rural areas in the USA had more than doubled since April 2021. On **March 10**, the CDC recommended the Transportation Security Administration (TSA) extend the mask requirement for all public transportation and transportation hubs through April 18. On **March 11**, the CDC updated the "COVID-19 Community Level" status, showing more than 98% of the US populations was in a location with either a low or medium COVID-19 community transmission level. The CDC published data in MMWR showing that during August–October 2021 school districts in Arkansas that had a universal mask requirement had a 23% lower incidence of COVID-19 among staff and students, compared to districts without mask requirements. On **March 12**, the CDC estimated 23% of COVID-19 infections in the USA were caused by Omicron B4.2 variant, with initial data suggesting BA.2 appeared to be more transmissible than the Omicron BA.1 variant. On **March 15**, the CDC published data in MMWR showing COVID-19 hospitalization rate among infants and children aged 4 ears and younger was five times higher during the peak of the Omicron variant surge compared to the Delta variant—63% of those children hospitalized had no underlying medical conditions. The CDC outlined its goals for the third year of global COVID-19 response, namely to increase vaccination rates around the world, reduce spread of the COVID-19 virus, and strengthen public health leadership, while improving long-term health security worldwide. On **March 18**, the CDC published data in two MMWR showing adults who received three doses of a COVID-19 mRNA vaccine were 94% less likely to require a ventilator or die from COVID-19 during the Omicron surge, compared to unvaccinated adults in the USA, and that Black adults were four times more likely to be hospitalized than White adults (March 2022). On **March 24**, the US Census Bureau released data showing deaths in the USA, between 2019 and 2020, increased by approximately 19% after onset of the COVID-19 pandemic in March, 2020, citing it as the largest spike in mortality in the USA in 100 years. On **March 26**, the CDC estimated approximately 55% of all COVID-19 cases in the USA had been caused by the Omicron BA.2 subvariant. On **March 29**, the CDC and FDA recommended a second mRNA COVID-19 booster for immunocompromised individuals and all adults aged 50 years and older to be given 4 months after the last booster dose. The CDC recommended all adults who received a primary vaccine series and booster dose of the Johnson & Johnson COVID-19 vaccine should receive a second booster dose with an mRNA COVID-19 vaccine. On **March 31**, the CDC released data from the Adolescent Behaviors and Experiences Survey showing from January 2021 to June 2021, among high school aged adolescents, 44% reported feeling persistently sad or hopeless, 55% reported emotional abuse in the home and lesbian, gay, bisexual, and female youth reported the poorest overall mental health, the most emotional abuse by a parent or caregiver, and attempted suicide more than their peers during the COVID-19 pandemic.

Federal Actions Taken on April 2022 [4]

On April 1, the CDC announced termination of Title 42, an Order that suspended the right to introduce migrants into the USA due to the public health risk of COVID-19, effective May 23, 2022. On **April 8**, the CDC published findings in MMWR showing the risk for cardiac complications in all age groups was significantly higher after COVID-19 infection than for those with mRNA COVID-19 vaccination. On **April 18**, the CDC mask mandate for indoor public transportation was struck down by a judge in Florida. On **April 19**, the CDC published data in MMWR showing that three in ten children aged between 5 and 11 years hospitalized with COVID-19 in the first few months of the Omicron surge had no underlying medical conditions, of these 9 in 10 were unvaccinated and 2 in 10 required ICU care. On **April 21**, the Department of Homeland Security extended the COVID-19 vaccine requirement for all non-citizens entering the USA and also required to have had a negative antigen test taken within 1 day of departure. On **April 26,** the White House purchased 20 million doses of the Pfizer oral antiviral drug Paxlovid as part of the Test to Treat initiative in 2200 locations where Americans could receive free COVID-19 testing and, if needed, treatment. This doubled the number of pharmacies stocked with Paxlovid from 20,000 to 40,000. On **April 29,** data from the CDC National Commercial Laboratory Seroprevalence Study estimated that, as of February 2022, approximately 75% of children and adolescents showed infection-induced antibodies to SARS-CoV-2 (evidence of prior COVID-19 infection) and that, since December 2021, about 33% had become newly seropositive. On **April 30**, the CDC reported the proportion of the US population that had been fully vaccinated against COVID-19, by age group, was 28% (aged 5–11); 59% (aged 12–17); 69% (aged 18–49); 80% (aged 50–64); and 90% (aged 65 and older).

Federal Actions Taken on May 2022 [4]

On **May 3**, the CDC recommended everyone continue to wear masks while in indoor transportation hubs to prevent spread of COVID-19, but this was no longer legally enforceable. On **May 12**, the CDC reported the number of recorded deaths attributed to COVID-19 in the USA had reached 1,000,000. On **May 25**, the CDC published data in MMWR showing COVID-19 survivors were twice as likely to develop a pulmonary embolism or respiratory condition and that approximately 1 in 5 had at least one health condition that could be attributable to previous COVID-19 infection. On **May 26**, the CDC updated the "COVID-19 Community Level" status showing 71.52% of the US population was in a location with low COVID-19 community transmission, 20.73% with medium, and 7.7% high COVID-19 community level. On **May 31**, the US Department of Justice requested a Federal appeals court overturn the order declaring the CDC mandate requiring individuals to wear masks on public transportation unlawful. On **May 28**, the CDC reported the weekly average of new COVID-19 infections (119,725) in the USA was six times higher than in 2021 (17,887).

Federal Actions Taken on June 2022 [4]

On **June 1**, the CDC reported the USA had recorded a total of 84,145,569 COVID-19 infections and more than a million deaths caused by COVID-19 (1,003,571). On **June 18,** the ACIP recommended Moderna and Pfizer-BioNTech COVID-19 vaccines for everyone aged 6 months—5 years, expanding vaccine eligibility to more than 20 million additional children in the USA. All individuals aged 6 months and older were eligible for COVID-19 vaccination in the USA, as of 18 June 2022. On **June 24**, the ACIP recommended the Moderna COVID-19 vaccine for everyone aged 6–17. On **June 30**, the FDA called for Omicron-specific updates to COVID-19 vaccine boosters from Pfizer-BioNTech and Moderna in the fall of 2022.

Federal Actions Taken on July 2022

On **July 6,** the CDC data showed that the Omicron subvariants BA.4 and BA.5 were dominant in the USA, comprising more than 70% of new COVID-19 infections. On **July 8**, the FDA approved the Pfizer-BioNTech COVID-19 vaccine for everyone aged 12–15 years.

3 Conclusion

The Federal government took many actions during the two and a half years of the COVID-19 pandemic to support state, local, and tribal responses to the global pandemic that impacted far more than the health of our citizens. The USA, like many other countries, faced immense difficulties in dealing with the rapidly evolving public health emergency. The pandemic exposed significant gaps in the ability of Federal and state governments to communicate and collaborate effectively, ranging from securing essential medical supplies to implementing expanded surveillance and reporting systems to identify and track the spread of the virus. Actions to develop diagnostic tools, test protocols, and validate accuracy and reliability of testing methods were taken but not timely in implementation. With a more robust national network for laboratory testing and validation, the early efforts would have been better aligned and coordinated for implementation. Billions of dollars were provided to augment and fill gaps in medical and public health capacity and capability in the US states, territories, and localities.

There are many lessons to learn from the COVID-19 pandemic, but it is too soon to obtain an accurate and clear picture of what went right, wrong, and why. Shortages of hospitalists, intensivists, epidemiologists, nurses, and infectious disease scientists before the pandemic made the on the ground responses during the pandemic consistently an uphill battle. Health professionals coping with the demands of COVID-19 infections and deaths without sufficient staff, supplies, or protective equipment created frustration and burnout. Lack of trust and resulting hesitance to comply with conflicting information about effectiveness of diagnostic tests, treatments, vaccines, and nonclinical prevention protocols (masking, social distancing, and isolation /quarantine policies) severely impacted efforts to stem the spread of the virus before vaccines became available. Preventable deaths occurred because political leaders and evidence-based scientists did not interact effectively, especially in an environment saturated with 24/7 media coverage, that undermined the nation's confidence in both. Differences in political leadership actions at the Federal, state, and local levels complicated the ability to achieve unity of effort in executing a national response strategy.

Lack of standardized reporting procedures and systems made it extremely difficult to exchange timely and accurate data and information, especially in the early months of the pandemic, when a cohesive response was critical to address the crisis effectively. The Federal government took many good actions, but it was not prepared, as it could have been, to leverage a strong national response framework of standard methods, practices, and resources to enable timely and effective collaborative actions by states and localities. Instead, lacking common data systems and supply chain visibility for timely and effective decision-making, many key players had to work strenuously to understand data that were not timely or accurate and find fast, but suboptimal ways to integrate local public health measures into healthcare delivery systems.

To repair and build the nation's capacity to respond efficiently and effectively to both public health and healthcare needs of our citizens— all citizens—we need to optimize our capacity, capability, and efficiency in supporting critical services, regardless of who is in charge and who will pay for it. Envisioning the end state and outcomes needed will help define and characterize a national framework that clearly delineates stakeholders, authorities, roles and responsibilities, governance models, capabilities, resources, benefits, and costs savings.

The positive note is that we have an opportunity to leverage new technologies, data platforms, and informatics to enable more effective, timely, and responsive decision-making. A coherent, fully integrated, national public health emergency response requires aligned, supported (continuously funded), and linked community-based elements of the currently fragmented health sectors. Once established, an integrated framework for Federal, state, and local collaboration, data sharing, and evidence-based response focused on social determinants of health and a One Health model of health services will be possible.

References

1. National Academy of Medicine (2023) In: Anise A, Adams L, Ahmed M, Bailey A, Chua PS, Chukwurah CS, Cocchiola M, Cupito A, Kadakia K, Lee J, Williams A (eds) Emerging stronger from COVID-19: priorities for health system transformation, NAM Special Publication. The National Academies Press, Washington, DC. https://doi.org/10.17226/26657
2. Epané JP, Zengul F, Ramamonjiarivelo Z, McRoy L, Weech-Maldonado R (2023) Resources availability and COVID-19 mortality among US counties. Public Health 11:1098571. https://doi.org/10.3389/fpubh.2023.1098571
3. U.S. Department of Homeland Security, National Response Framework, October 28, 2019 National Response Framework (fema.gov)
4. CDC Museum COVID-19 Timeline | David J. Sencer CDC Museum | CDC. https://www.cdc.gov/museum/timeline/covid19.html
5. COVID Relief Spending | USAspending. https://www.usaspending.gov/disaster/covid-19

Polydemic-Wise Resilient and Regenerative Communities: Enhancing Health and Human Security During Pandemics

Michael D. McDonald

Parsimoniously defined, the polydemic approach empowers a pandemic-threatened community by enabling truthful and trusted unity of effort to provide scientifically based information, tools, and methods ahead of the spread of the disease agent to all the resident, collectively intelligent decision-makers in the community. To prevent and manage SARS-COV-2 and other highly infectious disease agents in times of compromised information, both cybernetic and biological influencers within and across connected communities must act with collective intelligence and more rapidly than pathogen transmission within and between polydemic-wise communities. This was extremely difficult to achieve, if not impossible, in the absence of accurate and reliable scientific knowledge. Today, advanced computing and communication allow the rapid creation of pandemic-informed community clusters armed with appropriate surveillance, critical analysis, and unity of effort to assert nonviolent, noncoercive transformative power, what Kenneth Boulding defines as "integrative power."[1]

The ability to achieve this kind of positive response depends on biologically and cybernetically linked communities to collectively discern and spread essential science-based information, risk communication, and algorithms, with sensitivity to local needs, culture, knowledge, and networks. Understandably, communities of place and similar beliefs (sometimes independent of place) can have diverse perceptions of need, potential rigidity of cultural identity, social associations, currency of misinformation, and human confusion between epistemology and ontology. Some

[1] Kenneth E. Boulding, *Three Faces of Power*, (Newbury Park, California: Sage Publications, 1989).

M. D. McDonald (✉)
Health Initiatives Foundation, Inc., Washington, DC, USA

OVIAR Global Resilience Systems, Washington, DC, USA

Resilient American Communities (RAC), Washington, DC, USA

J. M. Rosen, R. R. Colwell (eds.), *The COVID-19 Pandemic*,
https://doi.org/10.1007/978-3-031-62772-9_14

individuals and communities—even when provided with health-preserving and life-saving information, tools, and methods—may be unable or unwilling to make appropriate individual and collective decisions to take essential action in a timely manner to protect themselves, their loved ones, and others from the consequences of pandemic conditions and emerging severe polycrises.[2] However, even they will benefit from those around them lowering their risk of infection and transmission by reducing the probability of pathological biological and cybernetic transmission within their communities.

There is now a rich set of data available from 2020 to 2023 years of COVID-19 that can help determine which communities performed above and which below average, in terms of pandemic resilience and regeneration. This thread of research is valuable to understand resilience and regeneration in pandemics. It is also important to establish standards-based communities of practice to catalyze collective intelligence within and between communities collaboratively striving to be resilient and regenerative in facing a pandemic.

Unexpected elements arose damaging prevention, response, and recovery during the COVID-19 pandemic including the tendency to divisively influence policies and culture that elevated risk and vulnerability. All too often hierarchical command and control systems were unable or unwilling to provide sustainable health-preserving and life-sustaining solutions during rapidly changing conditions for all communities under their governance or management control. In the best of circumstances within democratic societies, hyper-local and local groups step in with collective intelligence to combat a pathogen in their midst when higher-level intervention is failing.[3,4]

There is evidence that polydemic intervention[5] can significantly reduce morbidity and mortality of infectious disease conditions in a community and between connected communities acting in a unified effort with additional improvements to the quality of life. The key to success in dynamically building resilience and regenerative capacity in polydemic communities is the adoption of a co-creative process to collectively examine scientifically valid, evidence-based information, and train community members to fully engage and rapidly spread effective collective decision-making among all truthful and trusted colleagues. When collective intelligence is engaged bottom up and in unity of effort before and ongoing during exposure to potentially lethal pathogens, polydemic-wise communities can achieve far better outcomes.

---

[2] For example, simultaneous health crises, climate change, economic decline, political upheaval, and social crises including but not limited to violence and war, in which information warfare is used by domestic and/or foreign perpetrators further spread by those unwilling or unable to care about and address individual and communitarian interests.

[3] https://www.washingtonpost.com/graphics/2020/politics/trump-covid-pandemic-dark-winter/

[4] https://www.newyorker.com/magazine/2023/01/16/the-dire-aftermath-of-chinas-untenable-zero-covid-policy#:~:text=Last month, China, after suppressing, by an enormous viral surge

[5] Networked interventions within and between communities that provide information and decision support inspiring unity of effort to raise resilience and regenerative capacities at hyper-local, local, and bioregional levels, independent of political affiliation to address problems in complex adaptive systems—even within communities historically disadvantaged by social inequities, injustices, and other factors impacting their social determinants of health.

By using advanced distributed collectively intelligent (DCI) grids[6] associated with a resilience system to monitor and manage mission critical functions in a community, it is hypothesized that societies comprised of polydemic-wise communities will fare far better in dealing with complex and novel threats, based on 2020 pandemic norms. It is asserted that when organized within societies supporting participatory democracy, with tools and methods validating and verifying scientifically derived policies, algorithms, knowledge, information, data, and convergent best practices, polydemic-wise communities (PWCs) will achieve health, human security, resilience, and regenerative outcomes far better than 2020 norms.

When PWCs operate within co-creative cultures that value the pursuit of truth through open discourse and critical analysis following the rule of law (sometimes operating in limited ways with waivers under emergency powers) in an agile, truth-seeking society with effective management and governance, extraordinarily effective results can be achieved. Even when under the stress of fast-moving novel crises, PWCs with well-coordinated, self-triggering action teams trained and trusted to act at the hyper-local level with focus, agility, and convergence succeed better than communities highly dependent on hierarchical command and control systems alone, especially in times of conflicting authorities.[7]

Resilient American Communities (RAC) and Alliance for Global Resilience and Regeneration (AGRR) sites are positioned to codesign and implement public health and biosecurity further upstream at hyper-local and bioregional levels to prevent and manage infectious disease outbreaks. RAC and AGRR community health workers (CHWs) and community resilience workers (CRWs) play an important role in achieving results. When these action teams are given defined authority to self-trigger with specified focus, agility, and convergent capacities within and at the edges of their polydemic communities with proper situational awareness and sensing capacities, members within their resilience capacity zones, resilience networks, and resilience systems can gain and collectively share substantial resistance to bio-cyber pathogens.

In theory and in practice, there is evidence from the COVID-19 pandemic (and other infectious disease and bio-cybersecurity events) that severe impacts of a pandemic or polycrises (cluster of multiple often synergistic crises) can be reduced, if not avoided or curtailed in polydemic-wise communities. The hypothesis is that we are now entering a time in which both biological agents (e.g., SARS-CoV-2) and cybernetic (e.g., disinformation, misinformation, and malicious algorithms) pathogens are able to be rapidly detected, identified, and neutralized. It also assumes the possibility that we may already be dealing with and will have to reduce the impacts

[6] Distributed collectively intelligent (DCI) grids are human/machine systems in complex adaptive systems that sense, analyze, anticipate, and report on vulnerabilities, resilience, and regenerative capacities of organizations, communities and societies acting in unity of effort to achieve positive outcomes in health, human security, and prosperity, based on valid and verified evidence and best scientific practices, technological approaches, with adherence to ethical principles.

[7] David Alberts, The Agility Advantage: A Survival Guide for Complex Enterprises and Endeavors, Command and Control and Cyber Research Portal, 2011.

of dynamic intelligent perpetrators within our midst purposely increasing chaos in our communities with escalating deleterious bio-cyber capacities. Therefore, we have to ensure destructive bio-cyber actors can be controlled and, in some cases, eliminated from distributed collectively intelligent grids.

There is growing evidence that some networked infrastructures, tools, and methods that were used to engage prevention and management of bio-cyber events may also be used to manage and govern social and economic trajectories of communities in biological and cybernetic security events, whether naturally occurring or human-induced. Resilience systems and distributed collectively intelligent (DCI) grids operating truthfully and with trust can lessen, if not prevent, otherwise accelerating and cascading impacts in hyper-local, local, and bioregional polydemic communities. To act wisely, PWCs using DCI grids must elevate their ability to anticipate (and articulate openly the level of uncertainty inherent in their modeling efforts) the possibilities of cascading impacts from errors of omission and errors of commission.

RAC and AGRR communities strive for improved health and human conditions as "polydemic-wise communities (PWCs)" when they join with other PWCs to achieve better collectively intelligent outcomes. PWCs deviate from traditional levels of prevention and is response dependent only on individual risk reduction and hierarchically organized emergency management by encouraging and improving constructive participatory management and governance in community resilience and regeneration with outcomes often exceeding those of professionals and representative leaders taking charge under the unity of command often with limited pertinent domain knowledge.

Methods to collectively reduce risk in PWCs include convergent interventions to address:

- *Syndemic effects within pandemics* associated with social determinants of health that create inordinate suffering and poor outcomes for people of color and other minority communities, as well as in states and communities highly influenced by disinformation and misinformation. Syndemic effects occur when two or more factors contribute to ill health and create diverse and chaotic outcomes in populations. For example, early in the COVID-19 pandemic, historically disadvantaged communities were more likely to suffer severe illness, hospitalization, and death, far exceeding the average.
- *Bio-cyber disinformation and misinformation* encouraging counterproductive behavior and social processes worsening transmission of disease as a result of undermining adaptive public health and biosecurity responses, leading to escalating morbidity, mortality, and socioecological degradation within communities and the broader society.
- *Formation of networks of resilient and regenerative communities* to withstand adverse conditions in times of accelerating complex change by encouraging co-creative, transformative processes within and between communities designed to effectively evolve as complex adaptive systems with enhanced collective intelligence.

At this time in history, important progress is being made in developing metagenomic, genetic, and memetic tools and methods. They have contributed significantly to improved health and quality of life. The focus here is primarily on how polydemic networks of communities can use vaccines and memes (e.g., physical distancing, masks, MERV 13 air filtration, immune system enhancements, and safe online services) to reduce the impact of bio-cyber conditions and improve functional capacity.[8]

Bio-cyber events, designed to propagate intentional disinformation or unintended negative misinformation, can cause severe biological and social damage and be further amplified as biological and cybernetic replicators accelerate and serve as catalysts and cofactors for exponential transmission. In the case of COVID-19, false information was introduced as "alternative facts," traveling at astounding rates via mass media, social media, and political persuasion by highly connected transmitters of misinformation. Simultaneously, SARS-CoV-2 infections were spreading globally, transmitted by infected travelers exposing the virus to communities and societies worldwide at an accelerated rate. Unfortunately, willing agents of disinformation and carriers of misinformation introduce rapidly replicating falsehoods for political, social, and/or economic gain. During a time of highly lethal infectious disease, the corrosive effects on communities and societies can spread at an even more accelerated rate, exemplified by the COVID-19 pandemic effect on the world body politic and the world economy in 2019 and 2020. Unfortunately, COVID-19 infections continue, albeit no longer in raging pandemic form (Fig. 1).

By approximately the end of 2021, the United States suffered over one million deaths. In contrast, former US partners in the Global Health Security Agenda (GHSA) continued to carry out good governance for pandemic management and were able to contain COVID-19 deaths per capita to a fraction of that in the United States. Six of the East Asian GHSA partners formerly collaborating with the United States, namely, the Philippines, Vietnam, Malaysia, Thailand, Cambodia, and Laos, with more population collectively than the United States, had less than 17% of the US COVID-19 deaths in the same period, 2020–2023.

A comparison of death rates in states, with respect to voting behavior, in the United States during COVID-19 is shown in Fig. 2. Anti-vaccine disinformation and misinformation are considered to have played a significant role.[9]

Death rates for Quarter 1, 2021, and Quarter 1, 2022, show stark contrast collectively, in terms of the COVID-19 death rate. Death rates for states with widespread behaviors resonant with messaging of anti-COVID-19 vaccination were

---

[8] During severe and life-threatening illness, such as Long COVID, individuals are compromised in their ability to function fully. Similarly, during pandemics and bio-cyber events affecting populations, the capacities of communities and societies are also diminished when key mission critical functions (e.g., business transactions, transportation, availability of health services, food availability, education, day care) are compromised. Even well individuals' abilities to function are diminished within their impacted social structures leading to economic decline and increased social conflicts. PWCs are more agile in engaging work-arounds, such as use of online digital services for food delivery, telehealth, tele-education, and safer movement within public spaces.

[9] Marazitti, Donatella, "Brainwashing by Social Media: A Threat to Freedom, a Risk for Dictatorship", Clin Neuropsychiatry. 2022 Oct; 19(5): 277–279. (https://www.ncbi.nlm.nih.gov/pmc/articles/PMC9597644/).

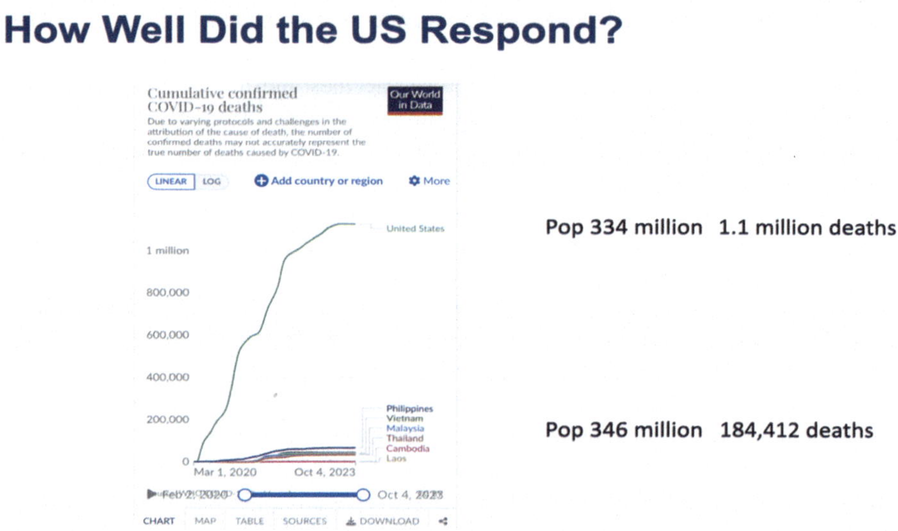

Fig. 1 The failure of U.S. Pandemic management during early COVID-19 pandemic

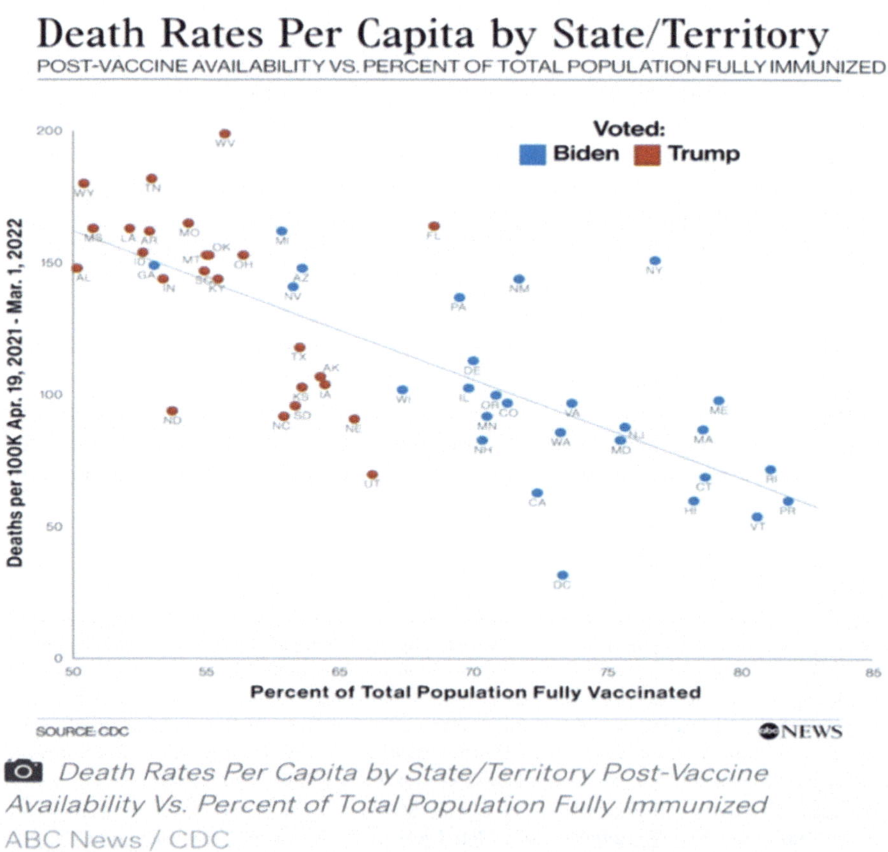

Fig. 2 Political rhetoric, mass behavior, and COVID mortality based on voting records by state

more than 38% higher than in states with significant vaccination.[10] "Hospitalization rates in the bottom 10 states are nearly four times higher and death rates are more than 5.5 times higher than in the top 10 states, according to analysis of federal data."[11]

1 Variable Risks and Syndemic Impacts

When impacted by pathogens, patterns of health and disease in populations most likely will be stochastic. Mathematically, they may also be more accurately depicted in simulation as interactive, chaotic, or complex.[12] If stochastic, they will reflect statistical patterns of varying probabilities determined by many dynamic variables, rather than being deterministic by nature (in which the causes and cures are entirely predictive of outcomes based on simple cause and effect interventions). Unfortunately, when large portions of specific communities are largely innumerate and do not think critically about the uncertainties involved, they tend to fall victim to misinformation and disinformation.

Given the number of variables that influence health and disease in communities, which are complex adaptive systems, many variables have to be considered, both holistically and influenced systematically with appropriate knowledge, to maintain and improve health status. This is particularly important in open communities (e.g., tourist towns) within the rapidly changing conditions associated with infectious disease spread, such as occurred during the evolving COVID-19 pandemic/endemicity.

Application of a model designed to optimize disease prevention and management in a polydemic-wise community requires continual monitoring of health status patterns. To serve its populations most effectively in infectious disease outbreaks, interventions must address differences in biological factors within and between individuals. They must also reduce inequities within and between communities and subpopulations, as well as differences in socioecological and environmental factors that can act as deleterious adjuvants or catalysts of biological and cybernetic pathogens.

A more comprehensive model of biological, lifestyle, medical, and environmental factors should be included in improving impact of the public health and related interventions to prevent and manage outbreaks, epidemics, pandemics, endemic conditions, and other biosecurity events.[13] Historically, in modern societies, this capability has been located only within communities of practice, namely health professionals working in public health and medical institutions. However, today, this knowledge is also directly accessible to the general public in communities out-

---

[10] https://abcnews.go.com/Health/red-blue-america-glaring-divide-covid-19-death/story?id=83649085

[11] https://www.cnn.com/2021/08/18/health/us-coronavirus-wednesday/index.html

[12] Four Ways of Thinking: Statistical, Interactive, Chaotic and Complex—David Sumpter https://www.youtube.com/watch?v=PPCfDe8TfJQ Oxford Mathematics

[13] Blum, Henrik L. Planning for Health: Development and Application of Social Change Theory, Human Sciences Press, 1974.

side clinical settings through the Internet and other media. The challenge is to build distributed collectively intelligent grids, resilience systems, resilience networks, and resilience capacity zones that directly enable the general public to embrace and execute polydemic-wise behavior.

Community risks vary with time, as disease agents mutate and population patterns change. Similarly, individual and group perceptions, decisions, and actions evolve relative to commonly accepted, scientifically based interventions to curb novel pathogenic agents and their variants. Advanced technological societies with strong and agile scientific communities of practice can shift rapidly to reduce the risk of biological and memetic threats by supporting polydemic-wise tools and methods (e.g., effective biological tests, vaccines, masks, air filters, risk communication, self-sequestering behaviors, quarantining, and isolation techniques, shaping individual and group behavior) in their own and other more vulnerable communities seeking assistance.

New practices that improve public perception and commitment to risk management and public health can address emerging challenges. However, if societal governance or public resistance inhibits adaptation, the outcome of risk communication can be minimal or negative. Executed properly with good governance, adept leadership, and strong community acceptance, new collectively intelligent, participatory approaches can transform norms and common practices to deal with increasing complexity threatening health and human security.

Unfortunately, in the United States during early 2020, the highest levels of societal governance crippled pandemic response. US public health institutions were defunded, and the United States suffered some of the highest levels of COVID-19 infection, morbidity, and mortality in the world during the first two waves of the pandemic. Highly infectious biological pathogens readily passed from one individual in their social groups to other social groups in the first year of the pandemic.

Once arriving in American communities, SARS-CoV-2 transmitted rapidly, with a pandemic impact, expanding rapidly, especially in super-spreader sites (e.g., schools, daycare, long-term care facilities, and healthcare facilities) and events (e.g., political rallies or large social gatherings like motorcycle rallies, weddings, and funerals). The destructive impact of the airborne disease agent was elevated by escalating disinformation and misinformation, crippling under-funded prevention and response activities. Less understood were confounding factors on the cybernetic side when during early waves of the pandemic disinformation, misinformation, or a medical center information system-disrupting computer virus or AI worm could damage community health and reduce societal pandemic resilience by diminishing collective resistance to both novel and the otherwise routine health threats, while escalating the negative impact on other mission critical functions.

Several years into the COVID-19 pandemic, individuals and small groups (in homes, childcare centers, schools, workplaces, multi-family buildings, and neighborhoods) are realizing the significant power at the hyper-local level to act with effective biological and cybernetic tools and methods to reduce risk at the point of potential infection. The benefits of wastewater testing, gathering reliable data on infections from reported PCR tests, home antigen tests, daily reports on hospitalizations and deaths, and employing effective air filters, masks, and vaccination are being recognized. The explosion of data and knowledge directly accessible to the general public at the community level derived from specialized mobile devices and apps in real time and near-real time with the added benefit of machine learning algorithms and AI to provide unprecedented situational awareness is just beginning to be appreciated. Accurate and comprehensive situational awareness can lead to reducing the chance of infections and adverse cascading effects if and when community members agree to alerts, warnings, notifications, and decision support to respond accordingly with collective sensitivity, without fear of privacy violation or manipulation by disinformation, misinformation, coercion, social ostracism, or violence.

Compare current information systems and the decision-making environment in the twenty-first century to the paucity of information available during the H1N1 influenza pandemic in 1918 and 1919. At that time, there were no computers, mass electronic media, Internet, or social media. From 2020 through 2023 at the time of each COVID-19 peak, the United States had rapid access to information on infection rate and the efficacy of public health knowledge and methods, as well as hospitalization, morbidity, and mortality rates on a daily and weekly basis for COVID-19. In many advanced PWCs, this information awareness was available at the census block level in hot spots through RAC and AGRR-related resilience networks and other sources.

It is abundantly clear why infection, morbidity, and mortality rates in the rapidly advancing RAC sites with near-ubiquitous access to scientifically verified and validated information and decision support during COVID-19 had vastly better outcomes in avoiding severe infection and premature death compared to the 1918 and 1919 flu pandemic. These variations are also demonstrated in the 2020–2023 COVID-19 data between PWCs and their sociodemographic peers in the twenty-first century that lacked access to pertinent evidence-based information in truthful and trusted exchanges.

The H1N1 and SARS-COV-2 public health/biosecurity crises, a century apart, were catalyzed by highly infectious, novel zoonotic viruses causing very high rates of severe illness and premature death, especially among specific cohorts at greatest risk, as the pathogens took on attributes of rapid human-to-human transmission through waves of viral mutation. In contrast, in the twenty-first century, communities with members willing to engage in science-based distributed collective intelligence fared far better in avoiding infection when also caring deeply about not engaging in behavior that would infect others. Their ability to avoid infection and transmission was vastly improved over those exposed to and accepting memes sowing distrust and promulgating dissension from effective public health measures, for example, masks, vaccines, and more comprehensive situational awareness augmented by truthful and trusted evidence-based information and decision support.

Knowledge and actions based on the best scientific public health knowledge, tools, and methods available to the general public in communities during the early twenty-first century saved millions of lives and provided ample social and economic benefit in free and democratic societies. It was unevenly distributed among

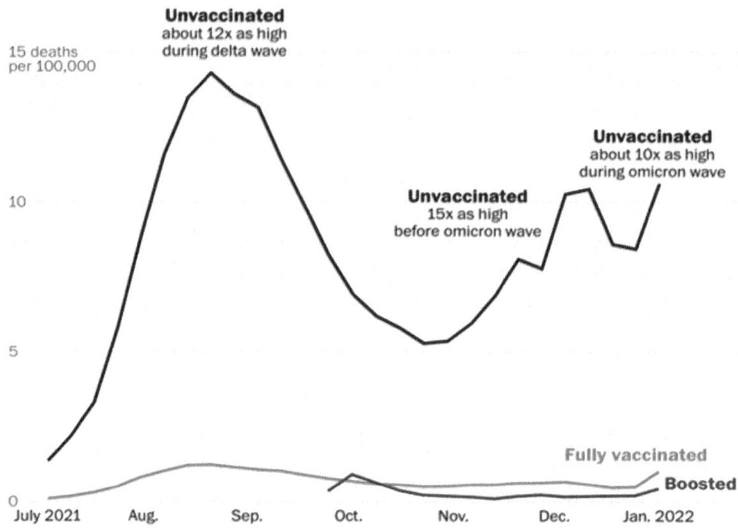

Fig. 3 Excess deaths associated with anti-vaccination

communities due to varying collective understanding and willingness to act for the common good (Fig. 3).

The patterns of infection, severe illness, and premature death worsened among anti-vaccine/anti-mask cohorts with poor levels of collective intelligence regarding COVID-19 risk and safeguards. Those who conformed to non-efficacious vaccine "alternatives" (e.g., ivermectin, hydroxychloroquine, chloroquine) were more likely to support and transmit misinformation and disinformation.[14] By mid- to late 2021, those adhering to anti-COVID-19 vaccine beliefs died from SARS-CoV-2 infection at a rate 10–15 times the rate of those wearing masks with mRNA vaccine-induced immunity to COVID-19.

Early on in the COVID-19 pandemic, many dependent elderly, those with co-morbidities hence elevated risk (overweight, diabetes, immune system suppression), and frontline workers (e.g., nurses and ambulance personnel) who could not control exposure to the threat of SARS-CoV-2 infection had extremely high rates of infection, severe illness, and premature death before mRNA and other effective COVID-19 vaccines were available. In contrast, those who wore quality masks, had better collective information, and followed effective self-sequestering, honoring public health mandates (such as improved hygiene, isolation, quarantine, and active avoidance of super-spreader sites and events) had better outcomes through non-pharmaceutical means. This new level of situational awareness for those cohorts that had locus of control to avoid risk of infection early on in the pandemic fared far better, with reduced COVID-19 risk.

[14] Eysenbach, Gunther, "Impact of Trump's Promotion of Unproven COVID-19 Treatments and Subsequent Internet Trends: Observational Study", J Med Internet Res. 2020 Nov; 22(11): e20044.

About 6–12 months into the pandemic, significantly different morbidity and mortality rates were noted among communities. Starting in this time, Resilient American Communities (RAC) initiated activities in association with the US Resilience System and utilized its dashboards and convocations to explore emerging equity issues, with better data on subpopulation infection, hospitalization, severe illness, and excess death rates. Within the first pandemic year (2020) in the United States, Native American, African American, Latin American Communities, and other minority groups showed significantly elevated morbidity and mortality associated with SARS-CoV-2 infection. Elevated morbidity and mortality continued into the second year of the pandemic for BIPOC communities. Focusing on addressing social, climate, economic, and environmental injustice and social determinants of health leading up to the third wave of COVID-19, COVID-19 morbidity and mortality rates began to improve in some of the most impacted communities following the first two waves of COVID-19.

By the summer and fall of 2020, a type of collective intelligence began to emerge in American Communities among those actively striving to build resilience. In 2021, and into 2022 and 2023, a more unusual and unexpected pattern emerged as rates of vaccination with mRNA COVID-19 vaccines induced vaccine-acquired immunity for those who chose to be vaccinated for free. Yet, sources of disinformation and networks of misinformation proliferated in populations associated with specific political and religious beliefs, even while acceptance of vaccination, masking, and social distancing was building rapidly in the broader population, including historically marginalized and disenfranchised populations.

Some modelers asserted in 2022 that Sweden had achieved herd immunity twice.[12] According to the models, Sweden reached herd immunity to the Wuhan strain in early December 2020 and again to the alpha strain at the beginning of April in 2021. These reports stimulated debate, especially on whether solely bottom-up pandemic practices were more advantageous than strict top down interventions when situational awareness is readily available to populations having strong trust in good participatory governance in a more homogenous population as in Sweden.

Public health institutions have long based infectious disease response on the belief that outbreaks, epidemics, and pandemics can be controlled, at least to some degree, by preventing the spread of the disease agent. The value of public health derives from its ability to reduce adverse effects on individuals, communities, and societies through preventive actions. In 2021, 2022, and 2023, science-based public health practices provided even more power directly to the general public in elevating their role in limiting negative impacts during the COVID-19 pandemic. "A new analysis finds that in the two years since the first US patient received a COVID-19 vaccine, the country's vaccination program has prevented more than 18 million hospitalizations and more than 3 million deaths ..." (in the United States).[15] The US COVID-19 vaccination program has saved the United States more than $1 trillion in medical costs and preserved hospital resources, kept children in school, and allowed

[15] https://www.statnews.com/2022/12/13/covid-vaccines-prevented-3-million-deaths/

reopening businesses and other activities. A key factor in the success of vaccination and other preventive measures is the public's trust and willingness to engage them.

Factors elevating morbidity and mortality correlate with "social determinants of health," varying by socioeconomic status and community affiliation, and other memetic factors. These emerged during the COVID-19 pandemic as a "syndemic," that is, synergistic epidemic with aggregation of two or more concurrent or sequential epidemics or disease clusters in a population, with biological and social interactions exacerbating prognosis of illness.[16] As RAC and other groups focused on prevention and management of syndemic effects by addressing social determinants of health emphasizing social equity, patterns of COVID-19 infection, morbidity, and mortality rates began to shift in more positive directions for vulnerable populations.

Sources and patterns of disinformation and misinformation simultaneously became clear as social network analysis, graph mathematics, and machine learning algorithms were applied to analyze and improve datasets and data management systems by considering variables designed to understand syndemic effects in communities as a unit of analysis. Where simpler, traditional, unifactorial, and deterministic biological views of pandemics and epidemics caused by a single pathogen were prominent, outcomes were poor especially when influenced by divisive political, economic, and religiously associated influencers.[17] Communities fared better when patterns of health and disease were understood in the context of complex adaptive systems informed by complexity science, for example, in PWCs.

Community-based complex adaptive systems aided by health informatics, ubiquitous computing, and proliferating networks with improving levels of community-based risk data (e.g., COVID-19 wastewater analysis, COVID-19 nasal tests) in an environment of trust and collective intelligence provided better situational awareness. Polydemic-wise bottom-up communities of practice improve compliance with public health practices to suppress COVID-19, hence better health, social, and economic outcomes.

During 2002, the Disaster Knowledge Management System (DKMS) gathered information about outbreaks in China through word-of-mouth reports, requiring months to be confirmed by the World Health Organization as the pathogen spread to other countries. "SARS" (SARS-CoV) was recognized as a potential pandemic agent through school attendance records in Guangdong Province and other syndromic data, taking months for the alarm to be raised. However, there was limited

[16] Merrill, S. (2009) Introduction to Syndemics: A Critical Systems Approach to Public and Community Health *Wiley* p. 304.

[17] Misinformation Research is Buckling Under GOP Legal Attacks, https://www.washingtonpost.com/technology/2023/09/23/online-misinformation-jim-jordan/?pwapi_token=eyJ0eXAiOiJKV1QiLCJhbGciOiJIUzI1NiJ9.eyJyZWFzb24iOiJnaWZ0IiwibmJmIjoxNjk1NDQxNjAwLCJpc3MiOiJzdWJzY3JpcHRpb25zIiwiZXhwIjoxNjk2ODIzOTk5LCJpYXQiOjE2OTU0NDE2MDAsImp0aSI6IjViNGIyNTliLWQyMDYtNDVjMy04NDA2LTM3YmQ5MTliMDM0YiIsInVybCI6Imh0dHBzOi8vd3d3Lndhc2hpbmd0b25wb3N0LmNvbS90ZWNobm9sb2d5LzIwMjMvMDkvMjMvb25saW51LW1pc21uZm9ybWF0aW9uLWppbS1qb3JkYW4vIn0.HHCodmnsq8WQRY7fd6zeHYg_lRFgvNyGqrpIIYT97Bs

sharing of data and information at the hyper-local and local levels. CDC and WHO took months to alert the world of the spread of this airborne human pathogen, given the inclination of China to suppress early indicators of the emerging disease.[18]

In December 2019, cases of the disease were spreading in Wuhan, unidentified by almost all global health specialists outside of China. In those early days before definitive genetic analysis, many potential causes were considered. However, the emerging new cases in 2019 had strong similarities to those identified in the early findings of the highly pathogenic agent now commonly known as severe acute respiratory syndrome (SARS) that emerged during November of 2002 in China.

In the 2020s, as news of an emerging disease was coming to the attention of international scientists supported by disease surveillance systems, those working within communities of practice with previous experience with SARS in 2002 became increasingly alarmed. For months in both 2002 and 2019, the Chinese government tried to suppress information about the emerging SARS-CoV outbreaks, but in early 2020 with improved disease surveillance networks and increased density of social network penetration globally, suspicions of an emerging pandemic exploded around the world. While China shut down domestic airline traffic, international flights out of China continued unabated.

Given the need for rapid response to large-scale, high-severity biosecurity and disaster management events, RAC was formed at the beginning of the COVID-19 pandemic to engage bottom-up disease surveillance and response to SARS-COV-2. RAC coordinated early rapid response efforts at the local resilience network level (county and community levels) and at the hyper-local resilience capacity zone (i.e., household, neighborhood/village/organization/census tract) level in a series of communities with high syndemic vulnerabilities. For example, where these subsystems of the Florida Resilience Systems and their tools and methods have been engaged fully and equitably with collective intelligence down to neighborhood and household levels, Resilient American Communities have made important contributions to the prevention and management of pandemic and syndemic impacts throughout the COVID-19 pandemic, with progressively increasing influences from 2021 forward.

The Resilience Systems' focus on scientific principles, critical analysis, the pursuit of truth, requirements of evidence, and equitable care for the most vulnerable populations reinforced by equitable availability of cutting-edge data management and knowledge management have played crucial roles in benefiting communities and their members. Enhanced situational awareness based in the sciences of complexity associated with the use of the resilience systems and their nested subsystems (Resilience Network dashboards, serious games, and mission critical function assessments in resilience capacity zones) provided in association with Resilient American Communities (RAC) and the Alliance for Global Resilience and Regeneration (AGRR) further elevated positive outcomes. The collective intelligence within RAC and AGRR unities of effort engage the benefits of biological and cybernetic advancements to lower risks while elevating resilience and regeneration, especially among the most vulnerable parts of the society where unity of command

[18] https://www.ncbi.nlm.nih.gov/pmc/articles/PMC8814736/

has been suboptimal or has failed. When working with nascent RAC and AGRR distributed collectively intelligent grids, PWCs can now achieve far more positive outcomes consistent with the expectations of the public—in part due to transcending their focus on distrust in mainstream institutions that have historically treated them poorly. Ultimately, this type of unity of effort has been proven to provide progressively better results through RAC collective intelligence, even before their distributed collectively intelligent (DCI) grids have the substantial benefits expected with rapidly improving enhancements in association with machine learning and artificial intelligence.

In 2020, 2021, 2022, 2023, and into 2024, the struggle to suppress SARS-CoV-2 infections and COVID-19 morbidity and mortality (as well as other elements of intensifying polycrises) in communities striving for resilience and regeneration has intensified utilizing a proliferation of more powerful networked and inter-networked information technologies grounded in science as well as with validated and verified data and evidence. On the one hand, the emergence of more democratized citizen science tools and methods available through social media with rapidly proliferating machine learning and AI provide new power at the local, hyper-local, and bioregional levels. However, the democratized production of knowledge has led not only to new powers in distributed public sense-making with better decision-making benefiting the public.

There is evidence that purveyors of false information about COVID-19 vaccines and infection-protecting mask use have contributed to more than three times the deaths of Americans in the first 3 years of the COVID-19 pandemic than from the American casualties in the Vietnam war.[19] Such are the casualties of modern culture wars within 24-hour media cycles and social media, when they are not able to be overcome by evidence-based distributed collective intelligence and the reestablishment of democratic norms and principles at all levels of management and governance. Although it has not yet been definitively established in a court of law as scientifically true regarding COVID-19, there is ample evidence that repeated lies and propaganda (e..g., "COVID-19 mRNA vaccines are unproven to be safe and efficacious") by trusted yet untrustworthy authorities in emergent conditions may be as damaging to the human mind's discernment — and the collective intelligence of communities — as repeated exposure to close explosions are to the brain.[20] [21]

Negative outcomes of the COVID-19 pandemic are attributable not only to the super-spreading biological replication of SARS-CoV-2 but also to the cybernetic replication of algorithms and memes shaping "alternative facts" and behaviors inconsistent with resilience and regeneration in a rapidly changing world. The rudimentary manipulation of memes is not new, but it is beginning to be better understood in the scientific disciplines of memetics, knowledge science, cognitive psychology, and the sciences of complexity. With solid science and evidence-based

[19] Hotez, Peter J., The Deadly Rise of Anti-science: A Scientist's Warning, JHU Press, 2023 https://www.youtube.com/watch?v=XGhBTKzrJEY&t=112s

[20] https://www.netreputation.com/the-power-and-impact-of-propaganda/

[21] https://disinformation-nation.org/why-propaganda-matters/

facts in hand, information warfare can be managed and litigated against upstream by strengthening distributed collectively intelligent (DCI) grids at the hyper-local, local, and bioregional levels identifying sources of information and supporting resilience and regenerative systems.[22,23,24]

The United States entered into the fifth wave of the COVID-19 pandemic in the summer and fall of 2023 with the vast majority of public health and medical professionals holding the following facts as true, helpful, and supported by ample evidence:

- SARS-CoV-2 continues to mutate remaining a public health concern.
- COVID-19 vaccines create safe and efficacious immune response to dominant SARS-CoV-2 variants and subvariants, reducing the likelihood of severe illness and premature death.
- Well-fitted, high-quality masks help protect against SARS-CoV-2 transmission.
- Aging populations and those with co-morbid conditions are at higher risk. Family, friends, and caregivers need to make collective efforts to be especially careful to suppress transmission of SAR-CoV-2 in their presence.
- High-quality MERV 13 filtration systems are effective in reducing transmission in spaces where individuals infected with SARS-CoV and other airborne pathogens may be shedding virus.
- COVID-19 testing results from wastewater, home antigen tests, and PCR tests should be ubiquitously accessible for proof of COVID-19 infection and transmission.
- Communities will benefit from consensus about self-sequestering, quarantines, and isolation practices in COVID-19, flu, RSV, and other infectious disease hotspots.
- Super-spreader sites and events should be avoided when SARS-CoV-2 infections have been identified in those sites. Full disclosure, while protecting privacy, is recommended to avoid disease transmission.
- Disinformation should be identified and disclosed.
- Frontline workers must be trained to reduce risk to themselves and others while being suitably acknowledged and rewarded for their public service.

2 With Hindsight and Looking Forward (Fig. 4)

The general public and health professionals have a better perspective on the emergence and ongoing patterns of COVID-19 since the first known cases occurred in Wuhan, China, during late 2019. Vaccines are available against COVID-19 and will remain critical for community health, resilience, and regeneration, especially as

[22] Folkenflik, David, Yang, Mary, "Fox News settles blockbuster defamation lawsuit with Dominion Voting Systems," NPR, April 23, 2023.

[23] https://www.wnycstudios.org/podcasts/otm/episodes/exxonmobil-big-tobacco1 (legal suits against EXXON Mobil for falsifying climate information Merchants of Doubt).

[24] Hals, Tom and Stempel, Jonathan, "Alex Jones files for bankruptcy following $1.5 billion Sandy Hook verdicts," Reuters, December 2, 2022.

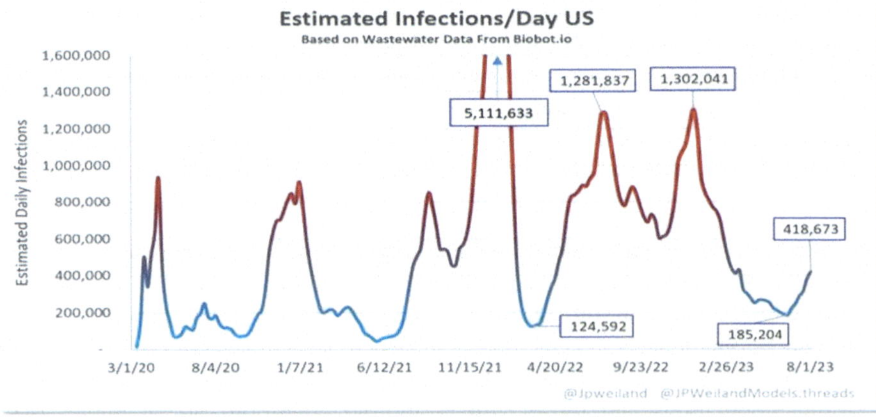

Fig. 4 Monitoring waves of COVID infections to reduce mortality and morbidity

new COVID-19 subvariants emerge. Collective efforts will likely take years if not decades to mature with an incorporation of improvements in the sciences of complexity and transformative technologies relative to PWCs. Yet, immediate value is available today in nascent RAC and AGRR sites within fast-evolving resilience capacity zones and resilience networks, linked to metroplex, national, and global resilience systems.

Unity of effort, based on improvements in distributed collectively intelligent (DCI) grids associated with resilience systems, where the public has direct access to validated and verified real-time information grounded in geospatial intelligence and decision support empowered with well-refined ML, AI, and graph mathematics is crucially important. Public understanding of protection from bio-cyber events and crises at the hyper-local, local, and bioregional levels will play an increasingly important collective role in the suppression of anti-science disinformation and misinformation to public health through verified and validated crowd-sourcing (Fig. 5).

Health and human security in the State of New Hampshire, for example, has been negatively impacted by extremist coercion and violent threats. These negative impacts had become so extreme in 2021 that acceptance of CDC funds by the Governor's Executive Council was blocked temporarily by armed protesters outside and entering the chambers in which the routine processing of public health funds and other funding bills were to be enacted, including but not limited to vaccination programs.[25][26][27] With public understanding that the SARS-COV-2 virus is not going

[25] https://www.nhpr.org/nh-news/2021-09-29/nh-executive-council-meeting-postponed-due-to-safety-concerns

[26] https://www.wmur.com/article/new-hampshire-executive-council-meeting-postponed/37790951#

[27] https://www.thetruthaboutguns.com/new-hampshire-governor-cancels-outdoor-inauguration-over-armed-protesters/

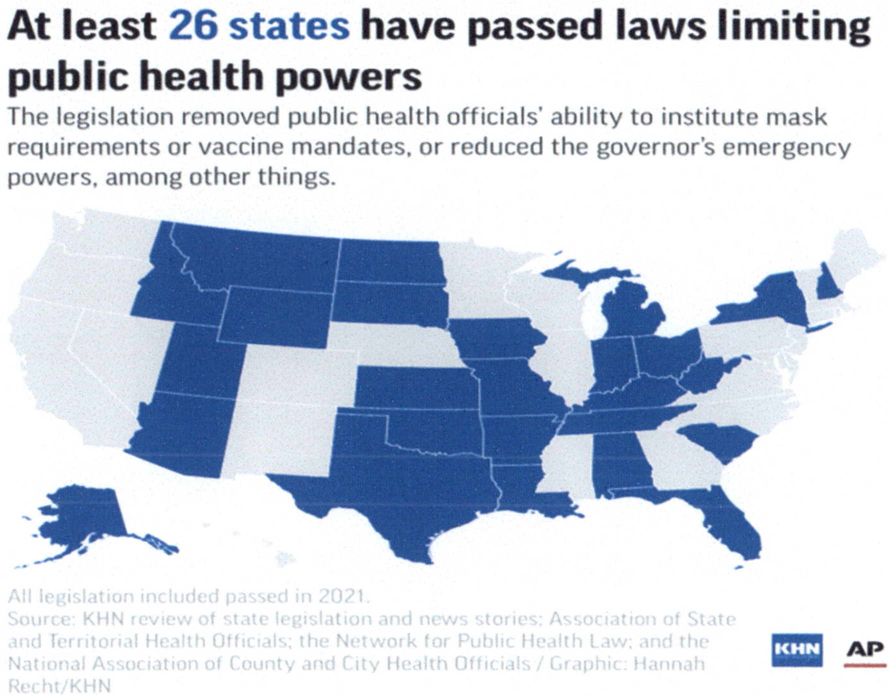

Fig. 5 Political disruption of public health emergency power

away and is continuing to mutate, new patterns of individual and collective behaviors will need to be engaged to lower risk with deepening understanding of how to advance protection against COVID-19 in the midst of efforts to create barriers to effective public health practices.

3 Conclusion

Community-based systems are needed to guide human populations with collective intelligence at hyper-local, local, and bioregional levels. Polydemic-wise communities (PWCs) can play a role going forward empowered with distributed collectively intelligent (DCI) grids interoperating within resilience systems, resilience networks, and resilience capacity zones. Communities striving to become more resilient and regenerative will need to refine and enhance their tele-cognitive infrastructures and embrace skills that improve the health and security of their residents with far less damage from disinformation and misinformation. Lessons learned by RAC and AGRR sites about bio-cyber events during the COVID-19 pandemic can be helpful in addressing new bio-cybersecurity threats. DCI grids can play an important role by replicating and scaling the influences of PWCs.

Economic Consequences of Pandemics and How to Prevent Them

Reda Cherif, Fuad Hasanov, and Henry Chen

1 Introduction

The COVID-19 pandemic has shown that a pandemic can create sudden and severe supply and demand disruptions and alter the structure of economies, unleashing a force that is difficult to control using standard policy tools (see [15, 16]). There are simultaneous shifts in the demand structure and labor supply and a multiplication of shocks affecting tastes and production. Moreover, amid heightened uncertainty the extent of policy intervention and its timing become difficult to assess.

The economic and social costs of the COVID-19 pandemic in the USA have been unprecedented. The country has witnessed a colossal number of deaths, extraordinary strains on the healthcare system and medical personnel, national lockdowns, severe social, learning, and health disruptions, and unparalleled economic fallout. As of December 2022, more than one million people died [63], while estimated excess deaths are over 1.2 million [91].

Prepared by Reda Cherif (IMF, University of Cambridge), Fuad Hasanov (Corresponding author: IMF, Georgetown University, University of Cambridge), and Henry Chen (Georgetown University). The views expressed in this paper are those of the authors and do not necessarily represent the views of the IMF, its executive board, or IMF management.

R. Cherif
IMF, Washington, DC, USA

Bennett Institute for Public Policy, University of Cambridge, Cambridge, UK

F. Hasanov (✉)
Bennett Institute for Public Policy, University of Cambridge, Cambridge, UK

IMF and Georgetown University, Washington, DC, USA
e-mail: FHasanov@imf.org

H. Chen
Georgetown University, Washington, DC, USA

J. M. Rosen, R. R. Colwell (eds.), *The COVID-19 Pandemic*,
https://doi.org/10.1007/978-3-031-62772-9_15

The pandemic's toll on the US economy, even if measured in temporary losses of output and fiscal support provided, is staggering, on the order of seven trillion dollars in 2020–21, about a third of the US GDP [30]. In the second quarter of 2020, as the country went into the lockdown, real GDP contracted by about 9%, or about 1.8 trillion dollars, a whopping 150 billion dollars a month (on an annual basis, about a 30% drop). As lockdowns eased, output recovered, but nonetheless, by the end of the year, output fell by 3.4% due to a fall in demand and trade and supply chain disruptions [60]. It was a larger decline in output and income than the fall witnessed during the Great Recession of 2009 (see Fig. 1). Before the pandemic in January 2020, the IMF projected a 2% growth rate for the US economy [59]. The resulting loss in output was more than a trillion dollars in 2020 (about 5% of 2019 real GDP). Although the US economy had rebounded strongly due to the unprecedented fiscal stimulus—about 5.8 trillion dollars in 2020–21, more than 25% of GDP [61]—and large-scale monetary expansion, the estimated gap in 2021 was still about 150 billion dollars (0.7% of 2019 real GDP).

A prolonged pandemic with long periods of lockdowns or social distancing, business and school closures, and heightened uncertainty would not only be economically costly but could also entail unintended economic and social consequences. If the pandemic is protracted, it could result in permanent income losses, discouraging workers, eroding human capital, slowing down investment and research, and worsening the future job prospects of young workers [24, 25, 69, 70]. The strain on the healthcare systems and medical personnel, foregone medical procedures, premature deaths, mental well-being of population, civil strife, and increased inequality given that many low-income workers cannot work from home are other costs of the pandemic that do not much show up in the GDP calculations. One study estimates health costs of the pandemic at about 8 trillion dollars [36], bringing the 2020–21 total economic and health costs to about 15 trillion dollars (about 75% of GDP). Another study puts the total economic cost of the pandemic, which included standard economic costs such as those resulting from business closures and direct health costs but excluded long-term health and learning costs, at about 14 trillion dollars for 2020–2023 [96]. Human capital losses are even more staggering, between 14 and 28 trillion dollars in terms of foregone GDP [56].

Fig. 1 US real GDP (2007–2020) and forecast (2020–2021). (Source: World Development Indicators and IMF World Economic Outlook)

The cost-benefit analysis of fighting the pandemic with all means possible is clear. Even if the cost of tests, vaccines, and increased medical facilities and personnel were in the hundreds of billions of dollars a year, it would still be a fraction of the exorbitant cost of the pandemic measured in trillions of dollars. A simple example illustrates this point. Universal testing (15% of the population daily, or about once a week, and 5 dollars per test), costing less than 100 billion dollars annually, would still be worth it even if the chance of success were about 1% [30].

The pandemic clearly illustrated that minimizing the profound consequences of pandemics requires a preventive risk-based approach. Rather than devising optimal, or largest bang for the buck, economic policies to mop up after the pandemic hits, an optimal economic policy is to prevent pandemics. Akin to the crisis prevention framework for the financial industry implemented after the 2008 financial crisis, fighting pandemics requires a pandemic prevention infrastructure. This infrastructure includes monitoring and surveillance of potential pathogens, platforms to develop tests, vaccines, and cures ex ante, the network to produce, scale up, and distribute the medical solutions globally, and an institution to coordinate across major stakeholders. This risk-based approach is minuscule in cost compared to the pandemic fallout.

In this chapter, we give an overview of the effects of the pandemic on the US economy. In particular, we give an account of the short-term and potential long-term effects. We also study the inflationary pressures that appeared during the pandemic and the difficulty of conducting macroeconomic policy amid such large shocks, representing a major cost on the economy. We then turn to proposing a risk-based framework to tackle future pandemics based on three key principles: surveillance, prevention, and coordination.

2 Colossal Impact of the COVID-19 Pandemic

2.1 Sectoral Effects

The COVID-19 pandemic led to sharp declines in economic sectors dependent on in-person and close interactions such as retail, aviation, and leisure and hospitality. Other sectors such as e-commerce saw a boom for precisely the same reasons, that is, a substitute for these services.

The retail, aviation, and hospitality sectors suffered substantially when the pandemic hit. Retail sales fell by 20% in the first quarter of the pandemic although they recovered within 6 months [17]. In contrast, during the 2008 financial crisis, retail sales bottomed after more than a year, falling by 12%, and only recovered their precrisis level after 60 months. Air transportation is another sector that suffered severely from the pandemic [5]. The volume of supply relative to the size of the economy declined by more than 30% in 2020 compared to the 2010–2019 average, and it was slightly below the average in 2022. Airlines suffered from a combination of resource constraints, labor shortages, and input costs. Leisure/hospitality is another notable sector given the extent of its decline and its contribution to job losses. At the height of the pandemic, from February 2020 to April 2020, the sector made up 40% of all job losses, and despite the recovery, only a fraction of the jobs lost were recovered in 2022.

Other sectors benefitted from the pandemic, especially those leveraging digital technologies and replacing in-person interactions. In particular, retail e-commerce, as a share of total retail sales, jumped in the second quarter of 2020 from about 10% to 15% and remained around that level until the third quarter of 2022, indicating a permanent increase [94]. This marks a stark acceleration compared to the linear pre-pandemic trend when the share increased by about 0.5% per year since 2013. There was also a rise in digital services. For example, mental health specialists estimated that close to 40% of their 2020 revenue came from telemedicine services [95]. Indeed, Haltiwanger [55] finds that the non-store retail industry accounted for nearly 33% of the entrepreneurial surge that began in the second half of 2020, by far the largest share.

The pandemic greatly disrupted the agricultural sector, leading to a paradoxical situation of simultaneous food shortages and waste. Many farmers were unable to distribute their produce to their clients in the food services industry (e.g., restaurants), and due to the compartmentalized nature of the food market, they could not redistribute the produce to retail stores, which experienced shortages. This forced both producers and government to quickly coordinate a response. On the one hand, lockdown measures implemented during the pandemic reduced consumer expenditures on food service meals by over 50% [26], causing monthly revenues for food services at restaurants and bars to drop by 47.5 billion dollars year over year [62]. Chenarides et al. [26] argue that this disruption exposed what was already a critically weak food supply chain, as many firms either relied predominantly on food service distribution or retail distribution, instead of a flexible mixture of both. This resulted in staggering levels of food waste among parts of the supply chain, as food services-reliant firms saw greatly reduced demand [98].

On the flip side of the supply chain, lockdown measures and trade restrictions may have exacerbated food insecurity. The lockdown caused retail food supply shortages and increased grocery prices, which were especially severe in the meat market [84], leading to a spike in food insecurity that affected nearly 15% of households [80]. Ziliak [100] finds that pandemic-era food insecurity more than doubled relative to the Great Recession of 2009, with senior citizens especially hard hit. Interestingly, stock prices of major food supply chain companies reflected the mismatch between food distributors and food retailers, with the distributors witnessing much higher price volatility and perceived risk than retailers [57]. There is evidence that trade restrictions may have also played a role, as restrictions applied during the pandemic on agricultural products, particularly by China and the USA, increased global agricultural price volatility, leading to higher prices and greater food insecurity [99].

To help address the food supply chain imbalance, the federal and state governments instituted various programs. One such program, the Farm to Families Food Box Program, connected producers with large surpluses with families suffering from food insecurity [42, 90]. Moreover, Tian et al. [92] find that community food services, such as food banks and pantries, also played an important role in alleviating food insufficiency, especially for lower- and middle-class families.

2.2 Labor Market

The effects of the COVID-19 pandemic on the US economy were significant and widespread. One area especially affected was the labor market, where the labor force participation rate and employment-to-population ratio both fell to historic lows and have yet to return to their pre-pandemic levels after 3 years. Research suggests that this decrease was explained to a large extent by three factors: an increase in early retirements, more workers staying home to care for children and relatives [58], and an increase in substance abuse among prime-aged workers [51].

A sizable share of workers who left the labor market retired. Compared to the trend, an extra 2.8 million workers retired over the first 18 months of the pandemic [19]. The factors contributing to their decision included fear of COVID-19, which is more lethal for older populations, and a lower level of spending because of savings on services and higher asset prices, including those on housing, equity, and bonds. Albanesi and Ki [8] find that the workers who left for childcare reasons tended to be women, especially those with children, causing women's employment to drop by much larger amounts than men's employment. Alongside the drop in the number of workers, Faberman et al. [45] find substantial labor market underutilization in terms of aggregate hours, with the number of desired labor hours decreasing across all workers, including those who remained in the labor force, suggesting that the labor market is even tighter than headline statistics imply.

The lagging effect of policy further contributed to the slower rebound of the employment levels. Dergiades et al. [37] also find that, due to the lagging effect of unemployment relief, the early stimulus was only able to partially offset the initial employment shock felt during the pandemic, a gap that may have caused some labor market scarring. Indeed, Chetty et al. [32] also find that expansionary fiscal policies, while boosting consumer spending, were unable to restore full employment.

In terms of workforce composition, real-time private sector data indicate that lower-wage workers may have experienced more persistent job losses than higher-wage ones [32]. This finding is consistent with Ma et al. [69, 70], who found that recent pandemics caused persistent unemployment among less educated workers. Chetty et al. [32] also find that low-wage employment rates were significantly lower in areas that were initially harder hit by the pandemic, suggesting geographic differences. Moreover, Deitz [38] finds that there are one million newly disabled workers in the workforce since the pandemic, which could be caused by the prevalence of long COVID-19 or other pandemic-related health problems.

In the aftermath of the past pandemics, pandemic-induced labor market effects have caused long-run labor scarcity and reduced real rates of return, consistent with the neoclassical model [64]. However, Gagnon et al. [49] use a similar model to predict that, unlike other pandemics, the COVID-19 pandemic will not affect the long-run economic capacity of the USA, given that mortalities were not concentrated among those in their working or reproductive prime age. Nonetheless, overall, the literature provides evidence that the COVID-19 pandemic caused significant short-term negative labor market effects.

2.3 Mysterious Sources of Inflationary Pressure

The pandemic shock was a primary driver in creating the inflationary environment affecting the economy as it started rebounding in 2021. Many studies attempted to understand the main factors behind the surge in inflation. However, there were also a variety of conflicting conclusions. Ball et al. [14] decompose US inflation to find that labor market tightening contributed significantly to core inflation, with energy and supply chain shocks explaining the rise in headline inflation. Using a structural factor model and an extensive set of variables measuring inflation and activity, Eickmeier and Hofmann [43] identify the contribution of demand and supply factors to inflation. They find that the inflation surge since mid-2021 has been driven by a combination of highly unusual expansionary demand conditions and tight supply conditions. Santacreu and LaBelle [86], using an industry-fixed effects model, find that global supply chain disruptions are highly correlated with the U.S. Producer Price Index, with the inflationary effects largely present among industries with large amounts of foreign exposure. Among such industries, there was the motor vehicle industry, which was a key driver of Consumer Price Index (CPI) inflation early in the pandemic due to long-lasting semiconductor shortages and increased demand for used cars as demand for public transportation fell [34]. In addition, the comparatively large fiscal stimulus pursued by the USA in response to the pandemic shock, while boosting domestic consumer demand, did so in a way that outpaced global supply adjustments, resulting in a stronger inflationary outburst in the USA compared to other advanced countries [40]. U.S. Census Bureau research indeed shows nominal goods and services revenues at historic highs by late 2021, well exceeding pre-pandemic trends and largely driven by e-commerce sales [85].

Monetary factors and skewed demand for housing could have added to the inflationary pressure. Allen [9] finds that the short-term debt needed to fund this stimulus was largely covered by the Federal Reserve but resulted in the central bank over-compensating for pandemic-induced financial market illiquidity and implementing a round of quantitative easing that could have contributed to inflationary risks. Moreover, the rapid rise in demand for single-family residential space caused rental rates to rise, which further added to headline inflation [76]. Unsurprisingly, rising rents along with rising consumer prices meant that a sizeable proportion of Americans, upward of 35%, had difficulty covering their usual expenses in late 2020 and early 2021 [23].

The dynamics of inflation can also be explained *ex post* as the result of a rapid shift in demand, sustained by fiscal and monetary policy interventions, faced with a supply struggling to keep up with the change in demand while facing a series of disruptions. In the beginning of the pandemic, the spread of the virus shifted the demand from services to goods as a large number of consumers could not access services with a high risk of contagion as a result of fear or lockdowns (e.g., restaurants, cinemas, cruises, and travel).

Households coped with the pandemic by building an at-home leisure and work infrastructure by investing in video game consoles, large TV sets, and at-home fitness equipment such as stationary bikes, computers, and furniture. The need for larger space and single-family housing away from urban centers also increased as families shied away from dense urban areas. In other words, most of the initial increase in demand concerned durable goods [89]. In addition, stable or increasing disposable income and low interest rates further increased the demand for durable consumer goods. Meanwhile, labor supply shocks affected sectors where remote work was not feasible, including manufacturing, at a time of spike in demand. In 2020–2021, most sectors witnessed extensive factory closures as a result of value chain disruptions, new safety regulations, COVID-19 illnesses, and a shortage of workers [85]. Labor-intensive sectors such as apparel and complex value chain sectors such as transportation manufacturing faced the most severe disruption with 38 days and 16 days of closure throughout 2020, respectively. In comparison, "basic" heavy industries such as chemicals and paper manufacturing were closed only about 3 days during the same year. However, in 2022, value chain disruptions were being resolved. Although maritime shipping prices skyrocketed during the pandemic, they collapsed to around their pre-pandemic levels while inventories of microchips, cars, and housing started rising rapidly.

It is not implausible that a period of rapid increase in the durable good demand above its pre-pandemic trend could be followed by a rapid convergence back toward the trend. Real durable goods consumption grew by 34% between the second quarter of 2020 and the second quarter of 2021, much above trend growth. Subsequently, it started stagnating and then declining toward its trend level. This is illustrated by the initial pent-up demand for electronics, cars, and furniture, followed by the emergence of supply glut, including that of microchips, which seemed unimaginable in 2020–21. Growth of real durable consumption expenditure was −3% between the second quarter of 2021 and the third quarter of 2022 [93].

Meanwhile, real personal consumption expenditure on services fell by 14% between the last quarter of 2019 and the second quarter of 2020, when it bottomed as a result of lockdowns and fear of the virus. However, it quickly recovered and reached its trend level by the second quarter of 2022. Consumption of non-durable goods saw a smaller decline in the second quarter of 2020, then grew faster than its trend, albeit to a much lesser extent compared to durable goods, between the second and last quarters of 2021 and, afterward, fell below its trend by the second quarter of 2022.

In parallel, the dynamics of personal disposable income per capita and real wages are not consistent with the wage-price spiral narrative. Indeed, after several spikes above trend, including in the second quarter of 2020 and the first quarter of 2021, which could be explained by the extraordinary fiscal stimulus (transfers to households), disposable income per capita grew by only 2% over the first three quarters of 2022. In other words, wages and transfers were growing at real negative rates in 2022. In addition, Autor et al. [11] find that real wage growth was positive only for workers at the bottom 30% of hourly earnings distribution between the third quarter of 2020 and the third quarter of 2022. They argue that labor market tightness has added about one percentage point to inflation since the start of the pandemic, driven by intensified competition in the labor market rather than an increase in labor demand.

If the picture painted above shows demand below or close to its pre-pandemic trend, especially in terms of durable goods, if value chain disruptions seemed to be largely resolved or at least were less severe, and if wages were not spiraling upward, what can explain the persistence of inflation in 2022?

A plausible explanation is that a key contributor to high inflation was the increase in profit margins during 2020–2022 [21]. This can be seen through the contribution of profit margins to unit price growth. Indeed, unit price growth can be decomposed into three elements: unit labor cost, non-labor input cost, and corporate profits. During the pandemic, between the second quarter of 2020 and the last quarter of 2021, corporate profits contributed to more than half the unit price growth, while unit labor cost contributed less than 8%. In contrast, on average over the previous 40 years, corporate profits only contributed 11.4%, while unit labor cost contributed 61.8%. In other words, instead of the traditional wage-price spiral, what happened during the pandemic can be described as profit-price spiral, where an increase in some prices led other firms to raise their prices much more than what was needed to cover the increase in their input and labor costs. This could be the result of the combination of two forces: the rise in market power observed over the past decades, giving firms more bargaining power and resulting in potentially opportunistic behavior as prices started rising, and the relatively healthy balance sheets of households built during the pandemic, allowing households to absorb the price increase.

Interpreting inflation in the post-pandemic period, when inflation is spiking, is particularly problematic because of inherent lags and imprecision in the measurement of the rental cost. The rental category represents more than 40% of the CPI basket. A key component of this category is the "owner's equivalent rent," the indirect cost of living in one's own house, which is estimated by the Bureau of Labor Statistics (BLS) using the cost of renting in equivalent units. Beyond the issue of defining equivalent units, rents are based on a stock of leases, often multi-year contracts, which means that changes in the cost of housing in the CPI will only reflect the change in the cost for new renters (or owners), which in turn will affect the average rents and thus CPI with a sizable lag (see [1]). There is a risk of being backward-looking in deciding policy as new tenants are facing falling rents (as the data in the last quarter of 2022 suggested) while CPI indicates a rising cost of average rents. This may be particularly problematic for the conduct of monetary policy tasked with fighting inflation..

The financial markets also saw asset price inflation and volatility, increasing wealth inequality that contributed to the consumption boom. Aziz et al. [13] find that the pandemic dramatically increased asset and oil market volatility, which led to speculation and numerous asset bubbles in the months following the pandemic. Interestingly, they find that equities reacted more closely to changes in the oil price than to the severity of COVID-19, a finding that is corroborated by Managi et al. [72]. However, given that oil price volatility was largely pandemic-induced, the researchers suggest that the pandemic was still the major driver behind increased financial volatility and the increased business uncertainty that followed. Ferguson and Storm [46] argue that the richest 10% of US households saw a huge increase in household wealth as a result of asset inflation, contributing substantially to an aggregate consumption increase from July 2021.

All in all, a confluence of various pandemic-induced factors has been at play, pushing inflation to unprecedented levels since the 1980s. A spike in energy and food prices, including imported prices, and huge supply and labor market disruptions as the pandemic unfolded, contributed substantially to an increase in and persistence of inflation. A compositional shift of demand toward durables supported by fiscal packages, monetary easing, and stock and house price jumps was another key factor. Changes in work patterns, increased remote work, and worker movements out of and in the urban centers pushed up housing and rental costs. Inherent lags in rental inflation measurements and pass-through effects contributed to the inflation persistence. The tight labor markets are driven by increased competition, contributing only a fraction to inflation, while the wage-price spiral has not materialized. Instead, what could be called a profit-price spiral could be an overlooked but significant factor driving high and persistent inflation. With many of these factors working in reverse by the end of 2022, inflation has been on a declining path.

2.4 Poverty and Inequality

The COVID-19 pandemic was a massive negative economic shock, which affected the poorest workers in terms of market outcomes more severely than the Great Recession of 2009 in the aftermath of the 2008 financial crisis. However, the massive progressive package of aid in response to the pandemic led to a historic drop in poverty levels, especially for minorities and children. Although the official poverty rates did not show a significant change since 2019, an extension of the measure by the U.S. Bureau of Labor Statistics, including previously unaccounted government programs and federal and state taxes as well as regional variations, shows a dramatic fall in poverty [35]. It is remarkable that between 2019 and 2021, the overall poverty rate fell from 11.8% to 7.8%. Over this period, more than eight million Americans exited poverty. The decline was the most marked among minorities and children. In 2021, the child poverty rate fell to 5.2%, its lowest level since 2009 and more than half its level in 2019 (12.6%). This sharp decrease was mostly explained by the federal government's expansion of anti-poverty programs in the face of the COVID-19 pandemic. In contrast, the poverty rates were much higher during the 2008 financial crisis. For example, in 2011, it peaked for the whole population at 16% and 18% for children. In other words, while the recession of 2009 led to a historic spike in poverty, the response to the pandemic reduced it to a historically low level.

In terms of market outcomes, excluding taxes and transfers, the COVID-19 recession had a much more regressive effect on income compared to the 2008 financial crisis [66]. In other words, excluding policy intervention, the pandemic recession hit the income of the poorest workers harder than in 2008–2009 while it had a milder effect after taxes and transfers. The poorest workers (bottom quintile) were more likely to face a large drop in real annual income, of 10% or more, in 2020 than in 2009 (a 17% higher chance). In contrast, the richest workers (top quintile) were 10% less likely to face a similar drop in income during the pandemic.

In parallel, the policy package adopted in response to the COVID-19 was much more ambitious both in scale and in the extent of progressivity compared to the package passed during the 2009 recession. A majority of adults received stimulus checks of $1800 and $1100 per child. In addition, unemployment insurance benefits were increased by $600 per week. As argued by Larrimore et al. [66], the flat structure of the transfers and their extent relative to incomes received meant that not only large declines in income, after tax and transfers, were less frequent during the pandemic than during the Great Recession of 2009, but they were also less frequent than in 2019.

The exceptional levels of government transfers combined with consumption retrenchment on services led to a historical spike in saving rates in 2020–21 although it was highly concentrated among the top half of households in terms of income [7]. The accumulation of excess savings during the first year of the pandemic dwindled rapidly in 2022. Aladangady et al. [7] estimate that US households accumulated about $2.3 trillion in excess savings between 2020 and summer 2021. However, they also show that subsequently households have decumulated about one-quarter of these excess savings by mid-2022 as the saving rate has dropped below its pre-pandemic trend.

2.5 Acceleration of Preexisting Trends

The longer-term effects of the pandemic on the US economy remain to be seen; however, the existing research points us in a few directions. The pandemic acted as an accelerator of preexisting trends in terms of remote work, robotization, and migration away from large urban centers [82].

While some research finds reduced migration at the beginning of the pandemic [47, 67], most research suggests that a large wave of migration occurred during the pandemic. According to consumer data, the pandemic accelerated preexisting migration trends by increasing the rate of migration away from larger urban areas and toward small and medium urban areas [22]. Indeed, Cohen et al. [33] analyze real estate data to find that the housing market in New York City has shrunk dramatically, as both sale volumes and sale prices declined, suggesting a large migration out of the city. Moreover, this reduction in asset values was larger in lower-income communities, which may also have long-term implications for wealth inequality. Similarly, using U.S. Census Bureau data, Frey [48] finds that metro areas saw significant population loss during the pandemic, a combination of both increased outward migration and reduced levels of natural births. The effect of such migration on local economies was sizable. Not only were real estate prices elevated in areas where remote work was prevalent [77, 82], but municipal bond yields increased in cities suffering from outward migration, suggesting increased risk and reduced investor confidence in these cities [53].

Much of this migration was likely the result of new technologies enabling workplace mobility and shifts in the preferences of workers in favor of remote work and more flexibility, which are likely to persist. Aksoy et al. [6] show that most people around the world desire to work about 2 days a week from home. They also show that workers are ready to forego 5% pay raise to be able to work from home 2–3 days a week. Barrero et al. [15, 16] estimate that 20% of workers plan to continue some form of working from home after the pandemic.

Research finds that the increased digitalization and technological investments induced by the pandemic also improved within-sector productivity in the US economy, an improvement that has not been witnessed in other advanced economies [41]. Nodari [79] suggests a similar point, finding that post-pandemic GDP growth trends will in part be determined by countries' ability to leverage technological improvements to better adjust to pandemic-induced changes.

There has been a trend toward more automation. However, the pandemic resulted in global value chain disruptions, which in turn could have triggered a wave of onshoring. Increased automation helps compensate for the increased cost of labor. The scarcity of workers, followed by the wage increases, has further encouraged automation [68].

However, there could also be a negative effect on long-run productivity driven by learning losses induced by the pandemic. During the COVID-19 pandemic, national test results showed a significant decrease in math scores across almost all states, which research attributed to prolonged online learning, with the decline in passing rates for math being much larger in school districts with less in-person instruction [54]. These changes were also larger in districts with larger minority and lower-income populations, which could further increase inequality in the aftermath of the pandemic. In total, Patrinos et al. [81] estimate that students in the USA suffered half a year's worth of learning loss during the pandemic, which could result in 4.6% lower lifetime earnings, especially for students from disadvantaged backgrounds [56]. Similarly, research using historical data estimates that yearlong school closures reduce the present value of a student's lifetime income by up to $40,000, affecting especially secondary school students [20]. For the nation's economy as a whole, the present value of lost GDP could be between $14 and $28 trillion in value, much more than economic and health impact in 2020–23 [56].

2.6 Back to Normal: Dangers of Policy Uncertainty Amid Rapid Structural Shifts

Policymakers have faced the challenge of assessing the size of the fiscal stimulus and monetary easing needed to avoid a major crisis in the face of a pandemic. Given the extent of the crisis with millions of workers and thousands of firms in the services sector facing unemployment and closure, the safest policy consists in casting a wide safety net. Indeed, the risks of seeing the economy spiraling into a major recession are large. However, a generous rescue fiscal package and expansionary monetary policy can compound the inflationary forces at work during a pandemic. More importantly, a key difficulty in macroeconomic fine-tuning in the context of a post-pandemic world is the possibility of rapid structural shifts, triggered by changes in macroeconomic policy itself. In other words, a tight labor market and excess savings, for example, could potentially be swiftly reversed in the case of a policy reversal to fight inflation, which the central banks pursued in 2022.

An economy that is sustained by generous fiscal and monetary easing could be overheating during the pandemic, in part due to the spike in demand for durable goods and tightness in the labor market due to health-related and dependent care disruptions. The demand for durable goods could be initially sustained by fiscal stimulus and low interest rates but eventually as demand for these goods is satiated while pandemic restrictions ease, demand shifts back toward services such as travel and restaurants. Eventually, pre-pandemic trends become dominant.

The tightness in the labor market could also potentially be reversed rapidly. Indeed, a large number of workers left for early retirement, career change, or care for young children, but the forces driving those changes could also work in reverse to bring the workers back. Indeed, in 2022, 1.5 million retired workers were already back in the labor market, pulled by a combination of high prices, especially health insurance premia for retirees under 65 who do not qualify for Medicare, and the prospects of a downturn. Many workers who left to start their own endeavors, typically in digital or high-end personal services, face a demand with high-income elasticity and are potentially sensitive to a downturn. The reopening of daycare centers allows more parents to come back to the labor market if need be. Meanwhile, remote work could also dwindle if and when labor market conditions change, leading workers to give up the ability to work from home, or from another city, when fear of unemployment starts spreading.

3 Preventive Risk-Based Framework for Fighting Pandemics

As the economic and social fallout from pandemics is colossal, the optimal policy for fighting pandemics should include a massive investment in prevention to avert it. The preventive approach is vital not only from public health and social perspectives but also from an economic standpoint. With economic and health costs of about 70% of GDP in 2020–21, millions of deaths, and potentially large and long-term health and labor scarring effects, the cost of pandemic preparedness is minuscule in comparison with pandemic costs. Glennerster et al. [50] estimate that the global expected losses (economic, mortality, and learning) from future pandemics, accounting for intensity and frequency with the data from the past pandemics since 1600, amount to about 800 billion dollars per year or more than 150 billion for the USA alone. Even with a conservative estimate of the COVID-19-like pandemic event of one every 138 years [73], the 70% of GDP fallout (economic and health) in the USA is 0.5% annually, or about 100 billion dollars (in 2019 pre-pandemic prices), suggesting at least that level of spending on preparedness would be justified. The required spending in the USA could be orders of magnitude less than the annual expected loss. Indeed, the World Bank and World Health Organization put forth an annual estimate of about 10 billion dollars needed for the global pandemic preparedness [74].

A perpetual pandemic preparedness program is worth the annual cost with expected net present value of benefits in hundreds of billion dollars globally [50]. Glennerster et al. [50] calculate that an advanced investment program of 60 billion dollars upfront to create sufficient vaccine production capacity followed by a five billion dollar annual expenditure to maintain it would support the production of 24 billion vaccine doses to vaccinate 70% of the global population within 6 months. Even if the USA alone were to engage in the advanced investment program (proportionate to the global program in terms of cost) to create vaccine production capacity to vaccinate its own population, the expected net present value of benefits would amount to tens of billion dollars. The above numbers suggest that the annual expenditure needed to maintain the US pandemic preparedness is in the order of a few billion dollars per year.

Not surprisingly, many have called for the investment in pandemic preparedness. Esvelt [44] calls for delay, detect, and defend approach. Increasing security protocols in virus research and screening DNA synthesis machines to preclude printing viruses at home would mitigate potential leaks of dangerous pathogens. Detecting viruses requires monitoring tools like wastewater surveillance. In fact, Esvelt [44] argues that the detection of any patterns of exponential growth of nucleic acid fragments would sound an alarm about potential biological threats. Cheap and fast sequencing of genetic material makes it possible, and Esvelt [44] suggests that the untargeted sequencing of wastewater at all 328 US entry points would cost less than a billion dollars a year [83]. Finally, a quick distribution of personal protective equipment (PPE), coupled with resilient supply chains and local production, would help mitigate the spread of the virus. Esvelt [44] also suggests using low-wavelength light fixtures to kill the virus in closed spaces. This further calls for increased innovation in PPE and other monitoring and defensive tools to fight off the virus. If expensive screening procedures were implemented at the US airports to screen for potential terrorist threats after the 9/11 attacks, pandemic preparedness screening should be a much easier and cheaper task [83].

The pandemic preparedness would also include investment in environmental measures to prevent the spread of viruses from wildlife to humans. Bernstein et al. [18] argue that investing tens of billions of dollars in measures precluding tropical deforestation, wildlife trafficking and meat trade, and disease surveillance and control in domestic and wildlife animals could halve the risk of zoonotic pandemics and save about a million and one-half lives a year. The authors also suggest training more veterinarians and wildlife biologists and creating a global database of virus genomics. The worsening climate change resulting in more frequent and severe droughts, increased deforestation, and destructive weather events would only exacerbate the arrival and impact of novel infectious diseases and epidemics [3].

3.1 Applying Lessons from the 2008 Financial Crisis

The pandemic preparedness program is akin to what economists advocate for financial crisis preparedness, especially in the aftermath of the 2008 financial crisis. The crisis exposed a costly central bank strategy of mopping up after the burst of the asset price bubble. The crisis preparedness was clearly less costly and disruptive. The lessons learned did help maintain financial stability when the pandemic hit in 2020. The 2008 financial crisis made it very clear that markets on their own, no matter how sophisticated and perhaps especially when they are sophisticated, are not stable and that actions of one can affect many (externalities). It was a massive market failure made possible by lax regulation [87, 88]. Economists would argue that incentives are important to encourage "good" behavior, and in the build-up to the financial crisis, incentives for the financial sector to provide funds to productive enterprises, minimize risk, and reduce transaction costs were largely missing. Instead, banks engaged in risky behavior, creating dubious financial instruments, lending to, and borrowing from each other within an international banking network. As some banks got into trouble, the shock propagated through the network, further exacerbated by the lack of information of which banks owed what assets and were thus solvent. In short, regulators were blindly trusting in the self-correcting forces and incentives of the market while the financial sector was busy making quick profits and disregarding systemic risks.

The lessons of the 2008 financial crisis for crisis preparedness can be summarized with three key principles—surveillance, prevention, and coordination. First, the crisis revealed that banks' balance sheets were too opaque not only for regulators but also for banks themselves. The risks taken were not obvious to many as even the rating agencies were complacent in the risk ratings of the financial instruments issued. In addition, as the crisis unfolded, the lack of information intensified the crisis as lending stopped and financial markets froze. The knowledge and flow of information such as banks' balance sheets, bank networks, and linkages with other financial institutions are important both before and during the crisis to mitigate the impact of the crisis. Second, surveillance is not sufficient without the tools to prevent the crisis. Building capital buffers, prohibiting activities, and changing risk incentives were needed in addition to capital injections during the crisis. Third, the coordination among the Treasury, the central bank, banks, and other financial participants was important to share information and enact a plan of action as the crisis unfolded. The crisis committee to assess risks and information flows and engage in "war games" is also needed for the crisis preparedness.

Similarly, the COVID-19 pandemic was exacerbated by massive market failures. Governments scrambled to produce PPE, tests, vaccines, and cure, resorting to prolonged lockdowns to protect public health. Eventually, the cost of lockdowns, as was clearly visible in the second quarter of 2020, grew too large, and keeping the economy operating at some capacity became important too. The slow production and distribution of test kits and PPE in the beginning of the pandemic and, once vaccines were rolled out, vaccine hesitancy and misinformation have made the pandemic deadlier and costly than it needed to be. Even in 2022, 2 years after the start of the pandemic, despite the vaccination campaign, the emergence of new variants continuing to reinfect and kill people, and long-term health effects of the COVID-19 disease, the funds needed for the government-provided sick leave, child support, test kits, vaccines, and cure have dried up. More importantly, the drop in research funding, combined with legal and regulatory bottlenecks, has stalled the development and distribution of improved solutions for fighting the virus. With the pandemic fatigue setting in, the private sector cannot squash the pandemic, and it is the government that must coordinate the efforts among government agencies, firms, and citizens to minimize the pandemic costs.

The principles from the financial crisis preparedness can be applied to the pandemic preparedness. First, surveillance entails building an early warning system to monitor risk and survey potential pathogens. Collecting and sharing data and conducting analysis are an important component of the surveillance. For instance, sewage monitoring systems have proven to be very effective [44]. Second, on the prevention tools, developing medical solutions such as vaccines, tests, and cure ex ante, building easy-to-scale-up facilities, and setting up advance market commitments to nudge high supply when needed would insure resilience. Like the capital buffers for banks, a risk-based approach calls for redundancy in terms of monitoring, research, and facilities, providing resilience, a more important feature of the robust system than efficiency. Finally, the pandemic preparedness board with all the key stakeholders such as regulators, researchers, manufacturers, and distributors would have a mandate for pandemic preparedness and help coordinate the work needed.

3.2 Building a Testing Infrastructure Against Pandemics

We illustrate these three principles on building a testing infrastructure against pandemics. Whether effective vaccines and therapeutics are available, a viable way to squash a pandemic is to complement these tools with a universal and continuous testing and isolation policy. Lockdowns are costly, while effective and safe vaccines take time and more investment (e.g., clinical trials) to develop and may not fully protect against the emergence of new variants. Rather than relying solely on vaccines to fight pandemics, including the COVID-19 pandemic, testing policy should be a key component of the strategy. One can argue that lockdowns are medieval, vaccines are Victorian, and testing is modern.

Building a testing infrastructure would constitute the first line of defense against pandemics. The cities of the twenty-first century need a testing infrastructure to fight pandemics in the same way cities of the nineteenth century built their sanitation systems to eliminate cholera. This testing infrastructure consists of not only rapid, do-it-yourself (DIY), easily accessible, and cheap test kits but also built-in testing equipment in public buildings and transport, and other places of congregation to monitor for and identify various pathogens and extinguish or isolate them quickly.

State intervention for development, regulation, production, purchase, and coordination and oversight at the highest level, similar to the Second World War effort, is required to implement this policy. The intervention to provide test kits includes support for the development of tests, regulation for emergency approval of tests, industrial policy for mass production and purchase, use of distribution networks across health clinics and pharmacies, and coordination among public and private players. This cost is minuscule compared to the human, social, and economic losses of the pandemic. Implementing universal testing through a massive scale-up of rapid and cheap test kits would be worth it even if the odds of success were less than one in a hundred. For instance, testing everyone once a week with a cost of 5 dollars per test would cost billions, compared to the trillions of economic and health losses the USA has incurred in 2020–21.

3.2.1 Surveillance

Surveillance, as the first key principle in the pandemic preparedness, requires an early warning system. This system would consist of both physical infrastructure for testing and the development of tests needed to identify different pathogens. In addition, the availability of rapid, cheap, DIY, and accessible test kits would also be an important part of the prevention leg of the pandemic preparedness. The early warning system would require timely and comprehensive data collection, analysis, and dissemination. R&D spending needs to be ramped up, and barriers to entry for biotech firms need to be lifted for competition to take hold. Regulations and codes need to be revamped to be ready for new pandemics.

The physical infrastructure for testing would play an important role in pandemic preparedness. Adding the testing equipment to existing buildings and means of transport would be less costly than overhauling the whole system. Sewage monitoring, air virus detection equipment, and thermal screening could be a good place to start. In fact, sewage monitoring was used in the past to monitor a polio vaccination campaign [71]. However, the development of new construction and transportation would pave the way for a seamless design of built-in testing equipment. This dual approach calls for innovation in testing equipment for both existing and new buildings and means of transport. Eventually, the real challenge is to install this equipment in the existing facilities, but with planning and investment, it could be done in relatively short periods of time.

Once the data are collected, it needs to be analyzed and disseminated. This requires creating a database of pathogens and various computer, statistical, and biological tools for the analysis of a vast amount of collected data. Like economic data are collected, analyzed, and shared, the data generated by the early warning system could follow a similar template. The summary information could be shared across various research and government agencies while making sure the sensitive data stay private. Visualization of the data such as regional and neighborhood maps of the spread of flu or COVID-19 could provide useful aggregate information to the general public.

Monitoring risks would also involve the development of test kits and screening test systems. Testing equipment could be developed for health clinics or points of care in addition to the places of congregation, buildings, and means of transport. More importantly, DIY, rapid, cheap, and easily accessible test kits would be key to identifying and arresting the spread of potential pathogens. The dual approach of developing test kits for existing pathogens and innovating to design testing equipment for novel pathogens needs to be pursued. Developing universal tests and prototype test kits for various pathogens would allow public health officials to stay one step ahead of the epidemic. These tests could also take a form of a screening based on smell, cough, and other main symptoms of the disease that are analyzed using machine learning methods. Some of these screening systems based on cough and smell have been in development and have been shown to be relatively precise in identifying COVID-19 [12].

Research and development of tests and prototypes require increased investment in R&D. Public spending in health R&D is the second largest after defense but is only about 44 billion dollars (2020), or about 0.2% of GDP, about a quarter of the total R&D and about one-half of defense R&D [78]. The call for more public R&D spending, including on health, has been loud and clear (e.g., [52]), and from the COVID-19 pandemic experience, it is hard to deny that the returns to basic and applied science are enormous. It is the development of mRNA technology prior to the pandemic that helped create effective mRNA vaccines against the COVID-19 within weeks of the virus genome sequencing.

To increase the odds of success, both public research institutes and pharmaceutical and biotechnology firms and startups should engage in research and development. The barriers to entry, such as the availability of financing or labs, should be minimized to allow all researchers and innovators to be part of the process. In addition, a portfolio approach for various test technologies needs to be adopted to maximize the return-risk tradeoff. For instance, Athey et al. [10] argue that given the enormous costs of the COVID-19 pandemic, a much larger portfolio of vaccine candidates with a much larger spending envelope should have been tried. According to Ahuja et al. [4], the US government program to develop and produce vaccines, Operation Warp Speed (OWS), should have tried 27 vaccine candidates and spent more than 50 billion dollars, compared to the six vaccine candidates it approved for a cost of 18 billion dollars.

In addition to innovation, building testing infrastructure requires a new set of regulations. Like energy efficiency standards, building standards and codes need to be improved to encourage the development and provision of built-in testing systems. The drug regulator needs to update regulations to expedite and streamline approval of tests for emergency use when time is of the essence. Moreover, in times of pandemics, an epidemiological approach rather than a clinical approach to testing needs to take precedence, in which test accuracy could be sacrificed for scale, ease of use, and availability [28, 65]. More importantly, these new rules and standards need to be analyzed and put in place before another pandemic hits.

3.2.2 Prevention

The prevention component of the pandemic preparedness requires a wide penetration of testing equipment in public places and availability of rapid, DIY, and cheap test kits. Since testing equipment may be costly to install in the short run, test kits provide a cheap alternative. In times of pandemic, test kits would also provide a viable way to squash the pandemic if they are widely available [28]. The readiness for this eventuality involves building and maintaining easy-to-scale-up facilities and stockpiling inputs for mass production of test kits, having ready standard advance market commitments for large purchases in times of pandemics, and setting up distribution networks for fast dissemination across the country.

In normal times, the market can take the lead on the production of test kits, but certain spare capacity should be built into the system. Once test kits leave the public or private laboratories, private firms can produce the test kits demanded. If the cost of these tests turns out to be large, also potentially due to lower demand, the government could provide the needed support, whether directly to producers or indirectly through the insurance system, to keep the costs down and make the kits widely available. In addition, creating spare capacity whether in firms' manufacturing facilities or in dedicated government-owned contractor-operated (GOCO) manufacturing plants like those during the Second World War would be a critical part of the pandemic preparedness infrastructure. Although it may be costly to maintain, it acts precisely as an insurance policy against pandemics. The spare capacity would allow production to be scaled up easily in times of need. The spare capacity needs to be applied to the whole supply chain, which requires additional investment, stockpiling, and maintenance. This approach could also be applied to other pandemic preparedness programs such as PPEs, vaccines, or therapeutics.

Creating a supply chain map for testing would save time and resources when the next pandemic hits. The information would include relevant firms, facilities, staff and raw materials, and geography of production and distribution along the whole chain. Identifying key vulnerabilities ex ante would help create redundancies in the network and resilient supply chain. Knowing which facilities to repurpose when needed, how to do this, and having the necessary regulations in place, such as the updated Defense Production Act, would expedite the increase in production.

Once facilities to produce are available, incentives need to be provided to firms to ramp up production when needed. One of these tools that proved to be useful in the COVID-19 pandemic is advance market commitments (AMCs). AMCs were an indispensable part of the OWS to give incentives to private firms to commit to much higher production of vaccines once they were approved for use. These contracting tools need to be designed in advance. Athey et al. [10] discuss how AMCs could be used to support a ramp-up in production. They argue for investments in capacity at risk, emphasizing that high capacity increases the speed of supply provision, which in times of pandemics is crucial. AMCs allow a transfer of investment risk from firms to the government, while various funding mechanisms could be designed to provide incentives for speed and quality. Contracting on capacity instead of doses/kits would further ameliorate potential hoarding and trade restrictions observed during the COVID-19 pandemic.

The efficient dissemination of test kits is as important as quick and mass production, highlighting the two crucial principles to beat the pandemic—scale and speed. Like supply chain maps, the distribution network of pharmacies, clinics, and other locations (e.g., grocery stores) needs to be known in advance. The database on shipping and trucking companies, travel routes, and maximum operational capacity of the delivery companies and final sellers of the product, including personnel, would help flag various bottlenecks and direct at-risk investments. Essentially, being able to ramp up not only production but also distribution very quickly would keep the pandemic at bay.

Once the pandemic hits, scale and speed become critical for the testing policy to be effective. To squash the curve, testing needs to be continuous and periodic, that is, testing everyone, for instance, weekly. In fact, depending on the testing strategy adopted, simulations show that the needed testing rates for the COVID-19 pandemic are in the range of 5–15% of the population daily [30, 31]. The needed weekly testing rates imply a large number of tests. Smart testing strategies like group testing, other screening systems such as "symptom-based" tests, travel restrictions to the hub of urban centers, or testing on a geographical grid, by hubs, or places of congregation, can minimize the number of tests needed, especially in the production ramp-up stage [28]. At the initial stages of the pandemic, if the tests do not have high sensitivity and specificity, more frequent testing rates may be required to control the spread [28, 65]. Eventually, developing precise tests should be important to contain the spread and minimize the isolation of healthy individuals.

Experimentation with mass testing during the COVID-19 pandemic has shown that the spread of the virus was largely contained. South Korea started mass testing very early on and, with contact tracing, kept the epidemic spread in check [27]. In Italy, the region of Veneto has fared much better than Lombardy as it implemented a large-scale testing of the population, isolating the infected. On the level of a town in Veneto, the virus was squashed with universal testing before and after a two-week lockdown [101].

An important component of the universal or mass testing strategy is the industrial policy for a production ramp-up of test kits. Relying on the market to produce the needed quantities would take too long, and even with enough time, the final production may not be sufficient given the uncertainties of the demand for tests. Other market failures such as capacity constraints, coordination failures among firms, market power of firms, and externalities would only exacerbate the ramp-up challenge. Clearing all the bottlenecks along the whole supply chain requires state intervention [29]. Being prepared for the ramp-up in production before the pandemic hits would make the whole enterprise much easier and faster than what was witnessed during the COVID-19 pandemic. Whether these are expedited regulatory approval of tests, waivers of intellectual property rights, or "forced entry" through the Defense Production Act, the rules and procedures need to be decided in advance.

Although the challenge of a production ramp-up may seem insurmountable even if the necessary rules are in place, the fast mobilization of resources, akin to the war effort, as observed during the Second World War, is possible. A famous challenge to the nation by President Roosevelt in 1940 stated that the USA needed to produce more combat airplanes than it cumulatively produced in the previous 40 years, ever since the first flight of the Wright brothers in 1903. By 1943, the country produced more than 50,000 airplanes annually, a 30-fold increase. As President Roosevelt said, "Powerful enemies must be out-fought and out-produced" [29].

The COVID-19 pandemic showed that a resilient industrial ecosystem to ramp up production of test kits and other essential items would be an insurance against future pandemics. During the pandemic, the mobilization of resources for vaccines was made possible by the government's OWS. The program was instrumental in not only the development of a portfolio of vaccines and advance purchases but also in clearing the bottlenecks for the production ramp-up (Adler, [29]). However, tests and vaccines should not be separate items on the list of priorities. In fact, building the industrial ecosystem for health should help create public sector capabilities and needed capacities and strengthen the national and local health system [75].

3.2.3 Coordination

Public-private coordination is key to putting the pandemic preparedness strategy into action and implementing it during the pandemic. The coordination can be done by a task force or committee set up at the high level of the government and composed of key public and private sector stakeholders. This task force will hold regular meetings and issue communications to the public. It can also direct the key government agency, for example, the U.S. Biomedical Advanced Research and Development Authority (BARDA), to lead the effort on a day-to-day basis in the preparation and implementation of the strategy [2].

The task force needs to set up relevant objectives, clear existing bottlenecks, and coordinate across government agencies, manufacturers, and distributors. The state support needs to be directed to ease bottlenecks in regulation, intellectual property rights, supply chain and final production, distribution, or communication. For instance, during the COVID-19 pandemic, the EU vaccine task force mapped, tracked, and cleared bottlenecks in vaccine production in Europe. As an example, a German dengue vaccine bottling factory was repurposed for Johnson & Johnson's vaccine [39].

In setting up objectives, the task force needs to embrace clarity and ambition. For instance, to successfully implement universal testing, the objective would specify the testing rate of the population daily (e.g., a 15% or a weekly testing rate) and gather resources to implement it. This would involve experimenting with various testing technologies, ramping up production, and providing testing and quarantine facilities.

The task force would provide the necessary incentives and enforce accountability. It would facilitate sharing information and production know-how, harnessing the capabilities of large firms and the innovativeness of smaller ones, and enforcing strict accountability for the funds received. The incentives for the private sector should be aligned to engage in this endeavor. For instance, public procurement and advance purchases, provision of intellectual property licenses, and access to cheap financing would make it worthwhile for firms to invest. Essentially, in times of uncertainty, the state takes up this role to resolve uncertainty and acts as a provider of the first resort. The state could guarantee the demand for tests and provide intellectual property licensing agreements for more firms to participate in production.

Overall, the task force would coordinate across firms in the value chain and government agencies, facilitate information sharing, and clear all the resulting hurdles.

4 Concluding Remarks

This chapter gives an overview of the costs of the COVID-19 pandemic on the economy. It shows that the direct costs were colossal. Moreover, the pandemic had deep and potentially long-lasting effects in terms of income (scarring) and on the demographics of cities, automation, remote work, learning at school, health outcomes (e.g., mental health), and the labor market in general. We have argued that a major cost of a pandemic could consist in the difficulty of conducting monetary and fiscal policy amid major and rapid shifts in the economy. Indeed, the COVID-19 unleashed an unprecedented spike in inflation and required an exceptional level of fiscal support. However, the specific conditions, including global value chain disruptions, shifts in demand from services to durable goods, and the resilience of industry where remote work is an option, have made interpreting inflation data and predicting the effects of policy extremely difficult.

Given the colossal short- and long-term costs, we argue that putting together a pandemic preparedness program is orders of magnitude cheaper than the expected costs of pandemics. We propose a risk-based framework to tackle future pandemics. It is based on three key principles: surveillance, prevention, and coordination. Similar to the financial crisis prevention rules implemented after the 2008 financial crisis, a pandemic prevention infrastructure is needed. Surveillance involves monitoring for and analyzing potential pathogens and disseminating information. Prevention calls for developing platforms and capacities to design and produce en masse medical solutions such as tests, vaccines, and cure, while efficiently distributing the products across the country. Finally, an institution or a task force at the high levels of the government will be tasked with preparedness, coordinating across both public and private sectors.

The whole approach to health and pandemics needs to be rethought with prerogatives given to prevention and preparedness as well as equity and inclusion. Indeed, as the chair of the World Health Organization's (WHO) Council on the "Economics of Health for All" established during the pandemic in November 2020, Professor Mazzucato of the University College London has stated, "Health for all must be at the heart of government investment and innovation decisions—and it must be governed with the common good in mind" [97].

References

1. Adams B, Loewenstein L, Montag H, Verbrugge R (2022) Disentangling rent index differences: data, methods, and scope. BLS working paper 555. Available: https://www.bls.gov/osmr/research-papers/2022/pdf/ec220100.pdf
2. Adler D, Cherif R, Hasanov F (2021) How to deliver 10 billion COVID-19 vaccines at warp speed. World Economic Forum. Available: https://www.weforum.org/agenda/2021/08/how-to-deliver-10-billion-covid-19-vaccines-at-warp-speed/

3. Afelt A, Frutos R, Devaux C (2018) Bats, coronaviruses, and deforestation: toward the emergence of novel infectious diseases? Front Microbiol 9:702
4. Ahuja A, Athey S, Baker A, Budish E, Castillo JC, Glennerster R, Kominers SD, Kremer M, Lee J, Prendergast C et al (2021) Preparing for a pandemic: accelerating vaccine availability. Am Econ Assoc Papers Proc 111:331–335
5. Airlines for America (2022) Impact of COVID-19: data updates. Available: https://www.airlines.org/dataset/impact-of-covid19-data-updates/
6. Aksoy CG, Barrero JM, Bloom N, Davis SJ, Dolls M, Zarate P (2022) Working from home around the World. NBER Working Paper 30446. Available: https://www.nber.org/papers/w30446
7. Aladangady A, Cho D, Feiveson L, Pinto E (2022) Excess savings during the COVID-19 pandemic, FEDS Notes. Board of Governors of the Federal Reserve System. Available: https://doi.org/10.17016/2380-7172.3223
8. Albanesi S, Ki J (2021) Effects of the COVID-19 recession on the US labor market: occupation, family, and gender. J Econ Perspect. Available: https://pubs.aeaweb.org/doi/pdfplus/10.1257/jep.35.3.3
9. Allen WA (2021) Monetary policy and government debt management during the coronavirus pandemic. Natl Inst Econ Rev 255:79–84. Available: https://doi.org/10.1017/nie.2021.1
10. Athey, Susan, Juan Camilo Castillo, Esha Chaudhuri, Michael Kremer, Alexandre Simoes Gomes, and Christopher Snyder (2022) "Expanding capacity for vaccines against COVID-19 and future pandemics: a review of economic issues." NBER working paper 30192. Available: http://www.nber.org/papers/w30192
11. Autor D, Dube A, McGrew A (2022) The unexpected compression: competition at work in the low wage economy, Working paper, presented at Markus's Academy. Princeton University. Available: https://bcf.princeton.edu/events/david-autur-on-the-unexpected-compression-competition-at-work-in-the-low-wage-economy/
12. Azeli Y, Fernández A, Capriles AF et al (2022) A machine learning COVID-19 mass screening based on symptoms and a simple olfactory test. Sci Rep 12:15622. Available: https://doi.org/10.1038/s41598-022-19817-x
13. Aziz A, Izraf M, Azhari A, Ashraful Mobin M (2022) Detecting asset Price bubbles during the COVID-19 crisis and its implications: evidence from the stock and oil market. Int J Bank Financ 17(2):91–114. Available: https://doi.org/10.32890/ijbf2022.17.2.4
14. Ball, Laurence, Daniel Leigh, Prachi Mishra (2022) Understanding U.S. inflation during the COVID Era." International Monetary Fund Working Paper No. 2022/208. Available: https://www.imf.org/en/Publications/WP/Issues/2022/10/28/Understanding-U-S-525200.
15. Barrero, Jose Maria, Nicholas Bloom, Steven J. Davis (2021a) Why working from home will stick." NBER working paper 28731. Available: https://www.nber.org/papers/w28731.
16. Barrero JM, Bloom N, Davis SJ, Meyer BH (2021b) COVID-19 is a persistent reallocation shock. AEA Papers Proc 111:287–291. Available: https://www.aeaweb.org/articles?id=10.1257/pandp.20211110
17. Bauer L, Broady K, Edelberg W, O'Donnell J (2020) Ten facts about COVID-19 and the U.S. economy. Brookings Institution. Available: https://www.brookings.edu/research/ten-facts-about-covid-19-and-the-u-s-economy/
18. Bernstein AS, Ando AW, Loch-Temzelides T, Vale MM, Li BV, Li H, Busch J, Chapman CA, Kinnaird M, Nowak K, Castro MC, Zambrana-Torrelio C, Ahumada JA, Xiao L, Roehrdanz P, Kaufman L, Hannah L, Daszak P, Pimm SL, Dobson AP (2022) The costs and benefits of primary prevention of zoonotic pandemics. Sci Adv. https://doi.org/10.1126/sciadv.abl4183
19. Bhattarai A (2022) Millions retired early during the pandemic. Many are now returning to work, new data shows. The Washington Post. Available: https://www.washingtonpost.com/business/2022/05/05/retirement-jobs-work-inflation-medicare/
20. Bilo S, Ajwad MI, AlAnsari E, AlHumaidan L, Alrashidi F (2022) School is closed: simulating the long-term impact of the COVID-19 pandemic-related school disruptions in Kuwait, Policy research working paper 10035. The World Bank. Available: https://openknowledge.world-

bank.org/bitstream/handle/10986/37411/IDU0e7f7494c051340403109d2300111bd4dbbc2.pdf?sequence=1&isAllowed=y
21. Bivens J (2021) Corporate profits have contributed disproportionately to inflation. How should policymakers respond? Working economics blog. Economic Policy Institute. Available: https://www.epi.org/blog/corporate-profits-have-contributed-disproportionately-to-inflation-how-should-policymakers-respond/
22. Brown JP, Tousey C (2021) How the pandemic influenced trends in domestic migration across U.S. urban areas. Econ Rev (Federal Reserve Bank of Kansas City) 106(4):39–58. Available: https://www.kansascityfed.org/research/economic-review/how-the-pandemic-influenced-trends-in-domestic-migration-across-us-urban-areas/
23. Center of Budget and Policy Priorities (2022) The COVID-19 economy's effects on food, housing, and employment hardships. Available: https://www.cbpp.org/research/poverty-and-inequality/tracking-the-covid-19-economys-effects-on-food-housing-and
24. Cerra, Valerie, Antonio Fatas, and Sweta Saxena (2020) "Hysteresis and business cycles," CEPR discussion paper No. DP14531.
25. Cerra V, Fatas A, Saxena SC (2021) Fighting the scarring effects of COVID-19. Ind Corp Chang 30(2):459–466. Available: https://doi.org/10.1093/icc/dtab030
26. Chenarides L, Manfredo M, Richards TJ (2021) COVID-19 and food supply chains. Appl Econ Perspect Policy 43:270–279. Available: https://doi.org/10.1002/aepp.13085
27. Cheong I (2020) The experience of South Korea with COVID-19. In: Baldwin R, di Mauro BW (eds) Mitigating the COVID economic crisis: act fast and do whatever it takes, a VoxEU.org eBook. CEPR. Available: https://voxeu.org/content/mitigating-covid-economic-crisis-act-fast-and-do-whatever-it-takes
28. Cherif R, Hasanov F (2022) Testing our way out of pandemics. Health Technol Policy 12:100714. Available: https://doi.org/10.1016/j.hlpt.2022.100714
29. Cherif R, Hasanov F (2021) Industrial policy against pandemics. Ind Corp Chang:1–15. Available: https://doi.org/10.1093/icc/dtab042
30. Cherif R, Hasanov F (2020) This is why we should test everyone for COVID-19. World Economic Forum. Available: https://www.weforum.org/agenda/2020/12/universal-testing-best-hedge-covid-19-heres-why/
31. Cherif R, Hasanov F, Mina M (2020) Universal testing to end the pandemic. Health Affairs Blog. https://doi.org/10.1377/hblog20201102.521193
32. Chetty, Raj, John N. Friedman, Nathaniel Hendren, Michael Stepner, The Opportunity Insights Team (2022) "The economic impacts of COVID-19: evidence from a new public database built using private sector data." NBER working paper 27431. Available: http://www.nber.org/papers/w27431.
33. Cohen JP, Friedt FL, Lautier JP (2022) The impact of the coronavirus pandemic on New York City real estate: first evidence. J Reg Sci 62(3):858–888. Available: https://doi.org/10.2139/ssrn.3959105
34. Council of Economic Advisers, Executive Office of the President (2022) Economic report of the president (2022). U.S. Government Publishing Office. Available: https://www.govinfo.gov/app/details/ERP-2021
35. Creamer J, Shrider EA, Burns K, Chen F (2022) Poverty in the United States: 2021. U.S. Census Bureau. Available: https://www.census.gov/library/publications/2022/demo/p60-277.html
36. Cutler D, Summers L (2020) The COVID-19 pandemic and the $16 trillion virus. JAMA 324(15):1495–1496
37. Dergiades T, Milas C, Panagiotidis T (2022) Unemployment claims during COVID-19 and economic support measures in the U.S. Econ Model 113. Available: https://doi.org/10.1016/j.econmod.2022.105891
38. Deitz R (2022) Long COVID appears to have led to a surge of the disabled in the workplace. Federal Reserve Bank of New York, Liberty Street Economics. Available: https://libertystreeteconomics.newyorkfed.org/2022/10/long-covid-appears-to-have-led-to-a-surge-of-the-disabled-in-the-workplace/

39. De Gruyter C (2021) Industrial policy saved Europe's vaccine drive. Foreign Policy. Available: https://foreignpolicy.com/2021/05/10/industrial-policy-saved-europes-vaccine-drive/
40. De Soyres F, Santacreu AM, Young H (2022) Fiscal policy and excess inflation during Covid-19: a cross-country view, FEDS notes. Board of Governors of the Federal Reserve System. Available: https://doi.org/10.17016/2380-7172.3083
41. de Vries K, Erumban A, van Ark B (2021) Productivity and the pandemic: short-term disruptions and long-term implications: the impact of the COVID-19 pandemic on productivity dynamics by industry. IEEP 18(3):541–570. Available: https://doi.org/10.1007/s10368-021-00515-4
42. Dorfman JH, Grant JD, Worley JM (2022) Using truck-to-trunk programs to increase food security and help support farm incomes. Appl Econ Perspect Policy 44:1440–1447. Available: https://doi.org/10.1002/aepp.13234
43. Eickmeier S, Hofmann B (2022) What drives inflation? Disentangling demand and supply forces. BIS working papers no 1047. Available: https://www.bis.org/publ/work1047.htm.
44. Esvelt, Kevin (2022) "Delay, detect, defend: preparing for a future in which thousands can release new pandemics." Geneva paper 29/22
45. Faberman, Jason, Andreas I. Mueller, and Aysegul Sahin (2022) "Has the willingness to work fallen during the Covid pandemic?" NBER working paper 29784. Available: https://www.nber.org/papers/w29784.
46. Ferguson, Thomas and Servaas Storm (2023) "Myth and reality in the great inflation debate: supply shocks and wealth effects in a multipolar world economy." INET Working Paper 196. Available: https://doi.org/10.36687/inetwp196.
47. Frey WH (2021) Pandemic population change across metro America: Accelerated migration, less immigration, fewer births and more deaths. Brookings Institution. Available: https://www.brookings.edu/research/despite-the-pandemic-narrative-americans-are-moving-at-historically-low-rates/
48. Frey WH (2022) New census data shows a huge spike in movement out of big metro areas during the pandemic. Brookings Institution. Available: https://www.brookings.edu/blog/the-avenue/2022/04/14/new-census-data-shows-a-huge-spike-in-movement-out-of-big-metro-areas-during-the-pandemic/
49. Gagnon E, Johannsen B, López-Salido D (2020) Supply-side effects of pandemic mortality: insights from an overlapping-generations model, Finance and economics discussion series 2020-060. Board of Governors of the Federal Reserve System. Available: https://doi.org/10.17016/FEDS.2020.060
50. Glennerster, Rachel, Christopher Snyder, and Brandon Joel Tan (2022) "Calculating the costs and benefits of advance preparations for future pandemics." NBER working paper 30565. Available: http://www.nber.org/papers/w30565.
51. Greenwood J, Guner N, Kopecky KA (2022) Substance abuse during the pandemic: implications for labor-force participation. HSOA J Addict Addict Dis. Available: http://www.jeremygreenwood.net/papers/ggkLFP.pdf
52. Gruber J, Johnson S (2019) Jump-starting America: how breakthrough science can revive economic growth and the American dream. Public Affairs, New York
53. Gustafson M, Haslag PH, Weagley D, Ye Z (2022) A flash in the Pan(demic)? Migration risks and municipal bonds. Georgia Tech Scheller College of Business research paper no. 4029984. Available: https://doi.org/10.2139/ssrn.4029984
54. Halloran, Clare, Rebecca Jack, James C. Okun, and Emily Oster (2021) "Pandemic schooling mode and student test scores: evidence from US States." NBER working paper 29497. Available: http://www.nber.org/papers/w29497.pdf.
55. Haltiwanger, John C (2021) "Entrepreneurship during the COVID-19 pandemic: evidence from the business formation statistics" NBER working paper 28912. Available: https://www.nber.org/papers/w28912.
56. Hanushek, E. and L. Woessmann (2020) "The economic impacts of learning losses." OECD education working papers 225. Available: https://doi.org/10.1787/21908d74-en.

57. Hohler J, Lansink AO (2021) Measuring the impact of COVID-19 on stock prices and profits in the food supply chain. Agribusiness 37(1):171–186. Available: https://doi.org/10.1002/agr.21678
58. Hornstein A, Kudlyak M (2022) The Pandemic's impact on unemployment and labor force participation trends, Economic brief, no. 22-12. Federal Reserve Bank of Richmond. Available: https://www.richmondfed.org/publications/research/economic_brief/2022/eb_22-12
59. IMF (2020) World economic outlook update. Available: https://www.imf.org/en/Publications/WEO/Issues/2020/01/20/weo-update-january2020
60. IMF (2021a) World economic outlook. Available: https://www.imf.org/en/Publications/WEO/Issues/2021/10/12/world-economic-outlook-october-2021
61. IMF (2021b) Fiscal monitor. Database of Country Fiscal Measures in Response to the COVID-19 Pandemic. Available: https://www.imf.org/en/Topics/imf-and-covid19/Fiscal-Policies-Database-in-Response-to-COVID-19
62. Johansson R (2021) America's farmers: resilient throughout the COVID pandemic. US Department of Agriculture. Available: https://www.usda.gov/media/blog/2020/09/24/americas-farmers-resilient-throughout-covid-pandemic#:~:text=Overall%2C%20agricultural%20exports%20during%20the,percent%20for%20non%2Dag%20exports
63. Johns Hopkins University (2022) Coronavirus resource center. The Center for Systems Science and Engineering. Available: https://coronavirus.jhu.edu/map.html
64. Jordà, Òscar, Sanjay R. Singh, Alan M. Taylor (2020) "Longer-run economic consequences of pandemics." Federal Reserve Bank of San Francisco Working Paper 2020-09. Available: https://doi.org/10.24148/wp2020-09.
65. Larremore DB, Wilder B, Lester E, Shehata S, Burke JM, Hay JA, Tambe M, Mina MJ, Parker R (2021) Test sensitivity is secondary to frequency and turnaround time for COVID-19 screening. Sci Adv 7(1). https://doi.org/10.1126/sciadv.abd5393
66. Larrimore J, Mortenson J, Splinter D (2022) Income declines during COVID-19, FEDS notes. Board of Governors of the Federal Reserve System. Available: https://doi.org/10.17016/2380-7172.3063
67. Lavelle M, Kepner E (2022) U.S. Migration patterns before and after the start of the Covid-19 pandemic, Midwest Economy Blog. Federal Reserve Bank of Chicago. Available: https://www.chicagofed.org/publications/blogs/midwest-economy/2022/migration-before-and-during-pandemic
68. Lee D (2021) As COVID-19 wanes, employers are accelerating the use of robots. Where does that leave workers? Los Angeles Times. Available: https://www.latimes.com/politics/story/2021-05-04/covid-automation-robots-trends-effects-on-workers
69. Ma, Chang, John Rogers, and Sili Zhou (2020a) "Global economic and financial effects of modern pandemics and epidemics," Working paper.
70. Ma C, Rogers J, Zhou S (2020b) Modern pandemics: recession and recovery, International finance discussion papers 1295. Board of Governors of the Federal Reserve System. Available: https://doi.org/10.17016/IFDP.2020.1295
71. Mallapaty S (2020) How sewage could reveal true scale of coronavirus outbreak. Nature 580:176–177. Available: https://www.nature.com/articles/d41586-020-00973-x
72. Managi S, Yousfi M, Zaied YB, Mabrouk NB, Lahouel BB (2022) Oil Price, US stock market and the US business conditions in the era of COVID-19 pandemic outbreak. Econ Anal Policy 73:129–139. Available: https://doi.org/10.1016/j.eap.2021.11.008
73. Marani M, Katul GG, Pan WK, Parolari AJ (2021) Intensity and frequency of extreme novel epidemics. Proc Natl Acad Sci 118(35):e2105482118
74. Mazzucato M, Donnelly A (2022) The world is still failing at pandemic preparedness and response. Project Syndicate. Available: https://www.project-syndicate.org/commentary/health-investment-failing-at-pandemic-preparedness-response-by-mariana-mazzucato-and-alan-donnelly-2022-11

75. Mazzucato M, Cherif R, Hasanov F (2020) How to end the pandemic this year. Project Syndicate. Available: https://www.project-syndicate.org/commentary/universal-testing-to-end-covid19-pandemic-by-reda-cherif-et-al-2020-10
76. Molly B, Chen S, Nothaft FE (2021) Housing preferences during the pandemic: effect on home price, rent, and inflation measurement. Bus Econ 56:200–211. Available: https://doi.org/10.1057/s11369-021-00241-4
77. Mondragon, John A. and Johannes Wieland (2022) "Housing demand and remote work." NBER working paper 30041. Available: https://www.nber.org/papers/w30041.
78. National Science Foundation (2022) "Federal R&D funding, by budget function: fiscal years 2020–22." NSF 22-316, Table 23. Available: https://ncses.nsf.gov/pubs/nsf22316.
79. Nodari G (2020) Global growth: drivers and post-pandemic prospects. BIS Q Rev:45–60. Available: https://www.bis.org/publ/qtrpdf/r_qt2203b.htm
80. Parekh N, Ali SH, O'Connor J, Tozan Y, Jones AM, Capasso A, Foreman J, DiClemente RJ (2021) Food insecurity among households with children during the COVID-19 pandemic: results from a study among social media users across the United States. Nutr J 20:73. Available: https://doi.org/10.1186/s12937-021-00732-2
81. Patrinos HA, Vegas E, Carter-Rau R (2022) An analysis of COVID-19 student learning loss, Policy research working paper no. 10033. The World Bank. Available: https://openknowledge.worldbank.org/handle/10986/37400
82. Pickert R, Tanzi A, Gu J (2022) How Covid-19 transformed the U.S. economy. Bloomberg. Available: https://www.bloomberg.com/graphics/2022-us-economy-pandemic-recovery/?leadSource=uverify%20wall
83. Piper K (2022) How to make COVID the last pandemic. Vox. Available: https://www.vox.com/future-perfect/2022/11/23/23475085/covid-pandemic-infectious-disease-virus-pathogen-illness-kevin-esvelt-genetic-sequencing
84. Ramsey AF, Goodwin BK, Hahn WF, Holt MT (2021) Impacts of COVID-19 and Price transmission in U.S. meat markets. Agric Econ 52(3):441–458. Available: https://doi.org/10.1111/agec.12628
85. Roman S, Cooke-Hull S, Dunfee M, Flaherty M, Haskell J, Holland V, Jackson C, Shevlin C (2022) The coronavirus pandemic's economic impact. US Census Bureau. Available: https://www.census.gov/library/publications/2022/econ/coronavirus-pandemics-economic-impact.html
86. Santacreu AM, LaBelle J (2022) Global supply chain disruptions and inflation during the COVID-19 pandemic. Fed Reserve Bank St Louis Rev 104(2):78–91. Available: https://doi.org/10.20955/r.104.78-91
87. Stiglitz, Joseph (2018) "Ten years later." Roosevelt Institute working paper
88. Stiglitz J (2010) Lessons from the global financial crisis of 2008. Seoul J Econ 23(3):321–339
89. Tauber K, Van Zandweghe W (2021) Why has durable goods spending been so strong during the COVID-19 Pandemic? Economic commentary 2021-16. Federal Reserve Bank of Cleveland. Available: https://doi.org/10.26509/frbc-ec-202116
90. Taylor DE, Farias LM, Kahan LM, Talamo J, Surdoval A, McCoy ED (2022) Understanding the challenges faced by Michigan's family farmers: race/ethnicity and the impacts of a pandemic. Agric Hum Values 39(3):1077–1096. Available: https://doi.org/10.1007/s10460-022-10305-6
91. The Economist (2022) Tracking COVID-19 excess deaths across countries. Available: https://www.economist.com/graphic-detail/coronavirus-excess-deaths-tracker
92. Tian Z, Schmidt C, Goetz SJ (2022) The role of community food Services in Reducing US food insufficiency in the COVID-19 pandemic. J Agric Resour Econ 47(3):580–597. Available: https://jareonline.org/articles/the-role-of-community-food-services-in-reducing-us-food-insufficiency-in-the-covid-19-pandemic/
93. U.S. Bureau of Economic Analysis (2022) Real personal consumption expenditures: durable goods. Retrieved from FRED, Federal Reserve Bank of St. Louis. Available: https://fred.stlouisfed.org/series/PCDGCC96

94. U.S. Census Bureau (2022a) U.S. Census Bureau News. Available: https://www.census.gov/retail/mrts/www/data/pdf/ec_current.pdf
95. U.S. Census Bureau (2022b) Census Bureau Releases COVID-19 economic brief. Available: https://www.census.gov/newsroom/press-releases/2022/covid-19-economic-brief.html
96. Walmsley T, Rose A, John R, Wei D, Hlávka JP, Machado J, Byrd K (2023) Macroeconomic consequences of the COVID-19 pandemic. Econ Model 120:106147. Available: https://doi.org/10.1016/j.econmod.2022.106147
97. WHO (2022) Council on the economics of health for all. Main webpage. Available: https://www.who.int/groups/who-council-on-the-economics-of-health-for-all
98. Yaffe-Bellany D, Corkery M (2020) Dumped milk, smashed eggs, plowed vegetables: food waste of the pandemic. The New York Times. Available: https://www.nytimes.com/2020/04/11/business/coronavirus-destroying-food.html
99. Yan W, Cai Y, Lin F, Ambaw DT (2021) The impacts of trade restrictions on world agricultural Price volatility during the COVID-19 pandemic. China World Econ 29(6):139–158. https://doi.org/10.1111/cwe.12398
100. Ziliak JP (2021) Food hardship during the COVID-19 pandemic and great recession. Appl Econ Perspect Policy 43:132–152. Available: https://doi.org/10.1002/aepp.13099
101. Zingales L (2020) Why mass testing is crucial: the US should study the Veneto model to fight Covid-19. The Stigler Center at the University of Chicago Booth School of Business, blog. Available: https://promarket.org/why-mass-testing-is-crucial-the-us-should-study-the-veneto-model-to-fight-covid-19

SARS-CoV-2 Induced Cellular Adaptations to Avoid Humoral Immunity Countermeasures

Antonio Merolli

1 Introduction

1.1 From the "Common Cold" to the "Severe Acute Respiratory Syndrome"

A "severe acute respiratory syndrome" (SARS) caused by a coronavirus called "SARS-associated coronavirus (SARS-CoV)" was reported in Asia in February 2003. The severity and fatality of the syndrome in humans were a clinical novelty for a member of the Coronaviridae virus family, popularly associated with the mild and benign "common cold." A few years later, in 2012, a new severe acute respiratory syndrome, again caused by a coronavirus, was identified in Saudi Arabia. It was called "Middle East respiratory syndrome" (MERS), and it was caused by the "Middle East respiratory syndrome coronavirus (MERS-CoV)." Both SARS and MERS triggered respective one-year self-contained epidemics. They both prompted a large number of basic and clinical studies that unveiled the mechanisms of infection and explored possible therapeutic strategies. These studies were deemed useful to contrast a possible future third outbreak of a new severe respiratory coronavirus infection expected to occur sooner or later. As years passed, however, the number of published studies on SARS and MERS steadily declined.

At the end of 2019, a syndrome characterized by severe respiratory symptoms and an unusual lethality was detected in the Chinese mega-city of Wuhan. A new coronavirus was quickly identified and sequenced as the causative agent [62]. The World Health Organization (WHO) called this new virus "SARS-CoV-2" and the disease "COVID-19."

A. Merolli (✉)
Department of Chemistry and Chemical Biology – Department of Physics and Astronomy, Rutgers University, Piscataway, NJ, USA

J. M. Rosen, R. R. Colwell (eds.), *The COVID-19 Pandemic*,
https://doi.org/10.1007/978-3-031-62772-9_16

1.2 SARS and MERS were Not Enough to Prepare for an Effective Response to COVID-19

It was soon evident that COVID-19 was spreading quickly all over the world. COVID-19 became a pandemic, and this was a major difference with the older SARS and MERS. Old published studies on SARS and MERS were revisited. New ultrastructural studies on SARS-CoV-2, however, reported specific differences with SARS-CoV and MERS-CoV. For example, a study showed that SARS-CoV-2 had a reduced number of "spike" proteins on the surface compared to SARS-CoV [28]. Another study showed that VeroE6 cells, upon expressing the SARS-CoV-2 spike protein, could form syncytia as long as the ACE2 is present, but they do not form syncytia if transfected with the SARS-CoV spike protein [64], implying that cell fusion capability may be specific to SARS-CoV-2 but not SARS-CoV. In our opinion, cellular ultrastructural studies were somewhat overlooked; they are more expensive, time consuming, and technically demanding than most of the standardized molecular biology techniques.

The WHO and the national health organizations directed significant efforts to contrast the new virus, founding their strategies largely on the knowledge acquired on the older SARS-CoV. This seemed reasonable because SARS-CoV-2 shared a lot of SARS-CoV structure and employed the same "ACE2 docking" mechanism of SARS-CoV to enter and infect the target cell. The extracellular virion docks the cellular ACE2, thanks to its "spike" protein (Fig. 1). The virions share the extracellular space with the extracellular vesicles (Fig. 2).

There was a global effort to develop a vaccine able to provide robust antibody protection in the extracellular space to block both the circulating virions and those freshly released by the infected cells. A number of vaccines became available, and a mass vaccination was undertaken globally. The SARS-CoV-2, however, soon exhibited an unexpected clinical response that hinted at the existence of

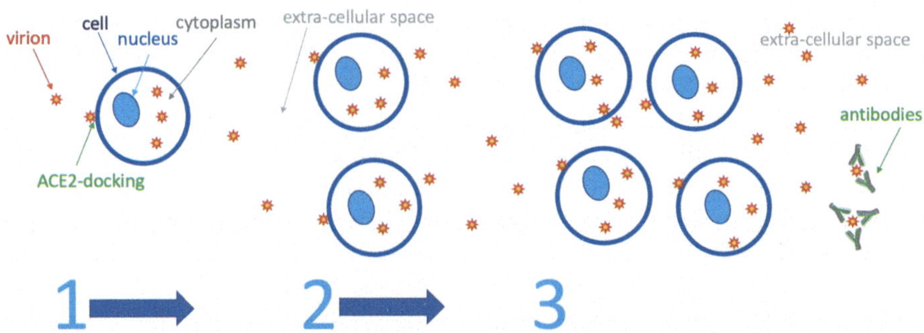

Fig. 1 SARS-CoV-2 extra-cytoplasmic "virion–ACE2 docking" mechanism of transmission. (SARS-CoV-2 virions enter the host cells by docking their spike glycoproteins to the membrane-bound angiotensin-converting enzyme 2. After an intracellular replication and assembly, the newly formed virions are released by the infected cells into the extracellular space. Circulating antibodies from the host can effectively suppress this extra-cytoplasmic route of transmission)

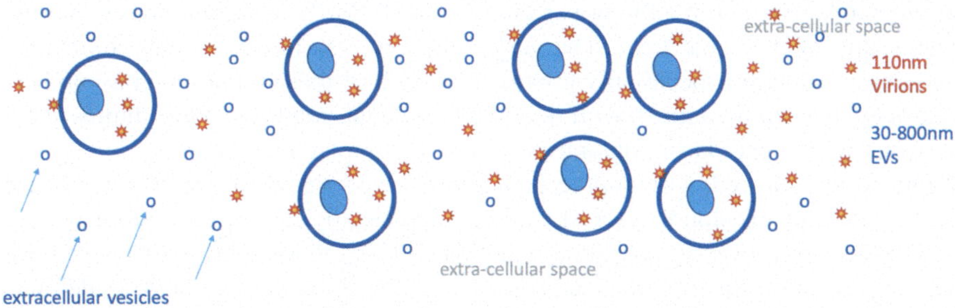

Fig. 2 SARS-CoV-2 virions share the extracellular space with the extracellular vesicles. (The extracellular space contains a variable amount of extracellular vesicles (EVs). For a long time considered as "cellular dust," EVs are now recognized as key players in cell–cell communication; they allow cells to exchange proteins, lipids, and genetic material. EVs and virions are difficult or impossible to be detected and differentiated by optical microscopy, but they share the same extracellular space)

mechanisms of immune evasion. It was more and more common to observe fully vaccinated responding individuals experiencing a second infection or even multiple reinfections [7]. Immune evasion was also suspected with the appearance of the so-called long-COVID syndrome, the persistence of the infection in immunized individuals. The Centers for Disease Control (CDC) defined the long-COVID as "the presence of signs, symptoms and conditions that continue, or develop, after the initial COVID-19 infection" (https://www.cdc.gov/coronavirus/2019-ncov/long-term-effects/index.html).

2 Ultrastructural Adaptations in Infected VeroE6 Cells configure an Evasion Route from Antibodies

In January 2020, protocols were developed at the helium-ion microscopy (HeIM) facility at Rutgers University to image biological samples with this exotic technology. The little diffused HeIM had already been shown to provide specific advantages in the imaging of biological samples when compared with standard scanning electron microscopy (SEM), most notably by avoidance of the electroconductive coating that covers and masks like a snow mantle the tiniest surface features of an object [39, 49]. Responsible for the HeIM facility, Prof. Leonard Feldman prompted the use of HeIM for the analysis of samples containing SARS-CoV-2. This was the beginning of a long journey to obtaining all the necessary safety clearances and, most importantly, infected samples from the few available sources (early cases in the New York area were recorded only in March 2020). After weeks of initial observations, a key collaboration with the Public Health Research Institute (PHRI) of Rutgers University in Newark, NJ, allowed the examination of cultures of VeroE6 cells infected with mNeonGreen-expressing SARS-CoV-2. These cultures were provided by Dr. Selvakumar Subbian and Dr. Theresa Chang. With these cultures, we were able to perform a three-way correlative HeIM/fluorescence/phase-contrast microscopy of the same field. The very same cell was identified and studied by merging information specifically provided by each of these three imaging modalities, namely, ultrastructure with HeIM, functional markers with fluorescence microscopy, and cellular topography with phase-contrast microscopy.

Our results showed, with unprecedented resolution, three key features that could shield viral transmission from host antibody defense, by sharing the cytoplasm via (1) the formation of long tunneling nanotubes (TNT) that connected two or more cells over a submillimetric distance (Fig. 3); (2) the occurrence of large-scale multicellular fusion (syncytia) (Fig. 4); and (3) the abundant presence of extracellular vesicles (EVs) of various sizes (Fig. 5).

These three ultrastructural features taken together (TNT, fusion, and EVs) can provide the mechanisms for an intracytoplasmic (or intra-cytosolic) transmission of SARS-CoV-2. Such possible transmission will not require the constant release of new virions in the extracellular space. This intracytoplasmic route may be an

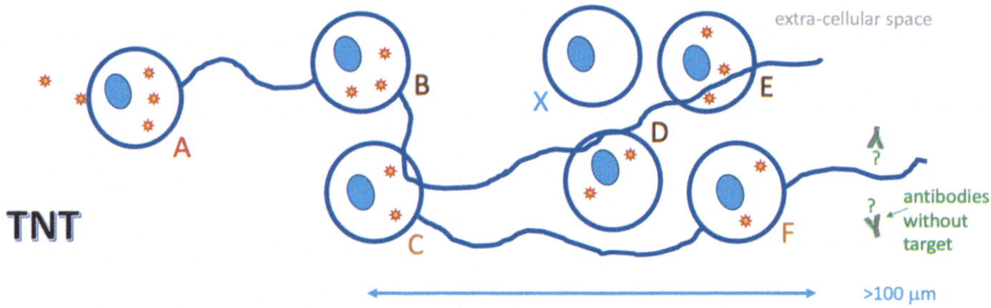

Fig. 3 Sharing of cytoplasm via tunneling nanotubes. (TNTs are tubular structures of nanometer-to-micrometer diameters that connect the cytoplasm of adjacent or distant cells. A cell infected by circulating virions may transmit the viral genome, viral components, or even the assembled virion via a TNT that attaches to another cell. Multiple cells can be connected (A-B-C-D-E or A-B-C-F). The presence of circulating antibodies in the extracellular space will not protect non-infected cells from being infected via TNTs)

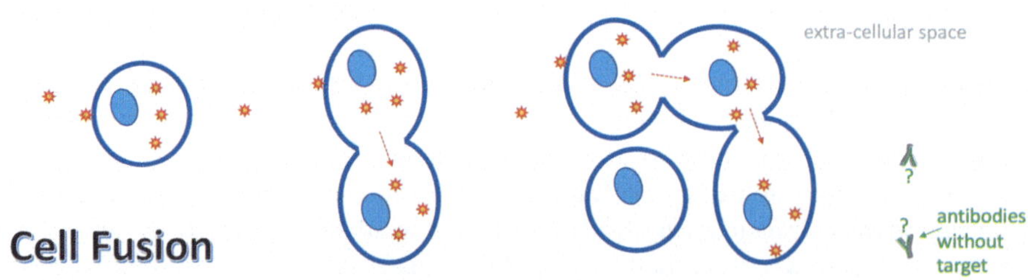

Fig. 4 Sharing of cytoplasm via cell fusion (syncytia). (SARS-CoV-2-infected cells are able to fuse together with adjacent cells. Cell fusion may transmit the viral genome, or viral components, or even the assembled virion to another cell. The fusion event may sequentially involve multiple cells. The presence of circulating antibodies in the extracellular space will not protect the non-infected cells from being infected via fusion with an infected cell)

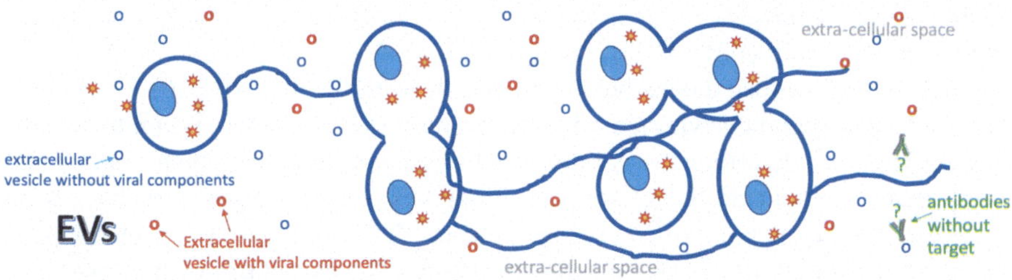

Fig. 5 Sharing of cytoplasm via extracellular vesicles. (The viral genome, viral components, or even an assembled virion can be transmitted to another cell over a long distance via extracellular vesicles (EVs). The presence of circulating antibodies in the extracellular space will not protect the non-infected cells from being infected by extracellular vesicles containing viral components)

alternative to the well-known extra-cytoplasmic ACE2 docking route. The intracytoplasmic mechanism could also contribute to the elusiveness of SARS-CoV-2 from antibody-based immune surveillance and response. After the publication of our results, articles appeared that specifically identified the SARS-CoV-2 transmission via intracytoplasmic modalities, TNT in particular [6, 45, 47].

3 SARS-COV-2 Induced Tunneling Nanotubes, Cell Fusion, and Extracellular Vesicles

3.1 Tunneling Nanotubes

Early in COVID-19 spreading, it was reported that SARS-CoV-2 infected VeroE6 cells produced more filopodia than usual and that these filopodia were "longer" than those present in uninfected cells [5, 10]. Observing these "very long filopodia" with correlative HeIM/fluorescence/phase-contrast microscopy, it was clear that they were not physiological filopodia but rather pathological "tunneling nanotubes" (TNTs) [52]. They were not present in control uninfected VeroE6 cells. These TNTs connect two or more infected host cells over submillimetric distances. Other authors have reported on the role of TNTs in SARS-COV-2 infection [55, 63].

TNTs are tubular structures of nanometer-to-micrometer diameter that connect the cytoplasm of adjacent or distant cells, thus providing an intracytoplasmic passage to exchange and/or transport components as small as ions, lipids, nucleic acids, microRNA, and cytoplasmic proteins, or as large as organelles such as endosomes, lysosomes, mitochondria, and portions of the endoplasmic reticulum or Golgi apparatus [1, 21, 26, 31, 44, 48, 55, 63, 65]. TNTs are reported to be transient structures and can form or disintegrate in a matter of minutes. They are known to promote the spread of various pathogens, including viruses [17, 33], prions [19], fungi [9], and mycoplasma [30]. TNTs are categorically different from filopodia, in length (they are longer than filopodia), diameter (they can be as thin as filopodia but as large as axons), composition, and function. Filopodia are made of F-actin, while TNTs, despite being composed of F-actin in their vast majority, can also incorporate microtubules or be mainly composed of microtubules [59].

SARS-CoV-2-infected cells bulge and detach from the floor (Fig. 6) while emitting TNTs in a specific asymmetric bipolar fashion (Fig. 7). TNTs retain their individual structure when they cross other TNTs but may fuse their membrane when overlapping another cell. Structures resembling vesicles or mitochondria have been imaged at this fusion point of entering/exiting a cell. Clear images of spherical bulging along the length of TNTs were also observed. The cytoplasm of multiple cells may be connected along a single TNT (Fig. 8). Some TNTs appear to reveal the mechanical tension between two cells (Fig. 9); in such cases, we documented TNTs pulling on both cells resulting in a stretched morphology. TNTs showed regions of bulging content, and this feature is reminiscent of pictures of axonal transport. With their growth cone, length, potential microtubule skeleton, and intraluminal transport, TNTs resemble neuronal axons [42]. As in axonal transport, the TNT diameter enlarges when bulky content is transported inside.

TNT-mediated intracellular spread can protect viruses from circulating immune surveillance and possible viral-neutralization activity in the extracellular matrix. Intercellular viral spread via TNTs avoids virus–cell interaction that may initiate host defense signaling and mount the antiviral response. Many viruses, such as the

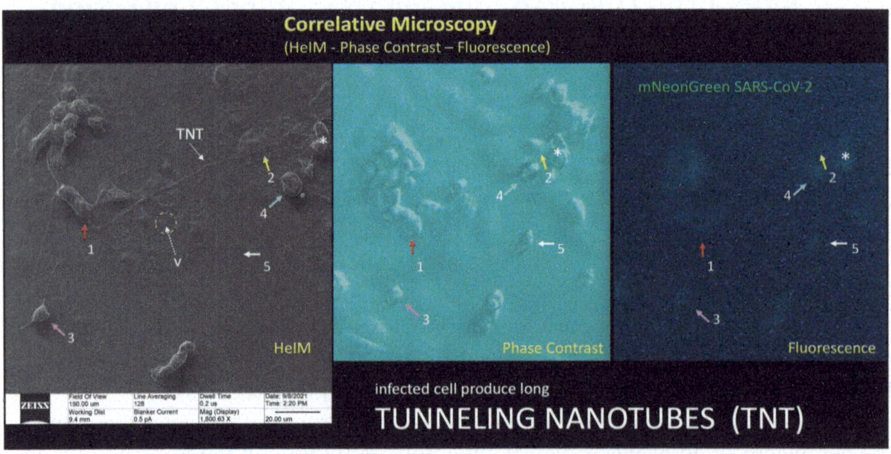

Fig. 6 Correlative microscopy. (A 150-micron square area imaged by HeIM, phase-contrast microscopy, and fluorescence microscopy. HeIM and phase-contrast microscopy show the topography of the cells. Fluorescence microscopy highlights their infectious status because only infected cells fluoresce in green. In this area, HeIM shows that two cells (1 and 2) are connected via a TNT of about 100 microns in length. There is also a rounded bipolar cell (3) emitting TNTs, and correlative microscopy shows that it is infected; this cell (3) did not change its position with respect to the initial phase-contrast image. Rounded infected cells, however, tend to detach from the floor. The correlation of phase-contrast microscopy and HeIM shows cell 4 changed its position slightly and cell 5 moved out of the field prior to HeIM imaging. Only HeIM was able to detect the presence of extracellular vesicles, and the one indicated by "v" is shown at higher magnification in Fig. 11d (HeIM 1800 X; Fluorescence and Phase Contrast 400X). (Reproduced from Ref. [39]))

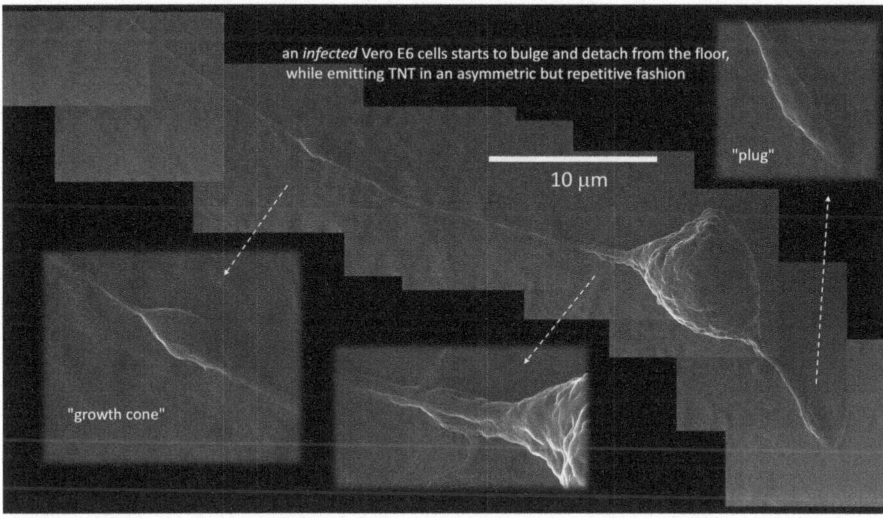

Fig. 7 Details of an infected cell emitting TNTs. (Cell 3 of Fig. 6 is visualized in a mosaic of HeIM photomicrographs taken at 27,000X. SARS-CoV-2-infected cells bulge and detach from the floor while emitting TNTs in a specific asymmetric bipolar fashion. TNTs generally stem with a "growth cone" of specific morphology, and a smaller diameter tube is sometimes observed stemming from it. It is common to observe an opposite TNT with a structure that the TEM literature describes as a "plug" for a cellular socket. (Reproduced from Ref. [39]))

influenza virus, human immunodeficiency virus (HIV), and herpes simplex virus (HSV), are known to evade host immunity and avoid pharmaceutical targeting by using TNTs to transmit their genomes to naïve/new cells [55, 57].

3.2 Cell Fusion

Cell–cell fusion (also known as syncytia) can be induced by certain types of viral infection, such as HIV, respiratory syncytial virus, and HSV [36]. Syncytia formation has been reported in the literature associated with SARS-COV-2 infection [11, 38, 64], most notably in histopathologic lung sections from patients who died from COVID-19 [11]. The literature reports a relevant finding that VeroE6 cells, upon expressing the SARS-CoV-2 spike protein, form syncytia as long as ACE2 is present but cannot when transfected with SARS-CoV spike protein [64], implying that cell fusion capability may be specific to SARS-CoV-2.

We reported how SARS-CoV-2-infected VeroE6 cells can form large-scale (submillimeter) clusters of multiple fused cells (syncytia) 24 h post-infection. This type of fusion is not observed in uninfected cells. It is reasonable to hypothesize that virus-induced cell fusion can facilitate the transfer of the viral genome to neighboring cells [38] by sharing the cytoplasm between cells. It is useful to remember, however, that replication of new viruses "highjacks" several components of the

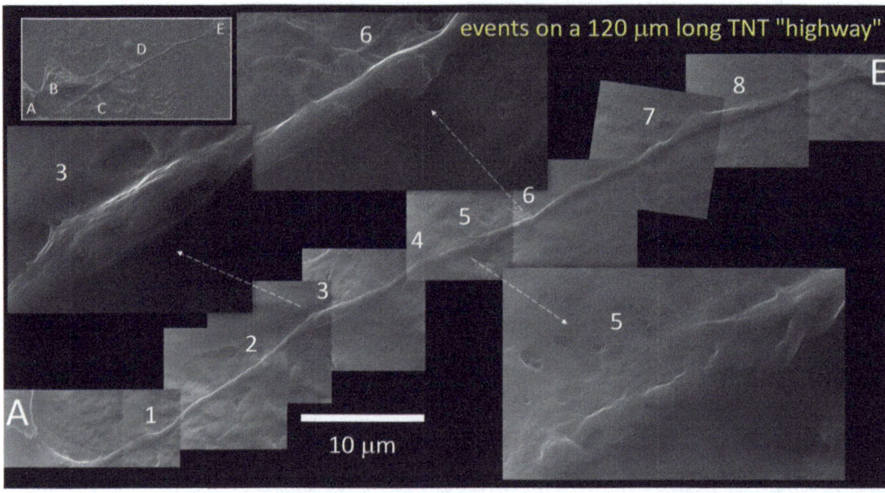

Fig. 8 Tunneling nanotube "highway." (The long TNT that connects cell A to B (as shown in Fig. 6, labeled 1 and 2) interacts with at least three other cells along its path (B, C, and D, in the top left inset). A mosaic reconstruction of HeIM micrographs (27,000X) highlights 8 notable events, some of which are magnified at a higher definition (54,000X). (1) Spherical bulging; (2) spherical bulging and membrane fusion, in a region rich in mitochondria; (3) diameter enlargement by cargo content; (4) spherical bulging close to (5) epi-nuclear membrane fusion area (detailed); (6) well-defined spherical bulging; (7) two close spherical bulging; (8) diameter enlargement. (Reproduced from Ref. [39]))

intracellular vesicle machinery, and newly replicated intracellular coronaviruses are found inside larger vesicular/vacuolar structures, not as free single virions inside the cytoplasm [12]. HeIM provided images of the cellular membrane continuity of the syncytia (Fig. 10, B1 and B2), while correlative fluorescence/phase-contrast microscopy showed the presence of multiple nuclei, identified by the DAPI-specific fluorescent blue marker, in these syncytia (Fig. 10 PC and M).

3.3 Extracellular Vesicles

The general term extracellular vesicles (EVs) refers to any membrane vesicle released into extracellular space. For a long time, EVs were considered to be "cellular dust" and did not attract much attention from researchers [41]. However, EVs have recently been identified as key players in cell–cell communication, allowing cells to exchange proteins, lipids, and genetic material [50, 58]. An accepted classification definition is vesicles generated inside the cell and released into extracellular space as "exosomes" (diameter range 30–150 nm). In contrast, vesicles pinched off from the plasma membrane are referred to as "microvesicles" (diameter range 150–1000 nm) [15, 41]. Some authors consider apoptotic bodies (bulk protrusions from dying cells that may end up in extracellular vesicles 800 nm diameter or larger)

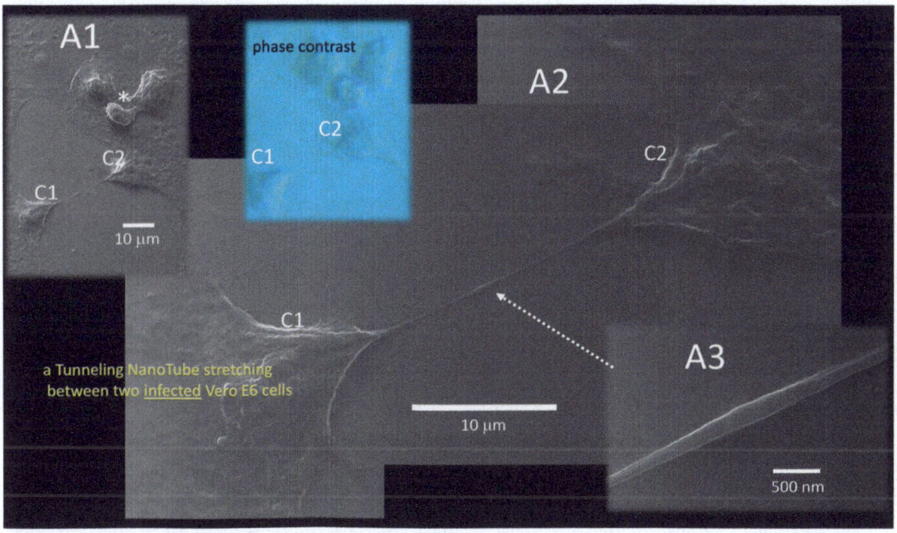

Fig. 9 TNT stretched between two cells. (A TNT under mechanical tension between two cells (C1 and C2) lies on the flat featureless floor of the culture dish. Fully flat healthy cells and a trio of bulging and detaching infected cells (asterisk in A1) can be seen nearby. This stretched TNT pulls on both cells, shaping them with a triangular morphology (A2). We ruled out the possibility that this tension could be caused by dehydration because correlative microscopy showed the same stretched morphology in the hydrated state in phase contrast, prior to HeIM imaging (blue inset). Stretched TNT between two cells often present bulging mid-way (dotted arrow). (magnification in A1 2700X; in A2 10,800X; in A3 77,000). (Reproduced from Ref. [39])

as part of the EV family too. The great similarity or even close relationship between EVs and some viruses has raised the question of how much they are related [41].

EVs are involved in SARS-CoV-2 infection [20, 46, 53] and could be used as biomarkers of disease severity [13]. SARS-CoV-2 RNA has been identified in the exosomal cargo of samples from patients with COVID-19, but not in healthy subjects, suggesting the virus uses endocytosis to spread infection [4].

The majority of EVs are below the resolution of optical microscopy. Their dimensions made the smallest EVs difficult or impossible to be differentiated from coated artifacts in scanning electron microscopy. HeIM is a powerful tool that can be used to detect EVs in a practical and unambiguous way. Many spherical vesicles whose diameter range from about 30 nm to more than 1 micron have been observed in both SARS-CoV-2-infected and uninfected VeroE6 cells, apparently in a similar number. We documented extracellular vesicles budding from the cell membrane, lying on the cell membrane, or floating as free objects in extracellular space (Fig. 11). Apoptotic bodies were observed budding from infected cells; some of them were captured just before leaving a disintegrating cell (Fig. 12).

A finding in relation to the presence of extracellular vesicles is that both infected and uninfected VeroE6 cells showed significant presence of caveolae (Fig. 11f). These "holes" in the cell membrane have previously been observed in

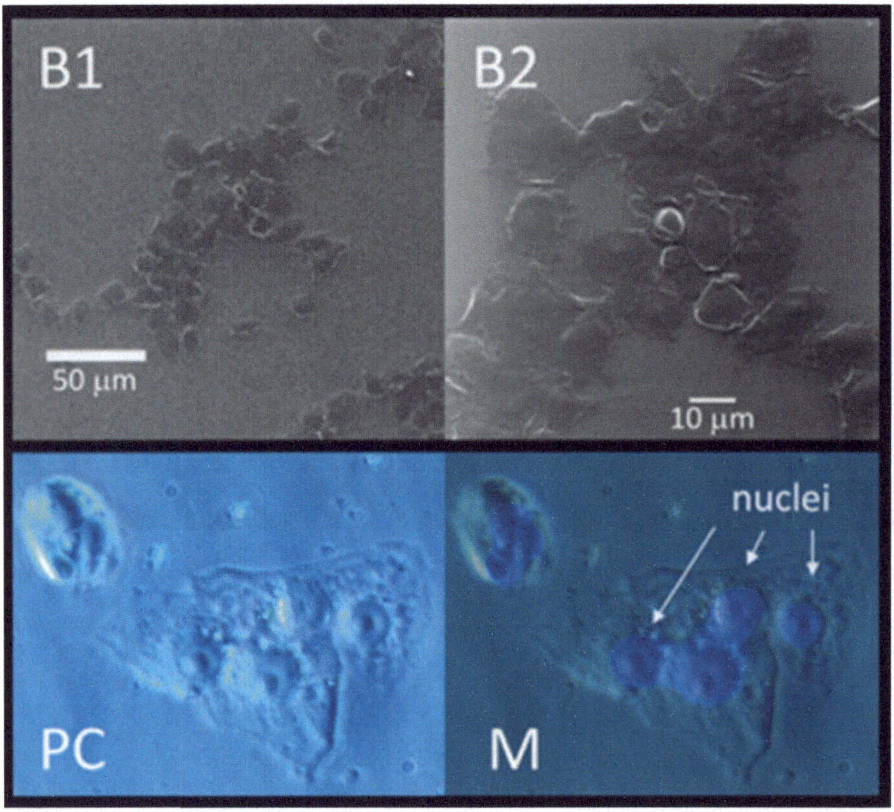

Fig. 10 Fusion of SARS-CoV-2-infected VeroE6 cells. (Infected cells form syncytia of multiple fused cells 24 h post-infection. HelM highlights the continuity of the cellular membranes (B1 at 450X and B2 at 2700X). Fluorescence microscopy shows the presence of multiple DAPI-stained fluorescent blue nuclei (M) in clusters of cells imaged using phase-contrast (PC) microscopy. (Reproduced from [39]))

not-electroconductive-coated mammalian cells; they are the circular aperture of larger spherical cavities formed by scaffolding "caveolin" proteins [40]. They are easily recognized by HelM, and prior studies using HelM have reported the caveolae to be lipid nanodomains [51]. Importantly, caveolae can be both entry or exit points for vesicles in the diameter range of the SARS-CoV-2 [14, 58]. It is also very intriguing that SARS-CoV-2 viral protein 3a viroporin has a specific caveolin-binding site [8]. We speculate if this ultrastructural feature plays a role in the susceptibility of VeroE6 cells to SARS-CoV-2 infection and in the mechanism of SARS-CoV-2 infection overall.

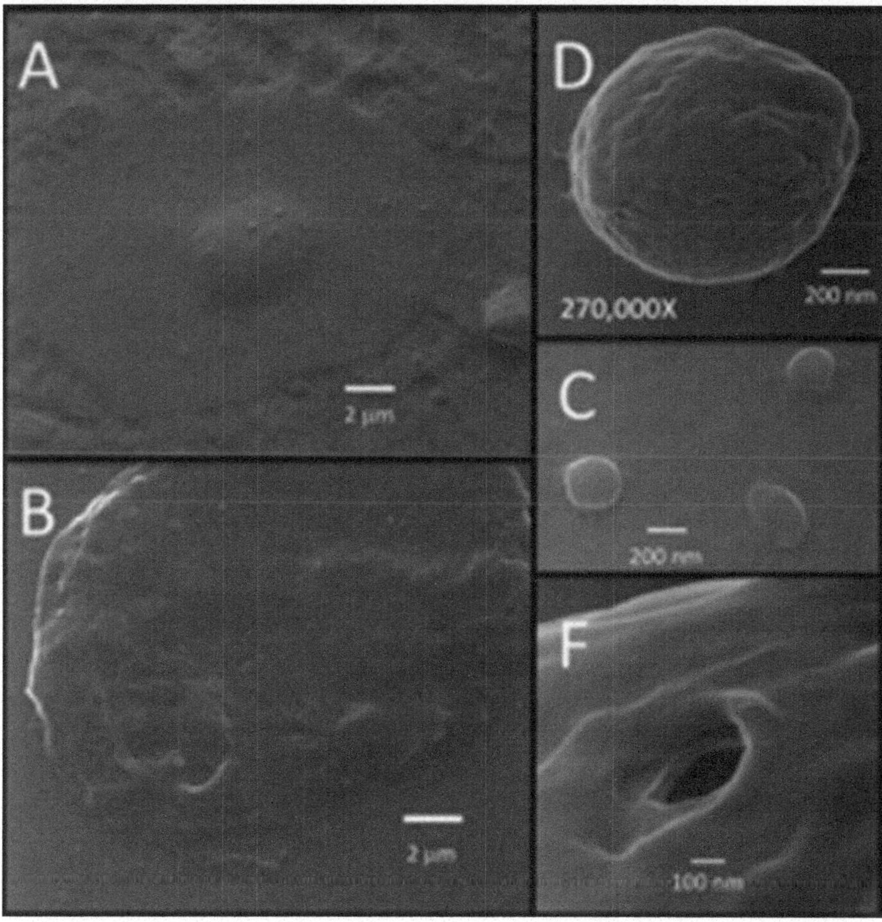

Fig. 11 Extracellular vesicles and caveolae in SARS-CoV-2-infected VeroE6 cells. (Extracellular vesicles and caveolae are present in large numbers in infected VeroE6 cells (magnification in A and B 10,800X; C135,000X; D 180,000X; F 270,000X). (Reproduced from [39]))

4 Virus Immune Evasion

4.1 Evasion from the Innate and the Adaptative Immune Systems

As mass vaccination took place, the occurrence of new infections in fully vaccinated individuals was reported worldwide. While vaccination significantly reduced the severity of the infection, a fraction of vaccinated individuals reinfected with SARS-CoV-2 progressed to extremely severe illness and even death [3]. The new symptomatic infections were labeled "breakthrough infections," and the huge number of such infections raised the question whether immune evasion was related to

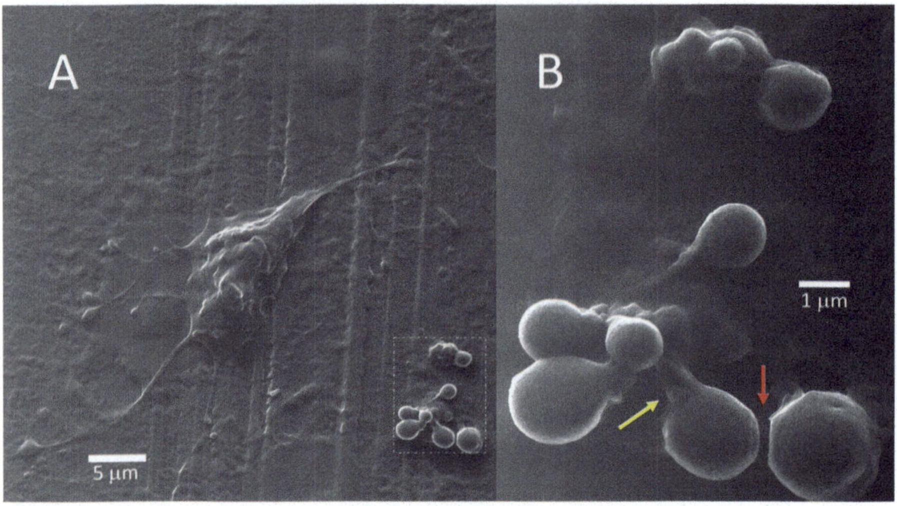

Fig. 12 Apoptosis of SARS-CoV-2-infected VeroE6 cells. (Budding apoptotic bodies captured just before leaving two disintegrating cells (**a**). Apoptotic bodies magnified in (**b**) show a clear stalk (yellow arrow) and a close membrane apposition (red arrow; detachment possibly due to dehydration). The parallel lines are the glass carving of the reference frame used for correlative microscopy (magnification in A 5400X; in B 27,000X). (Reproduced from [39]))

the renewed disease progression of SARS-CoV-2 [7]. It is still little known how the virus transmits, infects, and propagates in a host with a robust immune response.

Virus immune evasion is not a new concept. Multiple mechanisms of immune evasion have been discovered in the past three decades [2]. Viruses have developed strategies to cope with both innate and adaptative immune responses [23]. Human respiratory RNA viruses have developed a well-known immune evasion toward the innate immune system [29]. SARS-CoV, MERS-CoV, and SARS-CoV-2 all have demonstrated a capacity to evade multiple innate immune sensing ([27, 47]). However, SARS-CoV-2 evasion of humoral response (antibody-based) hints at specific evasion of the adaptative immune system.

While molecular mechanisms are today at the core of research attention, our study has showed how ultrastructural subcellular and cellular features (TNT and cell fusion, respectively) can elucidate immune evasion mechanisms at the cellular level. Ultrastructural research should be a part of investigations of the immune evasion mechanisms, unknown today but important, if not critical, in a future pandemic.

4.2 Extracellular Vesicles and Immune Evasion

Extracellular vesicles play a role in immune evasion but are difficult to study because of their size. Viruses may be using EVs to infect naïve/new cells [56]. The physical and chemical characteristics of many EVs, as well as biogenesis pathways, resemble those of retroviruses. EVs generated by virus-infected cells can incorporate viral proteins and fragments of viral RNA, similar to defective (noninfectious) retroviruses. EVs are known to facilitate HIV-1 infection and dissemination; HIV-1 has been reported as "entrapped" in exosome aggregates [25]. "Trojan" exosomes might provide retroviruses the ability to take advantage of cell-encoded intercellular vesicle traffic [18, 22, 25]. HIV-1 exploits the surface properties of exosomes to facilitate rapid infection of progeny virus and, in so doing, camouflages the virus from immune surveillance. Surrounding itself with exosomes, HIV-1 can accelerate infection and dissemination [25].

It has been hypothesized that SARS-CoV-2-infected cells can release EVs with viral antigens or cargo. EVs acting as a "Trojan horse" could explain the reappearance of the viral RNA in patients recovered from COVID-19 [14, 16]. Extracellular vesicles can convey pathogen molecules that serve as antigens or agonists of innate immune receptors to induce host defense and immunity or serve as regulators of host defense and mediators of immune evasion [25, 50]. We speculate that the mechanism of camouflaging the virus from immune surveillance might also trigger an autoimmune response from the host in those instances where close association of viral and host antigens promotes cross-recognition.

Several drugs are known to interfere with the metabolism of EVs, and their importance in fighting future pandemics should not be overlooked.

4.3 Intracytoplasmic Route of SARS-COV-2 Transmission as Humoral Immunity Evasion

The usual field of observation in our study was on regions 250 nm^2 or larger. In this way, we could map hundreds of cells, but, at the same time, we could easily zoom-in at the nanometer scale. In a large field, several cells could be seen to be connected, either by TNT or by fusion, while at the same time extracellular vesicles were scat tered around; this is interpreted as the three features coexist in space and time. In all three instances, the cytoplasmic content is shared among host cells. This may be an alternative route of transmission and possibly infection, clearly distinct from the well-known extra-cytoplasmic ACE2 (or other receptor-mediated) docking mechanism. Humoral immune evasion can be achieved using this route.

Even if the close phylogenetic relationship between SARS-CoV and SARS-CoV-2 makes it reasonable to translate much of our knowledge of SARS-CoV to SARS-CoV-2, we should focus on the differences between SARS-CoV and SARS-CoV-2 to explain the divergent clinical patterns of disease caused by these two viruses. We have already cited the fusogenic potential specific to the SARS-CoV-2 spike protein, but not the SARS-CoV spike protein [64]. Furthermore, SARS-CoV-2 has a lower density of spikes and can produce a high number of defective copies (with little or no spikes) released outside the infected cell [28, 35], but still remain very infective. These observations provide strong evidence, even if indirect, that an alternative route to ACE2 docking must exist for SARS-CoV-2. The intracytoplasmic route can hide the virus from host immune surveillance and potential antiviral response based on the detection of extracellular virions. By the intracytoplasmic route, naïve cells could be infected by viral mRNA transmitted free or in the form of cargo in microvesicles. Defective virions, presenting little or no spikes, which would be ineffective in infecting other cells via the extra-cytoplasmic route, might become effective when transmitted via the intracytoplasmic route.

5 Considerations for the Next Pandemic

5.1 Drugs Effective against SARS-CoV-2 that interfere with Tunneling Nanotubes, Cell Fusion, and Extracellular Vesicles

Specific drugs that target the mechanisms of the described intracytoplasmic route (TNTs, fusion, and EVs) will theoretically be promising candidates for treatment of COVID-19. The literature reports a broad spectrum of possible therapeutic molecules against each of these three mechanisms. Thus, if drugs already in use were effective against SARS-CoV-2, this could be taken as indirect evidence that the intracytoplasmic route plays a role in the infection.

A quick search of the literature showed that several drugs interfering with the TNTs, fusion, or EVs had been empirically tested and proven effective. A few interesting examples follow.

(A) The proven clinical efficacy of colchicine, a well-known microtubule inhibitor, might interfere with TNT microtubule formation [32, 37]. The clinical efficacy of colchicine suggests putative evidence of the role of TNTs in COVID-19 pathogenesis.
(B) Drugs that suppress cell fusion have been tested in vitro, including niclosamide, which was effective in cell protection against virus-induced cell death [11, 38].
(C) Drugs that interfere with EV machinery were tested before the COVID-19 pandemic, namely, during the SARS and MERS emergencies. Chloroquine elevates the pH in endosomal vesicles and is used to treat malaria and has shown to be effective against SARS-CoV [60].

Timing of administration of these drugs will be very important, in the sense that any candidate drug will likely need to be administered during early stages of SARS-CoV-2, when TNT formation, cell fusion, and EV release occur.

5.2 Clinical Observation Should Lead the Fight Against a New Disease

It is reasonable to consider a new pathogen will be able to develop an immune evasion mechanism due to the pressure of the natural and/or artificial (pharmacological) host defense. In the case of SARS-CoV-2, it would have been useful to look for mechanisms of immune evasion immediately after the occurrence of COVID-19 infections in vaccinated individuals. This occurrence was labeled "breakthrough infection" and served as a reminder that no vaccine is 100% effective. However, it was not investigated from the viewpoint of a source of new information. Breakthrough infections can have many causes but should be investigated for two important reasons. First, there is a duty to treat patients who do not respond to prevalent therapy, and second, breakthrough infections can disclose a mechanism not previously known.

Clinical observation is key to medical knowledge. The relevance of the old-fashioned daily "record of signs and symptoms" (... including breakthrough infection ...) and the "response of the patient to the therapy" cannot be downplayed. Observation of the response to therapy is, actually, done very early when a new disease breaks out. This is because we use (or we should use) every possible therapeutic strategy when we still do not know the causative mechanism. Therapeutic protocols are established based on symptoms, or on similarity with other known diseases, or just because a drug already in use for another concurrent treatment seems to be effective against the unknown pathogen as well. This was the exemplary case here for the microtubule-inhibitor "colchicine." Its efficacy in COVID-19 was reported with increasing evidence during the year before the publication of the research reported here. The possible mechanism of action against SARS-CoV-2 was purely speculative. The discovery of microtubule-rich TNT induction in infected cells suggested an explanation of why a drug interfering with microtubule formation could affect the course of COVID-19.

The method of investigation plays a role. Molecular analysis requires available and affordable benchtop equipment, whereas ultrastructural imaging is far more expensive, time consuming, and less available, and correlative microscopy is still in development. Bias related to social factors cannot be overlooked. Nevertheless, it is important to be honest and open-minded when evaluating results, and flexible enough to follow new paths when unexpected findings point in directions different from the expected.

6 Methods

6.1 VeroE6 Cell Culture and SARS-CoV-2 Infection

The African green monkey kidney epithelial cell line (VeroE6) supports SARS-CoV-2 replication in culture, while many cell lines are reported to be refractory to SARS-CoV-2 infection [43, 54]. All of the work involving infectious SARS-CoV-2 was performed in the biosafety level 3 facilities at Rutgers University. The SARS-CoV-2-expressing mNeonGreen was provided by Dr. Theresa Chang (Rutgers University) and Dr. Pei-Yong Shi (University of Texas). VeroE6 cells were grown in Dulbecco's modified essential medium (DMEM) (Cat# D6429, Sigma-Aldrich, USA) containing 10% fetal bovine serum (FBS) (Sigma-Aldrich, USA). For SARS-CoV-2 infection, 0.3×10^6 VeroE6 cells were seeded onto a 13-mm-diameter glass coverslip with carved reference frames, in a six-well cell-culture plate with 2 mL of DMEM + FBS medium. At 18–24 h post-seeding, the spent medium was aspirated.

The cells were infected with SARS-CoV-2-expressing mNeonGreen at a multiplicity of infection (MOI) of 0.1 (i.e., 1 viral particle for 10 host cells) in 400 μL of FBS-free DMEM. The plates were incubated at 37 °C for 1 h with intermittent rocking for infection. The wells were replenished with 2 mL of DMEM + FBS. At 24 h post-infection, the cell culture supernatants were removed, and the cells were fixed with 4% paraformaldehyde (Cat# 19943-k2, Thermo Scientific, USA) [34].

6.2 Helium-Ion Microscopy

The principle of HeIM operation is very similar to that of SEM, except that HeIM utilizes a beam of positively charged helium ions (He+) instead of negatively charged electrons to excite and detect secondary electrons from the sample surface [24, 61]. Due to the high brightness and low energy spread of its atomically sharp gas-field ion source, the smallest attainable focused spot size is about 0.3 nm. With its significantly smaller convergence angle compared to SEM, HeIM achieves a much larger depth of field, which is particularly useful for imaging three-dimensional structures. Due to their higher mass, He+ ions do not spread as much as electrons, resulting in a smaller escape volume of the secondary electrons and a higher surface resolution of the HeIM than SEM. A further benefit of HeIM is its charge compensation capability during secondary electron detection. In the HeIM, charges that accumulate on insulating samples are positive, so a low-energy electron "flood gun," which irradiates the sample with a diffuse beam of electrons, is used to achieve charge neutralization. This method eliminates the need for a conductive coating of samples and allows for direct visualization of their morphology at a nanoscale level [39, 51] (Fig. 13).

6.3 Correlative Microscopy

Reference frames carved on glass coverslips allowed us to perform correlative HeIM/fluorescence/phase-contrast microscopy (Zeiss Axio Observer, Zeiss Jena D). The ability to identify and study the same cell using the three different imaging modalities was extremely useful, as we could merge information specifically provided by each. As HeIM does not require any electroconductive coating, samples

Fig. 13 (Helium-ion microscopy images the biological sample (blue) while Scanning electron microscopy images the electroconductive coating of the biological sample (orange). HeIM has a specific advantage in avoiding the need of an electroconductive coating that masks, like a snow mantle, the surface features embedded in the coating thickness)

imaged by HeIM could be re-imaged with phase-contrast or fluorescence microscopy, an option not available using SEM because of the coating [39]. The routine sequence of imaging usually began with phase-contrast microscopy, followed by fluorescence microscopy. VeroE6 cells infected with SARS-CoV-2 were identified by the green fluorescence of the mNeonGreen protein expressed by the virus intracellularly.

Acknowledgement Our study was largely supported by the National Science Foundation and by the Department of Physics and Astronomy of Rutgers University.

References

1. Abounit S, Zurzolo C (2012) Wiring through tunneling nanotubes–from electrical signals to organelle transfer. J Cell Sci 125:1089–1098. https://doi.org/10.1242/jcs.083279
2. Alcami A, Koszinowski UH (2000) Viral mechanisms of immune evasion. Trends Microbiol 8(9):410–418. https://doi.org/10.1016/s0966-842x(00)01830-8. PMID: 10989308; PMCID: PMC7141568
3. Bahl A et al (2021) Vaccination reduces need for emergency care in breakthrough COVID-19 infections: a multicenter cohort study. Lancet Reg Health Am:100065. https://doi.org/10.1016/j.lana.2021.100065
4. Barberis E et al (2021) Circulating exosomes are strongly involved in SARS-CoV-2 infection. Front Mol Biosci 8:632290. https://doi.org/10.3389/fmolb.2021.632290
5. Barreto-Vieira DF et al (2021) Morphology and morphogenesis of SARS-CoV-2 in Vero-E6 cells. Mem Inst Oswaldo Cruz 116:e200443. https://doi.org/10.1590/0074-02760200443
6. Baselga M, Moreo E, Uranga-Murillo I et al (2023) Ultrastructural analysis and three-dimensional reconstruction of cellular structures involved in SARS-CoV-2 spread. Histochem Cell Biol 159:47–60. https://doi.org/10.1007/s00418-022-02152-7
7. Bergwerk M et al (2021) Covid-19 breakthrough infections in vaccinated health care workers. N Engl J Med. https://doi.org/10.1056/NEJMoa2109072
8. Bianchi M, Borsetti A, Ciccozzi M, Pascarella S (2021) SARS-Cov-2 ORF3a: Mutability and function. Int J Biol Macromol 170:820–826. https://doi.org/10.1016/j.ijbiomac.2020.12.142. Epub 2021 Jan 8. PMID: 33359807; PMCID: PMC7836370
9. Bloemendal S, Kück U (2013) Cell-to-cell communication in plants, animals, and fungi: a comparative review. Naturwissenschaften 100:3–19. https://doi.org/10.1007/s00114-012-0988-z
10. Bouhaddou M et al (2020) The global phosphorylation landscape of SARS-CoV-2 infection. Cell 182:685–712.e619. https://doi.org/10.1016/j.cell.2020.06.034
11. Braga L et al (2021) Drugs that inhibit TMEM16 proteins block SARS-CoV-2 spike-induced syncytia. Nature 594:88–93. https://doi.org/10.1038/s41586-021-03491-6
12. Bullock HA, Goldsmith CS, Miller SE (2021) Best practices for correctly identifying coronavirus by transmission electron microscopy. Kidney Int 99:824–827. https://doi.org/10.1016/j.kint.2021.01.004
13. Cappellano G et al (2021) Circulating platelet-derived extracellular vesicles are a Hallmark of SARS-CoV-2 infection. Cells 10. https://doi.org/10.3390/cells10010085
14. Chatterjee V, Yang X, Ma Y, Wu MH, Yuan SY (2020) Extracellular vesicles: new players in regulating vascular barrier function. Am J Physiol Heart Circ Physiol 319:H1181–H1196. https://doi.org/10.1152/ajpheart.00579.2020
15. Colombo M, Raposo G, Théry C (2014) Biogenesis, secretion, and intercellular interactions of exosomes and other extracellular vesicles. Annu Rev Cell Dev Biol 30:255–289. https://doi.org/10.1146/annurev-cellbio-101512-122326

16. Elrashdy F, Aljaddawi AA, Redwan EM, Uversky VN (2020) On the potential role of exosomes in the COVID-19 reinfection/reactivation opportunity. J Biomol Struct Dyn:1–12. https://doi.org/10.1080/07391102.2020.1790426
17. Eugenin EA, Gaskill PJ, Berman JW (2009) Tunneling nanotubes (TNT) are induced by HIV infection of macrophages: a potential mechanism for intercellular HIV trafficking. Cell Immunol 254:142–148. https://doi.org/10.1016/j.cellimm.2008.08.005
18. Gould SJ, Booth AM, Hildreth JE (2003) The Trojan exosome hypothesis. Proc Natl Acad Sci USA 100:10592–10597. https://doi.org/10.1073/pnas.1831413100
19. Gousset K, Zurzolo C (2009) Tunnelling nanotubes: a highway for prion spreading? Prion 3:94–98. https://doi.org/10.4161/pri.3.2.8917
20. Guervilly C et al (2021) Dissemination of extreme levels of extracellular vesicles: tissue factor activity in patients with severe COVID-19. Blood Adv 5:628–634. https://doi.org/10.1182/bloodadvances.2020003308
21. Gurke S, Barroso JF, Gerdes HH (2008) The art of cellular communication: tunneling nanotubes bridge the divide. Histochem Cell Biol 129:539–550. https://doi.org/10.1007/s00418-008-0412-0
22. Hildreth JEK (2017) HIV as Trojan exosome: immunological paradox explained? Front Immunol 8:1715. https://doi.org/10.3389/fimmu.2017.01715
23. Hilleman MR (2004) Strategies and mechanisms for host and pathogen survival in acute and persistent viral infections. Proc Natl Acad Sci USA 101(Suppl 2):14560–14566. https://doi.org/10.1073/pnas.0404758101. Epub 2004 Aug 5. PMID: 15297608; PMCID: PMC521982
24. Joens MS et al (2013) Helium ion microscopy (HIM) for the imaging of biological samples at sub-nanometer resolution. Sci Rep 3:3514. https://doi.org/10.1038/srep03514
25. Kadiu I, Narayanasamy P, Dash PK, Zhang W, Gendelman HE (2012) Biochemical and biologic characterization of exosomes and microvesicles as facilitators of HIV-1 infection in macrophages. J Immunol 189:744–754. https://doi.org/10.4049/jimmunol.1102244
26. Karlikow M et al (2016) Drosophila cells use nanotube-like structures to transfer dsRNA and RNAi machinery between cells. Sci Rep 6:27085. https://doi.org/10.1038/srep27085
27. Kasuga Y, Zhu B, Jang KJ, Yoo JS (2021) Innate immune sensing of coronavirus and viral evasion strategies. Exp Mol Med 53(5):723–736. https://doi.org/10.1038/s12276-021-00602-1. Epub 2021 May 6. PMID: 33953325; PMCID: PMC8099713
28. Ke Z et al (2020) Structures and distributions of SARS-CoV-2 spike proteins on intact virions. Nature 588:498–502. https://doi.org/10.1038/s41586-020-2665-2
29. Kikkert M (2020) Innate immune evasion by human respiratory RNA viruses. J Innate Immun 12(1):4–20. https://doi.org/10.1159/000503030. Epub 2019 Oct 14. PMID: 31610541; PMCID: PMC6959104
30. Kim BW, Lee JS, Ko YG (2019) Mycoplasma exploits mammalian tunneling nanotubes for cell-to-cell dissemination. BMB Rep 52:490–495. https://doi.org/10.5483/BMBRep.2019.52.8.243
31. Kimura S, Hase K, Ohno H (2012) Tunneling nanotubes: emerging view of their molecular components and formation mechanisms. Exp Cell Res 318:1699–1706. https://doi.org/10.1016/j.yexcr.2012.05.013
32. Knack RS, Losso LC, Knack RS, Hanada T (2021) Preliminary results of clinical use of colchicine in COVID-19 treatment. BMJ Case Rep 14. https://doi.org/10.1136/bcr-2021-244482
33. Kumar A et al (2017) Influenza virus exploits tunneling nanotubes for cell-to-cell spread. Sci Rep 7:40360. https://doi.org/10.1038/srep40360
34. Kumar R, Kolloli A, Subbian S (2021) Inactivation and elimination of SARS-CoV-2 in biosamples using simple fixatives and ultrafiltration. Methods Protoc 4. https://doi.org/10.3390/mps4010018
35. Laue M et al (2021) Morphometry of SARS-CoV and SARS-CoV-2 particles in ultrathin plastic sections of infected Vero cell cultures. Sci Rep 11:3515. https://doi.org/10.1038/s41598-021-82852-7
36. Leroy H et al (2020) Virus-mediated cell-cell fusion. Int J Mol Sci 21. https://doi.org/10.3390/ijms21249644

37. Lien CH et al (2021) Repurposing colchicine in treating patients with COVID-19: a systematic review and meta-analysis. Life (Basel) 11. https://doi.org/10.3390/life11080864
38. Lin L, Li Q, Wang Y, Shi Y (2021) Syncytia formation during SARS-CoV-2 lung infection: a disastrous unity to eliminate lymphocytes. Cell Death Differ 28:2019–2021. https://doi.org/10.1038/s41418-021-00795-y
39. Merolli A, Kasaei L, Ramasamy S et al (2022) An intra-cytoplasmic route for SARS-CoV-2 transmission unveiled by helium-ion microscopy. Sci Rep 12:3794. https://doi.org/10.1038/s41598-022-07867-0
40. Nassoy P, Lamaze C (2012) Stressing caveolae new role in cell mechanics. Trends Cell Biol 22:381–389. https://doi.org/10.1016/j.tcb.2012.04.007
41. Nolte-'t Hoen E, Cremer T, Gallo RC, Margolis LB (2016) Extracellular vesicles and viruses: are they close relatives? Proc Natl Acad Sci USA 113:9155–9161. https://doi.org/10.1073/pnas.1605146113
42. Nussenzveig HM (2019) Are cell membrane nanotubes the ancestors of the nervous system? Eur Biophys J 48:593–598. https://doi.org/10.1007/s00249-019-01388-x
43. Ogando NS et al (2020) SARS-coronavirus-2 replication in Vero E6 cells: replication kinetics, rapid adaptation and cytopathology. J Gen Virol 101:925–940. https://doi.org/10.1099/jgv.0.001453
44. Osswald M et al (2015) Brain tumour cells interconnect to a functional and resistant network. Nature 528:93–98. https://doi.org/10.1038/nature16071
45. Pepe A, Pietropaoli S, Vos M, Barba-Spaeth G, Zurzolo C (2022) Tunneling nanotubes provide a route for SARS-CoV-2 spreading. Sci Adv 8(29):eabo0171. https://doi.org/10.1126/sciadv.abo0171. Epub 2022 Jul 20. PMID: 35857849; PMCID: PMC9299553
46. Rosell A et al (2020) Patients with COVID-19 have elevated levels of circulating extracellular vesicle tissue factor activity that is associated with severity and mortality. Arterioscler Thromb Vasc Biol 41:Atvbaha120315547. https://doi.org/10.1161/atvbaha.120.315547
47. Rubio-Casillas A, Redwan EM, Uversky VN (2022) SARS-CoV-2: a master of immune evasion. Biomedicines 10(6):1339. https://doi.org/10.3390/biomedicines10061339. PMID: 35740361; PMCID: PMC9220273
48. Rustom A, Saffrich R, Markovic I, Walther P, Gerdes HH (2004) Nanotubular highways for intercellular organelle transport. Science 303:1007–1010. https://doi.org/10.1126/science.1093133
49. Schmidt M, Byrne JM, Maasilta IJ (2021) Bio-imaging with the helium-ion microscope: a review. Beilstein J Nanotechnol 12:1–23. https://doi.org/10.3762/bjnano.12.1
50. Schorey JS, Harding CV (2016) Extracellular vesicles and infectious diseases: new complexity to an old story. J Clin Invest 126:1181–1189. https://doi.org/10.1172/jci81132
51. Schürmann M et al (2015) Helium ion microscopy visualizes lipid Nanodomains in mammalian cells. Small 11:5781–5789. https://doi.org/10.1002/smll.201501540
52. Sherer NM, Mothes W (2008) Cytonemes and tunneling nanotubules in cell-cell communication and viral pathogenesis. Trends Cell Biol 18:414–420. https://doi.org/10.1016/j.tcb.2008.07.003
53. Sur S et al (2021) Exosomes from COVID-19 patients carry tenascin-C and fibrinogen-? In triggering inflammatory signals in cells of distant organ. Int J Mol Sci 22. https://doi.org/10.3390/ijms22063184
54. Takayama K (2020) In vitro and animal models for SARS-CoV-2 research. Trends Pharmacol Sci 41:513–517. https://doi.org/10.1016/j.tips.2020.05.005
55. Tiwari V, Koganti R, Russell G, Sharma A, Shukla D (2021) Role of tunneling nanotubes in viral infection, neurodegenerative disease, and cancer. Front Immunol 12:680891. https://doi.org/10.3389/fimmu.2021.680891
56. Urbanelli L et al (2019) The role of extracellular vesicles in viral infection and transmission. Vaccines (Basel) 7. https://doi.org/10.3390/vaccines7030102
57. Valdebenito S, Ono A, Rong L, Eugenin EA (2023) The role of tunneling nanotubes during early stages of HIV infection and reactivation: implications in HIV cure. NeuroImmune Pharmacol Ther 2(2):169–186. https://doi.org/10.1515/nipt-2022-0015

58. van Niel G, D'Angelo G, Raposo G (2018) Shedding light on the cell biology of extracellular vesicles. Nat Rev Mol Cell Biol 19:213–228. https://doi.org/10.1038/nrm.2017.125
59. Vignais ML, Caicedo A, Brondello JM, Jorgensen C (2017) Cell connections by tunneling nanotubes: effects of mitochondrial trafficking on target cell metabolism, homeostasis, and response to therapy. Stem Cells Int 2017:6917941. https://doi.org/10.1155/2017/6917941
60. Vincent MJ et al (2005) Chloroquine is a potent inhibitor of SARS coronavirus infection and spread. Virol J 2:69. https://doi.org/10.1186/1743-422x-2-69
61. Wirtz T, De Castro O, Audinot JN, Philipp P (2019) Imaging and analytics on the helium ion microscope. Annu Rev Anal Chem (Palo Alto, Calif) 12:523–543. https://doi.org/10.1146/annurev-anchem-061318-115457
62. Wu F, Zhao S, Yu B et al (2020) A new coronavirus associated with human respiratory disease in China. Nature 579:265–269. https://doi.org/10.1038/s41586-020-2008-3
63. Zhang S, Kazanietz MG, Cooke M (2020) Rho GTPases and the emerging role of tunneling nanotubes in physiology and disease. Am J Physiol Cell Physiol 319:C877–c884. https://doi.org/10.1152/ajpcell.00351.2020
64. Zhang Z et al (2021) SARS-CoV-2 spike protein dictates syncytium-mediated lymphocyte elimination. Cell Death Differ 28:2765–2777. https://doi.org/10.1038/s41418-021-00782-3
65. Zhu H et al (2016) Rab8a/Rab11a regulate intercellular communications between neural cells via tunneling nanotubes. Cell Death Dis 7:e2523. https://doi.org/10.1038/cddis.2016.441

Deploying Biotechnology at Scale Through Systems Integration to Combat COVID-19

Matthew McKnight and Swati Sureka

1 Executive Summary

Large-scale solutions to respond to and ultimately prevent biological crises like COVID-19 and any future pandemic necessitate comprehensive and integrated approaches. As in other domains of national security, a nimble response to a biological threat requires innovation and a massive operational scale. Achieving an effective early warning capability [1] against an infectious disease will require a trusted partnership between public and private sectors to link data curation and analytics with operations and signal collection at scale.

During the COVID-19 pandemic, a new category of biosecurity operations emerged: large-scale biosecurity system integration of physical assets, for example, tests, staff, couriers, and laboratories, new and emerging technologies, including rapid testing, genomic sequencing, variant determination, and their digital output to ensure optimal operational execution to provide an effective early warning system. This systems integrator function must flow between the public and private sectors and provide, in addition to platforms for end users, supply chain management for laboratories and manufacturers—while providing relevant data to public health officials.

COVID-19 accelerated creation of such a capability in Concentric by Ginkgo (Concentric), the outbreak suppression offering of Ginkgo Bioworks (Ginkgo), where a platform with related services was constructed that focused on integrating

Disclaimer: Ginkgo Bioworks is a publicly traded company. Matthew McKnight and Swati Sureka have an equity interest in Ginkgo Bioworks. Ginkgo Bioworks may financially benefit from the research presented here.

M. McKnight · S. Sureka (✉)
Ginkgo Bioworks, Boston, MA, USA
e-mail: ssureka@ginkgobioworks.com

J. M. Rosen, R. R. Colwell (eds.), *The COVID-19 Pandemic*,
https://doi.org/10.1007/978-3-031-62772-9_17

capabilities for a scaled system to detect, intercept, and mitigate biothreats [2]. This capability is critical for large-scale response to biological incidents in the future.

Testing in K-12 schools (a community sentinel population [3]) was established across the country in partnership with federal, state, and local public health authorities, and wastewater and air monitoring pilots were launched, strengthening the whole genome sequencing network. The system was also applied through a strong public-private partnership between CDC, Ginkgo, and XpresCheck to achieve viral genomic surveillance of international passengers, which was the first to detect Omicron subvariants BA.2 and BA.3 in the USA [4] and has since persisted as sustained biosecurity infrastructure for detecting a variety of biological threats.

This systems integration was built on a foundation of deep understanding of biology and biotechnology, in partnership with a trusted and credible partner in biosecurity. The vision of Concentric is to build a platform for the twenty-first century biosecurity that brings together component subsystems and ensures the subsystems are resilient, functioning together to meet national security, biosecurity, and public health needs of the US and allied stakeholders.

2 Specific Challenges to Public Health and Pandemic Response Leadership

The repeated surges of the COVID-19 pandemic made clear that a biosecurity and pandemic response required multifaceted action, including coordination and manufacturing and supply chain management. Furthermore, public health planning has to be inclusive to avoid excluding subsections of communities. The following describes the challenges and provides an overview of how Concentric addressed them in the system design:

2.1 Coordination, Demand Management, and Efficient Collection for Laboratory Testing

PCR testing is delivered by thousands of laboratories using different assays and collection devices, all of which depend on a wide variety of supply chain participants [5]. Moreover, variation in data reporting systems and lack of integration across systems can lead to data transfer delays, especially during surges of the biothreat agents. The Concentric system design resulted in optimal sample throughput at scale by managing front end support, collection supply chain, and logistics. Ginkgo established laboratory standards for validation, performance, and reporting across the laboratory network, ensuring an integrated system with rapid turnaround and real-time reporting.

2.2 Supply Chain Management and Onsite Delivery/Reporting for Antigen Testing

Currently, antigen testing relies on a fragmented and international supply chain, onsite medical assistance requirements, and variation in results reporting. Concentric provided the capability to serve as distributor—partnering with distributors of multiple types of rapid antigen tests, and a diversified offering to mitigate supply shortages. Concentric also integrated new test technologies rapidly. Concentric partnered with major national and multinational suppliers, recognized for quality and reliability, for sample collection materials, for example, kits, swabs, and tubes. Multiple suppliers participated in each procurement category. The multi-sourcing strategy enabled flexibility, scalability, and risk mitigation. The team shared values, platform, and mission, and engaged, promoted, and championed diverse suppliers.

2.3 Implementing Large-Scale Testing in Nontraditional Settings (Encouraging Equitable Access)

Testing large groups of individuals regularly or providing testing on demand requires purpose-built software, seamless reporting to both public health and community decision makers, decentralized supply chain management and optimization, quality verification, and inclusive stakeholder communication.

An effective biosecurity platform must also provide value to the end users and constituents, whether individuals receiving testing services or communities running testing programs. Enhanced end-user engagement increases opportunities for data collection and provides a foundation for diagnostic, screening, or surveillance programs. Toward that end, we implemented a feature set for managing "test takers" and test taker communities. We built and continually adapted an accessible, easy-to-use test platform for individual end users.

Concentric provided support for administering community test programs at a variety of scales and durations. Reporting and analytics dashboards were provided to communities operating at multiple jurisdictional levels, for example, individual schools, districts, and airports. Dashboards consolidated signals from different modalities, creating a "pathogen weather map" to assist public health officials in making informed decisions.

3 These Challenges Require a Systems Integrator

The federal government is responsible for providing excellent and reliable pandemic response capabilities for the United States and, ideally, the global community. It has at its discretion the powerful levers of financial resources, purchasing authority, and regulatory guidance, as well as a reservoir of technology, talent, and healthcare/public health infrastructure. However, the massive capabilities can be fragmented and decentralized.

A next generation pandemic response must be underpinned with a strong bioeconomy [6], that is sustained in part with public funding to ensure capacity and flexibility for the peaks and valleys of demand in testing and biosecurity services. To achieve these goals, trusted private sector partners will be needed to (1) be guided by policy, (2) evaluate and integrate the industry, and (3) deliver quickly on a national scale.

A national operator able to consolidate data streams coming from many dozens of systems into a single stream and effectively integrate state-level operations (as was done by Concentric) enriches and empowers the public health system. With accessible data, public health officials in local communities and the federal government can take informed, decisive action to prevent further spread. A national systems integrator should have agility and the capability to work across technologies and partners, essentially supercharging innovation, with significant impact on public health.

As the Department of Defense relies on major systems integrators to implement policy and operational plans, so should there be systems integrators to enhance biosecurity when a requirement arises to deliver pandemic response, namely pandemic surveillance, at scale, consistently and equitably.

4 Features of Concentric by Ginkgo as a Systems Integrator

The Concentric solution began with the design of a purpose-built and user-centric software platform with the following features:

4.1 Test-Agnostic

In biosecurity applications, tests are "data generators." Whereas each test generates specific data, current and future tests should be available to the government for the purpose of deployment in pursuit of its objectives. At Concentric, qualified tests are integrated into the platform and, in accordance with regulatory requirements specific to each test, used for diagnostics, screening, or surveillance. To accomplish this objective, the following were developed and integrated:

- *Laboratory integration API* that allows any qualified laboratory to manage samples and upload results.
- *Result entry portal* that allows healthcare providers or other qualified personnel (including those using over the counter (OTC) tests) to upload results of tests in real time.
- *Streamlined physician ordering* via private physician ordering services and/or officials designated by public health authorities.
- *Real-time 50-state and territory public health reporting* for all diagnostic and screening tests.

4.2 Data Standardization, Reporting, and Predictive Analytics

- *Standardization*: Useful data for Public health require inclusion of multiple data types, each with specific standardization norms. Concentric regularly shared structured data from different public health reporting information systems, including HHS Protect through Operation Expanded Testing [7] and other state reporting systems.
- *Reporting dashboards*: Aggregate-level dashboards were shared with federal, state, and district officials, allowing assessment of the performance of their biosurveillance, for example, testing, sequencing, and meta-data, employing metrics such as pool positivity rate and number of tests per week. Data are easily exported at the program level, allowing data-driven decisions. The dashboards enabled officials to manage hundreds of individual testing sites in a single centralized data source.
- *Demand planning for antigen tests/pool positivity rates*: Trends such as pool positivity rates and behaviors around material ordering were leveraged to manage demand surges amidst supply constraints. This allowed better forecasting time and location of potential surges and anticipated demand for testing supplies, as well as laboratory capacity and local labor availability, including healthcare staffing.

4.3 Capacity Planning and Allocation Management

- *Laboratories*: Concentric actively tested in partnership with 23 laboratories and developed a working relationship with approximately 40 laboratories, enabling a capacity to test more than 400,000 samples per day. The pooling product enabled a potential maximum capacity to collect samples from over 50 million individuals per week.
- *Medical professionals and nurses*: Concentric established partnerships with ten medical staffing agencies, hiring ca. 3000 healthcare providers and administrative coordinators to facilitate sample collection for weekly testing programs. The ability to add new partners, expanding the provider team as quickly as 2 weeks and across nearly all of the USA was established during the pandemic period.
- *Couriers*: Concentric established relationships with a total of five logistics providers at the national and local levels, coordinating sample pickup at ca. 3000 locations weekly. Samples were delivered to the local laboratory the same day or next morning. Logistics partners had the ability to expand coverage with 2–3 weeks' notice.

4.4 Technical Validation of Lab Partner Quality and Scalability

- *Ginkgo focuses on science,* and new solutions are assessed on underlying science and gap analyses, a core operational model.

- *Partner laboratories across the USA were selected to be geographically, organizationally, and technologically diverse*. Performance is measured at a high standard with robust validation, including blinded validation and compliance verification.
- *National distribution and integrated technology infrastructure enable extensive scalability.*

4.5 Achieving Maximum Efficient Discovery and Manufacturing Processes for Medical Countermeasures

- *Systems integration enables connection throughout an entire biosecurity continuum.* Test data, passive or active, are essential, however the data also prime therapeutics and vaccine discovery and delivery.
- The Ginkgo platform was designed to *enable extensive, rapid prototyping in discovery and during manufacturing.* In response to COVID-19, Ginkgo initiated programs to support Moderna, providing process optimization for key raw materials used in manufacturing of mRNA vaccines—including mRNA-1273, the Moderna vaccine candidate against COVID-19 [8]—and Aldevron, yielding a manufacturing breakthrough in the vaccinia capping enzyme used to manufacture mRNA vaccines [9].

5 Nationwide K-12 School Testing

Concentric was able to recognize very early that COVID-19-related disruption of schooling would have serious effects on children and also economic recovery. An emergency COVID-19 response for schools was initiated across the USA, with a pioneering of classroom-based pooled testing backed by systems integration to provide end-to-end, community-centered services across the nation.

Lacking mitigation planning related to school reopening, repeated outbreaks, closures, and super-spreader events occurred. The testing programs carried out at ca. 4000 US schools provided school leaders with confidence in the COVID-19 status of their classrooms as they reopened, empowering them to focus on engaging their communities and mounting appropriate responses to cases that occurred [10]. Across the country, testing-based intervention, including Test to Stay programs, proved critical in preserving in-person learning days while managing transmission risk [11, 12].

An effective mitigation strategy is offering testing for every child and each staff member weekly, regardless of symptoms. To conduct this volume of testing with PCR required development of a method, namely pooled testing, [13] whereby individual nasal swabs of students and staff were grouped by classroom and tested simultaneously, at a cost of a single sample. When pooled results were negative, educators and parents had a high degree of confidence that the classroom was free of COVID-19. When the test was positive, quick action identified and mitigated cases and exposures.

Many US states—from California to Arizona to Maryland—adopted the pooled testing model during the height of the COVID-19 pandemic. In Baltimore, Maryland, over 40,000 students and staff were tested weekly across 125 testing locations, enabling restoration of in-person learning and extracurricular and support services, including free meals and school-based healthcare [14]. In Tucson, Arizona, participation exceeded 62,000 across 100 locations [15]. In addition to building for scale, the Concentric networked systems integration model enabled delivery of pooled testing to rural areas where access to public testing services was less available—such as Anderson Valley, California, for example, where the Concentric in-school testing program was able to report approximately 2–3× faster than local PCR testing sites [16].

Many were involved in this important endeavor. The COVID-19 pandemic triggered an unprecedented mobilization of public, private, academic, and community organizations to facilitate careful reopening of schools. The distribution of critical capabilities, supported by widespread commitment of communities to mutual support, drives the Concentric systems integration model for pandemic response.

Acknowledgment We would like to thank the Ginkgo Biosecurity team, past and present, who built the Concentric system and many of whom contributed ideas and language to previous versions of this chapter. We are also incredibly grateful to our ecosystem of public health leaders, technical and operational partners, and communities across the country who worked hand-in-hand with us to develop and deploy Concentric capabilities during the COVID-19 pandemic.

References

1. Kamalrathe T et al (2023) Need for effective detection and early warnings for epidemic and pandemic preparedness planning in the context of multi-hazards: lessons from the COVID-19 pandemic. Int Disaster Risk Reduct 92:103724. Published online 2023 Apr 29. https://doi.org/10.1016/j.ijdrr.2023.103724
2. Ginkgo Bioworks (2021) Concentric by Ginkgo stands ready to continue fighting pandemic, providing biosecurity and COVID testing nationwide to schools, workplaces, and other congregate settings. Available via PR Newswire. https://www.prnewswire.com/news-releases/concentric-by-ginkgo-stands-ready-to-continue-fighting-pandemic-providing-biosecurity-and-covid-testing-nationwide-to-schools-workplaces-and-other-congregate-settings-301373567.html
3. Hoch M et al (2021) Weekly SARS-CoV-2 sentinel surveillance in primary schools, kindergartens, and nurseries, Germany, June–November 2020. Emerg Infect Dis 27(8):2192–2196. https://doi.org/10.3201/eid2708.204859
4. Wegrzyn R et al (2023) Early detection of severe acute respiratory syndrome coronavirus 2 variants using traveler-based genomic surveillance at 4 US airports, September 2021–January 2022. Clin Infect Dis 76(3):e540–e543. https://doi.org/10.1093/cid/ciac461
5. Library of Congress (2021) Covid-19 testing supply chain—CRS reports. Congressional Research Service https://crsreports.congress.gov/product/pdf/IF/IF11774
6. Morhard R (2022) The bioeconomy revolution can end the panic-and-neglect cycle in health security: Think Global Health. Council on Foreign Relations https://www.thinkglobalhealth.org/article/bioeconomy-revolution-can-end-panic-and-neglect-cycle-health-security

7. Ginkgo Bioworks (2022) Clinical enterprise, Inc. and Ginkgo Bioworks to continue support of federal Covid-19 testing program through 2022. Business Wired https://www.businesswire.com/news/home/20220728005065/en/Clinical-Enterprise-Inc.-and-Ginkgo-Bioworks-to-Continue-Support-of-Federal-COVID-19-Testing-Program-Through-2022
8. Ginkgo Bioworks (2020) Ginkgo Bioworks provides support on process optimization to Moderna for COVID-19 response. PR Newswire. https://www.prnewswire.com/news-releases/ginkgo-bioworks-provides-support-on-process-optimization-to-moderna-for-covid-19-response-301040876.html
9. Ginkgo Bioworks (2021) Aldevron's collaboration with Ginkgo Bioworks yields manufacturing breakthrough for vaccinia capping enzyme used for manufacturing of mRNA vaccines. PR Newswire https://www.prnewswire.com/news-releases/aldevrons-collaboration-with-ginkgo-bioworks-yields-manufacturing-breakthrough-for-vaccinia-capping-enzyme-used-for-manufacturing-of-mrna-vaccines-301351786.html
10. Shah Family Foundation (n.d.) Case studies. Open & Safe Schools. https://www.openandsafeschools.org/case-studies-1
11. Harris-McCoy K et al (2021) Evaluation of a test to stay strategy in transitional kindergarten through grade 12 schools — Los Angeles County, California, august 16–October 31, 2021. MMWR Morb Mortal Wkly Rep 70(1773–1777):1773. https://doi.org/10.15585/mmwr.mm705152e1
12. Nemoto N et al (2021) Evaluation of test to stay strategy on secondary and tertiary transmission of SARS-CoV-2 in K–12 schools—Lake County, Illinois, august 9–October 29, 2021. MMWR Morb Mortal Wkly Rep 70:1778–1781. https://doi.org/10.15585/mmwr.mm705152e2
13. Mutesa L, Ndishimye P, Butera Y et al (2021) A pooled testing strategy for identifying SARS-CoV-2 at low prevalence. Nature 589:276–280. https://doi.org/10.1038/s41586-020-2885-5
14. Ginkgo Bioworks (2022) Case study: regular testing allows Baltimore City public school system to make informed public health decisions for kids and families. Ginkgo Bioworks. https://static1.squarespace.com/static/606b2f5bf2e9ed2df82167c5/t/63176480e1e343380a498508/1662477440722/CaseStudy_Public_BCPSS_One_Page_062722.pdf
15. Ginkgo Bioworks (2022) Case study: Tucson Unified School District's COVID-19 testing program establishes 100 locations in just 2 months. Ginkgo Bioworks. https://static1.squarespace.com/static/606b2f5bf2e9ed2df82167c5/t/631769db70b1da610c6f747b/1662478811384/CaseStudy-Public-TUSD-OnePage-050322.pdf
16. Ginkgo Bioworks (2022) Case study: Anderson Valley Unified School District brings the power of weekly testing to rural California. Ginkgo Bioworks. https://static1.squarespace.com/static/606b2f5bf2e9ed2df82167c5/t/631765c4bf2ae46e5570536f/1662477764601/CaseStudy-Public-Anderson+Valley+Case+Study+updated-080122-compressed.pdf

Conclusion: COVID-19 Pandemic—Science, Technology, and Healthcare Delivery

Joseph M. Rosen and Rita R. Colwell

1 Concluding Remarks

A comprehensive view of the many aspects of the COVID-19 pandemic has been provided in this book. The pandemic impacted the United States and the world in a manner not seen for more than a hundred years. It changed lives in dramatic ways. An entire generation has been affected in its educational systems, ways of working, and how its lives are lived, all of which have been changed forever.

The pandemic continues to evolve and has moved into an endemic state. The information provided in this book comprises a foundation for understanding the trials endured over the COVID-19 years. Each of the authors offers their view of the pandemic and its implications for the health sciences. A historical perspective of pandemics, the role of biodefense, and how its threads are intimately woven into this pandemic, as a survey of the pandemics of the past one hundred years provided lessons for the current pandemic.

Our intent has been from the first conception of this book to provide an in-depth understanding of the virology of the pandemic. It establishes the basis of knowing the virus and how to cope with it and its variants in the future. Details of the virus's behavior help to understand how to predict what it will do, how to evaluate and test for it, and how to provide treatment for COVID-19 infections. To this end, leading experts studying this virus have shared their knowledge.

Artificial intelligence (AI) and robotics help us to understand better those technologies that can be used to address demands on talent and resources to meet the pandemic, e.g., using AI to test for the virus and employing computer systems to

J. M. Rosen
Department of Surgery, Dartmouth-Hitchcock Medical Center, Lebanon, NH, USA

R. R. Colwell (✉)
Institute for Advanced Computer Studies, University of Maryland, College Park, MD, USA
e-mail: rcolwell@umd.edu

J. M. Rosen, R. R. Colwell (eds.), *The COVID-19 Pandemic*,
https://doi.org/10.1007/978-3-031-62772-9_18

replace humans when demands of the pandemic prove overwhelming to the healthcare system. Robotics can replace healthcare workers in many instances by providing substitutes for human healthcare workers, especially in intensive care units where the healthcare workers are highly vulnerable.

The economic impact of the pandemic, as discussed by experts from The World Bank, helps provide an understanding of the enormous burden of the pandemic on the world economy. An important conclusion is that resources should be made available today to prevent another such pandemic in the future because the cost of the COVID-19 pandemic has been overwhelming for the United States and the world economy.

Information is crucial during any pandemic and that point was brutally clear during this pandemic in the disinformation and misinformation disseminated during COVID-19. Resilience of communities makes it possible to prepare better to respond to future pandemics.

The many variants of the COVID-19 virus that emerged during the course of this pandemic and the chronology of the pandemic highlight the human impact of the virus throughout the United States and the world populations. How the pandemic affected specific countries outside the United States, with Vietnam as an example of how well a country can handle a pandemic, is instructive. China showed how an autocratic country can handle a pandemic, and Afghanistan provided an example of the difficulties in handling the added demands of a pandemic on a country already burdened by war.

Overall, there are some very positive lessons learned from COVID-19, one of which is in the genomic analysis of wastewater, providing a measure of pathogens circulating in a community. Many countries quickly adopted the analysis of wastewater and sludge for the COVID-19 virus, thereby having a risk prediction of the virus in the community, often up to a week to prepare for COVID-19 cases in the community. This technology is potentially a powerful new public health tool. Similarly, a better understanding of the environmental drivers of pandemics, e.g., ambient temperatures, humidity, population movement, and environmental and sociological factors, enables risk prediction by employing powerful satellite sensors and global public health thereby becoming powerfully enabled.

The new healthcare system, based on the diligent efforts of the former Surgeon General C. Everett Koop, provides a new way to approach pandemics, namely, a dual-use system integrating healthcare systems across all 50 states of the United States. It integrates new technologies and policies for an improved response to a pandemic and formulates a cybercare system bringing together both cybertechnologies and healthcare systems.

Finally, the new technologies of the twenty-first century, notably synthetic biology and biotechnology allowed mRNA to be used to create vaccines that proved dramatically effective and able to be employed rapidly and effectively. Synthetic biology allowed the manufacture of vaccines in larger quantities at a lower cost than any prior vaccine production technologies.

In summary, a valuable overview of the science and technology influencing the trajectory of the COVID-19 pandemic, i.e., genomics of the virus, epidemiology and healthcare, robotics, mode of the virus pathogenicity, how a virus can impact healthcare systems, and devastate an economy, has been deftly discussed in detail. Simulation modeling to predict the spread of the virus and risk of infection in the United States and globally, and highly effective communication geared to pandemic response provides a foundational framework, hopefully thwarting the next pandemic.

Index

J. M. Rosen, R. R. Colwell (eds.), *The COVID-19 Pandemic*,
https://doi.org/10.1007/978-3-031-62772-9

SPRINGER NATURE

GPSR Compliance

The European Union's (EU) General Product Safety Regulation (GPSR) is a set of rules that requires consumer products to be safe and our obligations to ensure this.

If you have any concerns about our products, you can contact us on ProductSafety@springernature.com

In case Publisher is established outside the EU, the EU authorized representative is:

Springer Nature Customer Service Center GmbH
Europaplatz 3
69115 Heidelberg, Germany

The manufacturer's authorised representative in the EU is Springer Nature Customer Service Centre GmbH, Europaplatz 3, 69115 Heidelberg, Germany. If you have any concerns regarding our products, please contact ProductSafety@springernature.com

Printed and bound by CPI Group (UK) Ltd, Croydon, CR0 4YY
01/10/2026
02225906-0015